D

Communications and Control Engineering

Graham C. Goodwin, María M. Seron and
José A. De Doná

Constrained Control and Estimation

An Optimisation Approach

With 109 Figures

 Springer

Graham C. Goodwin, PhD
María M. Seron, PhD
José A. De Doná, PhD
School of Electrical Engineering and Computer Science
The University of Newcastle, Australia

Series Editors
E. D. Sontag • M. Thoma • A. Isidori • J. H. van Schuppen

British Library Cataloguing in Publication Data
Goodwin, Graham C. (Graham Clifford), 1945–
    Constrained control and estimation: an optimisation
    approach. — (Communications and control engineering)
    1. Automatic control
    I. Title   II. Secron, M. (María), 1963–   III. De Doná, José A.
    629.8
    ISBN 1852335483

Library of Congress Cataloging-in-Publication Data
Goodwin, Graham C.
    Constrained control and estimation: and optimisation approach / Graham Goodwin,
    María Seron, and José De Doná.
        p. cm. — (Communications and control engineering; ISSN 0178-5354)
    ISBN 1-85233-548-3 (alk. paper)
    1. Predictive control.   2. Control theory.   I. Goodwin, Graham C. (Graham Clifford),
        1945–   II. Seron, M. (María), 1963–   III. Title.   IV. Series.
    TJ217.6.D4 2005
    629.8—dc21   2001049308

Printed in the United States of America.   (SBA)

9 8 7 6 5 4 3 2

Springer Science+Business Media, LLC

springer.com

Preface

This book gives an introduction to the fundamental principles underlying constrained control and estimation. This subject has a long history in practice, particularly in the control of chemical processes and in channel equalisation in digital communications. Recently, significant advances have also been made in the supporting theory. In this context, the objective of this book is to describe the foundations of constrained control and estimation. The treatment is aimed at researchers and/or practitioners and builds on core principles in signals and systems, optimisation theory and optimal control. We also emphasise the common links and connections that exist between estimation and control problems.

What Is Constrained Control?

It is generally true in control system design that higher levels of performance are associated with "pushing the system hard". The latter, however, is usually limited by the presence of physical constraints on system components. As a simple, common world example, consider the case of automobile control. It is well known that rapid acceleration and deceleration are associated with large throttle displacement and strong braking action, respectively. It is also known that there exist maximum and minimum available throttle displacements and braking capacity, that is, the input to the system is constrained. Furthermore, we might suspect that variables other than the system input are subject to constraints; for example, acceleration and deceleration have to be limited to prevent wheels from losing traction. These are constraints on the output or state of the system. Similar constraints arise in virtually all control problems. For example, valves in chemical process control have a maximum displacement (when fully open) and a minimum displacement (when fully closed). These are examples of *input constraints*. Also, for safety or other operational reasons, it is usual to impose limits on allowable temperatures, levels and pressures. These are examples of *state constraints*.

One possible strategy for dealing with constraints is to modify the design so that limits are never violated. However, it is heuristically reasonable that this may be counterproductive. Indeed, because it is usually true that higher performance levels are associated with pushing the limits, there is a strong incentive to operate the system on constraint boundaries. Within this context, the subject of optimal constrained control provides the necessary tools required to solve this class of problems. Specifically, the aim is to maximise performance whilst ensuring that the relevant constraints on both inputs (manipulated variables) and states (process variables) are not violated.

What Is Constrained Estimation?

Constraints occur in estimation problems for similar reasons as they do in control problems, save that in estimation, the constraints typically arise as a priori known conditions rather than as required conditions. For example, in estimating the concentration in a distillation column, it is typically known that the liquid levels in the trays must lie between given lower and upper limits. By enforcing these kinds of constraints during estimation, one should expect more accurate and realistic results. Another area in which constrained estimation occurs is the case where the signal to be estimated is known, a priori, to belong to a finite alphabet. This is the core problem, for example, in signal recovery in digital communications. Well-known constrained estimators used in the latter context include decision feedback equalisers and the Viterbi algorithm.

Why a Special Treatment of Constrained Control and Estimation?

Most of the existing literature on the topic of control and estimation deals with unconstrained problems. However, as discussed above, there are strong practical reasons why constraints cannot be ignored if one is seeking high performance. Also, from a theoretical perspective, constrained control and estimation represents the obvious "next step" beyond traditional linear theory. In fact we shall see that adding constraints to an otherwise linear control or estimation problem still leads to a problem that is computationally tractable. Indeed, at the present state of development, the theory of constrained control and/or estimation for linear systems is approaching the completeness of the traditional theory for the unconstrained case. Thus there remains no real impediment to teaching this theory alongside traditional treatments of linear estimation and control. In summary, constrained control and estimation lies at the junction of practical importance and theoretical tractability.

Why a Book at this Particular Time?

There has recently been a surge of interest in constrained estimation and control problems and many new results have appeared which underpin practical applications in many areas.

For example, constrained control has been utilised in industry for three or four decades, primarily in the area of process control where long-time constants of the systems facilitated the necessary calculations. However, there have recently been several advances that have significantly broadened the realm of application of constrained control. These advances include:

- Computer speeds have increased dramatically making it feasible to apply constrained control methods to high speed systems, including electromechanical and aerospace systems.
- New insights have been obtained into constrained control which show that, in many cases of practical interest, the necessary computations can often be significantly simplified. This has further enhanced the domain of potential application of the ideas.
- Theoretical support for the topic is growing. This gives increased confidence in the application of the methods.
- The topic builds on many core principles from mathematical systems theory. Thus, it is a useful vehicle by which neophyte researchers can be acquainted with a broad range of tools in systems theory, convex optimisation, and optimal control and estimation.

Book Philosophy

This book is aimed at going a step beyond traditional linear control theory to include consideration of constraints. Our premise is that one should accept the existence of constraints and deal with them rather than avoid them. Thus, this book addresses high performance control system design and signal estimation in the presence of constraints. We adopt an optimisation-based approach to these problems. The principal tools used are prediction and optimisation. Prime topics are receding horizon control and moving horizon estimation. We treat related approaches in so far that they can be viewed as special cases of this philosophy. For example, it has recently been shown that anti-windup methods can sometimes be viewed as simplified forms of receding horizon control. Also, decision feedback equalisers turn out to be a special case of a more general moving horizon optimisation problem.

Book Content

The book gives a comprehensive treatment of constrained control and estimation. Topics to be addressed include:

- an overview of optimisation;
- linear and nonlinear receding horizon control;
- links to classical optimal control theory, including the discrete minimum principle;
- input and state constraints in control system design;
- constrained control solutions having a finite parameterisation for specific classes of problems;
- stability of constrained controllers;
- numerical procedures for solving constrained optimisation problems;
- output feedback;
- an overview of Bayesian estimation theory;
- constrained state estimation;
- links between constrained estimation and constrained control.

Related Literature

The book includes a comprehensive set of references to contemporary literature. We also note that there have recently been several excellent books published that complement the material in the current book. In particular, we point to the books by Camacho and Bordons (1999), Maciejowski (2002), Borrelli (2003), and Rossiter (2003).

Intended Audience

The current book is aimed at those wishing to gain an understanding of the fundamental principles underlying constrained control and estimation. The book could be used as the basis of a junior level course for research students or as the basis of a self-study program by practising engineers.

Flavour and Structure of the Book

The book emphasises the mathematical underpinnings of the topic. It summarises and utilises core ideas from signals and systems, optimisation theory, classical optimal control and Bayesian estimation. Also, the book deals with dual problems that arise in control, state estimation and signal recovery. The book assumes that the reader has appropriate background in systems theory, including linear control theory, stability theory and state space methods. With this as background, the book is self-contained and encompasses all necessary material to understand constrained control and estimation, including:

- optimisation and quadratic programming;
- controller design in the presence of constraints;

- stability;
- Bayesian estimation;
- estimator design in the presence of constraints;
- optimisation with finite set constraints.

The book also contains three case studies. These case studies are intended to show how the theory described in the book can be put into practice on problems of practical relevance. The chosen case studies are:

- rudder roll stabilisation of ships;
- cross directional control;
- control over communication networks.

These applications are described in sufficient detail so that the reader can gain an appreciation of the practical issues involved.

The book is divided into three parts:

- Part I: Foundations
- Part II: Further Developments
- Part III: Case Studies

Part I was written by the principal authors. Parts II and III are based on contributions prepared by other authors within our working group. Note, however, that Parts II and III are not simply a collection of contributions; the contents have been carefully chosen, edited and arranged by the principal authors and thus form an integrated presentation in combination with Part I. The split into three parts is aimed at dividing the material into distinct areas (foundations, further developments and case studies) and at providing appropriate recognition to those who assisted with the overall book project.

The book is accompanied by a website, which contains related material such as papers by the authors, lecture slides, worked examples, Matlab routines, and so on (see http://murray.newcastle.edu.au/cce/).

Newcastle, Australia   *Graham C. Goodwin*
June 2004   *María M. Seron*
                                                                      José A. De Doná

Acknowledgements

The authors gratefully acknowledge input from many colleagues and friends who assisted with the development of this book. Special thanks go to Hernan Haimovich, Tristan Perez, Osvaldo Rojas, Daniel Quevedo and James Welsh, who each contributed material for Parts II and III. We also acknowledge input from other students and colleagues at the University of Newcastle, including Adrian Wills, Juan I. Yuz and Juan Carlos Agüero. In particular, we are grateful to Claus Müller for the stability results of Chapter 4, Sections 10.6 and 10.7 in Chapter 10, and his careful reading of other parts of the manuscript; we are also grateful to Xiang W. Zhuo for contributing simulation examples for Chapters 9 and 10. A special thanks is due to David Mayne who was instrumental in introducing the authors to this topic and who directly inspired much of the development. Also, parts of the book were inspired by the contributions of many others, including (but not restricted to) Manfred Morari, Jim Rawlings, Hannah Michalska, Peter Tøndel, Thor Johansen, Karl Åström, Alberto Bemporad, Mogens Blanke, Mike Cannon, Lucien Polak, David Clarke, Steven Duncan, Frank Algöwer, Thor Fossen, Elmer Gilbert, Basil Kouvaritakis, Jan Maciejowski, Wook Hyun Kwon, Christopher Rao, Edoardo Mosca, Rick Middleton, Bob Skelton, Arie Feuer, Greg Stewart, Arthur Jutan, Will Heath and many others too numerous to name. We also acknowledge Rosslyn, Jayne and Dianne for their generous support. Finally, the third author wishes to acknowledge his young son Stefano for the loss of shared moments, which were as missed by his father as they were for him.

Contents

Part II   Further Developments

Part III Case Studies

Part I

Foundations

1

Introduction

1.1 Overview

The goal of this chapter is to provide a general overview of constrained control and estimation. This is intended to motivate the material to follow. Section 1.2 treats constrained control, Section 1.3 deals with constrained estimation, and Section 1.4 draws parallels between these two problems.

1.2 Introduction to Constrained Control

Handling constraints in control system design is an important issue in most, if not all, real world problems.

It is readily appreciated that all real world control systems have an associated set of constraints; for example, inputs always have maximum and minimum values and states are usually required to lie within certain ranges. Of course, one could proceed by ignoring these constraints and hope that no serious consequences result from this approach. This simple procedure may be sufficient at times. On the other hand, it is generally true that higher levels of performance are associated with operating on, or near, constraint boundaries. Thus, a designer really cannot ignore constraints without incurring a performance penalty.

As an illustration of these facts consider a simple automobile control problem. We mentioned in the Preface that there exist maximum and minimum available throttle displacements, that is, the system input is constrained. Other variables are also subject to constraints; for example, acceleration and deceleration have to be limited to prevent the vehicle's wheels from loosing traction. These factors constitute a constraint on the state of the system. Thus, modern cars incorporate both traction control (for acceleration) and anti-skid braking [ABS] (for deceleration). Both mechanisms ensure safe operation when variables are pushed to their limits.

As another simple example, consider the problem of rudder roll stabilisation of ships. The prime function of the rudder is to maintain the ship's heading. However, the rudder also imparts a rolling moment to the ship. Thus, the rudder can be used to achieve a measure of roll stabilisation. Since the rolling moment induced by the rudder is relatively small, it can be appreciated that large rudder displacements will be called upon, especially under heavy sea conditions. Of course, practical rudders must operate subject to constraints on both their total displacement (typically ± 30 degrees) and slew rate (typically ± 15 degrees per second). Indeed, it is generally agreed that rudder roll stabilisation can actually be counterproductive unless appropriate steps are taken to adequately deal with the presence of constraints. We will devote Chapter 14 to a more comprehensive introduction to rudder roll stabilisation. Other practical problems are discussed in Chapters 15 and 16.

Most of the existing literature on control theory deals with unconstrained problems. Nonetheless, as discussed above, there are strong practical reasons why a system should be operated on constraint boundaries. Thus, this book is aimed at going a step beyond traditional linear control theory to include consideration of constraints.

Our view of the existing methods for dealing with constraints in control system design is that they can be broadly classified under four headings:

- cautious
- serendipitous
- evolutionary
- tactical

In the "cautious" approach, one aims to explicitly deal with constraints by deliberately reducing the performance demands until the point where the constraints are not met at all. This has the advantage of allowing one to essentially use ordinary unconstrained design methods and hence to carry out a rigorous linear analysis of the problem. On the other hand, this is achieved at the cost of a potentially important loss in achievable performance since we expect high performance to be associated with pushing the boundaries, that is, acting on or near constraints.

In the "serendipitous" approach, one takes no special precautions to handle constraints, and hence occasional violation of the constraints is possible (that is, actuators reach saturation, states exceed their allowed values, and so on). Sometimes this can lead to perfectly acceptable results. However, it can also have a negative impact on important performance measures, including closed loop stability, since no special care is taken of the constrained phase of the response.

In the "evolutionary" approach, one begins with an unconstrained design philosophy but then adds modifications and embellishments to ensure that the negative consequences of constraints are avoided, or at least minimised, whilst ensuring that performance goals are attained. Examples of evolutionary approaches include various forms of anti-windup control, high gain-low gain control, piecewise linear control and switching control.

One might suspect that, by careful design and appropriate use of intuition, one can obtain quite acceptable results from the evolutionary approach provided one does not push too hard. However, eventually, the constraints will override the usual linear design paradigm. Under these conditions, there could be advantages in "starting afresh". This is the philosophy of the so-called "tactical" approaches, in which one begins afresh with a formulation that incorporates constraints from the beginning in the design process. One way of achieving this is to set the problem up as a constrained optimisation problem. This will be the approach principally covered in this book.

Of course, the above classification does not cover all possibilities. Indeed, many methods fall into several categories.

To provide further motivation for this subject, we will present a simple example illustrating aspects of the cautious, serendipitous and tactical approaches.

We will base our design on linear quadratic regulator [LQR] theory. Thus, consider an objective function of the form:

$$V_N(\{x_k\}, \{u_k\}) \triangleq \frac{1}{2}x_N^T P x_N + \frac{1}{2}\sum_{k=0}^{N-1}(x_k^T Q x_k + u_k^T R u_k), \qquad (1.1)$$

where $\{u_k\}$ denotes the *control sequence* $\{u_0, u_1, \ldots, u_{N-1}\}$, and $\{x_k\}$ denotes the corresponding *state sequence* $\{x_0, x_1, \ldots, x_N\}$. In (1.1), $\{u_k\}$ and $\{x_k\}$ are related by the linear state equation:

$$x_{k+1} = Ax_k + Bu_k, \quad k = 0, 1, \ldots, N-1,$$

where x_0, the initial state, is assumed to be known.

In principle one can adjust the following parameters to obtain different manifestations of performance:

- the optimisation horizon N
- the state weighting matrix Q
- the control weighting matrix R
- the terminal state weighting matrix P

Actually, adjusting one or more of these parameters to manipulate key performance variables turns out to be one of the principal practical attributes of constrained linear control. We illustrate some of the basic features of constrained control using the objective function (1.1) via the following simple example.

Example 1.2.1. Consider the specific linear system:

$$x_{k+1} = Ax_k + Bu_k, \qquad (1.2)$$
$$y_k = Cx_k,$$

with

$$A = \begin{bmatrix} 1 & 1 \\ 0 & 1 \end{bmatrix}, \qquad B = \begin{bmatrix} 0.5 \\ 1 \end{bmatrix}, \qquad C = \begin{bmatrix} 1 & 0 \end{bmatrix},$$

which is the zero-order hold discretisation with sampling period 1 of the double integrator

$$\frac{d^2 y(t)}{dt^2} = u(t).$$

We take the initial condition (for illustrative purposes) to be $x_0 = \begin{bmatrix} -6 & 0 \end{bmatrix}^{\mathrm{T}}$ and suppose that the actuators have maximum and minimum values (saturation) so that the control magnitude is constrained such that $|u_k| \leq 1$ for all k. We will design cautious, serendipitous, and tactical feedback controllers for this system. A schematic of the feedback control loop is shown in Figure 1.1, where "sat" represents the actuator modelled by the *saturation function*

$$\mathrm{sat}(u) \triangleq \begin{cases} 1 & \text{if } u > 1, \\ u & \text{if } |u| \leq 1, \\ -1 & \text{if } u < -1. \end{cases} \tag{1.3}$$

Note that the section of Figure 1.1 in the dashed-line box is part of the physical reality and is not subject to change (unless, of course, the actuator is replaced).

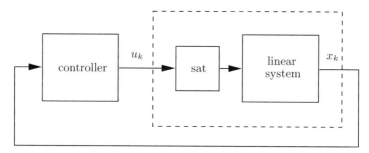

Figure 1.1. Feedback control loop for Example 1.2.1.

(i) Cautious Design

A *cautious* strategy would be, for example, to design a linear state feedback with *low gain* such that the control limits are never reached.

For example, using the objective function (1.1) with infinite horizon ($N = \infty$, $P = 0$) and weighting matrices $Q = C^{\mathrm{T}}C = \begin{bmatrix} 1 & 0 \\ 0 & 0 \end{bmatrix}$ and $R = 20$ gives the linear state feedback law:

$$u_k = -Kx_k = -\begin{bmatrix} 0.1603 & 0.5662 \end{bmatrix} x_k.$$

This control law *never violates the given physical limits* on the input *for the given initial condition*. The resulting input and output sequences are shown in Figure 1.2.

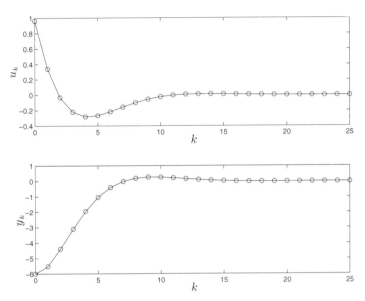

Figure 1.2. u_k and y_k for the cautious design $u_k = -Kx_k$ with weights $Q = C^{\mathrm{T}}C$ and $R = 20$.

We can see from Figure 1.2 that the input u_k has a maximum value close to 1 (achieved at $k = 0$) which clearly satisfies the given constraint *for this initial condition*. However, the achieved output response is rather slow. Indeed, it can be seen from Figure 1.2 that the "settling time" is of the order of eight samples.

(ii) Serendipitous Design

Now, suppose that for the same $Q = C^{\mathrm{T}}C$ in the infinite horizon objective function we try to obtain a faster response by reducing *the control weight to* $R = 2$. We expect that this will lead to a control law having "higher gain."

The resultant *higher gain control* would give the input and output sequences shown in dashed lines in Figure 1.3 *provided the input constraint could be removed* (that is, if the saturation block were removed from Figure 1.1). However, we can see that the *input constraints would have been violated* in the presence of actuator saturation. (The input at $k = 0$ is well beyond the allowed limit of $|u_k| = 1$.)

To *satisfy the constraints* we next incorporate the saturation function (1.3) in the controller and simply saturate the input signal when it violates the constraint. This leads to the control law:

$$u_k = \text{sat}(-Kx_k) = -\text{sat}(Kx_k).$$

Note that, in terms of performance, this is equivalent to simply letting the input saturate through the actuator in Figure 1.1. We call this control law *serendipitous* since no special considerations of the presence of the constraints have been made in the design calculations. The resulting input and output sequences are shown by circle-solid lines in Figure 1.3.

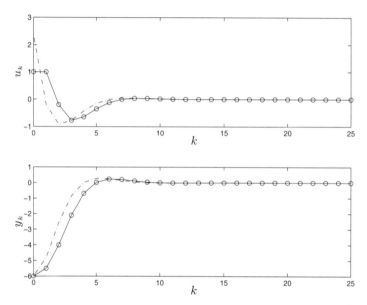

Figure 1.3. u_k and y_k for the unconstrained LQR design $u_k = -Kx_k$ (dashed line), and for the serendipitous strategy $u_k = -\text{sat}(Kx_k)$ (circle-solid line), with weights $Q = C^T C$ and $R = 2$.

We see from Figure 1.3 that the amount of overshoot is essentially the same whether or not the input is constrained. Of course, the response time achieved with a constrained input is longer than for the case when the input is unconstrained. However, note that the constraint is part of the physical reality and cannot be removed unless we replace the actuator. On the other hand, the serendipitous design (with $R = 2$) appears to be making better use of the available control authority than the cautious controller (with $R = 20$). Indeed, the settling time is now approximately five samples even when the input is constrained. This is approximately twice as fast as for the cautious controller, whose performance was shown in Figure 1.2.

Encouraged by the above result, we might be tempted to "push our luck" even further and aim for an even faster response by further reducing the weighting on the input signal. Accordingly, we *decrease the control weighting* in the LQR design even further, for example, to $R = 0.1$.

In Figure 1.4 we can see the resulting input and output sequences (when the input constraint, that is, the saturation block in Figure 1.1, is removed) for the linear controller $u_k = -Kx_k$ (dashed line). We now observe an *unconstrained* settling time of approximately three samples. However, when the input constraint is taken into account by setting $u_k = -\mathrm{sat}(Kx_k)$, then we see that significant overshoot occurs and the settling time "blows out" to 12 samples (circle-solid line).

Perhaps we should not be surprised by this result since no special care has been taken to tailor the design to deal with constraints, that is, the approach remains serendipitous.

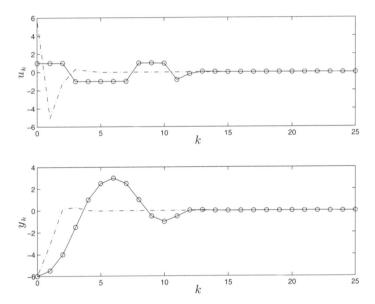

Figure 1.4. u_k and y_k for the unconstrained LQR design $u_k = -Kx_k$ (dashed line), and for the serendipitous strategy $u_k = -\mathrm{sat}(Kx_k)$ (circle-solid line), with weights $Q = C^{\mathrm{T}}C$ and $R = 0.1$.

We have seen above that as we try to push the system harder, the *serendipitous strategy ultimately fails to give a good result* leading to the output having large overshoot and long settling time. We can gain some insight into what has gone wrong by examining the state space trajectory corresponding to the serendipitous strategy. This is shown in Figure 1.5, where x_k^1 and x_k^2 denote the components of the state vector x_k in the discrete time model (1.2).

The control law $u = -\text{sat}(Kx)$ partitions the state space into three regions in accordance with the definition of the saturation function (1.3). Hence, the serendipitous strategy can be characterised as a *switched* control strategy in the following way:

$$u = \mathcal{K}(x) = \begin{cases} -Kx & \text{if } x \in R_0, \\ 1 & \text{if } x \in R_1, \\ -1 & \text{if } x \in R_2. \end{cases} \tag{1.4}$$

Notice that this is simply an alternative way of describing the serendipitous strategy since for $x \in R_0$ the input actually lies between the saturation limits. The partition is shown in Figure 1.5.

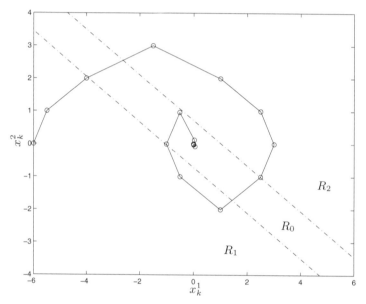

Figure 1.5. State space trajectory and space partition for the serendipitous strategy $u_k = -\text{sat}(Kx_k)$, with weights $Q = C^{\mathsf{T}}C$ and $R = 0.1$.

Examination of Figure 1.5 suggests a heuristic argument as to why the serendipitous control law may not be performing well in this case. We can think, in this example, of x^2 as "velocity" and x^1 as "position." Now, in our attempt to change the position rapidly (from -6 to 0), the velocity has been allowed to grow to a relatively high level ($+3$). This would be fine if the braking action were unconstrained. However, our input (including braking) is limited to the range $[-1, 1]$. Hence, the available braking is inadequate to "pull the system up", and overshoot occurs.

(iii) Tactical Design

Perhaps the above heuristic argument gives us some insight into how we could remedy the problem. A sensible idea would seem to be to try to "look ahead" and take account of future input constraints (that is, the limited braking authority available). To test this idea, we take the objective function (1.1) as a starting point.

We use a prediction horizon $N = 2$ and minimise, at each sampling instant i and for the current state x_i, the two-step objective function:

$$V_2(\{x_k\}, \{u_k\}) = \frac{1}{2} x_{i+2}^{\mathrm{T}} P x_{i+2} + \frac{1}{2} \sum_{k=i}^{i+1} (x_k^{\mathrm{T}} Q x_k + u_k^{\mathrm{T}} R u_k), \qquad (1.5)$$

subject to the equality and inequality constraints:

$$\begin{aligned} x_{k+1} &= A x_k + B u_k, \\ |u_k| &\le 1, \end{aligned} \qquad (1.6)$$

for $k = i$ and $k = i + 1$.

In the objective function (1.5), we set, as before, $Q = C^{\mathrm{T}} C$, $R = 0.1$. The terminal state weighting matrix P is taken to be the solution of the Riccati equation $P = A^{\mathrm{T}} P A + Q - K^{\mathrm{T}} (R + B^{\mathrm{T}} P B) K$, where $K = (R + B^{\mathrm{T}} P B)^{-1} B^{\mathrm{T}} P A$ is the corresponding gain.

As a result of minimising (1.5) subject to (1.6), we obtain an optimal fixed-horizon control sequence $\{u_i, u_{i+1}\}$. We then apply the resulting value of u_i to the system. The state evolves to x_{i+1}. We now shift the time instant from i to $i + 1$ and repeat this procedure. This is called *receding horizon control* [RHC] or *model predictive control*. RHC has the ability to "look ahead" by considering the constraints not only at the current time i but also at future times within the prediction interval $[i, i + N - 1]$. (This idea will be developed in detail in Chapter 4.)

The input and output sequences for the LQR design $u = -Kx$ (dashed line) that violates the constraints and the sequences for the receding horizon design (circle-solid line) are shown in Figure 1.6.

We can see from Figure 1.6 that the output trajectory with constrained input now has minimal overshoot. Thus, the idea of "looking ahead" and applying the constraints in a receding horizon fashion has apparently "paid dividends."

Actually, we will see in Chapter 6 that the receding horizon strategy described above also leads to a partition of the state space into different regions in which affine control laws hold. The result is shown (for interest) in Figure 1.7. The region R_2 corresponds to the region R_2 in Figure 1.5 and represents the area of state space where $u = -1$ is applied. Comparing Figure 1.5 and Figure 1.7 we see that the region R_2 has been "bent over" in Figure 1.7 so that $u = -1$ occurs at lower values of x^2 (velocity) than was the case in

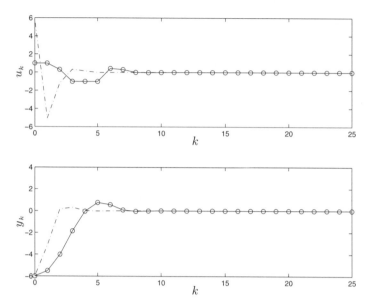

Figure 1.6. u_k and y_k for the unconstrained LQR design $u_k = -Kx_k$ (dashed line), and for the receding horizon design (circle-solid line), with weights $Q = C^{\mathsf{T}}C$ and $R = 0.1$.

Figure 1.5. This is in accordance with our heuristic argument about "needing to brake earlier."

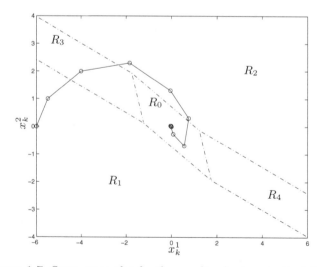

Figure 1.7. State space plot for the receding horizon tactical design.

Obviously we have not given full details of the above example especially in relation to the tactical approach; the example has been introduced only to "wet the readers' appetite" as to what might appear in the remainder of the book. Indeed, in forthcoming chapters, we will analyse, in some detail, the concepts raised in the above simple example.

1.3 Introduction to Constrained Estimation

Constraints are also often present in estimation problems. A classical example of a constrained estimation problem is the case in which *binary* data (say ± 1) are transmitted through a communication channel where it suffers dispersion causing the data to overlay itself. In the field of communications, this is commonly referred to as *intersymbol interference* [ISI]. The associated estimation problem is: Given the output of the channel, provide an estimate of the transmitted signal.

To illustrate some of the ideas involved in the above problem, let us assume, for simplicity, that the intersymbol interference produced by the channel can be modelled via a finite impulse response [FIR] model of the form:

$$y_k = \sum_{\ell=0}^{m} g_\ell u_{k-\ell} + n_k, \tag{1.7}$$

where y_k, u_k, n_k denote the channel output, input and noise, respectively. Also, $(g_0 \ldots g_m)$ denotes the (finite) impulse response of the channel. We assume here (for simplicity) that $g_0 \ldots g_m$ are known. Also, for simplicity, we assume that the channel is minimum phase (that is, has a stable inverse).

Now, heuristically, one might expect that one should "invert" the channel so as to recover the input sequence $\{u_k\}$ from a given sequence of output data $\{y_k\}$. Such an inverse can be readily found by utilising feedback ideas. Specifically, if we expand the channel transfer function as:

$$G(z) = g_0 + \ldots + g_m z^{-m} = g_0 + \tilde{G}(z),$$

then we can form an inverse by the feedback circuit shown in Figure 1.8.

To verify that the circuit of Figure 1.8 does, indeed, produce an inverse, we see that the transfer function from y_k to $\tilde{u}_k$ is

$$T(z) = \frac{\dfrac{1}{g_0}}{1 + \dfrac{\tilde{G}(z)}{g_0}} = \frac{1}{g_0 + \tilde{G}(z)} = \frac{1}{G(z)}.$$

Thus, we have generated an inverse to the system transfer function $G(z)$. Hence, in the absence of noise and other errors, we can expect that the signal $\tilde{u}_k$ in Figure 1.8 will converge to u_k following an initial transient (note that

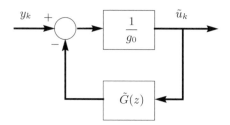

Figure 1.8. Feedback inverse circuit.

we have assumed that $G(z)$ has a stable inverse). Under ideal conditions this is exactly what does happen. However, in practice, the presence of the noise term in (1.7) will lead to estimation errors. Indeed, a little thought shows that $\tilde{u}_k$ may not even belong to the set $\{+1, -1\}$ even though we know, a priori, that the true transmitted signal, u_k, does.

An improvement seems to be to simply take the nearest value from the set $\{+1, -1\}$ corresponding to $\tilde{u}_k$. This leads to the circuit shown in Figure 1.9, where

$$\text{sign}(\tilde{u}_k) \triangleq \begin{cases} +1 & \text{if } \tilde{u}_k \geq 0, \\ -1 & \text{if } \tilde{u}_k < 0. \end{cases}$$

Comparing Figure 1.9 with Figure 1.8 may lead us to develop a further

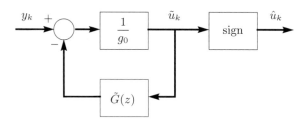

Figure 1.9. Constrained feedback inverse circuit.

embellishment of this simple idea. In particular, we see in Figure 1.8 that the feedback path through the transfer function $\tilde{G}(z)$ uses the estimated input $\tilde{u}_k$. Now, in Figure 1.9 our belief is that $\hat{u}_k$ should be a better estimate of the input than $\tilde{u}_k$ since we have *forced* the constraint $\hat{u}_k \in \{+1, -1\}$. This suggests that we could try feeding back $\hat{u}_k$ instead of $\tilde{u}_k$, as shown in Figure 1.10.

The arguments leading to Figure 1.10 are rather heuristic. Nonetheless, the constrained estimator in Figure 1.10 finds very widespread application in

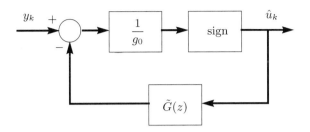

Figure 1.10. Constrained estimation with decision feedback, or "decision feedback equaliser [DFE]."

digital communications, where it is given a special name—decision feedback equaliser [DFE].

The reader might now be asking how one could improve on the circuit of Figure 1.10. We can gain some insight as to from where further improvements might come by expressing the result shown in Figure 1.10 as the solution to an optimisation problem. Specifically, assume that we are given (estimates of) past values of the input, $\{\hat{u}_{k-1}, \ldots, \hat{u}_{k-m}, \ldots\}$, and that we model the output $\hat{y}_k$ as

$$\hat{y}_k = g_0 u'_k + g_1 \hat{u}_{k-1} + \ldots + g_m \hat{u}_{k-m}.$$

We can now ask what value of u'_k causes $\hat{y}_k$ to be, at time k, as close as possible to the observed output y_k. We measure how close $\hat{y}_k$ is to y_k by the following one-step objective function:

$$V_1(\hat{y}_k, u'_k) = [y_k - \hat{y}_k]^2.$$

We also require that $u'_k \in \{+1, -1\}$. The solution to this constrained optimisation problem is readily seen to be:

$$\hat{u}_k = \text{sign}\left\{\frac{1}{g_0}[y_k - g_1 \hat{u}_{k-1} - \ldots - g_m \hat{u}_{k-m}]\right\}. \tag{1.8}$$

However, the reader can verify that this precisely corresponds to the arrangement illustrated in Figure 1.10. One might anticipate that by exploiting the connection with constrained optimisation one can obtain better performance, since more elaborate objective functions can be employed. How this might be achieved is discussed below, and a more detailed description will be given in Chapter 13.

The following example illustrates the above ideas.

Example 1.3.1. Consider the channel model

$$y_k = u_k - 1.7u_{k-1} + 0.72u_{k-2} + n_k,$$

where u_k is a random binary signal and n_k is an independent identically distributed [i.i.d.] noise having a Gaussian distribution of variance σ^2. We first assume the ideal situation in which the channel has no noise, $\sigma^2 = 0$. Since the channel model has a stable inverse, we implement the inversion estimator depicted in Figure 1.8. The result of the simulation is represented in Figure 1.11. Notice that the estimator yields perfect signal recovery. This

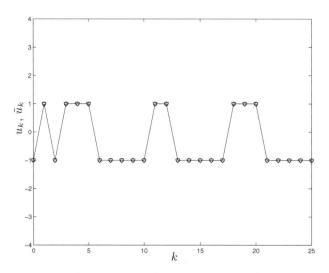

Figure 1.11. Data u_k (circle-solid line) and estimate $\tilde{u}_k$ (triangle-solid line) using the feedback inverse circuit of Figure 1.8. Noise variance: $\sigma^2 = 0$.

is very encouraging. However, this situation in which no noise is present is unrealistic.

Next, we simulate the inversion estimator of Figure 1.8 when the received signal is affected by noise n_k of variance $\sigma^2 = 0.1$. The result of the simulation is shown in Figure 1.12. Note that in this case the estimate $\tilde{u}_k$ differs from u_k and does not belong to the range ± 1. We conclude that not taking account of the constraints in the estimation leads to a poor result.

We next simulate the estimator represented in Figure 1.9 with noise of variance $\sigma^2 = 0.1$. In this implementation, the nearest value of the estimate of the previous scheme from the set $\{+1, -1\}$ is taken. The result is shown in Figure 1.13. It can be observed that now the estimate $\hat{u}_k$ belongs to the set $\{+1, -1\}$, but the result is still poor. The reason is that we have not "informed" the estimator about the constrained estimates but have simply applied the constraint $\hat{u}_k = \mathrm{sign}(\tilde{u}_k) \in \{+1, -1\}$ "after the event."

As a further improvement to our estimator, we next implement the estimator of Figure 1.10 (the DFE) where we now feed back $\hat{u}_k \in \{+1, -1\}$ instead of $\tilde{u}_k$, thereby informing the estimator about the presence of constraints. The

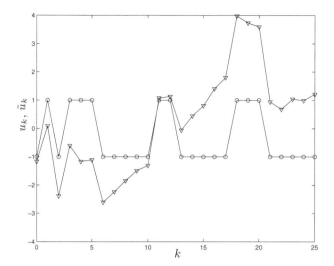

Figure 1.12. Data u_k (circle-solid line) and estimate $\tilde{u}_k$ (triangle-solid line) using the feedback inverse circuit of Figure 1.8. Noise variance: $\sigma^2 = 0.1$.

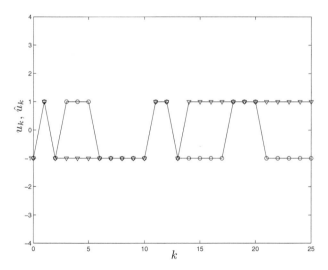

Figure 1.13. Data u_k (circle-solid line) and estimate $\hat{u}_k$ (triangle-solid line) using the constrained feedback inverse circuit of Figure 1.9. Noise variance: $\sigma^2 = 0.1$.

result of the simulation, for a noise of variance $\sigma^2 = 0.1$, is shown in Figure 1.14. Note that, despite the presence of noise in the channel, the DFE recovers the signal perfectly.

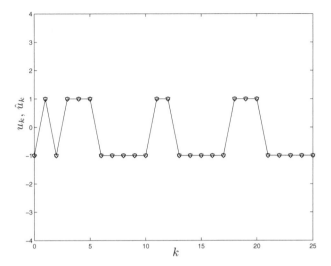

Figure 1.14. Data u_k (circle-solid line) and estimate $\hat{u}_k$ (triangle-solid line) using the DFE of Figure 1.10. Noise variance: $\sigma^2 = 0.1$.

One might wonder if the DFE circuit would always perform so well. We next investigate the performance of the DFE of Figure 1.10 when the noise variance is increased by a factor of 2; that is, $\sigma^2 = 0.2$. The result of the simulation is shown in Figure 1.15. Note that we have poor performance. The

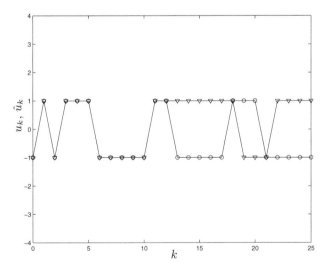

Figure 1.15. Data u_k (circle-solid line) and estimate $\hat{u}_k$ (triangle-solid line) using the DFE of Figure 1.10. Noise variance: $\sigma^2 = 0.2$.

reason is that we are only considering one observation at a time (as mentioned earlier, this scheme is equivalent to solving a one-step optimisation problem). It is a well-known phenomenon with this circuit that, once a detection error occurs, it may propagate. Thus, errors typically occur in "bursts." The reason for this error propagation is that previous estimates are assumed to be equal to the true signal. As we will study later in Chapter 13, other estimation mechanisms can be implemented in which multiple observations are considered simultaneously and some "degree of belief" in previous estimates is incorporated. We will give a first taste of these ideas below.   ○

The reader has probably noticed that there is a very close connection between the above problem and the tactical approach to the control problem discussed in Section 1.2. This suggests a route by which we may be able to improve the estimate given in (1.8). Referring to Example 1.2.1, we found that looking ahead so as to account for future consequences of current actions was helpful. Thus, we might be led to ask what would happen if we did not fix u'_k based only on the observation y_k but waited until we had observed both y_k and y_{k+1}. Of course, y_{k+1} also depends on u_{k+1}, but this consideration could be dealt with by asking that values of u'_k and u'_{k+1} belonging to the set $\{+1, -1\}$ be chosen such that the following two-stage objective function is minimised:

$$V_2(\hat{y}_k, \hat{y}_{k+1}, u'_k, u'_{k+1}) = [y_k - \hat{y}_k]^2 + [y_{k+1} - \hat{y}_{k+1}]^2, \qquad (1.9)$$

where

$$\hat{y}_k = g_0 u'_k + g_1 \hat{u}_{k-1} + \ldots + g_m \hat{u}_{k-m}, \qquad (1.10)$$

$$\hat{y}_{k+1} = g_0 u'_{k+1} + g_1 u'_k + g_2 \hat{u}_{k-1} + \ldots + g_m \hat{u}_{k-m+1}, \qquad (1.11)$$

and where the past estimates $\{\hat{u}_{k-1}, \hat{u}_{k-2}, \ldots\}$ are again assumed fixed and known.

The solution to the above problem can be readily computed by simple evaluation of V_2 for all possible constrained inputs; that is, for

$$\{u'_k, u'_{k+1}\} \in \{\{-1, -1\}, \{-1, 1\}, \{1, 1\}, \{1, -1\}\}. \qquad (1.12)$$

Notice that there are four possibilities and the optimal solution is simply the one that yields the lowest value of V_2. We could then fix the estimate of u_k (denoted $\hat{u}_k$) as the first element of the solution to this optimisation problem. We might then proceed to measure y_{k+2} and re-estimate u_{k+1}, plus obtain a fresh estimate of u_{k+2} by minimising:

$$V_2(\hat{y}_{k+1}, \hat{y}_{k+2}, u'_{k+1}, u'_{k+2}) = [y_{k+1} - \hat{y}_{k+1}]^2 + [y_{k+2} - \hat{y}_{k+2}]^2,$$

where

$$\hat{y}_{k+1} = g_0 u'_{k+1} + g_1 \hat{u}_k + \ldots + g_m \hat{u}_{k-m+1},$$

$$\hat{y}_{k+2} = g_0 u'_{k+2} + g_1 u'_{k+1} + g_2 \hat{u}_k + \ldots + g_m \hat{u}_{k-m+2},$$

and where $\{\hat{u}_k, \hat{u}_{k-1}, \ldots\}$ are now assumed fixed and known.

By the above procedure, we are already generating constrained estimates via a *moving horizon estimator* [MHE] subject to the constraint $u'_k \in \{+1, -1\}$. This kind of estimator will be studied in detail in Chapter 13.

Example 1.3.2. Consider again the channel model of Example 1.3.1. Here we implement a moving horizon estimator as described above. That is, we minimise, at each step, the two-stage objective function (1.9), subject to (1.10)–(1.11) and the constraints (1.12). We then take as the current estimate $\hat{u}_k$ the first value u'_k of the minimising sequence $\{u'_k, u'_{k+1}\}$.

The corresponding simulation results, for noise variance $\sigma^2 = 0.2$, are shown in Figure 1.16. We can see from this figure that the estimator recovers the signal perfectly. Comparing with Figure 1.15 (which shows the estimate provided by the DFE for the same noise variance), we can see that "looking ahead" two steps has been beneficial in this case.   ○

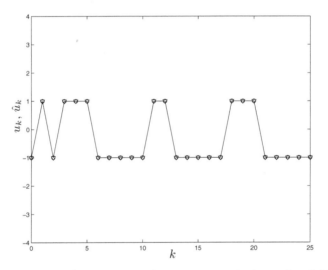

Figure 1.16. Data u_k (circle-solid line) and estimate $\hat{u}_k$ (triangle-solid line) using the moving horizon two-step estimator. Noise variance: $\sigma^2 = 0.2$.

1.4 Connections Between Constrained Control and Estimation

The brief introduction to constrained control and estimation given in Section 1.2 and Section 1.3 will have, no doubt, left the reader with the impression that these two problems are, at least, very *similar*. Indeed, both have been cast as finite horizon constrained optimisation problems. We will see later that these problems lead to the same underlying question, the only difference being a rather minor issue associated with the boundary conditions. Actually, we will show that a strong connection between constrained control and estimation problems is revealed when looked upon via a *Lagrangian duality* perspective. This will be the topic of Chapter 10.

1.5 The Remainder of the Book

The remainder of the book is devoted to expanding on the ideas introduced above. We will emphasise constrained optimisation approaches to the topics of control and estimation. Thus, we begin in the next chapter with a review of basic optimisation theory. This will be followed in Chapter 3 by a review of classical optimal control theory, including the discrete minimum principle. In Chapter 4, and following chapters, we will apply these ideas to the specific issues that arise in control and estimation problems.

1.6 Further Reading

For complete list of references cited, see References section at the end of book.

General

An introduction to unconstrained (linear) control can be found in a host of textbooks, such as Anderson and Moore (1989), Åström and Wittenmark (1990), Bitmead, Gevers and Wertz (1990), Goodwin, Graebe and Salgado (2001), Zhou, Doyle and Glover (1996).

The following books complement the material presented in the current chapter and give further information on receding horizon control: Camacho and Bordons (1999), Maciejowski (2002), Borrelli (2003), Rossiter (2003).

The book Proakis (1995) gives further background on channel equalisation in digital communications. See also the survey papers Qureshi (1985), Tugnait, Tong and Ding (2000).

2

Overview of Optimisation Theory

2.1 Overview

As foreshadowed in Chapter 1, the core idea underlying the approach described in this book to constrained control and estimation will be optimisation theory. This will be the topic of the current chapter. Optimisation theory has huge areas of potential application which extend well beyond the boundaries of control and estimation. However, control and estimation do present an ideal framework within which the basic elements of optimisation theory can be presented.

Key ideas that we present in this chapter include convexity, the Karush–Kuhn–Tucker optimality conditions and Lagrangian duality. These ideas will be drawn upon in following chapters when we apply them to the specific topics of constrained control and estimation. The material for this chapter has been extracted mainly from Bazaraa, Sherali and Shetty (1993). We refer the reader to this reference, as well as to the others mentioned in Section 2.8, for a more complete treatment of optimisation theory and a number of illustrative examples.

2.2 Preliminary Concepts

In this section we review some basic topological properties of sets that will be used throughout the book. Also, we review the definition of differentiability of real-valued functions defined on a subset S of $\mathbb{R}^n$.

2.2.1 Sets and Sequences

Given a point $x \in \mathbb{R}^n$, an *ε-neighbourhood* around x is defined as the set $N_\varepsilon(x) = \{y \in \mathbb{R}^n : ||y - x|| < \varepsilon\}$, for $\varepsilon > 0$, where $|| \cdot ||$ denotes the Euclidean norm of a vector in $\mathbb{R}^n$.

Let S be an arbitrary set in $\mathbb{R}^n$. A point x is said to be in the *closure* of S, denoted by cl S, if $S \cap N_\varepsilon(x) \neq \emptyset$ for every $\varepsilon > 0$. In other words, the closure of a set S is the set of all points that are arbitrarily close to S. If $S = \text{cl } S$, then S is called *closed*. A point $x \in S$ is in the *interior* of S, denoted by int S, if $N_\varepsilon(x) \subset S$ for some $\varepsilon > 0$. If $S = \text{int } S$, then S is called *open*.

A point x is in the *boundary* of S, denoted by ∂S, if $N_\varepsilon(x)$ contains at least one point in S and one point not in S for every $\varepsilon > 0$. Hence, a set S is closed if and only if it contains all its boundary points. Moreover, cl $S \equiv S \cup \partial S$ is the smallest closed set containing S. Similarly, a set S is open if and only if it does not contain any of its boundary points. Clearly, a set may be neither open nor closed, and the only sets in $\mathbb{R}^n$ that are both open and closed are the empty set and $\mathbb{R}^n$ itself. Also, note that any point $x \in S$ must be either an interior or a boundary point of S. However, in general, $S \neq \text{int } S \cup \partial S$, since S need not contain its boundary points. On the other hand, since int $S \subseteq S$, we have, int $S = S - \partial S$, whilst, in general, $\partial S \neq S - \text{int } S$.

A *sequence* of points, or vectors, $\{x_1, x_2, x_3, \ldots\}$, is said to *converge* to the *limit point* $\bar{x}$ if $||x_k - \bar{x}|| \to 0$ as $k \to \infty$; that is, if for any given $\varepsilon > 0$, there is a positive integer N such that $||x_k - \bar{x}|| < \varepsilon$ for all $k \geq N$. The sequence will be denoted by $\{x_k\}$, and the limit point $\bar{x}$ is represented by $x_k \to \bar{x}$ as $k \to \infty$. Any converging sequence has a unique limit point. By deleting certain elements of a sequence $\{x_k\}$, we obtain a *subsequence*, denoted by $\{x_k\}_K$, where K is a subset of all positive integers. To illustrate, let K be the set of all even positive integers, then $\{x_k\}_K$ denotes the subsequence $\{x_2, x_4, x_6, \ldots\}$.

An equivalent definition of closed sets, that is useful when demonstrating that a set is closed, is based on sequences of points contained in S. A set S is closed if and only if, for any convergent sequence of points $\{x_k\}$ contained in S with limit point $\bar{x}$, we also have $\bar{x} \in S$.

A set is *bounded* if it can be contained in a neighbourhood of sufficiently large but bounded radius. A *compact* set is one that is both closed and bounded. For every sequence $\{x_k\}$ in a compact set S, there is a convergent subsequence with a limit in S.

2.2.2 Differentiable Functions

We next investigate differentiability of a real-valued function f defined on a subset S of $\mathbb{R}^n$.

Definition 2.2.1 (Differentiable Function) *Let S be a set in $\mathbb{R}^n$ with a nonempty interior, and let $f : S \to \mathbb{R}$. Then, f is said to be* differentiable *at $\bar{x} \in \text{int } S$ if there exists a vector $\nabla f(\bar{x})^{\mathrm{T}} \in \mathbb{R}^n$, called the* gradient vector,[1] *and a function $\alpha : \mathbb{R}^n \to \mathbb{R}$, such that*

---

[1] Although nonstandard, here we will consider the gradient vector ∇f a row vector to be consistent with notation used in the remainder of the book.

$$f(x) = f(\bar{x}) + \nabla f(\bar{x})(x - \bar{x}) + \|x - \bar{x}\|\alpha(\bar{x}, x - \bar{x}) \quad \text{for all } x \in S, \quad (2.1)$$

where $\lim_{x \to \bar{x}} \alpha(\bar{x}, x - \bar{x}) = 0$. *The function f is said to be* differentiable on *the open set $S' \subseteq S$ if it is differentiable at each point in S'. The above representation of f is called a* first-order (Taylor series) expansion *of f at $\bar{x}$.*

○

Note that if f is differentiable at $\bar{x}$, then there can be only one gradient vector, and this vector consists of the partial derivatives, that is,

$$\nabla f(\bar{x}) = \left(\frac{\partial f(\bar{x})}{\partial x_1}, \frac{\partial f(\bar{x})}{\partial x_2}, \ldots, \frac{\partial f(\bar{x})}{\partial x_n} \right).$$

Definition 2.2.2 (Twice-Differentiable Function) *Let S be a set in $\mathbb{R}^n$ with a nonempty interior, and let $f : S \to \mathbb{R}$. Then, f is said to be* twice-differentiable *at $\bar{x} \in$ int S if there exists a vector $\nabla f(\bar{x})^{\mathrm{T}} \in \mathbb{R}^n$, and an $n \times n$ symmetric matrix $H(\bar{x})$, called the* Hessian matrix, *and a function $\alpha : \mathbb{R}^n \to \mathbb{R}$, such that*

$$f(x) = f(\bar{x}) + \nabla f(\bar{x})(x - \bar{x}) + \frac{1}{2}(x - \bar{x})^{\mathrm{T}} H(\bar{x})(x - \bar{x}) +$$
$$\|x - \bar{x}\|^2 \alpha(\bar{x}, x - \bar{x}) \quad \text{for all } x \in S,$$

where $\lim_{x \to \bar{x}} \alpha(\bar{x}, x - \bar{x}) = 0$. *The function f is said to be* twice-differentiable *on the open set $S' \subseteq S$ if it is twice-differentiable at each point in S'. The above representation of f is called a* second-order (Taylor series) expansion *of f at $\bar{x}$.*

○

For a twice-differentiable function, the Hessian matrix $H(\bar{x})$ comprises the second-order partial derivatives, that is, the element in row i and column j of the Hessian matrix is the second partial derivative $\partial^2 f(\bar{x})/\partial x_i \partial x_j$.

A useful theorem, which applies to differentiable functions defined on a convex set, is the *mean value theorem*, stated below. (Convex sets are formally defined in the next section.)

Theorem 2.2.1 (Mean Value Theorem) *Let S be a nonempty open convex set in $\mathbb{R}^n$, and let $f : S \longrightarrow \mathbb{R}$ be differentiable. Then, for every x_1 and x_2 in S, we must have*

$$f(x_2) = f(x_1) + \nabla f(x)(x_2 - x_1),$$

where $x = \lambda x_1 + (1 - \lambda)x_2$ for some $\lambda \in (0, 1)$.

○

2.2.3 Weierstrass' Theorem

The following result, based on the foregoing concepts, relates to the existence of a minimising solution for an optimisation problem. We shall say that $\min\{f(x) : x \in S\}$ exists if there exists a minimising solution $\bar{x} \in S$ such that $f(\bar{x}) \leq f(x)$ for all $x \in S$. On the other hand, we say that $\alpha = \inf\{f(x) : x \in S\}$ if α is the greatest lower bound of f on S. We now prove that if S is nonempty, closed and bounded, and if f is continuous on S, then a minimum exists.

Theorem 2.2.2 (Weierstrass' Theorem: Existence of a Solution)
Let $S \subset \mathbb{R}^n$ be a nonempty, compact set, and let $f : S \longrightarrow \mathbb{R}$ be continuous on S. Then $f(x)$ attains its minimum in S, that is, there exists a minimising solution to the problem $\min\{f(x) : x \in S\}$.

Proof. Since f is continuous on S, and S is both closed and bounded, f is bounded below on S. Consequently, since $S \neq \emptyset$, there exists a greatest lower bound $\alpha = \inf\{f(x) : x \in S\}$. Now, let $0 < \varepsilon < 1$, and consider the set $S_k = \{x \in S : \alpha \leq f(x) \leq \alpha + \varepsilon^k\}$ for $k = 1, 2, \ldots$. By the definition of an infimum, $S_k \neq \emptyset$ for each k, and so we can construct a sequence of points $\{x_k\} \subseteq S$ by selecting a point $x_k \in S_k$ for each $k = 1, 2, \ldots$. Since S is bounded, there exists a convergent subsequence $\{x_k\}_K \to \bar{x}$, indexed by the set K. By the closedness of S, we have $\bar{x} \in S$; and by the continuity of f, since $\alpha \leq f(x_k) \leq \alpha + \varepsilon^k$ for all k, we have $\alpha = \lim_{k \to \infty, k \in K} f(x_k) = f(\bar{x})$. Hence, we have shown that there exists a solution $\bar{x} \in S$ such that $f(\bar{x}) = \alpha = \inf\{f(x) : x \in S\}$, and so $\bar{x}$ is a minimising solution. This completes the proof. $\square$

2.3 Convex Analysis

One of the main concepts that underpins optimisation theory is that of *convexity*. Indeed, the *big divide* in optimisation is between convex problems and nonconvex problems, rather than between, say, linear and nonlinear problems. Thus, understanding the notion of convexity can be a crucial step in solving many real world problems.

2.3.1 Convex Sets

We have the following definition of a convex set.

Definition 2.3.1 (Convex Set) *A set $S \subset \mathbb{R}^n$ is* convex *if the line segment joining any two points of the set also belongs to the set. In other words, if $x_1, x_2 \in S$ then $\lambda x_1 + (1 - \lambda)x_2$ must also belong to S for each $\lambda \in [0, 1]$.* ○

Figure 2.1 below illustrates the notions of convex and nonconvex sets. Note that in Figure 2.1 (b), the line segment joining x_1 and x_2 does not lie entirely in the set.

The following are some examples of convex sets:

(i) **Hyperplane.** $S = \{x : p^\mathsf{T}x = \alpha\}$, where p is a nonzero vector in $\mathbb{R}^n$, called the *normal* to the hyperplane, and α is a scalar.

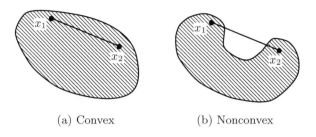

(a) Convex   (b) Nonconvex

Figure 2.1. Illustration of a convex and a nonconvex set.

(ii) **Half-space.** $S = \{x : p^{\mathrm{T}}x \leq \alpha\}$, where p is a nonzero vector in $\mathbb{R}^n$, and α is a scalar.

(iii) **Open half-space.** $S = \{x : p^{\mathrm{T}}x < \alpha\}$, where p is a nonzero vector in $\mathbb{R}^n$ and α is a scalar.

(iv) **Polyhedral set.** $S = \{x : Ax \leq b\}$, where A is an $m \times n$ matrix, and b is an m vector. (Here and in the remainder of the book the inequality should be interpreted *elementwise*.)

(v) **Polyhedral cone.** $S = \{x : Ax \leq 0\}$, where A is an $m \times n$ matrix.

(vi) **Cone spanned by a finite number of vectors.** $S = \{x : x = \sum_{j=1}^{m} \lambda_j a_j, \lambda_j \geq 0, \text{ for } j = 1, \ldots, m\}$, where $a_1, \ldots, a_m$ are given vectors in $\mathbb{R}^n$.

(vii) **Neighbourhood.** $N_\varepsilon(\bar{x}) = \{x \in \mathbb{R}^n : ||x - \bar{x}|| < \varepsilon\}$, where $\bar{x}$ is a fixed vector in $\mathbb{R}^n$ and $\varepsilon > 0$.

Some of the geometric optimality conditions presented in this chapter use *convex cones*, defined below.

Definition 2.3.2 (Convex Cone) *A nonempty set C in $\mathbb{R}^n$ is called a cone with vertex zero if $x \in C$ implies that $\lambda x \in C$ for all $\lambda \geq 0$. If, in addition, C is convex, then C is called a* convex cone. ○

Figure 2.2 shows an example of a convex cone and an example of a nonconvex cone.

2.3.2 Separation and Support of Convex Sets

Almost all optimality conditions and duality relationships use some sort of separation or support of convex sets. We begin by stating the geometric facts that, given a closed convex set S and a point $y \notin S$, there exists a unique point $\bar{x} \in S$ with minimum distance from y (Theorem 2.3.1) and a hyperplane that separates y and S (Theorem 2.3.2).

Theorem 2.3.1 (Closest Point Theorem) *Let S be a nonempty, closed convex set in $\mathbb{R}^n$ and $y \notin S$. Then, there exists a unique point $\bar{x} \in S$ with*

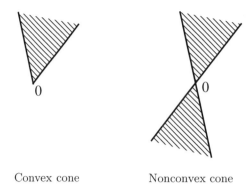

Convex cone   Nonconvex cone

Figure 2.2. Examples of cones.

minimum distance from y. Furthermore, $\bar{x}$ is the minimising point, or closest point to y, if and only if $(y - \bar{x})^{\mathrm{T}}(x - \bar{x}) \leq 0$ for all $x \in S$.

Proof. We first establish the existence of a closest point. Since $S \neq \emptyset$, there exists a point $\hat{x} \in S$, and we can confine our attention to the set $\bar{S} = S \cap \{x : \|y - x\| \leq \|y - \hat{x}\|\}$ in seeking the closest point. In other words, the closest point problem $\inf\{\|y - x\| : x \in S\}$ is equivalent to $\inf\{\|y - x\| : x \in \bar{S}\}$. However, the latter problem involves finding the minimum of a continuous function over a nonempty, compact set $\bar{S}$, and so by Weierstrass' theorem (Theorem 2.2.2) we know that there exists a minimising point $\bar{x}$ in S that is closest to the point y.

Next, we prove that the closest point is unique. Suppose that there is an $\bar{x}' \in S$ such that $\|y - \bar{x}\| = \|y - \bar{x}'\| = \gamma$. By convexity of S, $\frac{1}{2}\bar{x} + \frac{1}{2}\bar{x}' \in S$. By the triangle inequality, we obtain

$$\left\| y - \left(\frac{1}{2}\bar{x} + \frac{1}{2}\bar{x}' \right) \right\| \leq \frac{1}{2}\|y - \bar{x}\| + \frac{1}{2}\|y - \bar{x}'\| = \gamma.$$

If strict inequality holds, we have a contradiction to $\bar{x}$ being the closest point to y. Therefore, equality holds, and we must have $y - \bar{x} = \lambda(y - \bar{x}')$ for some λ. Since $\|y - \bar{x}\| = \|y - \bar{x}'\| = \gamma$, $|\lambda| = 1$. Clearly, $\lambda \neq -1$, because otherwise we would have $y = \frac{1}{2}\bar{x} + \frac{1}{2}\bar{x}' \in S$, contradicting the assumption that $y \notin S$. So, $\lambda = 1$, $\bar{x}' = \bar{x}$, and uniqueness is established.

Finally, we prove that $(y - \bar{x})^{\mathrm{T}}(x - \bar{x}) \leq 0$ for all $x \in S$ is both a necessary and sufficient condition for $\bar{x}$ to be the point in S closest to y. To prove sufficiency, let $x \in S$. Then,

$$\|y - x\|^2 = \|y - \bar{x} + \bar{x} - x\|^2 = \|y - \bar{x}\|^2 + \|\bar{x} - x\|^2 + 2(\bar{x} - x)^{\mathrm{T}}(y - \bar{x}).$$

Since $\|\bar{x} - x\|^2 \geq 0$ and $(\bar{x} - x)^{\mathrm{T}}(y - \bar{x}) \geq 0$ by assumption, $\|y - x\|^2 \geq \|y - \bar{x}\|^2$, and $\bar{x}$ is the minimising point. Conversely, assume that $\|y - x\|^2 \geq \|y - \bar{x}\|^2$

for all $x \in S$. Let $x \in S$ and note that $\bar{x} + \lambda(x - \bar{x}) \in S$ for all $0 \leq \lambda \leq 1$ by the convexity of S. Therefore,

$$\|y - \bar{x} - \lambda(x - \bar{x})\|^2 \geq \|y - \bar{x}\|^2. \tag{2.2}$$

Also

$$\|y - \bar{x} - \lambda(x - \bar{x})\|^2 = \|y - \bar{x}\|^2 + \lambda^2\|x - \bar{x}\|^2 - 2\lambda(y - \bar{x})^{\mathrm{T}}(x - \bar{x}). \tag{2.3}$$

From (2.2) and (2.3), we obtain

$$2\lambda(y - \bar{x})^{\mathrm{T}}(x - \bar{x}) \leq \lambda^2\|x - \bar{x}\|^2, \tag{2.4}$$

for all $0 \leq \lambda \leq 1$. Dividing (2.4) by any such $\lambda > 0$ and letting $\lambda \to 0^+$, the result follows. □

The above theorem is illustrated in Figure 2.3.

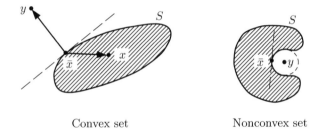

Convex set   Nonconvex set

Figure 2.3. Closest point to a closed convex set.

Definition 2.3.3 (Separation of Sets) *Let S_1 and S_2 be nonempty sets in $\mathbb{R}^n$. A hyperplane $H = \{x : p^{\mathrm{T}}x = \alpha\}$ separates S_1 and S_2 if $p^{\mathrm{T}}x \geq \alpha$ for each $x \in S_1$ and $p^{\mathrm{T}}x \leq \alpha$ for each $x \in S_2$. If, in addition, $p^{\mathrm{T}}x \geq \alpha + \varepsilon$ for each $x \in S_1$ and $p^{\mathrm{T}}x \leq \alpha$ for each $x \in S_2$, where ε is a positive scalar, then the hyperplane H is said to* strongly separate *the sets S_1 and S_2. (Notice that strong separation implies separation of sets.)*

Figure 2.4 illustrates the concepts of separation and strong separation of sets.

The following is the most fundamental separation theorem.

Theorem 2.3.2 (Separation Theorem) *Let S be a nonempty closed convex set in $\mathbb{R}^n$ and $y \notin S$. Then, there exists a nonzero vector p and a scalar α such that $p^{\mathrm{T}}y > \alpha$ and $p^{\mathrm{T}}x \leq \alpha$ for each $x \in S$.*

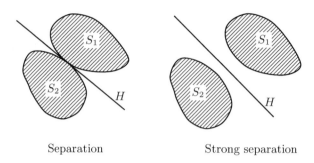

Separation   Strong separation

Figure 2.4. Separation and strong separation of sets.

Proof. S is a nonempty closed convex set and $y \notin S$. Hence, by Theorem 2.3.1 there exists a unique minimising point $\bar{x} \in S$ such that $(y - \bar{x})^{\mathrm{T}}(x - \bar{x}) \le 0$ for each $x \in S$. Letting $p = (y - \bar{x}) \ne 0$ and $\alpha = (y - \bar{x})^{\mathrm{T}}\bar{x} = p^{\mathrm{T}}\bar{x}$, we obtain $p^{\mathrm{T}}x \le \alpha$ for each $x \in S$. We also have $p^{\mathrm{T}}y - \alpha = (y - \bar{x})^{\mathrm{T}}(y - \bar{x}) = \|y - \bar{x}\|^2 > 0$ and, hence, $p^{\mathrm{T}}y > \alpha$. This completes the proof. $\qquad\square$

Closely related to the above concept is the notion of a *supporting hyper-plane*.

Definition 2.3.4 (Supporting Hyperplane at a Boundary Point)
Let S be a nonempty set in $\mathbb{R}^n$, and let $\bar{x} \in \partial S$. A hyperplane $H = \{x : p^{\mathrm{T}}(x - \bar{x}) = 0\}$ is called a supporting *hyperplane of S at $\bar{x}$ if either $p^{\mathrm{T}}(x - \bar{x}) \ge 0$ for each $x \in S$, or else, $p^{\mathrm{T}}(x - \bar{x}) \le 0$ for each $x \in S$.*

Figure 2.5 shows an example of a supporting hyperplane.

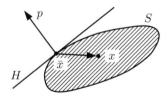

Figure 2.5. Supporting hyperplane.

The next result shows that a convex set has a supporting hyperplane at each boundary point. As a corollary, a result similar to Theorem 2.3.2, where S is not required to be closed, follows.

Theorem 2.3.3 (Supporting Hyperplane) *Let S be a nonempty convex set in $\mathbb{R}^n$, and let $\bar{x} \in \partial S$. Then there exists a hyperplane that supports S at*

$\bar{x}$; that is, there exists a nonzero vector p such that $p^{\mathrm{T}}(x - \bar{x}) \leq 0$ for each $x \in \mathrm{cl}\, S$.

Proof. Since $\bar{x} \in \partial S$, there exists a sequence $\{y_k\}$ not in cl S such that $y_k \to \bar{x}$. By Theorem 2.3.2, corresponding to each y_k there exists a p_k such that $p_k^{\mathrm{T}} y_k > p_k^{\mathrm{T}} x$ for each $x \in \mathrm{cl}\, S$. Without loss of generality, we can normalise the vector in Theorem 2.3.2 by dividing it by its norm, such that $\|p_k\| = 1$. Since $\{p_k\}$ is bounded, it has a convergent subsequence $\{p_k\}_K$ with limit p whose norm is also equal to 1. Considering this subsequence, we have $p_k^{\mathrm{T}} y_k > p_k^{\mathrm{T}} x$ for each $x \in \mathrm{cl}\, S$. Fixing $x \in \mathrm{cl}\, S$ and taking limits as $k \in K$ approaches ∞, we obtain, $p^{\mathrm{T}}(x - \bar{x}) \leq 0$. Since this is true for each $x \in \mathrm{cl}\, S$, the result follows. $\qquad\square$

Corollary 2.3.4 *Let S be a nonempty convex set in $\mathbb{R}^n$ and $\bar{x} \notin \mathrm{int}\, S$. Then there is a nonzero vector p such that $p^{\mathrm{T}}(x - \bar{x}) \leq 0$ for each $x \in \mathrm{cl}\, S$.*

Proof. If $\bar{x} \notin \mathrm{cl}\, S$, then the result follows from Theorem 2.3.2 choosing $y = \bar{x}$. On the other hand, if $\bar{x} \in \partial S$, the result follows from Theorem 2.3.3. $\qquad\square$

The next theorem shows that, if two convex sets are disjoint, then they can be separated by a hyperplane.

Theorem 2.3.5 (Separation of Two Disjoint Convex Sets) *Let S_1 and S_2 be nonempty convex sets in $\mathbb{R}^n$ and suppose that $S_1 \cap S_2$ is empty. Then there exists a hyperplane that separates S_1 and S_2; that is, there exists a nonzero vector p in $\mathbb{R}^n$ such that*

$$\inf\{p^{\mathrm{T}} x : x \in S_1\} \geq \sup\{p^{\mathrm{T}} x : x \in S_2\}.$$

Proof. Consider the set $S = S_1 \ominus S_2 \triangleq \{x_1 - x_2 : x_1 \in S_1 \text{ and } x_2 \in S_2\}$. Note that S is convex. Furthermore, $0 \notin S$, because otherwise $S_1 \cap S_2$ would be nonempty. By Corollary 2.3.4, there exists a nonzero $p \in \mathbb{R}^n$ such that $p^{\mathrm{T}} x \geq 0$ for all $x \in S$. This means that $p^{\mathrm{T}} x_1 \geq p^{\mathrm{T}} x_2$ for all $x_1 \in S_1$ and $x_2 \in S_2$, and the result follows. $\qquad\square$

The following corollary shows that the above result holds true even if the two sets have some points in common, as long as their interiors are disjoint.

Corollary 2.3.6 *Let S_1 and S_2 be nonempty convex sets in $\mathbb{R}^n$. Suppose that $\mathrm{int}\, S_2$ is not empty and that $S_1 \cap \mathrm{int}\, S_2$ is empty. Then, there exists a hyperplane that separates S_1 and S_2; that is, there exists a nonzero p such that*

$$\inf\{p^{\mathrm{T}} x : x \in S_1\} \geq \sup\{p^{\mathrm{T}} x : x \in S_2\}.$$

Proof. Replace S_2 by $\mathrm{int}\, S_2$, apply Theorem 2.3.5, and note that

$$\sup\{p^{\mathrm{T}} x : x \in S_2\} = \sup\{p^{\mathrm{T}} x : x \in \mathrm{int}\, S_2\}.$$

The result then follows. $\qquad\square$

2.3.3 Convex Functions

Convex functions have many important properties for optimisation problems. For example, any local minimum of a convex function over a convex set is also a global minimum. We present here some properties of convex functions, beginning with their definition.

Definition 2.3.5 (Convex Function) *Let* $f : S \rightarrow \mathbb{R}$, *where* S *is a nonempty convex set in* $\mathbb{R}^n$. *The function* f *is* convex *on* S *if*

$$f(\lambda x_1 + (1 - \lambda)x_2) \leq \lambda f(x_1) + (1 - \lambda)f(x_2)$$

for each x_1, $x_2 \in S$ *and for each* $\lambda \in (0, 1)$.

The function f *is* strictly convex *on* S *if the above inequality is true as a strict inequality for each distinct* x_1, $x_2 \in S$ *and for each* $\lambda \in (0, 1)$.

The function f *is* (strictly) concave *on* S *if* $-f$ *is (strictly) convex on* S.

○

The geometric interpretation of a convex function is that the value of f at the point $\lambda x_1 + (1 - \lambda)x_2$ is less than the height of the chord joining the points $[x_1, f(x_1)]$ and $[x_2, f(x_2)]$. For a concave function, the chord is below the function itself. Figure 2.6 shows some examples of convex and concave functions.

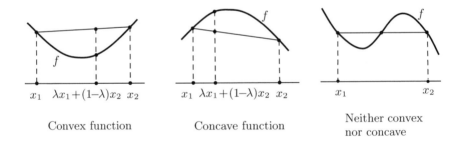

Convex function   Concave function   Neither convex nor concave

Figure 2.6. Examples of convex and concave functions.

The following are useful properties of convex functions.

(i) Let f_1, f_2, ..., $f_k : \mathbb{R}^n \rightarrow \mathbb{R}$ be convex functions. Then
  - $f(x) = \sum_{j=1}^{k} \alpha_j f_j(x)$, where $\alpha_j > 0$ for $j = 1, 2, \ldots k$, is a convex function;
  - $f(x) = \max\{f_1(x), f_2(x), \ldots, f_k(x)\}$ is a convex function.
(ii) Suppose that $g : \mathbb{R}^n \rightarrow \mathbb{R}$ is a concave function. Let $S = \{x : g(x) > 0\}$, and define $f : S \rightarrow \mathbb{R}$ as $f(x) = 1/g(x)$. Then f is convex over S.

(iii) Let $g : \mathbb{R} \to \mathbb{R}$ be a nondecreasing, univariate, convex function, and let $h : \mathbb{R}^n \to \mathbb{R}$ be a convex function. Then the composite function $f : \mathbb{R}^n \to \mathbb{R}$ defined as $f(x) = g(h(x))$ is a convex function.

(iv) Let $g : \mathbb{R}^m \to \mathbb{R}$ be a convex function, and let $h : \mathbb{R}^n \to \mathbb{R}^m$ be an affine function of the form $h(x) = Ax + b$, where A is an $m \times n$ matrix, and b is an $m \times 1$ vector. Then the composite function $f : \mathbb{R}^n \to \mathbb{R}$ defined as $f(x) = g(h(x))$ is a convex function.

Associated with a convex function f is the *level set* S_α defined as $S_\alpha = \{x \in S : f(x) \le \alpha\}$, $\alpha \in \mathbb{R}$. We then have:

Lemma 2.3.7 (Convexity of Level Sets) *Let S be a nonempty convex set in $\mathbb{R}^n$ and let $f : S \to \mathbb{R}$ be a convex function. Then the level set $S_\alpha = \{x \in S : f(x) \le \alpha\}$, where $\alpha \in \mathbb{R}$, is a convex set.*

Proof. Let $x_1, x_2 \in S_\alpha$. Thus, $x_1, x_2 \in S$, and $f(x_1) \le \alpha$ and $f(x_2) \le \alpha$. Now, let $\lambda \in (0, 1)$ and $x = \lambda x_1 + (1 - \lambda)x_2 \in S$ (by the convexity of S). Furthermore, by convexity of f,

$$f(x) \le \lambda f(x_1) + (1 - \lambda)f(x_2) \le \lambda \alpha + (1 - \lambda)\alpha = \alpha.$$

Hence, $x \in S_\alpha$, and we conclude that S_α is convex. $\qquad\square$

An important property of convex functions is that they are continuous on the interior of their domain, as we prove next.

Theorem 2.3.8 (Continuity of Convex Functions) *Let S be a nonempty convex set in $\mathbb{R}^n$ and let $f : S \to \mathbb{R}$ be a convex function. Then f is continuous on the interior of S.*

Proof. Let $\bar{x} \in$ int S. Hence, there exists a $\delta' > 0$ such that $\|x - \bar{x}\| \le \delta'$ implies that $x \in S$. Consider the vector $e_i \in \mathbb{R}^n$ having all elements equal to zero except for a 1 in the ith position. Now, construct

$$\theta \triangleq \max_{1 \le i \le n} \{\max [f(\bar{x} + \delta' e_i) - f(\bar{x}), f(\bar{x} - \delta' e_i) - f(\bar{x})]\}. \qquad (2.5)$$

Note, from the convexity of f, that we have:

$$f(\bar{x}) = f \left[\frac{1}{2}(\bar{x} + \delta' e_i) + \frac{1}{2}(\bar{x} - \delta' e_i) \right] \le \frac{1}{2}f(\bar{x} + \delta' e_i) + \frac{1}{2}f(\bar{x} - \delta' e_i),$$

for all $1 \le i \le n$, from where we conclude that $\theta \ge 0$.

Now, for any given $\epsilon > 0$, define:

$$\delta \triangleq \min \left\{ \frac{\delta'}{n}, \frac{\epsilon \delta'}{n\theta} \right\}. \qquad (2.6)$$

Choose an x with $\|x - \bar{x}\| \le \delta$. Let v_i denote the ith element of a vector v. If $x_i - \bar{x}_i \ge 0$, define $z_i = \delta' e_i$, otherwise define $z_i = -\delta' e_i$. Then, $x - \bar{x} = \sum_{i=1}^n \alpha_i z_i$, for some $\alpha_i \ge 0$, $1 \le i \le n$. Furthermore,

$$\|x - \bar{x}\| = \delta' \left(\sum_{i=1}^{n} \alpha_i^2 \right)^{\frac{1}{2}} \leq \delta. \tag{2.7}$$

It follows from (2.6) and (2.7) that $\alpha_i \leq 1/n$ and $\alpha_i \leq \epsilon/n\theta$, for $i = 1, 2, \ldots, n$. From the convexity of f, and since $0 \leq n\alpha_i \leq 1$, we obtain

$$f(x) = f\left(\bar{x} + \sum_{i=1}^{n} \alpha_i z_i \right) = f\left(\frac{1}{n} \sum_{i=1}^{n} (\bar{x} + n\alpha_i z_i) \right) \leq \frac{1}{n} \sum_{i=1}^{n} f(\bar{x} + n\alpha_i z_i)$$

$$= \frac{1}{n} \sum_{i=1}^{n} f[(1 - n\alpha_i)\bar{x} + n\alpha_i(\bar{x} + z_i)]$$

$$\leq \frac{1}{n} \sum_{i=1}^{n} [(1 - n\alpha_i)f(\bar{x}) + n\alpha_i f(\bar{x} + z_i)].$$

Therefore, $f(x) - f(\bar{x}) \leq \sum_{i=1}^{n} \alpha_i [f(\bar{x} + z_i) - f(\bar{x})]$. From (2.5) and the definition of z_i it follows that $f(\bar{x} + z_i) - f(\bar{x}) \leq \theta$ for each i; and since $\alpha_i \geq 0$, it follows that

$$f(x) - f(\bar{x}) \leq \theta \sum_{i=1}^{n} \alpha_i. \tag{2.8}$$

As noted above, $\alpha_i \leq \epsilon/n\theta$, for $i = 1, 2, \ldots, n$ and, thus, it follows from (2.8) that $f(x) - f(\bar{x}) \leq \epsilon$.

Now, let $y = 2\bar{x} - x$ and note that $\|y - \bar{x}\| \leq \delta$. Hence, as above, we have $f(y) - f(\bar{x}) \leq \epsilon$. But, $\bar{x} = \frac{1}{2}y + \frac{1}{2}x$, and by the convexity of f, we have $f(\bar{x}) \leq \frac{1}{2}f(y) + \frac{1}{2}f(x)$. Combining the last two inequalities, it follows that $f(\bar{x}) - f(x) \leq \epsilon$.

Summarising, we have shown that for any $\epsilon > 0$ there exists a $\delta > 0$ (defined as in (2.6)) such that $\|x - \bar{x}\| \leq \delta$ implies that $f(x) - f(\bar{x}) \leq \epsilon$ and that $f(\bar{x}) - f(x) \leq \epsilon$; that is, that $|f(x) - f(\bar{x})| \leq \epsilon$. Hence, f is continuous at $\bar{x} \in \text{int } S$, and the proof is complete. $\square$

2.3.4 Generalisations of Convex Functions

We present various types of functions that are similar to convex or concave functions but share only some of their desirable properties.

Definition 2.3.6 (Quasiconvex Function) *Let $f : S \to \mathbb{R}$, where S is a nonempty convex set in $\mathbb{R}^n$. The function f is* quasiconvex *if, for each $x_1, x_2 \in S$, the following inequality is true:*

$$f(\lambda x_1 + (1 - \lambda)x_2) \leq \max\{f(x_1), f(x_2)\} \quad \text{for each } \lambda \in (0, 1).$$

The function f is quasiconcave *if $-f$ is quasiconvex.*   ○

Note, from the definition, that a convex function is quasiconvex.

From the above definition, a function f is quasiconvex if, whenever $f(x_2) \geq f(x_1)$, $f(x_2)$ is greater than or equal to f at all convex combinations of x_1 and x_2. Hence, if f increases locally from its value at a point along any direction, it must remain nondecreasing in that direction. Figure 2.7 shows some examples of quasiconvex and quasiconcave functions.

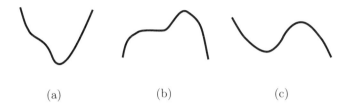

(a)   (b)   (c)

Figure 2.7. (a) Quasiconvex function. (b) Quasiconcave function. (c) Neither quasiconvex nor quasiconcave.

The following result states that a quasiconvex function is characterised by the convexity of its level sets.

Theorem 2.3.9 (Level Sets of a Quasiconvex Function) *Let $f : S \to \mathbb{R}$, where S is a nonempty convex set in $\mathbb{R}^n$. The function f is quasiconvex if and only if $S_\alpha = \{x \in S : f(x) \leq \alpha\}$ is convex for each real number α.*

Proof. Suppose that f is quasiconvex, and let $x_1, x_2 \in S_\alpha$. Therefore, $x_1, x_2 \in S$ and $\max \{f(x_1), f(x_2)\} \leq \alpha$. Let $\lambda \in (0,1)$, and let $x = \lambda x_1 + (1-\lambda)x_2$. By the convexity of S, $x \in S$. Furthermore, by the quasiconvexity of f, $f(x) \leq \max \{f(x_1), f(x_2)\} \leq \alpha$. Hence, $x \in S_\alpha$, and thus S_α is convex. Conversely, suppose that S_α is convex for each real number α. Let $x_1, x_2 \in S$ and take $\alpha = \max \{f(x_1), f(x_2)\}$. Hence, $x_1, x_2 \in S_\alpha$. Furthermore, let $\lambda \in (0,1)$ and $x = \lambda x_1 + (1-\lambda)x_2$. By assumption, S_α is convex, so that $x \in S_\alpha$. Therefore, $f(x) \leq \alpha = \max \{f(x_1), f(x_2)\}$. Hence, f is quasiconvex, and the proof is complete. $\square$

We will next define *strictly quasiconvex functions*.

Definition 2.3.7 (Strictly Quasiconvex Function) *Let $f : S \to \mathbb{R}$, where S is a nonempty convex set in $\mathbb{R}^n$. The function f is strictly quasiconvex if, for each $x_1, x_2 \in S$ with $f(x_1) \neq f(x_2)$, the following inequality is true*

$$f(\lambda x_1 + (1 - \lambda)x_2) < \max \{f(x_1), f(x_2)\} \quad \text{for each } \lambda \in (0,1).$$

The function f is strictly quasiconcave if $-f$ is strictly quasiconvex.   ○

Note from the above definition that a convex function is also strictly quasiconvex. Figure 2.8 shows some examples of quasiconvex and strictly quasiconvex functions.

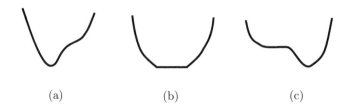

(a)   (b)   (c)

Figure 2.8. (a) Strictly quasiconvex function. (b) Strictly quasiconvex function. (c) Quasiconvex function but not strictly quasiconvex.

Notice that the definition precludes any *flat spots* from occurring anywhere except at extremising points. This, in turn, implies that a local optimal solution of a strictly quasiconvex function over a convex set is also a global optimal solution. (Local and global optima for constrained optimisation problems will be formally defined in Definition 2.5.1.)

We observe that strictly quasiconvex functions are not necessarily quasiconvex. However, if f is *lower semicontinuous*[2], then it can be shown that strict quasiconvexity implies quasiconvexity.

We will next introduce another type of function that generalises the concept of a convex function, called a *pseudoconvex function*. Pseudoconvex functions share the property of convex functions that, if $\nabla f(\bar{x}) = 0$ at some point $\bar{x}$, then $\bar{x}$ is a global minimum of f. (See Theorem 2.4.5.)

Definition 2.3.8 (Pseudoconvex Function) *Let S be a nonempty open set in $\mathbb{R}^n$, and let $f : S \to \mathbb{R}$ be differentiable on S. The function f is pseudoconvex if, for each $x_1, x_2 \in S$ with $\nabla f(x_1)(x_2 - x_1) \geq 0$, then $f(x_2) \geq f(x_1)$; or, equivalently, if $f(x_2) < f(x_1)$, then $\nabla f(x_1)(x_2 - x_1) < 0$.*

The function f is pseudoconcave if $-f$ is pseudoconvex.

The function f is strictly pseudoconvex if, for each distinct $x_1, x_2 \in S$ with $\nabla f(x_1)(x_2 - x_1) \geq 0$, then $f(x_2) > f(x_1)$; or, equivalently, if for each distinct $x_1, x_2 \in S$, $f(x_2) \leq f(x_1)$, then $\nabla f(x_1)(x_2 - x_1) < 0$.   ○

Note that the definition asserts that if the directional derivative of a pseudoconvex function at any point x_1 in the direction $x_2 - x_1$ is nonnegative,

---

[2] A function $f : S \to \mathbb{R}$, where S is a nonempty set in $\mathbb{R}^n$, is *lower semicontinuous* at $\bar{x} \in S$ if for each $\varepsilon > 0$ there exists a $\delta > 0$ such that $x \in S$ and $\|x - \bar{x}\| < \delta$ imply that $f(x) - f(\bar{x}) > -\varepsilon$. Obviously, a continuous function at $\bar{x}$ is also lower semicontinuous at $\bar{x}$.

then the function values are nondecreasing in that direction. Figure 2.9 shows examples of pseudoconvex and pseudoconcave functions.

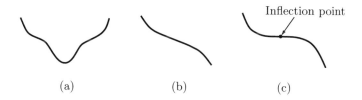

Figure 2.9. (a) Pseudoconvex function. (b) Both pseudoconvex and pseudoconcave. (c) Neither pseudoconvex nor pseudoconcave.

Several relationships among the different types of convexity can be established. For example, one of these relationships is that every pseudoconvex function is both strictly quasiconvex and quasiconvex. Figure 2.10 summarises the implications among the different types of convexity. In the particular case of a *quadratic* function f it can be shown that f is pseudoconvex if and only if f is strictly quasiconvex, which holds true if and only if f is quasiconvex.

Convexity at a Point

In some optimisation problems, the requirement of convexity may be too strong and not essential, and convexity at a point may be all that is needed. Hence, we present below several types of convexity at a point that are relaxations of the various forms of convexity presented so far.

Definition 2.3.9 (Various Types of Convexity at a Point) *Let S be a nonempty convex set in $\mathbb{R}^n$, and $f : S \to \mathbb{R}$. We then have the following definitions:*

Convexity at a point. The function f is said to be convex at $\bar{x} \in S$ if
$f(\lambda\bar{x} + (1 - \lambda)x) \leq \lambda f(\bar{x}) + (1 - \lambda)f(x)$ *for each $\lambda \in (0, 1)$ and each $x \in S$.*

Strict convexity at a point. The function f is strictly convex at $\bar{x} \in S$ if
$f(\lambda\bar{x} + (1 - \lambda)x) < \lambda f(\bar{x}) + (1 - \lambda)f(x)$ *for each $\lambda \in (0, 1)$ and each $x \in S$, $x \neq \bar{x}$.*

Quasiconvexity at a point. The function f is quasiconvex at $\bar{x} \in S$ if
$f(\lambda\bar{x} + (1 - \lambda)x) \leq \max\{f(\bar{x}), f(x)\}$ *for each $\lambda \in (0, 1)$ and each $x \in S$.*

Strict quasiconvexity at a point. The function f is strictly quasiconvex at $\bar{x} \in S$ if $f(\lambda\bar{x} + (1 - \lambda)x) < \max\{f(\bar{x}), f(x)\}$ for each $\lambda \in (0, 1)$ and each $x \in S$ such that $f(x) \neq f(\bar{x})$.

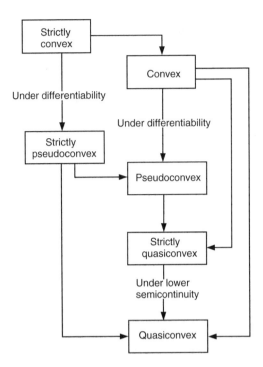

Figure 2.10. Relationship among various types of convexity. The arrows mean implications and, in general, the converses do not hold. (See Bazaraa et al. (1993) for a more complete picture of the relationships among the types of convexity.)

Pseudoconvexity at a point. Suppose f is differentiable at $\bar{x} \in \text{int } S$. Then f is pseudoconvex at $\bar{x}$ if $\nabla f(\bar{x})(x-\bar{x}) \geq 0$ for $x \in S$ implies that $f(x) \geq f(\bar{x})$. Strict pseudoconvexity at a point. Suppose f is differentiable at $\bar{x} \in \text{int } S$. Then f is strictly pseudoconvex at $\bar{x}$ if $\nabla f(\bar{x})(x - \bar{x}) \geq 0$ for $x \in S$, $x \neq \bar{x}$, implies that $f(x) > f(\bar{x})$. ○

Figure 2.11 illustrates some types of convexity at a point.

2.4 Unconstrained Optimisation

An unconstrained optimisation problem is a problem of the form

$$\text{minimise } f(x), \tag{2.9}$$

without any constraint on the vector x. Our ultimate goal in this book is constrained optimisation problems. However, we start by reviewing unconstrained problems because optimality conditions for constrained problems are a logical extension of the conditions for unconstrained problems.

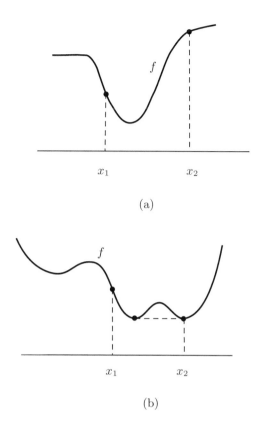

Figure 2.11. Convexity at a point. (a) f is quasiconvex but not strictly quasiconvex at x_1; f is both quasiconvex and strictly quasiconvex at x_2. (b) f is both pseudoconvex and strictly pseudoconvex at x_1; f is pseudoconvex but not strictly pseudoconvex at x_2.

Let us first define *local and global minima* for unconstrained problems.

Definition 2.4.1 (Local and Global Minima) *Consider the problem of minimising $f(x)$ over $\mathbb{R}^n$ and let $\bar{x} \in \mathbb{R}^n$. If $f(\bar{x}) \leq f(x)$ for all $x \in \mathbb{R}^n$, then $\bar{x}$ is called a* global minimum. *If there exists an ε-neighbourhood $N_\varepsilon(\bar{x})$ around $\bar{x}$ such that $f(\bar{x}) \leq f(x)$ for each $x \in N_\varepsilon(\bar{x})$, then $\bar{x}$ is called a* local minimum, *whilst if $f(\bar{x}) < f(x)$ for all $x \in N_\varepsilon(\bar{x})$, $x \neq \bar{x}$, for some $\varepsilon > 0$, then $\bar{x}$ is called a* strict local minimum. *Clearly, a global minimum is also a local minimum.* ○

Given a point $x \in \mathbb{R}^n$, we wish to determine, if possible, whether or not the point is a local or global minimum of a function f. For differentiable functions, there exist conditions that provide this characterisation, as we will see below. We will first present a result that allows the characterisation of *descent directions* of differentiable functions.

Theorem 2.4.1 (Descent Direction) *Suppose that $f : \mathbb{R}^n \to \mathbb{R}$ is differentiable at $\bar{x}$. If there exists a vector d such that $\nabla f(\bar{x})d < 0$, then there exists a $\delta > 0$ such that $f(\bar{x} + \lambda d) < f(\bar{x})$ for each $\lambda \in (0, \delta)$, so that d is a* descent direction *of f at $\bar{x}$.*

Proof. By the differentiability of f at $\bar{x}$, we have

$$f(\bar{x} + \lambda d) = f(\bar{x}) + \lambda \nabla f(\bar{x})d + \lambda \|d\| \alpha(\bar{x}, \lambda d),$$

where $\alpha(\bar{x}, \lambda d) \to 0$ as $\lambda \to 0$. Rearranging the terms and dividing by λ, $\lambda \neq 0$, we obtain

$$\frac{f(\bar{x} + \lambda d) - f(\bar{x})}{\lambda} = \nabla f(\bar{x})d + \|d\| \alpha(\bar{x}, \lambda d).$$

Since $\nabla f(\bar{x})d < 0$ and $\alpha(\bar{x}, \lambda d) \to 0$ as $\lambda \to 0$, there exists a $\delta > 0$ such that the right hand side above is negative for all $\lambda \in (0, \delta)$. The result then follows. $\qquad \square$

Corollary 2.4.2 *Suppose that $f : \mathbb{R}^n \to \mathbb{R}$ is differentiable at $\bar{x}$. If $\bar{x}$ is a local minimum, then $\nabla f(\bar{x}) = 0$.*

Proof. Suppose that $\nabla f(\bar{x}) \neq 0$. Then, letting $d = -\nabla f(\bar{x})^T$, we get $\nabla f(\bar{x})d = -\|\nabla f(\bar{x})\|^2 < 0$, and by Theorem 2.4.1 there is a $\delta > 0$ such that $f(\bar{x} + \lambda d) < f(\bar{x})$ for each $\lambda \in (0, \delta)$, contradicting the assumption that $\bar{x}$ is a local minimum. Hence, $\nabla f(\bar{x}) = 0$. $\qquad \square$

The above condition uses the gradient vector, whose components are the first partial derivatives of f; hence, it is called a *first-order condition*. Necessary conditions can also be stated in terms of the Hessian matrix H, which comprises the second derivatives of f, and are then called *second-order conditions*. One such condition is given below.

Theorem 2.4.3 (Necessary Condition for a Minimum) *Suppose that $f : \mathbb{R}^n \to \mathbb{R}$ is twice-differentiable at $\bar{x}$. If $\bar{x}$ is a local minimum, then $\nabla f(\bar{x}) = 0$ and $H(\bar{x})$ is positive semidefinite.*

Proof. Consider an arbitrary direction d. Then, since by assumption f is twice-differentiable at $\bar{x}$, we have

$$f(\bar{x} + \lambda d) = f(\bar{x}) + \lambda \nabla f(\bar{x})d + \frac{1}{2}\lambda^2 d^T H(\bar{x})d + \lambda^2 \|d\|^2 \alpha(\bar{x}, \lambda d), \qquad (2.10)$$

where $\alpha(\bar{x}, \lambda d) \to 0$ as $\lambda \to 0$. Since $\bar{x}$ is a local minimum, from Corollary 2.4.2 we have $\nabla f(\bar{x}) = 0$. Rearranging the terms in (2.10) and dividing by $\lambda^2 > 0$, we obtain

$$\frac{f(\bar{x} + \lambda d) - f(\bar{x})}{\lambda^2} = \frac{1}{2}d^{\mathrm{T}}H(\bar{x})d + \|d\|^2\alpha(\bar{x}, \lambda d). \tag{2.11}$$

Since $\bar{x}$ is a local minimum, $f(\bar{x} + \lambda d) \geq f(\bar{x})$ for sufficiently small λ. From (2.11), it is thus clear that $\frac{1}{2}d^{\mathrm{T}}H(\bar{x})d + \|d\|^2\alpha(\bar{x}, \lambda d) \geq 0$ for sufficiently small λ. By taking the limit as $\lambda \to 0$, it follows that $d^{\mathrm{T}}H(\bar{x})d \geq 0$; and, hence, $H(\bar{x})$ is positive semidefinite. $\square$

The conditions presented so far are necessary conditions for a local minimum. We now give a sufficient condition for a local minimum.

Theorem 2.4.4 (Sufficient Condition for a Local Minimum) *Suppose that* $f : \mathbb{R}^n \to \mathbb{R}$ *is twice-differentiable at* $\bar{x}$. *If* $\nabla f(\bar{x}) = 0$ *and* $H(\bar{x})$ *is positive definite, then* $\bar{x}$ *is a strict local minimum.*

Proof. Since f is twice-differentiable at $\bar{x}$, we must have, for each $x \in \mathbb{R}^n$,

$$f(x) = f(\bar{x}) + \nabla f(\bar{x})(x - \bar{x}) + \frac{1}{2}(x - \bar{x})^{\mathrm{T}}H(\bar{x})(x - \bar{x}) + \|x - \bar{x}\|^2\alpha(\bar{x}, x - \bar{x}), \tag{2.12}$$

where $\alpha(\bar{x}, x - \bar{x}) \to 0$ as $x \to \bar{x}$. Suppose, by contradiction, that $\bar{x}$ is not a strict local minimum; that is, suppose there exists a sequence $\{x_k\}$ converging to $\bar{x}$ such that $f(x_k) \leq f(\bar{x})$, $x_k \neq \bar{x}$, for each k. Considering this sequence, noting that $\nabla f(\bar{x}) = 0$ and $f(x_k) \leq f(\bar{x})$, and denoting $(x_k - \bar{x})/\|x_k - \bar{x}\|$ by d_k, (2.12) then implies that

$$\frac{1}{2}d_k^{\mathrm{T}}H(\bar{x})d_k + \alpha(\bar{x}, x_k - \bar{x}) \leq 0 \qquad \text{for each } k. \tag{2.13}$$

But $\|d_k\| = 1$ for each k; and, hence, there exists an index set K such that $\{d_k\}_K$ converges to d, where $\|d\| = 1$. Considering this subsequence, and the fact that $\alpha(\bar{x}, x_k - \bar{x}) \to 0$ as $k \in K$ approaches infinity, then (2.13) implies that $d^{\mathrm{T}}H(\bar{x})d \leq 0$. This contradicts the assumption that $H(\bar{x})$ is positive definite since $\|d\| = 1$. Therefore, $\bar{x}$ is indeed a strict local minimum. $\square$

As is generally the case with optimisation problems, more powerful results exist under (generalised) convexity conditions. The following result shows that the necessary condition $\nabla f(\bar{x}) = 0$ is also sufficient for $\bar{x}$ to be a global minimum if f is pseudoconvex at $\bar{x}$.

Theorem 2.4.5 (Necessary and Sufficient Condition for Pseudoconvex Functions) *Let* $f : \mathbb{R}^n \to \mathbb{R}$ *be pseudoconvex at* $\bar{x}$. *Then* $\bar{x}$ *is a global minimum if and only if* $\nabla f(\bar{x}) = 0$.

Proof. By Corollary 2.4.2, if $\bar{x}$ is a global minimum then $\nabla f(\bar{x}) = 0$. Now, suppose that $\nabla f(\bar{x}) = 0$, so that $\nabla f(\bar{x})(x - \bar{x}) = 0$ for each $x \in \mathbb{R}^n$. By the pseudoconvexity of f at $\bar{x}$, it then follows that $f(x) \geq f(\bar{x})$ for each $x \in \mathbb{R}^n$, and the proof is complete. $\square$

2.5 Constrained Optimisation

We now proceed to the main topic of interest in this book; namely, constrained optimisation. We first derive optimality conditions for a problem of the following form:

$$\text{minimise } f(x), \tag{2.14}$$

$$\text{subject to:}$$

$$x \in S.$$

We will first consider a general constraint set S. Later, the set S will be more explicitly defined by a set of equality and inequality constraints. For constrained optimisation problems we have the following definitions.

Definition 2.5.1 (Feasible and Optimal Solutions) *Let* $f : \mathbb{R}^n \rightarrow \mathbb{R}$ *and consider the constrained optimisation problem* (2.14), *where S is a nonempty set in* $\mathbb{R}^n$.

- *A point $x \in S$ is called a* feasible solution *to problem* (2.14).
- *If $\bar{x} \in S$ and $f(x) \geq f(\bar{x})$ for each $x \in S$, then $\bar{x}$ is called an* optimal solution, *a* global optimal solution, *or simply a* solution *to the problem.*
- *The collection of optimal solutions is called the set of* alternative optimal solutions.
- *If $\bar{x} \in S$ and if there exists an ε-neighbourhood $N_\varepsilon(\bar{x})$ around $\bar{x}$ such that $f(x) \geq f(\bar{x})$ for each $x \in S \cap N_\varepsilon(\bar{x})$, then $\bar{x}$ is called a* local optimal solution.
- *If $\bar{x} \in S$ and if $f(x) > f(\bar{x})$ for each $x \in S \cap N_\varepsilon(\bar{x})$, $x \neq \bar{x}$, for some $\varepsilon > 0$, then $\bar{x}$ is called a* strict local optimal solution.   ○

Figure 2.12 illustrates examples of local and global minima for problem (2.14).

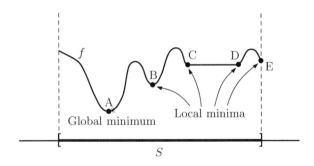

Figure 2.12. Local and global minima.

The function f and the constraint set S are shown in the figure. The points in S corresponding to A, B and E are also strict local minima, whereas those corresponding to the flat segment of the graph between C and D are local minima that are not strict.

In Chapter 13 we will treat a class of problems in which the constraint set S is not convex. However, in most of the book we will be concerned with problems in which the function f and set S in problem (2.14) are, respectively, a convex function and a convex set. Such a problem is known as a *convex programming problem*. The following result shows that each local minimum of a convex program is also a global minimum.

Theorem 2.5.1 (Local Minima of Convex Programs are Global Minima) *Consider problem* (2.14), *where S is a nonempty convex set in $\mathbb{R}^n$, and $f : S \to \mathbb{R}$ is convex on S. If $\bar{x} \in S$ is a local optimal solution to the problem, then $\bar{x}$ is a global optimal solution. Furthermore, if either $\bar{x}$ is a strict local minimum, or if f is strictly convex, then $\bar{x}$ is the unique global optimal solution.*

Proof. Since $\bar{x}$ is a local optimal solution, there exists an ε-neighbourhood $N_\varepsilon(\bar{x})$ around $\bar{x}$ such that

$$f(x) \geq f(\bar{x}) \quad \text{for each } x \in S \cap N_\varepsilon(\bar{x}). \tag{2.15}$$

By contradiction, suppose that $\bar{x}$ is not a global optimal solution so that $f(\hat{x}) < f(\bar{x})$ for some $\hat{x} \in S$. By the convexity of f, we have that:

$$f(\lambda \hat{x} + (1 - \lambda)\bar{x}) \leq \lambda f(\hat{x}) + (1 - \lambda)f(\bar{x}) < \lambda f(\bar{x}) + (1 - \lambda)f(\bar{x}) = f(\bar{x}),$$

for each $\lambda \in (0, 1)$. But, for $\lambda > 0$ and sufficiently small, $\lambda \hat{x} + (1 - \lambda)\bar{x} = \bar{x} + \lambda(\hat{x} - \bar{x}) \in S \cap N_\varepsilon(\bar{x})$. Hence, the above inequality contradicts (2.15), and we conclude that $\bar{x}$ is a global optimal solution.

Next, suppose that $\bar{x}$ is a strict local minimum. Then, as just proven, $\bar{x}$ is a global minimum. Now, suppose that $\bar{x}$ is not the unique global optimal solution. That is, suppose that there exist an $\hat{x} \in S$ such that $f(\hat{x}) = f(\bar{x})$. Then, defining $x_\lambda = \lambda \hat{x} + (1 - \lambda)\bar{x}$ for $0 \leq \lambda \leq 1$, we have, by the convexity of f and S, that $f(x_\lambda) \leq \lambda f(\hat{x}) + (1 - \lambda)f(\bar{x}) = f(\bar{x})$, and $x_\lambda \in S$ for all $0 \leq \lambda \leq 1$. By taking $\lambda \to 0^+$ we can make $x_\lambda \in N_\varepsilon(\bar{x}) \cap S$ for any $\varepsilon > 0$. However, this contradicts the strict local optimality of $\bar{x}$ and, hence, $\bar{x}$ is the unique global minimum.

Finally, suppose that $\bar{x}$ is a local optimal solution and that f is strictly convex. Since strict convexity implies convexity then, as proven earlier, $\bar{x}$ is a global optimal solution. By contradiction, suppose that $\bar{x}$ is not the unique global optimal solution so that there exists an $\tilde{x} \in S$, $\tilde{x} \neq \bar{x}$, such that $f(\tilde{x}) = f(\bar{x})$. By strict convexity, we have that $f(\frac{1}{2}\tilde{x} + \frac{1}{2}\bar{x}) < \frac{1}{2}f(\tilde{x}) + \frac{1}{2}f(\bar{x}) = f(\bar{x})$. Since S is convex, $\frac{1}{2}\tilde{x} + \frac{1}{2}\bar{x} \in S$, and the above inequality contradicts global optimality of $\bar{x}$. Hence, $\bar{x}$ is the unique global minimum, and this completes the proof. $\square$

2.5.1 Geometric Necessary Optimality Conditions

In this section we give a necessary optimality condition for problem (2.14) using the *cone of feasible directions* defined below. Note that, in the sequel and in Sections 2.5.2– 2.5.4, we do not assume problem (2.14) to be a convex program. As a consequence of this generality, only *necessary* conditions for optimality will be derived. In a later section, Section 2.5.5, we will impose suitable convexity conditions to the problem in order to obtain sufficiency conditions for optimality.

Definition 2.5.2 (Cones of Feasible Directions and of Improving Directions) *Let S be a nonempty set in $\mathbb{R}^n$ and let $\bar{x} \in \operatorname{cl} S$. The* cone of feasible directions *of S at $\bar{x}$, denoted by D, is given by*

$$D = \{d : d \neq 0, \text{ and } \bar{x} + \lambda d \in S \text{ for all } \lambda \in (0, \delta) \text{ for some } \delta > 0\}.$$

Each nonzero vector $d \in D$ is called a feasible direction. *Moreover, given a function $f : \mathbb{R}^n \to \mathbb{R}$, the* cone of improving directions *at $\bar{x}$, denoted by F, is given by*

$$F = \{d : f(\bar{x} + \lambda d) < f(\bar{x}) \text{ for all } \lambda \in (0, \delta) \text{ for some } \delta > 0\}.$$

Each direction $d \in F$ is called an improving direction, *or a* descent direction *of f at $\bar{x}$.*   ○

We will now consider the function f to be differentiable at the point $\bar{x}$. We can then define the sets

$$F_0 \triangleq \{d : \nabla f(\bar{x})d < 0\}, \tag{2.16}$$

$$F_0' \triangleq \{d \neq 0 : \nabla f(\bar{x})d \leq 0\}. \tag{2.17}$$

Observe that the set F_0 defined in (2.16) is an open half-space defined in terms of the gradient vector. Note also that, from Theorem 2.4.1, if $\nabla f(\bar{x})d < 0$, then d is an improving direction. It then follows that $F_0 \subseteq F$. Hence, the set F_0 gives an algebraic description of the set of improving directions F. Also, if $d \in F$, we must have $\nabla f(\bar{x})d \leq 0$, or else, analogous to Theorem 2.4.1, $\nabla f(\bar{x})d > 0$ would imply that d is an *ascent direction*. Hence, we have

$$F_0 \subseteq F \subseteq F_0'. \tag{2.18}$$

The following theorem states that a necessary condition for local optimality is that every improving direction in F_0 is not a feasible direction.

Theorem 2.5.2 (Geometric Necessary Condition for Local Optimality Using the Sets F_0 and D) *Consider the problem to minimise $f(x)$ subject to $x \in S$, where $f : \mathbb{R}^n \to \mathbb{R}$ and S is a nonempty set in $\mathbb{R}^n$. Suppose that f is differentiable at a point $\bar{x} \in S$. If $\bar{x}$ is a local optimal solution then $F_0 \cap D = \emptyset$, where $F_0 = \{d : \nabla f(\bar{x})d < 0\}$ and D is the cone of feasible directions of S at $\bar{x}$.*

Proof. Suppose, by contradiction, that there exists a vector $d \in F_0 \cap D$. Since $d \in F_0$, then, by Theorem 2.4.1, there exists a $\delta_1 > 0$ such that

$$f(\bar{x} + \lambda d) < f(\bar{x}) \quad \text{for each } \lambda \in (0, \delta_1). \tag{2.19}$$

Also, since $d \in D$, by Definition 2.5.2, there exists a $\delta_2 > 0$ such that

$$\bar{x} + \lambda d \in S \quad \text{for each } \lambda \in (0, \delta_2). \tag{2.20}$$

The assumption that $\bar{x}$ is a local optimal solution is not compatible with (2.19) and (2.20). Thus, $F_0 \cap D = \emptyset$. $\qquad\square$

The necessary condition for local optimality of Theorem 2.5.2 is illustrated in Figure 2.13, where the vertices of the cones F_0 and D are translated from the

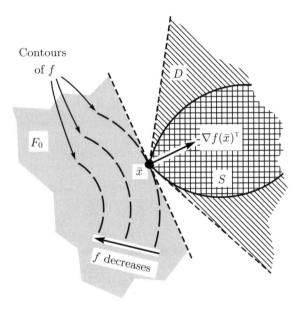

Figure 2.13. Illustration of the necessary condition for local optimality of Theorem 2.5.2: $F_0 \cap D = \emptyset$.

origin to $\bar{x}$ for convenience.

2.5.2 Problems with Inequality and Equality Constraints

We next consider a specific description for the feasible region S as follows:

$$S = \{x \in X : g_i(x) \leq 0, \ i = 1, \ldots, m, \ h_i(x) = 0, \ i = 1, \ldots, l\},$$

where $g_i : \mathbb{R}^n \to \mathbb{R}$ for $i = 1, \ldots, m$, $h_i : \mathbb{R}^n \to \mathbb{R}$ for $i = 1, \ldots, \ell$, and X is a nonempty open set in $\mathbb{R}^n$. This gives the following *nonlinear programming* problem with inequality and equality constraints:

$$
\begin{aligned}
&\text{minimise } f(x), \\
&\text{subject to:} \\
&g_i(x) \le 0 \quad \text{for } i = 1, \ldots, m, \\
&h_i(x) = 0 \quad \text{for } i = 1, \ldots, \ell, \\
&x \in X.
\end{aligned}
\tag{2.21}
$$

The following theorem shows that if $\bar{x}$ is a local optimal solution to problem (2.21), then either the gradients of the equality constraints are linearly dependent at $\bar{x}$, or else $F_0 \cap G_0 \cap H_0 = \emptyset$, where F_0 is defined as in (2.16) and the sets G_0 and H_0 are defined in the statement of the theorem.

Theorem 2.5.3 (Geometric Necessary Condition for Problems with Inequality and Equality Constraints) *Let X be a nonempty open set in $\mathbb{R}^n$, and let $f : \mathbb{R}^n \to \mathbb{R}$, $g_i : \mathbb{R}^n \to \mathbb{R}$ for $i = 1, \ldots, m$, $h_i : \mathbb{R}^n \to \mathbb{R}$ for $i = 1, \ldots, \ell$. Consider the problem defined in (2.21). Suppose that $\bar{x}$ is a local optimal solution, and let $I = \{i : g_i(\bar{x}) = 0\}$ be the index set for the* binding *or* active *constraints. Furthermore, suppose that each g_i for $i \notin I$ is continuous at $\bar{x}$, that f and g_i for $i \in I$ are differentiable at $\bar{x}$, and that each h_i for $i = 1, \ldots, \ell$ is continuously differentiable at $\bar{x}$. If $\nabla h_i(\bar{x})^\mathrm{T}$ for $i = 1, \ldots, \ell$ are linearly independent, then $F_0 \cap G_0 \cap H_0 = \emptyset$, where*

$$
\begin{aligned}
F_0 &= \{d : \nabla f(\bar{x})d < 0\}, \\
G_0 &= \{d : \nabla g_i(\bar{x})d < 0 \quad for\ i \in I\}, \\
H_0 &= \{d : \nabla h_i(\bar{x})d = 0 \quad for\ i = 1, \ldots, \ell\}.
\end{aligned}
\tag{2.22}
$$

Proof. We use contradiction. Suppose there exists a vector $y \in F_0 \cap G_0 \cap H_0$; that is, $\nabla f(\bar{x})y < 0$, $\nabla g_i(\bar{x})y < 0$ for each $i \in I$, and $\nabla h(\bar{x})y = 0$, where $\nabla h(\bar{x})$ is the $\ell \times n$ Jacobian matrix whose ith row is $\nabla h_i(\bar{x})$. We now construct a feasible arc from $\bar{x}$. For $\lambda \ge 0$, define $\alpha : \mathbb{R} \to \mathbb{R}^n$ by the following differential equation and boundary condition:

$$
\frac{d\alpha(\lambda)}{d\lambda} = \mathbf{P}(\lambda)y, \qquad \alpha(0) = \bar{x},
\tag{2.23}
$$

where $\mathbf{P}(\lambda)$ is the matrix that projects any vector into the null space of $\nabla h(\alpha(\lambda))$. For sufficiently small λ, the above equation is well-defined and solvable because $\nabla h(\bar{x})$ has full row rank and h_i, $i = 1, \ldots, \ell$, are continuously differentiable at $\bar{x}$, so that $\mathbf{P}$ is continuous in λ. Obviously, $\alpha(\lambda) \to \bar{x}$ as $\lambda \to 0^+$.

We now show that for sufficiently small $\lambda > 0$, $\alpha(\lambda)$ is feasible and $f(\alpha(\lambda)) < f(\bar{x})$, thus contradicting local optimality of $\bar{x}$. By the chain rule of differentiation and using (2.23), we obtain

$$\frac{dg_i(\alpha(\lambda))}{d\lambda} = \nabla g_i(\alpha(\lambda))\mathbf{P}(\lambda)y, \tag{2.24}$$

for each $i \in I$. In particular, y is in the null space of $\nabla h(\bar{x})$, and so for $\lambda = 0$, we have $\mathbf{P}(0)y = y$. Hence, from (2.24) and the fact that $\nabla g_i(\bar{x})y < 0$, we obtain

$$\left.\frac{dg_i(\alpha(\lambda))}{d\lambda}\right|_{\lambda=0} = \nabla g_i(\bar{x})y < 0, \tag{2.25}$$

for $i \in I$. Recalling that $g_i(\alpha(0)) = g_i(\bar{x}) = 0$ for all $i \in I$, this and (2.25) further imply that $g_i(\alpha(\lambda)) < 0$ for sufficiently small $\lambda > 0$, and for each $i \in I$. For $i \notin I$, $g_i(\bar{x}) < 0$, and g_i is continuous at $\bar{x}$, and thus $g_i(\alpha(\lambda)) < 0$ for sufficiently small λ. By the mean value theorem (Theorem 2.2.1), we have

$$h_i(\alpha(\lambda)) = h_i(\alpha(0)) + \lambda \left.\frac{dh_i(\alpha(\lambda))}{d\lambda}\right|_{\lambda=\mu} = \lambda \left.\frac{dh_i(\alpha(\lambda))}{d\lambda}\right|_{\lambda=\mu}, \tag{2.26}$$

for some $\mu \in (0, \lambda)$. However, by the chain rule of differentiation and similarly to (2.24), we obtain

$$\left.\frac{dh_i(\alpha(\lambda))}{d\lambda}\right|_{\lambda=\mu} = \nabla h_i(\alpha(\mu))\mathbf{P}(\mu)y.$$

By construction, $\mathbf{P}(\mu)y$ is in the null space of $\nabla h_i(\alpha(\mu))$ and, hence, from the above equation we obtain $\left.\dfrac{dh_i(\alpha(\lambda))}{d\lambda}\right|_{\lambda=\mu} = 0$. Substituting in (2.26), it follows that $h_i(\alpha(\lambda)) = 0$ for all i. Also, since X is open, $\alpha(\lambda) \in X$ for sufficiently small λ.

We have, so far, established that the arc $\alpha(\lambda)$ defined by (2.23) is a feasible solution to the problem (2.21) for each sufficiently small $\lambda > 0$, since $g_i(\alpha(\lambda)) < 0$ for all $i = 1, \ldots, m$, $h_i(\alpha(\lambda)) = 0$ for all $i = 1, \ldots, \ell$, and $\alpha(\lambda) \in X$. To complete the proof by contradiction we next prove that such a feasible arc $\alpha(\lambda)$ would constitute an arc of improving solutions. By an argument similar to that leading to (2.25), we obtain

$$\left.\frac{df(\alpha(\lambda))}{d\lambda}\right|_{\lambda=0} = \nabla f(\bar{x})y < 0,$$

and, hence, $f(\alpha(\lambda)) < f(\bar{x})$ for sufficiently small $\lambda > 0$. This contradicts local optimality of $\bar{x}$. Hence, $F_0 \cap G_0 \cap H_0 = \emptyset$, and the proof is complete. □

2.5.3 The Fritz John Necessary Conditions

In this section we express the geometric optimality condition $F_0 \cap G_0 \cap H_0 = \emptyset$ of Theorem 2.5.3 in a more usable algebraic form known as the Fritz John conditions.

Theorem 2.5.4 (The Fritz John Necessary Conditions) *Let X be a nonempty open set in $\mathbb{R}^n$, and let $f : \mathbb{R}^n \to \mathbb{R}$, $g_i : \mathbb{R}^n \to \mathbb{R}$ for $i = 1, \ldots, m$, $h_i : \mathbb{R}^n \to \mathbb{R}$ for $i = 1, \ldots, \ell$. Consider the optimisation problem defined in (2.21). Let $\bar{x}$ be a feasible solution, and let $I = \{i : g_i(\bar{x}) = 0\}$. Furthermore, suppose that each g_i for $i \notin I$ is continuous at $\bar{x}$, that f and g_i for $i \in I$ are differentiable at $\bar{x}$, and that each h_i for $i = 1, \ldots, \ell$ is continuously differentiable at $\bar{x}$. If $\bar{x}$ locally solves problem (2.21), then there exist scalars u_0 and u_i for $i \in I$, and v_i for $i = 1, \ldots, \ell$, such that*

$$u_0 \nabla f(\bar{x})^{\mathrm{T}} + \sum_{i \in I} u_i \nabla g_i(\bar{x})^{\mathrm{T}} + \sum_{i=1}^{\ell} v_i \nabla h_i(\bar{x})^{\mathrm{T}} = 0,$$

$$u_0, u_i \geq 0 \quad for\ i \in I, \tag{2.27}$$

$$(u_0, u_I, v) \neq (0, 0, 0),$$

where u_I and v are vectors whose components are u_i, $i \in I$, and v_i, $i = 1, \ldots, \ell$, respectively. Furthermore, if g_i, $i \notin I$ are also differentiable at $\bar{x}$, then the above conditions can be written as

$$u_0 \nabla f(\bar{x})^{\mathrm{T}} + \sum_{i=1}^{m} u_i \nabla g_i(\bar{x})^{\mathrm{T}} + \sum_{i=1}^{\ell} v_i \nabla h_i(\bar{x})^{\mathrm{T}} = 0,$$

$$u_i g_i(\bar{x}) = 0 \quad for\ i = 1, \ldots, m, \tag{2.28}$$

$$u_0, u_i \geq 0 \quad for\ i = 1, \ldots, m,$$

$$(u_0, u, v) \neq (0, 0, 0),$$

where u and v are vectors whose components are u_i, $i = 1, \ldots, m$, and v_i, $i = 1, \ldots, \ell$, respectively.

Proof. In the case where the vectors $\nabla h_i(\bar{x})^{\mathrm{T}}$ for $i = 1, \ldots, \ell$ are linearly dependent, then one can find scalars $v_1, \ldots, v_\ell$, not all zero, such that $\sum_{i=1}^{\ell} v_i \nabla h_i(\bar{x})^{\mathrm{T}} = 0$. Letting u_0 and u_i for $i \in I$ equal to zero, conditions (2.27) hold trivially.

Now suppose that $\nabla h_i(\bar{x})^{\mathrm{T}}$ for $i = 1, \ldots, \ell$ are linearly independent. Then, from Theorem 2.5.3, local optimality of $\bar{x}$ implies that the sets defined in (2.22) satisfy:

$$F_0 \cap G_0 \cap H_0 = \emptyset. \tag{2.29}$$

Let A_1 be the matrix whose rows are $\nabla f(\bar{x})$ and $\nabla g_i(\bar{x})$ for $i \in I$, and let A_2 be the matrix whose rows are $\nabla h_i(\bar{x})$ for $i = 1, \ldots, \ell$. Then, it is easy to see that condition (2.29) is satisfied if and only if the system:

$$A_1 d < 0,$$

$$A_2 d = 0,$$

is inconsistent. Now consider the following two sets:

$$S_1 = \{(z_1, z_2) : z_1 = A_1 d,\ z_2 = A_2 d,\ d \in \mathbb{R}^n\},$$
$$S_2 = \{(z_1, z_2) : z_1 < 0,\ z_2 = 0\}.$$

Note that S_1 and S_2 are nonempty convex sets and, since the system $A_1 d < 0$, $A_2 d = 0$ has no solution, then $S_1 \cap S_2 = \emptyset$. Then, by Theorem 2.3.5, there exists a nonzero vector $p^T = (p_1^T, p_2^T)$ such that

$$p_1^T A_1 d + p_2^T A_2 d \geq p_1^T z_1 + p_2^T z_2,$$

for each $d \in \mathbb{R}^n$ and $(z_1, z_2) \in \mathrm{cl}\, S_2$. Noting that $z_2 = 0$ and since each component of z_1 can be made an arbitrarily large negative number, it follows that $p_1 \geq 0$. Also, letting $(z_1, z_2) = (0, 0) \in \mathrm{cl}\, S_2$, we must have $(p_1^T A_1 + p_2^T A_2) d \geq 0$ for each $d \in \mathbb{R}^n$. Letting $d = -(A_1^T p_1 + A_2^T p_2)$, it follows that $-\|A_1^T p_1 + A_2^T p_2\|^2 \geq 0$, and thus $A_1^T p_1 + A_2^T p_2 = 0$. Summarising, we have found a nonzero vector $p^T = (p_1^T, p_2^T)$ with $p_1 \geq 0$ such that $A_1^T p_1 + A_2^T p_2 = 0$. Denoting the components of p_1 by u_0 and u_i, $i \in I$, and letting $p_2 = v$, conditions (2.27) follow. The equivalent form (2.28) is readily obtained by letting $u_i = 0$ for $i \notin I$, and the proof is complete. $\qquad\square$

In the Fritz John conditions (2.28) the scalars u_0, u_i for $i = 1, \ldots, m$, and v_i for $i = 1, \ldots, \ell$, are called the *Lagrange multipliers* associated, respectively, with the objective function, the inequality constraints $g_i(x) \leq 0$, $i = 1, \ldots, m$, and the equality constraints $h_i(x) = 0$, $i = 1, \ldots, \ell$. Observe that the v_i are *unrestricted* in sign. The condition that $\bar{x}$ be feasible for the optimisation problem (2.21) is called the *primal feasibility* [PF] condition. The requirements $u_0 \nabla f(\bar{x})^T + \sum_{i=1}^{m} u_i \nabla g_i(\bar{x})^T + \sum_{i=1}^{\ell} v_i \nabla h_i(\bar{x})^T = 0$, with $u_0, u_i \geq 0$ for $i = 1, \ldots, m$, and $(u_0, u, v) \neq (0, 0, 0)$ are called the *dual feasibility* [DF] conditions. The condition $u_i g_i(\bar{x}) = 0$ for $i = 1, \ldots, m$ is called the *complementary slackness* [CS] condition; it requires that $u_i = 0$ if the corresponding inequality is nonbinding (that is, $g_i(\bar{x}) < 0$), and allows for $u_i > 0$ only for those constraints that are binding. Together, the PF, DF and CS conditions are called the *Fritz John* [FJ] *optimality conditions*. Any point $\bar{x}$ for which there exist Lagrange multipliers $\bar{u}_0$, $\bar{u}_i$, $i = 1, \ldots, m$, $\bar{v}_i$, $i = 1, \ldots, \ell$, such that the FJ conditions are satisfied is called an *FJ point*.

The FJ conditions can also be written in vector form as follows:

$$\begin{aligned}
\nabla f(\bar{x})^T u_0 + \nabla g(\bar{x})^T u + \nabla h(\bar{x})^T v &= 0, \\
u^T g(\bar{x}) &= 0, \\
(u_0, u) &\geq (0, 0), \\
(u_0, u, v) &\neq (0, 0, 0),
\end{aligned} \qquad (2.30)$$

where $\nabla g(\bar{x})$ is the $m \times n$ Jacobian matrix whose ith row is $\nabla g_i(\bar{x})$, $\nabla h(\bar{x})$ is the $\ell \times n$ Jacobian matrix whose ith row is $\nabla h_i(\bar{x})$, and $g(\bar{x})$ is the m vector function whose ith component is $g_i(\bar{x})$. Also, u and v are, respectively, an m

vector and an ℓ vector, whose elements are the Lagrange multipliers associated with, respectively, the inequality and equality constraints.

At this point it is important to note that, given an optimisation problem, there might be points that satisfy the FJ conditions trivially. For example, if a feasible point $\bar{x}$ (not necessarily an optimum) satisfies $\nabla f(\bar{x}) = 0$, or $\nabla g_i(\bar{x}) = 0$ for some $i \in I$, or $\nabla h_i(\bar{x}) = 0$ for some $i = 1, \dots, \ell$, then we can let the corresponding Lagrange multiplier be any positive number, set all the other multipliers equal to zero, and satisfy conditions (2.27). In fact, given *any* feasible solution $\bar{x}$ we can always add a redundant constraint to the problem to make $\bar{x}$ an FJ point. For example, we can add the constraint $\|x - \bar{x}\|^2 \geq 0$, which holds true for all $x \in \mathbb{R}^n$, is a binding constraint at $\bar{x}$ and whose gradient is zero at $\bar{x}$.

2.5.4 Karush–Kuhn–Tucker Necessary Conditions

In the previous section we stated the FJ necessary conditions for optimality. We saw that the FJ conditions relate to the existence of scalars $u_0, u_i \geq 0$ and v_i, not all zero, such that the conditions (2.27) are satisfied. We also saw that there are instances where there are points that satisfy the conditions trivially, for example, when the gradient of some binding constraint (which might even be redundant) vanishes.

It is also possible that, at some feasible point $\bar{x}$, the FJ conditions (2.27) are satisfied with Lagrange multiplier associated with the objective function $u_0 = 0$. In such cases, the FJ conditions become virtually useless since the objective function gradient does not play a role in the optimality conditions (2.27) and the conditions merely state that the gradients of the binding inequality constraints and of the equality constraints are linearly dependent. Thus, when $u_0 = 0$, the FJ conditions are of no practical value in locating an optimal point. Under suitable assumptions, referred to as *constraint qualifications* [CQ], u_0 is guaranteed to be positive and the FJ conditions become the Karush–Kuhn–Tucker [KKT] conditions, which will be presented next. There exist various constraint qualifications for problems with inequality and equality constraints. Here, we use a typical constraint qualification that requires that the gradients of the inequality constraints for $i \in I$ and the gradients of the equality constraints at $\bar{x}$ be linearly independent.

Theorem 2.5.5 (Karush–Kuhn–Tucker Necessary Conditions)   *Let X be a nonempty open set in $\mathbb{R}^n$, and let $f : \mathbb{R}^n \to \mathbb{R}$, $g_i : \mathbb{R}^n \to \mathbb{R}$ for $i = 1, \dots, m$, $h_i : \mathbb{R}^n \to \mathbb{R}$ for $i = 1, \dots, \ell$. Consider the problem defined in (2.21). Let $\bar{x}$ be a feasible solution, and let $I = \{i : g_i(\bar{x}) = 0\}$. Suppose that f and g_i for $i \in I$ are differentiable at $\bar{x}$, that each g_i for $i \notin I$ is continuous at $\bar{x}$, and that each h_i for $i = 1, \dots, \ell$ is continuously differentiable at $\bar{x}$. Furthermore, suppose that $\nabla g_i(\bar{x})^{\mathrm{T}}$ for $i \in I$ and $\nabla h_i(\bar{x})^{\mathrm{T}}$ for $i = 1, \dots, \ell$ are linearly independent. If $\bar{x}$ is a local optimal solution, then there exist unique scalars u_i for $i \in I$, and v_i for $i = 1, \dots, \ell$, such that*

$$\nabla f(\bar{x})^{\mathrm{T}} + \sum_{i \in I} u_i \nabla g_i(\bar{x})^{\mathrm{T}} + \sum_{i=1}^{\ell} v_i \nabla h_i(\bar{x})^{\mathrm{T}} = 0, \tag{2.31}$$

$$u_i \geq 0 \quad for \ i \in I.$$

Furthermore, if g_i, $i \notin I$ are also differentiable at $\bar{x}$, then the above conditions can be written as

$$\nabla f(\bar{x})^{\mathrm{T}} + \sum_{i=1}^{m} u_i \nabla g_i(\bar{x})^{\mathrm{T}} + \sum_{i=1}^{\ell} v_i \nabla h_i(\bar{x})^{\mathrm{T}} = 0, \tag{2.32}$$

$$u_i g_i(\bar{x}) = 0 \quad for \ i = 1, \dots, m,$$

$$u_i \geq 0 \quad for \ i = 1, \dots, m.$$

Proof. We have, from the FJ conditions (Theorem 2.5.4), that there exist scalars $\hat{u}_0$ and $\hat{u}_i$, $i \in I$, and $\hat{v}_i$, $i = 1, \dots, \ell$, not all zero, such that

$$\hat{u}_0 \nabla f(\bar{x})^{\mathrm{T}} + \sum_{i \in I} \hat{u}_i \nabla g_i(\bar{x})^{\mathrm{T}} + \sum_{i=1}^{\ell} \hat{v}_i \nabla h_i(\bar{x})^{\mathrm{T}} = 0, \tag{2.33}$$

$$\hat{u}_0, \hat{u}_i \geq 0 \quad for \ i \in I.$$

Note that the assumption of linear independence of $\nabla g_i(\bar{x})^{\mathrm{T}}$ for $i \in I$ and $\nabla h_i(\bar{x})^{\mathrm{T}}$ for $i = 1, \dots, \ell$, together with (2.33) and the fact that at least one of the multipliers is nonzero, implies that $\hat{u}_0 > 0$. Then, letting $u_i = \hat{u}_i/\hat{u}_0$ for $i \in I$, and $v_i = \hat{v}_i/\hat{u}_0$ for $i = 1, \dots, \ell$ we obtain conditions (2.31). Furthermore, the linear independence assumption implies the uniqueness of these Lagrange multipliers. The equivalent form (2.32) follows by letting $u_i = 0$ for $i \notin I$. This completes the proof. □

As in the FJ conditions, the scalars u_i and v_i are called the *Lagrange multipliers*. Observe that the v_i are *unrestricted* in sign. The condition that $\bar{x}$ be feasible for the optimisation problem (2.21) is called the *primal feasibility* [PF] condition. The requirement that

$$\nabla f(\bar{x})^{\mathrm{T}} + \sum_{i=1}^{m} u_i \nabla g_i(\bar{x})^{\mathrm{T}} + \sum_{i=1}^{\ell} v_i \nabla h_i(\bar{x})^{\mathrm{T}} = 0, \text{ with } u_i \geq 0 \text{ for } i = 1, \dots, m$$

is called the *dual feasibility* [DF] condition. The condition $u_i g_i(\bar{x}) = 0$ for $i = 1, \dots, m$ is called the *complementary slackness* [CS] condition; it requires that $u_i = 0$ if the corresponding inequality is nonbinding (that is, $g_i(\bar{x}) < 0$), and it permits $u_i > 0$ only for those constraints that are binding. Together, the PF, DF and CS conditions are called the Karush–Kuhn–Tucker [KKT] optimality conditions. Any point $\bar{x}$ for which there exist Lagrange multipliers $\bar{u}_i$, $i = 1, \dots, m$, $\bar{v}_i$, $i = 1, \dots, \ell$, that, together with $\bar{x}$, satisfy the KKT conditions is called a *KKT point*.

The KKT conditions can also be written in vector form as follows:

$$\nabla f(\bar{x})^{\mathrm{T}} + \nabla g(\bar{x})^{\mathrm{T}} u + \nabla h(\bar{x})^{\mathrm{T}} v = 0,$$
$$u^{\mathrm{T}} g(\bar{x}) = 0, \qquad (2.34)$$
$$u \geq 0,$$

where $\nabla g(\bar{x})$ is the $m \times n$ Jacobian matrix whose ith row is $\nabla g_i(\bar{x})$, $\nabla h(\bar{x})$ is the $\ell \times n$ Jacobian matrix whose ith row is $\nabla h_i(\bar{x})$, and $g(\bar{x})$ is the m vector function whose ith component is $g_i(\bar{x})$. Also, u and v are, respectively, an m vector and an ℓ vector, whose elements are the Lagrange multipliers associated with, respectively, the inequality and equality constraints.

2.5.5 Karush–Kuhn–Tucker Sufficient Conditions

In the previous section we derived the KKT necessary conditions for optimality from the FJ optimality conditions. This derivation was done by asserting that the multiplier associated with the objective function is positive at a local optimum whenever a linear independence constraint qualification is satisfied. It is important to notice that the linear independence constraint qualification is only a *sufficient condition*[3] placed on the behaviour of the constraints to ensure that an FJ point (and hence, from Theorem 2.5.4, any local optimum) be a KKT point. Thus, the importance of the constraint qualifications is to guarantee that, by examining only KKT points, we do not lose out on optimal solutions. There is an important special case; namely, when the constraints are linear, in which case the KKT conditions are always necessary optimality conditions irrespective of the behaviour of the objective function. (Although we will not prove this result here, it comes from the fact that a more general constraint qualification to that of linear independence, known as *Abadie's constraint qualification*—see Abadie 1967—is automatically satisfied when the constraints are linear.) However, we are still left with the problem of determining, among all the points that satisfy the KKT conditions, which ones constitute local optimal solutions. The following result shows that, under moderate convexity assumptions, the KKT conditions are also sufficient for local optimality.

Theorem 2.5.6 (Karush–Kuhn–Tucker Sufficient Conditions) *Let X be a nonempty open set in $\mathbb{R}^n$, and let $f : \mathbb{R}^n \to \mathbb{R}$, $g_i : \mathbb{R}^n \to \mathbb{R}$ for $i = 1, \ldots, m$, $h_i : \mathbb{R}^n \to \mathbb{R}$ for $i = 1, \ldots, \ell$. Consider the problem defined in (2.21). Let $\bar{x}$ be a feasible solution, and let $I = \{i : g_i(\bar{x}) = 0\}$. Suppose that the KKT conditions hold at $\bar{x}$; that is, there exist scalars $\bar{u}_i \geq 0$ for $i \in I$, and $\bar{v}_i$ for $i = 1, \ldots, \ell$, such that*

$$\nabla f(\bar{x})^{\mathrm{T}} + \sum_{i \in I} \bar{u}_i \nabla g_i(\bar{x})^{\mathrm{T}} + \sum_{i=1}^{\ell} \bar{v}_i \nabla h_i(\bar{x})^{\mathrm{T}} = 0. \qquad (2.35)$$

---

[3] It is possible, in some optimisation problems, for a local optimum to be a KKT point and yet not satisfy the linear independence constraint qualification.

Let $J = \{i : \bar{v}_i > 0\}$ and $K = \{i : \bar{v}_i < 0\}$. Further, suppose that f is pseudoconvex at $\bar{x}$, g_i is quasiconvex at $\bar{x}$ for $i \in I$, h_i is quasiconvex at $\bar{x}$ for $i \in J$, and h_i is quasiconcave at $\bar{x}$ (that is, $-h_i$ is quasiconvex at $\bar{x}$) for $i \in K$. Then $\bar{x}$ is a global optimal solution to problem (2.21). In particular, if these generalised convexity assumptions on the objective and constraint functions are restricted to the domain $N_\varepsilon(\bar{x})$ for some $\varepsilon > 0$, then $\bar{x}$ is a local minimum for problem (2.21).

Proof. Let x be any feasible solution to problem (2.21). (In the case where we need to restrict the domain to $N_\varepsilon(\bar{x})$, then let x be a feasible solution to problem (2.21) that also lies within $N_\varepsilon(\bar{x})$.) Then, for $i \in I$, $g_i(x) \le g_i(\bar{x})$, since $g_i(x) \le 0$ and $g_i(\bar{x}) = 0$. By the quasiconvexity of g_i at $\bar{x}$ it follows that

$$g_i(\bar{x} + \lambda(x - \bar{x})) = g_i(\lambda x + (1 - \lambda)\bar{x}) \le \max\{g_i(x), g_i(\bar{x})\} = g_i(\bar{x}),$$

for all $\lambda \in (0, 1)$. This implies that g_i does not increase by moving from $\bar{x}$ along the direction $x - \bar{x}$. Thus, by an analogous result to that of Theorem 2.4.1, we must have

$$\nabla g_i(\bar{x})(x - \bar{x}) \le 0 \qquad \text{for } i \in I. \tag{2.36}$$

Similarly, since h_i is quasiconvex at $\bar{x}$ for $i \in J$ and h_i is quasiconcave at $\bar{x}$ for $i \in K$, we have

$$\nabla h_i(\bar{x})(x - \bar{x}) \le 0 \qquad \text{for } i \in J, \tag{2.37}$$
$$\nabla h_i(\bar{x})(x - \bar{x}) \ge 0 \qquad \text{for } i \in K. \tag{2.38}$$

Multiplying (2.36), (2.37) and (2.38) by $\bar{u}_i \ge 0$, $\bar{v}_i > 0$, and $\bar{v}_i < 0$, respectively, and adding the terms, we obtain

$$\sum_{i \in I} \bar{u}_i \nabla g_i(\bar{x})(x - \bar{x}) + \sum_{i \in J \cup K} \bar{v}_i \nabla h_i(\bar{x})(x - \bar{x}) \le 0. \tag{2.39}$$

Transposing (2.35), multiplying by $(x - \bar{x})$ and noting that $\bar{v}_i = 0$ for $i \notin J \cup K$, then (2.39) implies that

$$\nabla f(\bar{x})(x - \bar{x}) \ge 0.$$

By the pseudoconvexity of f at $\bar{x}$, we must have $f(x) \ge f(\bar{x})$, and the proof is complete.   □

An important point to note is that, despite the sufficiency of the KKT conditions under the generalised convexity assumptions of Theorem 2.5.6, the KKT conditions are *not necessary* for *optimality* for these problems. (This situation, however, only arises when the constraint qualification does not hold at a local optimal solution, and hence the local solution is not captured by the KKT conditions.)

2.5.6 Quadratic Programs

Quadratic programs represent a special class of nonlinear programs in which the objective function is quadratic and the constraints are linear. Thus, a quadratic programming [QP] problem can be written as

$$\text{minimise } \frac{1}{2}x^{\mathrm{T}}Hx + x^{\mathrm{T}}c, \tag{2.40}$$

$$\text{subject to:}$$
$$A_I^{\mathrm{T}}x \le b_I,$$
$$A_E^{\mathrm{T}}x = b_E,$$

where H is an $n \times n$ matrix, c is an n vector, A_I is an $n \times m_I$ matrix, b_I is an m_I vector, A_E is an $n \times m_E$ matrix and b_E is an m_E vector.

As mentioned in the previous section, since the constraints are linear we have that a constraint qualification, known as Abadie's constraint qualification, is automatically satisfied and, hence, a local minimum $\bar{x}$ is necessarily a KKT point. Also, since the constraints are linear, the constraint set $S = \{x : A_I^{\mathrm{T}}x \le b_I, A_E^{\mathrm{T}}x = b_E\}$ is a (polyhedral) convex set. Thus, the QP problem (2.40) is a convex program if and only if the objective function is convex; that is, if and only if H is symmetric and positive semidefinite. In this case we have, from Theorem 2.5.1, that $\bar{x}$ is a local minimum if and only if $\bar{x}$ is a global minimum. And, from Theorems 2.5.5 and 2.5.6 (and from the automatic fulfilment of Abadie's constraint qualification), we have that the above is true if and only if $\bar{x}$ is a KKT point. Furthermore, if H is positive definite, then we have that the objective function is strictly convex and we can conclude from Theorem 2.5.1 that $\bar{x}$ is the unique global minimum for problem (2.40).

The KKT conditions (2.34) for the QP problem defined in (2.40) are:

$$
\begin{array}{lll}
\text{PF:} & A_I^{\mathrm{T}}\bar{x} \le b_I, & \\
& A_E^{\mathrm{T}}\bar{x} = b_E, & \\
\text{DF:} & H\bar{x} + c + A_I u + A_E v = 0, & \tag{2.41}\\
& u \ge 0, & \\
\text{CS:} & u^{\mathrm{T}}(A_I^{\mathrm{T}}\bar{x} - b_I) = 0, &
\end{array}
$$

where u is an m_I vector of Lagrange multipliers corresponding to the inequality constraints and v is an m_E vector of Lagrange multipliers corresponding to the equality constraints.

2.6 Lagrangian Duality

In this section we present the concept of *Lagrangian duality*. Given a nonlinear programming problem, known as the *primal problem*, there exists another nonlinear programming problem, closely related to it, that receives the name of the *Lagrangian dual problem*. As we will see later in Section 2.6.3, under certain convexity assumptions and suitable constraint qualifications, the primal and dual problems have equal optimal objective values.

2.6.1 The Lagrangian Dual Problem

We will first define the primal and dual problems as separate optimisation problems. Later we will see that these two problems are closely related.

Thus, first consider the following nonlinear programming problem, called the *primal problem*.

Primal Problem P

$$\text{minimise } f(x), \tag{2.42}$$
$$\text{subject to:}$$
$$g_i(x) \leq 0 \quad \text{for } i = 1, \ldots, m,$$
$$h_i(x) = 0 \quad \text{for } i = 1, \ldots, \ell,$$
$$x \in X.$$

Then the *Lagrangian dual problem* is defined as the following nonlinear programming problem.

Lagrangian Dual Problem D

$$\text{maximise } \theta(u, v), \tag{2.43}$$
$$\text{subject to:}$$
$$u \geq 0,$$

where

$$\theta(u, v) = \inf\{f(x) + \sum_{i=1}^{m} u_i g_i(x) + \sum_{i=1}^{\ell} v_i h_i(x) : x \in X\} \tag{2.44}$$

is the *Lagrangian dual function*.

In the dual problem (2.43)–(2.44), the vectors u and v have as their components the Lagrange multipliers u_i for $i = 1, \ldots, m$, and v_i for $i = 1, \ldots, \ell$. Note that the Lagrange multipliers u_i, corresponding to the inequality constraints $g_i(x) \leq 0$, are restricted to be nonnegative, whereas the Lagrange multipliers v_i, corresponding to the equality constraints $h_i(x) = 0$, are unrestricted in sign.

Given the primal problem P (2.42), several Lagrangian dual problems D of the form of (2.43)–(2.44) can be devised, depending on which constraints are handled as $g_i(x) \leq 0$ and $h_i(x) = 0$, and which constraints are handled by the set X. Hence, an appropriate selection of the set X must be made, depending on the nature of the problem and the goal of formulating or solving the dual problem D.

The primal and dual problems can also be written in *vector* form. Consider the function $f : \mathbb{R}^n \to \mathbb{R}$ and the vector functions $g : \mathbb{R}^n \to \mathbb{R}^m$ and $h : \mathbb{R}^n \to \mathbb{R}^\ell$, whose ith components are g_i and h_i, respectively. Then, we can write:

Primal Problem P

$$\text{minimise } f(x), \tag{2.45}$$
$$\text{subject to:}$$
$$g(x) \le 0,$$
$$h(x) = 0,$$
$$x \in X.$$

Lagrangian Dual Problem D

$$\text{maximise } \theta(u, v), \tag{2.46}$$
$$\text{subject to:}$$
$$u \ge 0,$$

where $\theta(u, v) = \inf\{f(x) + u^{\mathsf{T}} g(x) + v^{\mathsf{T}} h(x) : x \in X\}$.

The relationship between the primal and dual problems will be explored below.

2.6.2 Geometric Interpretation of the Lagrangian Dual

An interesting geometric interpretation of the dual problem can be made by considering a simpler problem with only one inequality constraint and no equality constraint. Consider the following primal problem P:

Primal Problem P

$$\text{minimise } f(x), \tag{2.47}$$
$$\text{subject to:}$$
$$g(x) \le 0,$$
$$x \in X,$$

where $f : \mathbb{R}^n \to \mathbb{R}$ and $g : \mathbb{R}^n \to \mathbb{R}$, and define the following set in $\mathbb{R}^2$:

$$G = \{(y, z) : y = g(x), z = f(x) \text{ for some } x \in X\},$$

that is, G is the image of X under the (g, f) map. Figure 2.14 shows an example of the set G. Then, the primal problem consists of finding a point in G with $y \le 0$ that has minimum ordinate z. Obviously this point in Figure 2.14 is $(\bar{y}, \bar{z})$.

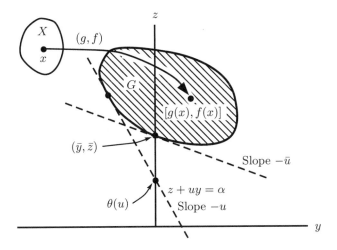

Figure 2.14. Geometric interpretation of Lagrangian duality: case with no duality gap.

Now, consider the Lagrangian dual problem D:

Lagrangian Dual Problem D

$$\text{maximise } \theta(u), \tag{2.48}$$

$$\text{subject to:}$$

$$u \geq 0.$$

The solution of the Lagrangian dual problem (2.48) requires one to first solve the following *Lagrangian dual subproblem*:

$$\theta(u) = \inf\{f(x) + ug(x) : x \in X\}. \tag{2.49}$$

Given $u \geq 0$, problem (2.49) is equivalent to minimise $z + uy$ over points (y, z) in G. Note that $z + uy = \alpha$ is the equation of a straight line with slope $-u$ that intercepts the z-axis at α. Thus, in order to minimise $z + uy$ over G we need to move the line $z + uy = \alpha$ parallel to itself as far down as possible, whilst it remains in contact with G. The last intercept on the z-axis thus obtained is the value of $\theta(u)$ corresponding to the given $u \geq 0$, as shown in Figure 2.14. Finally, to solve the dual problem (2.48), we have to find the line with slope $-u$ ($u \geq 0$) such that the last intercept on the z-axis, $\theta(u)$, is maximal. Such a line is shown in Figure 2.14. It has slope $-\bar{u}$ and supports the set G (recall Definition 2.3.4) at the point $(\bar{y}, \bar{z})$. Thus, the solution to the dual problem (2.48) is $\bar{u}$, and the optimal dual objective value is $\bar{z}$. It can be seen that, in the example illustrated in Figure 2.14, the optimal primal and dual objective values are equal. In such cases, it is said that there is no *duality gap*. In the next section we will develop conditions such that no duality gap exists.

2.6.3 Weak and Strong Duality

In this section we explore the relationships between the primal problem P and its Lagrangian dual problem D. In particular, we are interested in the conditions that the primal problem P must satisfy for the primal and dual objective values to be equal; this situation is known as *strong duality*. The first result shows that the objective value of any feasible solution to the dual problem constitutes a lower bound for the objective value of any feasible solution to the primal problem.

Theorem 2.6.1 (Weak Duality Theorem) *Consider the primal problem P given by (2.45) and its Lagrangian dual problem D given by (2.46). Let x be a feasible solution to P; that is, $x \in X$, $g(x) \leq 0$, and $h(x) = 0$. Also, let (u, v) be a feasible solution to D; that is, $u \geq 0$. Then:*

$$f(x) \geq \theta(u, v).$$

Proof. We use the definition of θ given in (2.44), and the facts that $x \in X$, $u \geq 0$, $g(x) \leq 0$ and $h(x) = 0$. We then have

$$\theta(u, v) = \inf\{f(\tilde{x}) + u^{\mathrm{T}}g(\tilde{x}) + v^{\mathrm{T}}h(\tilde{x}) : \tilde{x} \in X\}$$
$$\leq f(x) + u^{\mathrm{T}}g(x) + v^{\mathrm{T}}h(x) \leq f(x),$$

and the result follows.   □

Corollary 2.6.2

$$\inf\{f(x) : x \in X, g(x) \leq 0, h(x) = 0\} \geq \sup\{\theta(u, v) : u \geq 0\}. \qquad (2.50)$$

○

Note from (2.50) that the optimal objective value of the primal problem is greater than or equal to the optimal objective value of the dual problem. If (2.50) holds as a *strict* inequality, then it is said that there exists a *duality gap*. Figure 2.15 shows an example for the primal and dual problems defined in (2.47) and (2.48)–(2.49), respectively. Notice that, in the case shown in the figure, there exists a duality gap. We see, by comparing Figure 2.15 with Figure 2.14, that the presence of a duality gap is due to the nonconvexity of the set G. As we will see in Theorem 2.6.4 below, if some suitable convexity conditions are satisfied, then there is no duality gap between the primal and dual optimisation problems. Before stating the conditions that guarantee the absence of a duality gap, we need the following result.

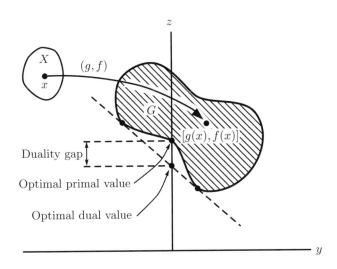

Figure 2.15. Geometric interpretation of Lagrangian duality: case with duality gap.

Lemma 2.6.3 *Let X be a nonempty convex set in $\mathbb{R}^n$. Let $\alpha : \mathbb{R}^n \to \mathbb{R}$ and $g : \mathbb{R}^n \to \mathbb{R}^m$ be convex,[4] and $h : \mathbb{R}^n \to \mathbb{R}^\ell$ be affine (that is, assume h is of the form $h(x) = Ax - b$). Also, let u_0 be a scalar, $u \in \mathbb{R}^m$ and $v \in \mathbb{R}^\ell$. Consider the following two systems:*

System 1: $\alpha(x) < 0, \quad g(x) \le 0, \quad h(x) = 0 \quad$ *for some $x \in X$.*

System 2: $u_0\alpha(x) + u^{\mathrm{T}}g(x) + v^{\mathrm{T}}h(x) \ge 0$ *for some $(u_0, u, v) \ne (0, 0, 0)$, $(u_0, u) \ge (0, 0)$ and for all $x \in X$.*

If System 1 has no solution x, then System 2 has a solution (u_0, u, v). Conversely, if System 2 has a solution (u_0, u, v) with $u_0 > 0$, then System 1 has no solution.

Proof. Assume first that System 1 has no solution. Define the set:

$$S = \{(p, q, r) : p > \alpha(x), q \ge g(x), r = h(x) \quad \text{for some } x \in X\}.$$

The reader can easily verify that, since X, α and g are convex and h is affine, the set S is convex. Since System 1 has no solution, we have that $(0, 0, 0) \notin S$. We then have, from Corollary 2.3.4, that there exists a nonzero vector (u_0, u, v) such that

$$(u_0, u, v)^{\mathrm{T}}[(p, q, r) - (0, 0, 0)] = u_0 p + u^{\mathrm{T}} q + v^{\mathrm{T}} r \ge 0, \qquad (2.51)$$

---

[4] That is, each component of the vector-valued function g is a convex function.

for each $(p, q, r) \in \mathrm{cl}\, S$. Now, fix an $x \in X$. Noticing, from the definition of S, that p and q can be made arbitrarily large, we have that in order to satisfy (2.51), we must have $u_0 \geq 0$ and $u \geq 0$. Also, note that $[\alpha(x), g(x), h(x)] \in \mathrm{cl}\, S$ and we have from (2.51) that

$$u_0 \alpha(x) + u^\mathsf{T} g(x) + v^\mathsf{T} h(x) \geq 0.$$

Since the above inequality is true for each $x \in X$, System 2 has a solution.

To prove the converse, assume that System 2 has a solution (u_0, u, v) such that $u_0 > 0$ and $u \geq 0$, and $u_0 \alpha(x) + u^\mathsf{T} g(x) + v^\mathsf{T} h(x) \geq 0$ for each $x \in X$. Suppose that $x \in X$ is such that $g(x) \leq 0$ and $h(x) = 0$. From the previous inequality we conclude that $u_0 \alpha(x) \geq -u^\mathsf{T} g(x) \geq 0$, since $u \geq 0$ and $g(x) \leq 0$. But, since $u_0 > 0$, we must then have that $\alpha(x) \geq 0$. Hence, System 1 has no solution. This completes the proof.   □

The following result, known as the *strong duality theorem*, shows that, under suitable convexity assumptions and under a constraint qualification, there is no *duality gap* between the primal and dual optimal objective function values.

Theorem 2.6.4 (Strong Duality Theorem) *Let X be a nonempty convex set in $\mathbb{R}^n$. Let $f : \mathbb{R}^n \to \mathbb{R}$ and $g : \mathbb{R}^n \to \mathbb{R}^m$ be convex, and $h : \mathbb{R}^n \to \mathbb{R}^\ell$ be affine. Suppose that the following constraint qualification is satisfied. There exists an $\hat{x} \in X$ such that $g(\hat{x}) < 0$ and $h(\hat{x}) = 0$, and $0 \in \mathrm{int}\, h(X)$, where $h(X) = \{h(x) : x \in X\}$. Then,*

$$\inf\{f(x) : x \in X, g(x) \leq 0, h(x) = 0\} = \sup\{\theta(u, v) : u \geq 0\}, \qquad (2.52)$$

where $\theta(u, v) = \inf\{f(x) + u^\mathsf{T} g(x) + v^\mathsf{T} h(x) : x \in X\}$. Furthermore, if the inf is finite, then $\sup\{\theta(u, v) : u \geq 0\}$ is achieved at $(\bar{u}, \bar{v})$ with $\bar{u} \geq 0$. If the inf is achieved at $\bar{x}$, then $\bar{u}^\mathsf{T} g(\bar{x}) = 0$.

Proof. Let $\gamma = \inf\{f(x) : x \in X, g(x) \leq 0, h(x) = 0\}$. By assumption there exists a feasible solution $\hat{x}$ for the primal problem and hence $\gamma < \infty$. If $\gamma = -\infty$, we then conclude from Corollary 2.6.2 that $\sup\{\theta(u, v) : u \geq 0\} = -\infty$ and, hence, (2.52) is satisfied. Thus, suppose that γ is finite, and consider the following system:

$$f(x) - \gamma < 0, \qquad g(x) \leq 0 \qquad h(x) = 0, \qquad \text{for some } x \in X.$$

By the definition of γ, this system has no solution. Hence, from Lemma 2.6.3, there exists a nonzero vector (u_0, u, v) with $(u_0, u) \geq (0, 0)$ such that

$$u_0[f(x) - \gamma] + u^\mathsf{T} g(x) + v^\mathsf{T} h(x) \geq 0 \qquad \text{for all } x \in X. \qquad (2.53)$$

We will next show that $u_0 > 0$. Suppose, by contradiction that $u_0 = 0$. By assumption, there exists an $\hat{x} \in X$ such that $g(\hat{x}) < 0$ and $h(\hat{x}) = 0$. Substituting in (2.53) we obtain $u^\mathsf{T} g(\hat{x}) \geq 0$. But, since $g(\hat{x}) < 0$ and $u \geq 0$, $u^\mathsf{T} g(\hat{x}) \geq 0$ is only possible if $u = 0$. From (2.53), $u_0 = 0$ and $u = 0$ imply that $v^{\mathrm{T}}h(x) \geq 0$ for all $x \in X$. But, since $0 \in \mathrm{int}\, h(X)$, we can choose an $x \in X$ such that $h(x) = -\lambda v$, where $\lambda > 0$. Therefore, $0 \leq v^{\mathrm{T}}h(x) = -\lambda\|v\|^2$, which implies that $v = 0$. Thus, it has been shown that $u_0 = 0$ implies that $(u_0, u, v) = (0, 0, 0)$, which is a contradiction. We conclude, then, that $u_0 > 0$. Dividing (2.53) by u_0 and denoting $\bar{u} = u/u_0$ and $\bar{v} = v/u_0$, we obtain

$$f(x) + \bar{u}^{\mathrm{T}}g(x) + \bar{v}^{\mathrm{T}}h(x) \geq \gamma \qquad \text{for all } x \in X. \tag{2.54}$$

This implies that $\theta(\bar{u}, \bar{v}) = \inf\{f(x) + \bar{u}^{\mathrm{T}}g(x) + \bar{v}^{\mathrm{T}}h(x) : x \in X\} \geq \gamma$. We then conclude, from Theorem 2.6.1, that $\theta(\bar{u}, \bar{v}) = \gamma$ and, from Corollary 2.6.2, that $(\bar{u}, \bar{v})$ solves the dual problem. Finally, to complete the proof, assume that $\bar{x}$ is an optimal solution to the primal problem; that is, $\bar{x} \in X$, $g(\bar{x}) \leq 0$, $h(\bar{x}) = 0$ and $f(\bar{x}) = \gamma$. From (2.54), letting $x = \bar{x}$, we get $\bar{u}^{\mathrm{T}}g(\bar{x}) \geq 0$. Since $\bar{u} \geq 0$ and $g(\bar{x}) \leq 0$, we get $\bar{u}^{\mathrm{T}}g(\bar{x}) = 0$. This completes the proof.   □

2.7 Multiconvex Problems

We have emphasised convex optimisation problems since these have many desirable properties, for example, all local minima are global minima, absence of duality gap, and so on. Sometimes a problem is nonconvex but can be partitioned into a finite number of subproblems, each of which is convex within a convex region. In this case, we can solve each of the convex problems using constraints to restrict the solution to the appropriate region. Then one can simply compare the resulting objective values and decide which is best. Of course, the disadvantage of this idea is that one has to solve as many convex problems as there are convex regions. Nonetheless, this is a useful strategy in many problems of interest in practice (see, for example, Chapter 9).

This completes our brief introduction to optimisation theory. Of course, this is a rich topic and many more results are available in the literature. We refer the reader to some of the books listed in Section 2.8. However, our brief introduction will suffice for the problems addressed here. Indeed, we will make extensive use of the concepts outlined in this chapter. As a prelude of what is to follow, we note that in Chapter 3 we will use the KKT optimality conditions in the context of nonlinear optimal control; and in Chapter 10 we will utilise strong Lagrangian duality to connect constrained control and estimation.

2.8 Further Reading

For complete list of references cited, see References section at the end of book.

General

This chapter is mainly based on Bazaraa et al. (1993).

The following books complement and extend the material presented in this chapter: Boyd and Vandenberghe (2003), Nocedal and Wright (1999), Nash and Sofer (1996), Floudas (1995), Fiacco and McCormick (1990), Fletcher (2000), Luenberger (1984), (1989), Fiacco (1983), Gill, Murray and Wright (1981), Abadie (1967).

3

Fixed Horizon Optimal Control
with Constraints

3.1 Overview

Chapter 2 dealt with rather general optimisation problems. As stated earlier, this theory finds widespread application across a large number of areas. Our goal in the present chapter is to begin to apply these ideas to one of the specific problems of interest in this book, namely that of constrained optimal control. In particular, our goal is to relate fixed horizon optimal control problems to the classical theory of optimal control. One of the main contributions of the current chapter is an introduction to the discrete minimum principle. This will turn out to be an elegant application of many of the concepts introduced in Chapter 2. Thus, this chapter serves to reinforce these ideas. However, some of this material is difficult. We thus suggest that some readers who are meeting this material for the first time might wish to proceed immediately, without loss of continuity, to Chapter 4 and return to this chapter at a later time.

3.2 Optimal Control Problems

We will begin with a rather general treatment using nonlinear state space models. However, in future chapters we will specialise the treatment to constrained linear systems since much more can be said about this case. In the deterministic discrete time case we will describe the system via a nonlinear state space model of the form

$$x_{k+1} = f(x_k, u_k), \quad k \geq i \geq 0, \qquad x_i = \bar{x}, \tag{3.1}$$

where $f : \mathbb{R}^n \times \mathbb{R}^m \to \mathbb{R}^n$ is some given nonlinear function, $x \in \mathbb{R}^n$ is the system state and $u \in \mathbb{R}^m$ is the control input. In the context of optimisation, we can think of the system of equations (3.1) as providing *equality constraints* that must be satisfied for all time instants k in an interval of interest. It is also common to assume that there exist additional constraints on the system state and input, which can be stated as *set constraints* of the form

$$u_k \in \mathbb{U} \quad \text{for } k = i, i+1, \ldots, i+N-1,$$
$$x_k \in \mathbb{X} \quad \text{for } k = i, i+1, \ldots, i+N, \tag{3.2}$$
$$x_{i+N} \in \mathbb{X}_f \subset \mathbb{X},$$

where $\mathbb{U} \subset \mathbb{R}^m$, $\mathbb{X} \subset \mathbb{R}^n$, and $\mathbb{X}_f \subset \mathbb{R}^n$ are some sets, and N is the *optimisation horizon*. Usually, $\mathbb{U}$ is compact, and $\mathbb{X}$ and $\mathbb{X}_f$ are closed.

The *fixed horizon optimal control* problem of interest here is the following:

$$\mathcal{P}_N(\bar{x}): \quad V_N^{\text{OPT}}(\bar{x}) \triangleq \min V_N(\{x_k\}, \{u_k\}),$$

subject to:
$$x_{k+1} = f(x_k, u_k) \quad \text{for } k = i, i+1, \ldots, i+N-1,$$
$$x_i = \bar{x}, \tag{3.3}$$
$$u_k \in \mathbb{U} \quad \text{for } k = i, i+1, \ldots, i+N-1,$$
$$x_k \in \mathbb{X} \quad \text{for } k = i, i+1, \ldots, i+N,$$
$$x_{i+N} \in \mathbb{X}_f \subset \mathbb{X},$$

where $\{x_k\} \triangleq \{x_i, \ldots, x_{i+N}\}$, $\{u_k\} \triangleq \{u_i, \ldots, u_{i+N-1}\}$ are the state and control sequences, and where $V_N(\{x_k\}, \{u_k\})$ is the objective function given by

$$V_N(\{x_k\}, \{u_k\}) \triangleq F(x_{i+N}) + \sum_{k=i}^{i+N-1} L(x_k, u_k). \tag{3.4}$$

F and L are some functions whose properties will be defined later.

The state and control sequences that attain the minimum in (3.3)–(3.4) are the *optimal sequences*, or *minimisers*. The value of the objective function at the minimisers is $V_N^{\text{OPT}}(\bar{x})$. The function $V_N^{\text{OPT}}(\cdot)$ is called the *value function*, and is a function of the initial state only.

The remainder of this chapter is concerned with necessary and sufficient conditions for the sequences $\{x_i, \ldots, x_{i+N}\}$ and $\{u_i, \ldots, u_{i+N-1}\}$ to be the minimisers of the optimisation problem (3.3)–(3.4) for particular instances of the constraints (3.2). We will build links to traditional optimal control theory, including the minimum principle and dynamic programming.

For the moment, we will treat general nonlinear systems. Later, in Chapter 5 to Chapter 8, we will focus on a special case in which the model (3.1) is linear and the constraint sets polyhedral. For this class of problems it will turn out that remarkable simple characterisations of the optimal control sequence can be given.

3.3 Necessary Conditions for Optimality

As discussed above, our approach to constrained control (and later constrained estimation) evolves from classical optimal control theory. Our goal in this section is to summarise some of the key features of the latter theory. We will give an outline presentation and refer the reader to associated literature for the full technical details.

We will treat the problem in two stages. Section 3.3.1 deals with the "unconstrained" case, that is, the case where only the equality constraints given by the state equations (3.1) are present; Section 3.3.2 deals with the "constrained" case, that is, where, in addition to the state equations (3.1), set constraints of the form (3.2) are present.

3.3.1 Necessary Conditions in the Absence of Set Constraints

We assume that the objective function and state equations do not depend explicitly on time. We can then take the initial time as $i = 0$ without loss of generality. We thus consider the following optimal control problem:

$$\mathcal{P}_N(\bar{x}) : \quad \text{minimise } V_N(\{x_k\}, \{u_k\}), \tag{3.5}$$

$$\text{subject to:}$$
$$x_{k+1} = f(x_k, u_k) \quad \text{for } k = 0, \ldots, N - 1, \tag{3.6}$$
$$x_0 = \bar{x}, \tag{3.7}$$

where

$$V_N(\{x_k\}, \{u_k\}) \triangleq F(x_N) + \sum_{k=0}^{N-1} L(x_k, u_k), \tag{3.8}$$

and $\{x_k\} \triangleq \{x_0, \ldots, x_N\}$, $\{u_k\} \triangleq \{u_0, \ldots, u_{N-1}\}$. We assume that $f : \mathbb{R}^n \times \mathbb{R}^m \to \mathbb{R}^n$, $F : \mathbb{R}^n \to \mathbb{R}$ and $L : \mathbb{R}^n \times \mathbb{R}^m \to \mathbb{R}$ are differentiable functions of their variables.

We will derive necessary optimality conditions for the sequences $\{x_0^*, \ldots, x_N^*\}$ and $\{u_0^*, \ldots, u_{N-1}^*\}$ to be minimisers of the optimisation problem (3.5)–(3.8) using the KKT necessary conditions (see Section 2.5.4 in Chapter 2). Note that problem (3.5)–(3.8) has equality constraints (given by the state equations (3.6)–(3.7)) and no inequality constraints. In order to apply the KKT optimality conditions of Theorem 2.5.5, we need to verify the constraint qualification that the gradients of the equality constraints are linearly independent when evaluated at the minimisers. To this end, let us define a new variable

$$\mathbf{x} \triangleq \begin{bmatrix} x_0^\mathrm{T} & \cdots & x_N^\mathrm{T} & u_0^\mathrm{T} & \cdots & u_{N-1}^\mathrm{T} \end{bmatrix}^\mathrm{T} \in \mathbb{R}^{(N+1)n+Nm}, \tag{3.9}$$

which comprises all the variables with respect to which the optimisation (3.5) is performed. We can then write the state equations (3.6)–(3.7) as $(N + 1)n$ equality constraints on $\mathbf{x}$ as follows:

$$h(\mathbf{x}) \triangleq \begin{bmatrix} \bar{x} - x_0 \\ f(x_0, u_0) - x_1 \\ \vdots \\ f(x_{N-1}, u_{N-1}) - x_N \end{bmatrix} = 0. \tag{3.10}$$

If we now let

$$x_k \triangleq \left[x_k^1 \cdots x_k^n\right]^\mathrm{T} \quad \text{for } k = 0, \ldots, N, \tag{3.11}$$

$$u_k \triangleq \left[u_k^1 \cdots u_k^m\right]^\mathrm{T} \quad \text{for } k = 0, \ldots, N-1, \tag{3.12}$$

$$f(x_k, u_k) \triangleq \left[f_1(x_k, u_k) \cdots f_n(x_k, u_k)\right]^\mathrm{T}, \tag{3.13}$$

and define

$$\frac{\partial f}{\partial x_k} \triangleq \begin{bmatrix} \dfrac{\partial f_1}{\partial x_k^1} & \cdots & \dfrac{\partial f_1}{\partial x_k^n} \\ \vdots & \ddots & \vdots \\ \dfrac{\partial f_n}{\partial x_k^1} & \cdots & \dfrac{\partial f_n}{\partial x_k^n} \end{bmatrix}, \qquad \frac{\partial f}{\partial u_k} \triangleq \begin{bmatrix} \dfrac{\partial f_1}{\partial u_k^1} & \cdots & \dfrac{\partial f_1}{\partial u_k^m} \\ \vdots & \ddots & \vdots \\ \dfrac{\partial f_n}{\partial u_k^1} & \cdots & \dfrac{\partial f_n}{\partial u_k^m} \end{bmatrix}, \tag{3.14}$$

for $k = 0, \ldots, N-1$, we can compute the $(N+1)n \times [(N+1)n + Nm]$ Jacobian matrix of the vector-valued function $h(\mathbf{x})$ in (3.10) as

$$\frac{\partial h}{\partial \mathbf{x}} = \begin{bmatrix} -I_n & 0 & 0 & \cdots & 0 & 0 & 0 & 0 & \cdots & 0 \\ \dfrac{\partial f}{\partial x_0} & -I_n & 0 & \cdots & 0 & 0 & \dfrac{\partial f}{\partial u_0} & 0 & \cdots & 0 \\ 0 & \dfrac{\partial f}{\partial x_1} & -I_n & \cdots & 0 & 0 & 0 & \dfrac{\partial f}{\partial u_1} & \cdots & 0 \\ \vdots & \vdots & \vdots & \ddots & \vdots & \vdots & \vdots & \vdots & \ddots & \vdots \\ 0 & 0 & 0 & \cdots & \dfrac{\partial f}{\partial x_{N-1}} & -I_n & 0 & 0 & \cdots & \dfrac{\partial f}{\partial u_{N-1}} \end{bmatrix}, \tag{3.15}$$

where 0 denotes zero matrices of appropriate dimensions, and I_n denotes the $n \times n$ identity matrix. Clearly, $\partial h / \partial \mathbf{x}$ in (3.15) has full row rank for all $\mathbf{x} \in \mathbb{R}^{(N+1)n + Nm}$, and hence the gradients of the equality constraints (3.10) are linearly independent in $\mathbb{R}^{(N+1)n + Nm}$. Thus, the constraint qualification required by the KKT optimality conditions of Theorem 2.5.5 holds for all $\mathbf{x} \in \mathbb{R}^{(N+1)n + Nm}$. Notice that this implies that the KKT conditions of Theorem 2.5.5 and the FJ conditions of Theorem 2.5.4 (the latter are necessary conditions for optimality without imposing any constraint qualification) are essentially the same for problem (3.5)–(3.8).

Next, we introduce Lagrange multipliers $\lambda_{-1} \in \mathbb{R}^n$ for the initial state equation (3.7), and $\{\lambda_k\} \triangleq \{\lambda_0, \ldots, \lambda_{N-1}\}$, $\lambda_k \in \mathbb{R}^n$ (usually referred to as *adjoint variables*) for the state equations (3.6), and form the (real valued) *Lagrangian function*

$$\mathcal{L}(\mathbf{x}, \boldsymbol{\lambda}) \triangleq F(x_N) + \sum_{k=0}^{N-1} L(x_k, u_k) + \lambda_{-1}^\mathrm{T}(\bar{x} - x_0) + \sum_{k=0}^{N-1} \lambda_k^\mathrm{T}[f(x_k, u_k) - x_{k+1}],$$

$$\tag{3.16}$$

where $\boldsymbol{\lambda} \triangleq \left[\lambda_{-1}^{\mathrm{T}} \ \lambda_0^{\mathrm{T}} \ \ldots \ \lambda_{N-1}^{\mathrm{T}} \right]^{\mathrm{T}}$. Let

$$\mathbf{x}^* \triangleq \left[(x_0^*)^{\mathrm{T}} \ \cdots \ (x_N^*)^{\mathrm{T}} \ (u_0^*)^{\mathrm{T}} \ \cdots \ (u_{N-1}^*)^{\mathrm{T}} \right]^{\mathrm{T}}$$

be the minimising vector corresponding to the sequences $\{x_0^*, \ldots, x_N^*\}$ and $\{u_0^*, \ldots, u_{N-1}^*\}$ that minimise problem (3.5)–(3.8). Observe that the dual feasibility condition in the KKT conditions (see (2.32) in Chapter 2) for the optimisation problem (3.5)–(3.8) is equivalent to the statement that there exists $\boldsymbol{\lambda}^* \triangleq \left[(\lambda_{-1}^*)^{\mathrm{T}} \ (\lambda_0^*)^{\mathrm{T}} \ \ldots \ (\lambda_{N-1}^*)^{\mathrm{T}} \right]^{\mathrm{T}}$ such that the partial derivative $\partial \mathcal{L} / \partial \mathbf{x}$ of the Lagrangian function (3.16) vanishes at $(\mathbf{x}^*, \boldsymbol{\lambda}^*)$. Hence, the following must hold:

$$
\begin{aligned}
\frac{\partial \mathcal{L}(\mathbf{x}^*, \boldsymbol{\lambda}^*)}{\partial x_k} &= 0 \quad \text{for } k = 0, \ldots, N, \\
\frac{\partial \mathcal{L}(\mathbf{x}^*, \boldsymbol{\lambda}^*)}{\partial u_k} &= 0 \quad \text{for } k = 0, \ldots, N-1.
\end{aligned}
\tag{3.17}
$$

In (3.17), $\dfrac{\partial \mathcal{L}}{\partial x_k}$ and $\dfrac{\partial \mathcal{L}}{\partial u_k}$ denote the row vectors of partial derivatives

$$\frac{\partial \mathcal{L}}{\partial x_k} \triangleq \left[\frac{\partial \mathcal{L}}{\partial x_k^1} \ \cdots \ \frac{\partial \mathcal{L}}{\partial x_k^n} \right],$$

$$\frac{\partial \mathcal{L}}{\partial u_k} \triangleq \left[\frac{\partial \mathcal{L}}{\partial u_k^1} \ \cdots \ \frac{\partial \mathcal{L}}{\partial u_k^m} \right],$$

where x_k and u_k have the form (3.11) and (3.12), respectively.

Before performing the differentiations in (3.17), we introduce the *Hamiltonian* $\mathcal{H} : \mathbb{R}^n \times \mathbb{R}^m \times \mathbb{R}^n \longrightarrow \mathbb{R}$ defined as

$$\mathcal{H}(x_k, u_k, \lambda_k) \triangleq L(x_k, u_k) + \lambda_k^{\mathrm{T}} f(x_k, u_k) \quad \text{for } k = 0, \ldots, N-1, \tag{3.18}$$

where $L(\cdot, \cdot)$ is the *per-stage* weighting in the objective function (3.8), and $f(\cdot, \cdot)$ is the vector-valued function on the right hand side of the state equations (3.6). Note that

$$\frac{\partial \mathcal{H}}{\partial x_k} = \frac{\partial L}{\partial x_k} + \lambda_k^{\mathrm{T}} \frac{\partial f}{\partial x_k},$$

$$\frac{\partial \mathcal{H}}{\partial u_k} = \frac{\partial L}{\partial u_k} + \lambda_k^{\mathrm{T}} \frac{\partial f}{\partial u_k},$$

where

$$\frac{\partial L}{\partial x_k} = \left[\frac{\partial L}{\partial x_k^1} \ \cdots \ \frac{\partial L}{\partial x_k^n} \right],$$

$$\frac{\partial L}{\partial u_k} = \left[\frac{\partial L}{\partial u_k^1} \ \cdots \ \frac{\partial L}{\partial u_k^m} \right],$$

and $\partial f/\partial x_k$, $\partial f/\partial u_k$ are defined in (3.14).

Then, if $\{x_0^*, \ldots, x_N^*\}$, $\{u_0^*, \ldots, u_{N-1}^*\}$ are the minimisers, and $\{\lambda_{-1}^*, \lambda_0^*, \ldots, \lambda_{N-1}^*\}$ are the corresponding Lagrange multipliers, we have, from (3.17), (3.16) and (3.18), that

$$\frac{\partial \mathcal{L}(\mathbf{x}^*, \boldsymbol{\lambda}^*)}{\partial x_k} = \frac{\partial \mathcal{H}(x_k^*, u_k^*, \lambda_k^*)}{\partial x_k} - (\lambda_{k-1}^*)^{\mathrm{T}} = 0 \quad \text{for } k = 0, \ldots, N-1, \quad (3.19)$$

$$\frac{\partial \mathcal{L}(\mathbf{x}^*, \boldsymbol{\lambda}^*)}{\partial x_N} = \frac{\partial F(x_N^*)}{\partial x_N} - (\lambda_{N-1}^*)^{\mathrm{T}} = 0, \quad (3.20)$$

$$\frac{\partial \mathcal{L}(\mathbf{x}^*, \boldsymbol{\lambda}^*)}{\partial u_k} = \frac{\partial \mathcal{H}(x_k^*, u_k^*, \lambda_k^*)}{\partial u_k} = 0 \quad \text{for } k = 0, \ldots, N-1. \quad (3.21)$$

Thus, from (3.19)–(3.21), and the state equations (3.6)–(3.7), a necessary condition for the sequences $\{x_0^*, \ldots, x_N^*\}$, $\{u_0^*, \ldots, u_{N-1}^*\}$ to be minimisers of (3.5)–(3.8) is that there exist a sequence of vectors $\{\lambda_{-1}^*, \lambda_0^*, \ldots, \lambda_{N-1}^*\}$ such that the following equations hold:

(i) State equations:

$$x_{k+1}^* = f(x_k^*, u_k^*) \quad \text{for } k = 0, \ldots, N-1,$$
$$x_0^* = \bar{x}.$$

(ii) Adjoint equations:

$$(\lambda_{k-1}^*)^{\mathrm{T}} = \frac{\partial \mathcal{H}(x_k^*, u_k^*, \lambda_k^*)}{\partial x_k} \quad \text{for } k = 0, \ldots, N-1.$$

(iii) Boundary condition:

$$(\lambda_{N-1}^*)^{\mathrm{T}} = \frac{\partial F(x_N^*)}{\partial x_N}.$$

(iv) Hamiltonian condition:

$$\frac{\partial \mathcal{H}(x_k^*, u_k^*, \lambda_k^*)}{\partial u_k} = 0 \quad \text{for } k = 0, \ldots, N-1. \quad (3.22)$$

○

Notice that (3.22) means that the minimising control u_k^* is a stationary point of the *restricted* Hamiltonian defined as[1]

$$\mathcal{H}(x_k^*, u_k, \lambda_k^*) \triangleq L(x_k^*, u_k) + (\lambda_k^*)^{\mathrm{T}} f(x_k^*, u_k) \quad \text{for } k = 0, \ldots, N-1.$$

---

[1] That is, the restricted Hamiltonian is the function of u_k obtained by setting $x_k = x_k^*$ and $\lambda_k = \lambda_k^*$ in the Hamiltonian (3.18).

3.3.2 Necessary Conditions in the Presence of Set Constraints

In the "constrained" case, meaning in the presence of set constraints of the form (3.2), necessary conditions for optimality are provided by the *discrete minimum principle*. Although well established for constrained *continuous time* systems, the validity of the minimum principle for constrained *discrete time* systems was a subject of great interest and controversy in the 1960s. For general discrete time systems, the minimum principle does not hold when constraints on the input are imposed. Indeed, its validity in this case relies upon a *convexity requirement* (see Assumption 3.1 for Theorem 3.3.1 that follows). Actually, this requirement is readily satisfied in the continuous time case but presents a nontrivial restriction in the discrete time case. We refer the interested reader to the papers listed in Section 3.6.

In Section 3.3.3 we will present a version of the discrete minimum principle based on Halkin (1966) where convexity is imposed as an assumption on some reachable sets of the system. As pointed out by Halkin, the required convexity condition is always justified in the case of linear difference equations or when the system of nonlinear difference equations approximates a system of nonlinear differential equations. Actually, the convexity requirement can be relaxed to that of "directional convexity," allowing more general discrete systems to be covered (see the papers in Section 3.6).

In Section 3.3.4, we will make different assumptions on the nature of the constraints. This will allow us to use the FJ and KKT optimality conditions of Sections 2.5.3 and 2.5.4 of Chapter 2 to derive a version of the discrete minimum principle under convexity assumptions.

3.3.3 A Discrete Minimum Principle

We consider the following optimisation problem.

$$\mathcal{P}_N(\bar{x}): \quad \text{minimise } V_N(\{x_k\}, \{u_k\}), \tag{3.23}$$

$$\text{subject to:}$$

$$x_{k+1} = f(x_k, u_k) \quad \text{for } k = 0, \ldots, N-1, \tag{3.24}$$

$$x_0 = \bar{x}, \tag{3.25}$$

$$u_k \in \mathbb{U} \subset \mathbb{R}^m \quad \text{for } k = 0, \ldots, N-1, \tag{3.26}$$

$$h_N(x_N) = 0, \tag{3.27}$$

where

$$V_N(\{x_k\}, \{u_k\}) \triangleq F(x_N) + \sum_{k=0}^{N-1} L(x_k, u_k). \tag{3.28}$$

As before, $\{x_k\} \triangleq \{x_0, \ldots, x_N\}$, $x_k \in \mathbb{R}^n$, and $\{u_k\} \triangleq \{u_0, \ldots, u_{N-1}\}$, $u_k \in \mathbb{R}^m$, are the state and control sequences, and (3.24)–(3.25) are the state equations. Now, in addition to the equality constraints given by the state

equations, the minimisation is performed subject to constraints (3.26) and
(3.27) on the input and terminal state, respectively. In (3.26), $\mathbb{U}$ is a given
set, and in (3.27), $h_N : \mathbb{R}^n \rightarrow \mathbb{R}^\ell$ is a vector-valued function representing ℓ
constraints on the terminal state.

The conditions required on the data of the optimisation problem (3.23)–
(3.28) are given in the following assumption.

Assumption 3.1

(i) The function $F(x)$ is twice-continuously differentiable.

(ii) For every $u \in \mathbb{U}$, the functions $f(x, u)$ and $L(x, u)$ are twice-continuously
differentiable with respect to x.

(iii) The terminal constraint function $h_N(x)$ is twice-continuously differentiable and satisfies the "constraint qualification" that the Jacobian matrix
$\partial h_N(x)/\partial x$ has full row rank for all $x \in \mathbb{R}^n$.

(iv) The functions $f(x, u)$ and $L(x, u)$, and all their first and second partial
derivatives with respect to x, are uniformly bounded on $A \times \mathbb{U}$ for any
bounded set $A \subset \mathbb{R}^n$.

(v) The matrix $\partial f(\,\cdot\,,\,\cdot\,)/\partial x$ is nonsingular on $\mathbb{R}^n \times \mathbb{U}$.

(vi) The set $\left\{ \begin{bmatrix} f(x, u) \\ L(x, u) \end{bmatrix} : u \in \mathbb{U} \right\}$ is convex for all $x \in \mathbb{R}^n$.   ○

Similarly to the unconstrained case, we define the Hamiltonian as

$$\mathcal{H}(x_k, u_k, \lambda_k, \eta) \triangleq \eta L(x_k, u_k) + \lambda_k^{\mathrm{T}} f(x_k, u_k) \quad \text{for } k = 0, \ldots, N - 1, \quad (3.29)$$

where η is a real number and λ_k, $k = 0, \ldots, N - 1$, are some vectors in $\mathbb{R}^n$.
We then have the following result.

Theorem 3.3.1 (Discrete Minimum Principle) *Subject to Assump-
tion 3.1, if the sequences* $\{x_0^*, \ldots, x_N^*\}$, $\{u_0^*, \ldots, u_{N-1}^*\}$ *are minimisers of
problem* $\mathcal{P}_N(\bar{x})$ *defined in (3.23)–(3.28), then there exist a sequence of vectors*
$\{\lambda_{-1}^*, \ldots, \lambda_{N-1}^*\}$, $\lambda_k^* \in \mathbb{R}^n$ *and a real number* η^* *such that the following
conditions hold:*

(i) *Adjoint equations:*

$$(\lambda_{k-1}^*)^{\mathrm{T}} = \frac{\partial \mathcal{H}(x_k^*, u_k^*, \lambda_k^*, \eta^*)}{\partial x_k} \quad \text{for } k = 0, \ldots, N - 1. \quad (3.30)$$

(ii) *Boundary conditions: There exists a real number* $\beta \geq 0$ *and a vector*
$\gamma \in \mathbb{R}^\ell$, *such that*

$$\lambda_{N-1}^* = \left[\frac{\partial h_N(x_N^*)}{\partial x} \right]^{\mathrm{T}} \gamma + \left[\frac{\partial F(x_N^*)}{\partial x} \right]^{\mathrm{T}} \beta, \quad (3.31)$$

$$\eta^* = \beta \geq 0, \quad (3.32)$$

where η^* *and* λ_{N-1}^* *are not simultaneously zero. Moreover, if* $\eta^* = 0$ *in
(3.32), then the vectors* $\{\lambda_{-1}^*, \ldots, \lambda_{N-1}^*\}$ *satisfying (3.30)-(3.31) are all
nonzero.*

(iii) Minimisation of the Hamiltonian:

$$\mathcal{H}(x_k^*, u_k^*, \lambda_k^*, \eta^*) \leq \mathcal{H}(x_k^*, u, \lambda_k^*, \eta^*), \tag{3.33}$$

for all $k = 0, \ldots, N - 1$ and all $u \in \mathbb{U}$.

Proof. (Outline) Assume $\{x_0^*, \ldots, x_N^*\}$, $\{u_0^*, \ldots, u_{N-1}^*\}$ are minimisers of (3.23)–(3.28). Define an auxiliary state $z_k \in \mathbb{R}$ satisfying

$$\begin{aligned} z_{k+1} &= z_k + L(x_k, u_k) \quad \text{for } k = 0, \ldots, N - 1, \\ z_0 &= 0, \end{aligned} \tag{3.34}$$

where $\{x_0, \ldots, x_{N-1}\}$, $\{u_0, \ldots, u_{N-1}\}$ satisfy the system equations (3.24). Note that $z_i = \sum_{k=0}^{i-1} L(x_k, u_k)$, and, from (3.28),

$$V_N(\{x_k\}, \{u_k\}) = z_N + F(x_N). \tag{3.35}$$

Consider the composite state $\begin{bmatrix} x_k^T & z_k \end{bmatrix}^T \in \mathbb{R}^{n+1}$, satisfying (3.24)–(3.25) and (3.34).

Comment on notation. The composite state $\xi_k \triangleq \begin{bmatrix} x_k^T & z_k \end{bmatrix}^T \in \mathbb{R}^{n+1}$ satisfies

$$\xi_{k+1} = \xi_k + \bar{f}(\xi_k, u_k),$$

where $\bar{f}(\xi_k, u_k) \triangleq \begin{bmatrix} f(x_k, u_k) - x_k \\ L(x_k, u_k) \end{bmatrix}$. The initial state ξ_0 must belong to the set

$$\{\xi : \bar{h}_i(\xi) = 0 \quad \text{for } i = 1, \ldots, n + 1\},$$

where $\bar{h}_i(\xi) \triangleq e_i^T \left(\xi - \begin{bmatrix} \bar{x} \\ 0 \end{bmatrix}\right)$, and where e_i is the vector having one as its ith component and all other components equal to zero. The terminal state ξ_N must belong to the set

$$\left\{\xi : \bar{h}_N(\xi) \triangleq h_N(x) = 0\right\}.$$

The control is constrained by (3.26). The objective function to be minimised, (3.35), is a function of the terminal state ξ_N only, that is,

$$V_N(\{x_k\}, \{u_k\}) = z_N + F(x_N) \triangleq g_0(\xi_N).$$

This formulation of the problem was used in Halkin (1966).   ∘

Define the set $\mathbb{W}$ as the set of reachable composite states $\begin{bmatrix} x_k^T & z_k \end{bmatrix}^T$ at time N. In other words, $\mathbb{W}$ is the set of all $[x_N^T \ z_N]^T$ corresponding to all sequences $\{x_0, \ldots, x_N\}$, $\{u_0, \ldots, u_{N-1}\}$, $\{z_0, \ldots, z_N\}$, satisfying (3.24)–(3.26) and (3.34).

Next, consider the optimal sequences $\{x_0^*, \ldots, x_N^*\}$, $\{u_0^*, \ldots, u_{N-1}^*\}$, and let $\{z_0^*, \ldots, z_N^*\}$ be the resulting sequence satisfying (3.34). We define the set

$\mathbb{S}_N^*$ as the set of all composite states $\begin{bmatrix} x^{\mathrm{T}} & z \end{bmatrix}^{\mathrm{T}}$ satisfying the terminal constraint (3.27), and for which the objective function takes a lesser value than the optimal value. Using (3.35), this set is given by

$$\mathbb{S}_N^* = \left\{ \begin{bmatrix} x \\ z \end{bmatrix} : h_N(x) = 0, \; z + F(x) < z_N^* + F(x_N^*) \right\}. \qquad (3.36)$$

We note that the sets $\mathbb{W}$ and $\mathbb{S}_N^*$ are disjoint—otherwise $\{x_0^*, \dots, x_N^*\}$, $\{u_0^*, \dots, u_{N-1}^*\}$ would not be optimal. Figure 3.1 illustrates these sets in the composite state space $\begin{bmatrix} x^{\mathrm{T}} & z \end{bmatrix}^{\mathrm{T}}$.

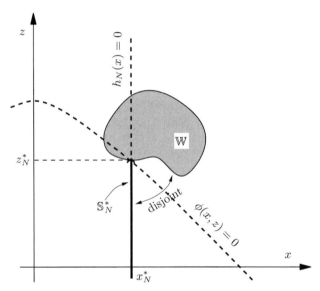

Figure 3.1. The composite state space $\begin{bmatrix} x^{\mathrm{T}} & z \end{bmatrix}^{\mathrm{T}}$ showing sets $\mathbb{W}$ and $\mathbb{S}_N^*$ for the original (nonlinear) problem. The curve $\phi(x, z) \triangleq [z - z_N^*] + [F(x) - F(x_N^*)] = 0$ is the set of composite states that give an objective value (3.35) equal to the optimal value $z_N^* + F(x_N^*)$.

In the case of linear systems, it is easy to prove that the sets $\mathbb{W}$ and $\mathbb{S}_N^*$ are convex, and hence separated since they are disjoint (see Theorem 2.3.5 in Chapter 2).

In the nonlinear case, however, the sets $\mathbb{W}$ and $\mathbb{S}_N^*$ are *not* necessarily convex, and hence not necessarily separated. This is illustrated in Figure 3.1, where no hyperplane separates the two sets. To circumvent this difficulty, we will consider a linearised problem using linear approximations with respect to x of the functions $f(x, u)$ in (3.24), $L(x, u)$ in (3.34), $h_N(x)$ in (3.27) and $F(x)$ in (3.35) around the optimal sequences. That is, we consider the system

$$x_{k+1} = f(x_k^*, u_k) + \frac{\partial f(x_k^*, u_k^*)}{\partial x_k}[x_k - x_k^*] \quad \text{for } k = 0, \dots, N-1, \tag{3.37}$$

$$z_{k+1} = z_k + L(x_k^*, u_k) + \frac{\partial L(x_k^*, u_k^*)}{\partial x_k}[x_k - x_k^*] \quad \text{for } k = 0, \dots, N-1, \tag{3.38}$$

$$x_0 = x_0^* = \bar{x}, \tag{3.39}$$

$$z_0 = 0, \tag{3.40}$$

with constraints

$$u_k \in \mathbb{U} \subset \mathbb{R}^m \quad \text{for } k = 0, \dots, N-1, \tag{3.41}$$

$$h_N(x_N^*) + \frac{\partial h_N(x_N^*)}{\partial x}[x_N - x_N^*] = \frac{\partial h_N(x_N^*)}{\partial x}[x_N - x_N^*] = 0, \tag{3.42}$$

and objective function

$$\hat{V}_N \triangleq z_N + F(x_N^*) + \frac{\partial F(x_N^*)}{\partial x}[x_N - x_N^*]. \tag{3.43}$$

We note that $\{x_0^*, \dots, x_N^*\}$, $\{u_0^*, \dots, u_{N-1}^*\}$ and $\{z_0^*, \dots, z_N^*\}$ are feasible sequences (but not necessarily optimal) for the linearised problem (3.37)–(3.43).

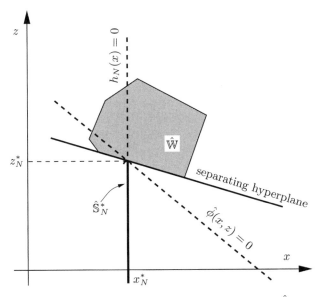

Figure 3.2. The composite state space $[x^{\mathrm{T}} \ z]^{\mathrm{T}}$ showing sets $\hat{\mathbb{W}}$ and $\hat{\mathbb{S}}_N^*$ for the linearised problem. The line $\hat{\phi}(x, z) \triangleq [z - z_N^*] + \frac{\partial F(x_N^*)}{\partial x}[x - x_N^*] = 0$ is the set of composite states that give a linearised objective value (3.43) equal to the optimal value $z_N^* + F(x_N^*)$.

We also define $\hat{\mathbb{W}}$, $\hat{\mathbb{S}}_N^*$ as we did for the sets $\mathbb{W}$ and $\mathbb{S}_N^*$, but for the linearised problem. Figure 3.2 illustrates these sets in the composite state space $[x^\mathrm{T} \ z]^\mathrm{T}$.

It is easy to show that the sets $\hat{\mathbb{W}}$ and $\hat{\mathbb{S}}_N^*$ are convex. It is not immediately obvious, but is proved in Halkin (1966), that the sets $\hat{\mathbb{W}}$ and $\hat{\mathbb{S}}_N^*$ are separated[2] by a hyperplane passing through the point $[(x_N^*)^\mathrm{T} \ z_N^*]^\mathrm{T}$ (see Definition 2.3.3 of Chapter 2). We conclude that there exists a *nonzero* vector $[\pi^\mathrm{T} \ \beta]^\mathrm{T} \in \mathbb{R}^{n+1}$ such that

$$[x - x_N^*]^\mathrm{T}\pi + [z - z_N^*]\beta \le 0 \quad \text{for all} \quad \begin{bmatrix} x \\ z \end{bmatrix} \in \hat{\mathbb{S}}_N^*, \tag{3.44}$$

$$[x - x_N^*]^\mathrm{T}\pi + [z - z_N^*]\beta \ge 0 \quad \text{for all} \quad \begin{bmatrix} x \\ z \end{bmatrix} \in \hat{\mathbb{W}}. \tag{3.45}$$

We define the sequence of nonzero vectors $\{\lambda_{-1}^*, \ldots, \lambda_{N-1}^*\}$, as the solution of (3.30) with boundary condition

$$\lambda_{N-1}^* = \pi, \tag{3.46}$$

and put

$$\eta^* = \beta. \tag{3.47}$$

We are now ready to show that conditions (3.30)–(3.33) are satisfied.

(i) Condition (3.30).

   This is satisfied by definition.

(ii) Conditions (3.31) and (3.32).

   First note from (3.46)–(3.47) that, since $[\pi^\mathrm{T} \ \beta]^\mathrm{T} \ne 0$, we have that η^* and λ_{N-1}^* are not simultaneously zero.

   Next, we have from (3.44) and (3.46) that

$$[x - x_N^*]^\mathrm{T}\lambda_{N-1}^* + [z - z_N^*]\beta \le 0 \tag{3.48}$$

   for all $[x^\mathrm{T} \ z]^\mathrm{T} \in \hat{\mathbb{S}}_N^*$, that is, all states x, z that satisfy the linearised terminal constraint (3.42) and such that the linearised objective function (3.43) has a lesser value than the optimal value $z_N^* + F(x_N^*)$. More precisely, (3.48) holds for all x, z such that

$$\frac{\partial h_N(x_N^*)}{\partial x}[x - x_N^*] = 0, \tag{3.49}$$

   and

---

[2] Halkin (1966) proves the following *linearisation lemma*: If the sets $\mathbb{W}$ and $\mathbb{S}_N^*$ are disjoint, then the sets $\hat{\mathbb{W}}$ and $\hat{\mathbb{S}}_N^*$ are separated.

$$[z - z_N^*] + \frac{\partial F(x_N^*)}{\partial x}[x - x_N^*] < 0. \qquad (3.50)$$

We will show that this implies that conditions (3.31) and (3.32) hold. Specifically, we define

$$A_1 = \begin{bmatrix} -\lambda_{N-1}^{*\mathrm{T}} & -\beta \\ \dfrac{\partial F(x_N^*)}{\partial x} & 1 \end{bmatrix}, \qquad A_2 = \begin{bmatrix} \dfrac{\partial h_N(x_N^*)}{\partial x} & 0 \end{bmatrix},$$

$$d = \begin{bmatrix} x - x_N^* \\ z - z_N^* \end{bmatrix}.$$

Now (3.48)–(3.50) imply that the system

$$A_1 d < 0, \quad A_2 d = 0$$

is inconsistent. Following the proof of Theorem 2.5.4 in Chapter 2, we have that there exists a nonzero vector $[q^{\mathrm{T}} \; v^{\mathrm{T}}]^{\mathrm{T}}$ with $q \in \mathbb{R}^2$, $v \in \mathbb{R}^{\ell}$, $q \geq 0$, such that

$$A_1^{\mathrm{T}} q + A_2^{\mathrm{T}} v = 0.$$

Letting $q = \begin{bmatrix} q_1 & q_2 \end{bmatrix}^{\mathrm{T}}$, the above equation implies

$$-\begin{bmatrix} \lambda_{N-1}^* \\ \beta \end{bmatrix} q_1 + \begin{bmatrix} \begin{bmatrix} \dfrac{\partial F(x_N^*)}{\partial x} \end{bmatrix}^{\mathrm{T}} & \begin{bmatrix} \dfrac{\partial h_N(x_N^*)}{\partial x} \end{bmatrix}^{\mathrm{T}} \\ 1 & 0 \end{bmatrix} \begin{bmatrix} q_2 \\ v \end{bmatrix} = 0.$$

or

$$-\lambda_{N-1}^* q_1 + \begin{bmatrix} \dfrac{\partial F(x_N^*)}{\partial x} \end{bmatrix}^{\mathrm{T}} q_2 + \begin{bmatrix} \dfrac{\partial h_N(x_N^*)}{\partial x} \end{bmatrix}^{\mathrm{T}} v = 0, \qquad (3.51)$$

$$-\beta q_1 + q_2 = 0. \qquad (3.52)$$

If $q_1 = 0$, then $q_2 = 0$ from (3.52), and by linear independence of the rows of $\partial h_N/\partial x$ we have from (3.51) that $v = 0$. However, this contradicts $[q^{\mathrm{T}} \; v^{\mathrm{T}}]^{\mathrm{T}} \neq 0$. Hence, $q_1 \neq 0$. Thus, from (3.52), we have $\beta = q_2/q_1 \geq 0$, which shows (3.32). Also, equation (3.31) follows from (3.51) substituting $\beta = q_2/q_1$ and choosing $\gamma = v/q_1$. Finally, if $\eta^* = \beta = 0$ in (3.32), then $\lambda_{N-1}^* = \pi \neq 0$ (since $[\pi^{\mathrm{T}} \; \beta]^{\mathrm{T}} \neq 0$). Hence, (3.30) yields

$$\lambda_{k-1}^* = \begin{bmatrix} \dfrac{\partial f(x_k^*, u_k^*)}{\partial x_k} \end{bmatrix}^{\mathrm{T}} \lambda_k^* \neq 0 \quad \text{for } k = 0, \dots, N-1,$$

since $\partial f(\cdot, \cdot)/\partial x$ is nonsingular on $\mathbb{R}^n \times \mathbb{U}$ by assumption.

(iii) Condition (3.33).

For any given $k \in [0, \dots, N]$ let $\hat{\mathbb{W}}_k$ be the set of all composite states $[x^{\mathrm{T}} \; z]^{\mathrm{T}}$ reachable at time k for the linearised system (3.37)–(3.40) with all admissible control sequences satisfying (3.41). Clearly $\hat{\mathbb{W}}_N = \hat{\mathbb{W}}$.

We shall prove by contradiction that for every $k \in [0, \ldots, N]$ and all $[x^{\mathrm{T}} \; z]^{\mathrm{T}} \in \hat{\mathbb{W}}_k$ we have

$$[x - x_k^*]^{\mathrm{T}} \lambda_{k-1}^* + [z - z_k^*]\beta \geq 0. \tag{3.53}$$

Indeed, let us assume that for some $j \in [0, \ldots, N]$, $[\tilde{x}_j^{\mathrm{T}} \; \tilde{z}_j]^{\mathrm{T}} \in \hat{\mathbb{W}}_j$ and $\varepsilon > 0$, we have

$$[\tilde{x}_j - x_j^*]^{\mathrm{T}} \lambda_{j-1}^* + [\tilde{z}_j - z_j^*]\beta = -\varepsilon < 0. \tag{3.54}$$

We show that this leads to a contradiction. We define $\{\tilde{x}_{j+1}, \ldots, \tilde{x}_N\}$, $\{\tilde{z}_{j+1}, \ldots, \tilde{z}_N\}$ as the solution of the linearised equations (3.37)–(3.38) starting at time $k = j$ with $x_j = \tilde{x}_j$, $z_j = \tilde{z}_j$ and using the optimal control sequence $\{u_{j+1}^*, \ldots, u_{N-1}^*\}$. That is,

$$\tilde{x}_{k+1} = f(x_k^*, u_k^*) + \frac{\partial f(x_k^*, u_k^*)}{\partial x_k}[\tilde{x}_k - x_k^*] \quad \text{for } k = j, \ldots, N-1,$$

$$\tilde{z}_{k+1} = \tilde{z}_k + L(x_k^*, u_k^*) + \frac{\partial L(x_k^*, u_k^*)}{\partial x_k}[\tilde{x}_k - x_k^*] \quad \text{for } k = j, \ldots, N-1.$$

Using the above equations and the fact that $x_{k+1}^* = f(x_k^*, u_k^*)$ and $z_{k+1}^* = z_k^* + L(x_k^*, u_k^*)$ we have that the difference between the quantity $[\tilde{x}_k - x_k^*]^{\mathrm{T}} \lambda_{k-1}^* + [\tilde{z}_k - z_k^*]\beta$ at step k and its subsequent value at step $k+1$ satisfies, for $k = j, \ldots, N-1$,

$$[\tilde{x}_k - x_k^*]^{\mathrm{T}} \lambda_{k-1}^* + [\tilde{z}_k - z_k^*]\beta - [\tilde{x}_{k+1} - x_{k+1}^*]^{\mathrm{T}} \lambda_k^* - [\tilde{z}_{k+1} - z_{k+1}^*]\beta =$$

$$[\tilde{x}_k - x_k^*]^{\mathrm{T}} \lambda_{k-1}^* + [\tilde{z}_k - z_k^*]\beta - \left\{ \frac{\partial f(x_k^*, u_k^*)}{\partial x_k}[\tilde{x}_k - x_k^*] \right\}^{\mathrm{T}} \lambda_k^*$$

$$-\beta \frac{\partial L(x_k^*, u_k^*)}{\partial x_k}[\tilde{x}_k - x_k^*] - [\tilde{z}_k - z_k^*]\beta =$$

$$[\tilde{x}_k - x_k^*]^{\mathrm{T}} \left\{ \lambda_{k-1}^* - \left[\frac{\partial f(x_k^*, u_k^*)}{\partial x_k} \right]^{\mathrm{T}} \lambda_k^* - \beta \left[\frac{\partial L(x_k^*, u_k^*)}{\partial x_k} \right]^{\mathrm{T}} \right\} =$$

$$[\tilde{x}_k - x_k^*]^{\mathrm{T}} \left\{ \lambda_{k-1}^* - \left[\frac{\partial \mathcal{H}(x_k^*, u_k^*, \lambda_k^*, \eta^*)}{\partial x_k} \right]^{\mathrm{T}} \right\} = 0, \tag{3.55}$$

where, in the last two equalities, we have also used (3.30), (3.47), and the definition of $\mathcal{H}(x_k, u_k, \lambda_k, \eta)$ from (3.29). Thus, due to hypothesis (3.54) we may conclude, using (3.55), that

$$[\tilde{x}_N - x_N^*]^{\mathrm{T}} \lambda_{N-1}^* + [\tilde{z}_N - z_N^*]\beta = -\varepsilon < 0. \tag{3.56}$$

However, this contradicts (3.45) since $\lambda_{N-1}^* = \pi$ (see (3.46)) and $[\tilde{x}_N^{\mathrm{T}} \; \tilde{z}_N]^{\mathrm{T}} \in \hat{\mathbb{W}}$ by construction. Thus (3.53) holds.

We now use (3.53) to construct a contradiction by assuming that for some $j \in [0, \ldots, N-1]$ there exists $\tilde{u}_j \in \mathbb{U}$ and an $\varepsilon > 0$ such that (3.33) is false, that is,

$$\mathcal{H}(x_j^*, \tilde{u}_j, \lambda_j^*, \eta^*) - \mathcal{H}(x_j^*, u_j^*, \lambda_j^*, \eta^*) = -\varepsilon. \tag{3.57}$$

Consider the composite state at the next time $j+1$, $[\tilde{x}_{j+1}^{\mathrm{T}} \; \tilde{z}_{j+1}]^{\mathrm{T}} \in \hat{\mathbb{W}}_{j+1}$, defined by

$$\begin{aligned}
\tilde{x}_{j+1} &= f(x_j^*, \tilde{u}_j), \\
\tilde{z}_{j+1} &= z_j^* + L(x_j^*, \tilde{u}_j),
\end{aligned} \tag{3.58}$$

which corresponds to the solution of the linearised equations (3.37)–(3.38) at time $j+1$ using at time $k = j$ the values $x_j = x_j^*$, $z_j = z_j^*$, and $u_j = \tilde{u}_j$. Note that the states (3.58) are also solution at time $j+1$ of the nonlinear equations (3.24), (3.34) starting from the same values at time $k = j$. Using (3.47) and the definition (3.29) of the Hamiltonian in (3.57), we obtain

$$[f(x_j^*, \tilde{u}_j)]^{\mathrm{T}}\lambda_j^* + L(x_j^*, \tilde{u}_j)\beta - [f(x_j^*, u_j^*)]^{\mathrm{T}}\lambda_j^* - L(x_j^*, u_j^*)\beta = -\varepsilon.$$

Substituting (3.58) and $x_{j+1}^* = f(x_j^*, u_j^*)$, $z_{j+1}^* = z_j^* + L(x_j^*, u_j^*)$ in the above equation yields

$$[\tilde{x}_{j+1} - x_{j+1}^*]^{\mathrm{T}}\lambda_j^* + [\tilde{z}_{j+1} - z_{j+1}^*]\beta = -\varepsilon.$$

However, this contradicts (3.53), which we have already established. We thus conclude that (3.57) is false, and hence the inequality (3.33) holds for $k = 0, \ldots, N-1$ and all $u \in \mathbb{U}$. This concludes the proof. $\qquad \square$

3.3.4 Connections Between the Minimum Principle and the Fritz John and Karush–Kuhn–Tucker Optimality Conditions

We consider the following optimisation problem:

$$\mathcal{P}_N(\bar{x}): \qquad \text{minimise } V_N(\{x_k\}, \{u_k\}), \tag{3.59}$$

$$\text{subject to:}$$
$$x_{k+1} = f(x_k, u_k) \quad \text{for } k = 0, \ldots, N-1, \tag{3.60}$$
$$x_0 = \bar{x}, \tag{3.61}$$
$$g_k(u_k) \leq 0 \quad \text{for } k = 0, \ldots, N-1, \tag{3.62}$$
$$g_N(x_N) \leq 0, \tag{3.63}$$
$$h_N(x_N) = 0, \tag{3.64}$$

where

$$V_N(\{x_k\}, \{u_k\}) \triangleq F(x_N) + \sum_{k=0}^{N-1} L(x_k, u_k). \tag{3.65}$$

As before, $\{x_k\} \triangleq \{x_0, \ldots, x_N\}$, $x_k \in \mathbb{R}^n$, and $\{u_k\} \triangleq \{u_0, \ldots, u_{N-1}\}$, $u_k \in \mathbb{R}^m$, are the state and control sequences, and (3.60)–(3.61) are the state equations. The functions $g_k : \mathbb{R}^m \to \mathbb{R}^r$, $k = 0, \ldots, N-1$, represent r (elementwise) inequality constraints on the input u_k (compare with (3.26) where constraints on the input were expressed as the set constraints $u_k \in \mathbb{U}$, $k = 0, \ldots, N-1$). The functions $g_N : \mathbb{R}^n \to \mathbb{R}^p$ and $h_N : \mathbb{R}^n \to \mathbb{R}^\ell$, represent, respectively, inequality and equality constraints on the terminal state (compare with (3.27) where only equality constraints on the terminal state were considered). We will assume that all functions in (3.59)–(3.65) are differentiable functions of their variables and that f and h_N are continuously differentiable at the optimal solution.

We will derive necessary optimality conditions for the sequences $\{x_0^*, \ldots, x_N^*\}$ and $\{u_0^*, \ldots, u_{N-1}^*\}$ to be minimisers of the optimisation problem (3.59)–(3.65) using the FJ necessary optimality conditions (see Section 2.5.3 in Chapter 2). We observe that, in contrast with the "unconstrained" case of Section 3.3.1, where the linear independence constraint qualification required by the KKT conditions holds for all feasible points, here a constraint qualification *would* need to be imposed if we were to use the KKT conditions as necessary conditions for optimality. On the other hand, the FJ conditions are always a necessary condition for optimality under the differentiability assumption, without requiring any constraint qualification.

Recalling the vector definition

$$\mathbf{x} \triangleq \begin{bmatrix} x_0^T & \cdots & x_N^T & u_0^T & \cdots & u_{N-1}^T \end{bmatrix}^T \in \mathbb{R}^{(N+1)n + Nm},$$

we can express problem (3.59)–(3.65) in the form

$$\begin{aligned} &\text{minimise } \phi(\mathbf{x}), \\ &\text{subject to:} \\ &g(\mathbf{x}) \leq 0, \\ &h(\mathbf{x}) = 0, \end{aligned} \tag{3.66}$$

where

$$\phi(\mathbf{x}) \triangleq F(x_N) + \sum_{k=0}^{N-1} L(x_k, u_k), \tag{3.67}$$

$$h(\mathbf{x}) \triangleq \begin{bmatrix} \bar{x} - x_0 \\ f(x_0, u_0) - x_1 \\ \vdots \\ f(x_{N-1}, u_{N-1}) - x_N \\ h_N(x_N) \end{bmatrix}, \tag{3.68}$$

$$g(\mathbf{x}) \triangleq \begin{bmatrix} g_0(u_0) \\ g_1(u_1) \\ \vdots \\ g_{N-1}(u_{N-1}) \\ g_N(x_N) \end{bmatrix}. \tag{3.69}$$

Suppose $\mathbf{x}^* = \left[(x_0^*)^{\mathrm{T}} \cdots (x_N^*)^{\mathrm{T}} (u_0^*)^{\mathrm{T}} \cdots (u_{N-1}^*)^{\mathrm{T}} \right]^{\mathrm{T}}$ is a minimiser of (3.66). Then the FJ conditions (see (2.30) in Chapter 2) hold for problem (3.66) at $\mathbf{x}^*$, that is, there exist a scalar η^* and vectors $\{\lambda_{-1}^*, \ldots, \lambda_{N-1}^*\}$, γ^*, $\{\nu_0^*, \ldots, \nu_N^*\}$ such that

$$\left[\frac{\partial \phi(\mathbf{x}^*)}{\partial \mathbf{x}} \right]^{\mathrm{T}} \eta^* + \left[\frac{\partial h(\mathbf{x}^*)}{\partial \mathbf{x}} \right]^{\mathrm{T}} \begin{bmatrix} \lambda_{-1}^* \\ \vdots \\ \lambda_{N-1}^* \\ \gamma^* \end{bmatrix} + \left[\frac{\partial g(\mathbf{x}^*)}{\partial \mathbf{x}} \right]^{\mathrm{T}} \begin{bmatrix} \nu_0^* \\ \vdots \\ \nu_N^* \end{bmatrix} = 0, \tag{3.70}$$

$$\begin{bmatrix} \nu_0^* \\ \vdots \\ \nu_N^* \end{bmatrix}^{\mathrm{T}} g(\mathbf{x}^*) = 0, \tag{3.71}$$

$$(\eta^*, \nu_0^*, \ldots, \nu_N^*) \geq 0, \tag{3.72}$$

$$(\eta^*, \lambda_{-1}^*, \ldots, \lambda_{N-1}^*, \gamma^*, \nu_0^*, \ldots, \nu_N^*) \neq 0, \tag{3.73}$$

where

$$\frac{\partial \phi}{\partial \mathbf{x}} = \begin{bmatrix} \dfrac{\partial L}{\partial x_0} & \cdots \cdots \cdots & \dfrac{\partial L}{\partial x_{N-1}} & \dfrac{\partial F}{\partial x_N} & \dfrac{\partial L}{\partial u_0} & \cdots \cdots & \dfrac{\partial L}{\partial u_{N-1}} \end{bmatrix},$$

$$
\frac{\partial h}{\partial \mathbf{x}} =
\begin{bmatrix}
-I_n & 0 & 0 & \cdots & 0 & 0 & 0 & 0 & \cdots & 0 \\
\frac{\partial f}{\partial x_0} & -I_n & 0 & \cdots & 0 & 0 & \frac{\partial f}{\partial u_0} & 0 & \cdots & 0 \\
0 & \frac{\partial f}{\partial x_1} & -I_n & \cdots & 0 & 0 & 0 & \frac{\partial f}{\partial u_1} & \cdots & 0 \\
\vdots & \vdots & \vdots & \ddots & \vdots & \vdots & \vdots & \vdots & \ddots & \vdots \\
0 & 0 & 0 & \cdots & \frac{\partial f}{\partial x_{N-1}} & -I_n & 0 & 0 & \cdots & \frac{\partial f}{\partial u_{N-1}} \\
0 & 0 & 0 & \cdots & 0 & \frac{\partial h_N}{\partial x_N} & 0 & 0 & \cdots & 0
\end{bmatrix},
$$

$$
\frac{\partial g}{\partial \mathbf{x}} =
\begin{bmatrix}
0 & \cdots & \cdots & \cdots & \cdots & 0 & \frac{\partial g_0}{\partial u_0} & 0 & \cdots & 0 \\
\vdots & \ddots & & & & \vdots & 0 & \frac{\partial g_1}{\partial u_1} & \cdots & 0 \\
\vdots & & \ddots & & & \vdots & \vdots & & \ddots & \vdots \\
\vdots & & & \ddots & & \vdots & 0 & 0 & \cdots & \frac{\partial g_{N-1}}{\partial u_{N-1}} \\
0 & \cdots & \cdots & \cdots & \cdots & \frac{\partial g_N}{\partial x_N} & 0 & 0 & \cdots & 0
\end{bmatrix}.
$$

We will next use the Hamiltonian

$$
\mathcal{H}(x_k, u_k, \lambda_k, \eta) \triangleq \eta L(x_k, u_k) + \lambda_k^{\mathrm{T}} f(x_k, u_k) \quad \text{for } k = 0, \ldots, N-1, \quad (3.74)
$$

in the dual feasibility condition (3.70), and write the FJ conditions (3.70)–(3.73) component-wise. We thus conclude that a necessary condition for the sequences $\{x_0^*, \ldots, x_N^*\}$, $\{u_0^*, \ldots, u_{N-1}^*\}$ to be minimisers of (3.59)–(3.65) is that there exist a scalar η^* and vectors $\{\lambda_{-1}^*, \ldots, \lambda_{N-1}^*\}$, γ^*, $\{\nu_0^*, \ldots, \nu_N^*\}$, not all zero, such that the following conditions hold:

(i) Adjoint equations:

$$
(\lambda_{k-1}^*)^{\mathrm{T}} = \frac{\partial \mathcal{H}(x_k^*, u_k^*, \lambda_k^*, \eta^*)}{\partial x_k} \quad \text{for } k = 0, \ldots, N-1. \quad (3.75)
$$

(ii) Boundary conditions:

$$
\lambda_{N-1}^* = \left[\frac{\partial h_N(x_N^*)}{\partial x_N} \right]^{\mathrm{T}} \gamma^* + \left[\frac{\partial F(x_N^*)}{\partial x_N} \right]^{\mathrm{T}} \eta^* + \left[\frac{\partial g_N(x_N^*)}{\partial x_N} \right]^{\mathrm{T}} \nu_N^*, \quad (3.76)
$$

$$
(\nu_N^*)^{\mathrm{T}} g_N(x_N^*) = 0, \quad (3.77)
$$

$$
\eta^* \geq 0, \ \nu_N^* \geq 0. \quad (3.78)
$$

(iii) Hamiltonian conditions:

$$\left[\frac{\partial \mathcal{H}(x_k^*, u_k^*, \lambda_k^*, \eta^*)}{\partial u_k}\right]^{\mathrm{T}} + \left[\frac{\partial g_k(u_k^*)}{\partial u_k}\right]^{\mathrm{T}} \nu_k^* = 0, \tag{3.79}$$

$$(\nu_k^*)^{\mathrm{T}} g_k(u_k^*) = 0, \tag{3.80}$$

$$\nu_k^* \geq 0, \tag{3.81}$$

for $k = 0, \ldots, N - 1$.   ∘

Now consider the following related condition:

$$\mathcal{H}(x_k^*, u_k^*, \lambda_k^*, \eta^*) \leq \mathcal{H}(x_k^*, u_k, \lambda_k^*, \eta^*) \text{ for all } u_k \text{ such that } g_k(u_k) \leq 0, \tag{3.82}$$

for $k = 0, \ldots, N - 1$, where $\mathcal{H}(x_k^*, u_k, \lambda_k^*, \eta^*)$ is the restricted Hamiltonian

$$\mathcal{H}(x_k^*, u_k, \lambda_k^*, \eta^*) \triangleq \eta^* L(x_k^*, u_k) + (\lambda_k^*)^{\mathrm{T}} f(x_k^*, u_k). \tag{3.83}$$

Notice that the KKT conditions for (3.82) (that is, for the problem: minimise $\mathcal{H}(x_k^*, u_k, \lambda_k^*, \eta^*)$ subject to $g_k(u_k) \leq 0$) coincide with (3.79)–(3.81) (compare with (2.34) in Chapter 2). However, in order to guarantee that (3.82) is a necessary condition for (3.79)–(3.81), and hence for the original problem (3.59)–(3.65), we need additional mild convexity assumptions (see Definition 2.3.9 in Chapter 2). Suppose now that $\mathcal{H}(x_k^*, u_k, \lambda_k^*, \eta^*)$ is *pseudoconvex* at u_k^*, and the constraint function $g_k(u_k)$ in (3.82) is *quasiconvex* at u_k^*. We can then apply the KKT sufficient optimality conditions of Theorem 2.5.6 in Chapter 2 to conclude that conditions (3.79)–(3.81) imply (3.82).

Thus, for the *original* optimisation problem (3.59)–(3.65), under the above (generalised) convexity assumptions, a necessary condition for the sequences $\{x_0^*, \ldots, x_N^*\}$, $\{u_0^*, \ldots, u_{N-1}^*\}$ to be minimisers is that there exist a scalar η^* and vectors $\{\lambda_{-1}^*, \ldots, \lambda_{N-1}^*\}$, γ^*, ν_N^*, not all zero, such that conditions (3.75)–(3.78) hold, and, furthermore, u_k^* minimises the restricted Hamiltonian for $k = 0, \ldots, N - 1$; that is, condition (3.82) holds.

Finally, we observe that, if the functions $h(\mathbf{x})$ in (3.68) and $g(\mathbf{x})$ in (3.69), which define the constraints for the (vector form of the) original problem (3.66), satisfy a constraint qualification (see discussion preceding Theorem 2.5.5 in Chapter 2), then we can apply the KKT necessary optimality conditions to the original problem, that is, we can set $\eta = 1$ in the FJ conditions (3.70)–(3.73) and in the Hamiltonian (3.74).

3.4 Sufficient Conditions for Optimality Using Dynamic Programming

The necessary conditions for optimality developed in Section 3.3 are not, in general, sufficient. However, it is possible to obtain sufficient conditions by using the following *principle of optimality*: "Any part of an optimal trajectory must itself be optimal." This principle is captured in the idea of dynamic programming (Bellman 1957). To develop this idea further, let $V_{N-k}^{\mathrm{OPT}}(x_k)$ be the *partial value function* at time k *assuming we are in state x_k at time k.* Of course, the catch here is that we do not know a priori which state we will be in at time k—more will be said on this point later.

We consider the optimisation problem (3.23)–(3.28). To illustrate the idea, we assume that, in addition to the equality constraints provided by the state equations, only the input is constrained to belong to a set $\mathbb{U} \subset \mathbb{R}^m$, that is, the optimisation is performed subject to (3.24)–(3.26) (no equality constraint on the terminal state). Clearly, from (3.28), the partial value function at time N is

$$V_0^{\mathrm{OPT}}(x_N) = F(x_N), \tag{3.84}$$

by definition of the objective function. The principle of optimality states that we must satisfy the following sufficient condition at each time k:

$$V_{N-k}^{\mathrm{OPT}}(x_k) = \min_{u_k \in \mathbb{U}} \left\{ V_{N-(k+1)}^{\mathrm{OPT}}(f(x_k, u_k)) + L(x_k, u_k) \right\}. \tag{3.85}$$

Unfortunately, we will not know, in advance, which value of x_k will be optimal at time k. For some simple problems, such as for linear quadratic unconstrained optimal control, $V_{N-k}^{\mathrm{OPT}}(x_k)$ has a finite parameterisation. However, in general, $V_{N-k}^{\mathrm{OPT}}(x_k)$ will be a complicated function of the states. In the latter case, we really have no option but to store $V_{N-k}^{\mathrm{OPT}}(x_k)$ for all possible values of x_k. We begin with $V_0^{\mathrm{OPT}}(x_N)$. Then using (3.85), we can evaluate the optimiser u_{N-1}^* and hence $V_1^{\mathrm{OPT}}(x_{N-1})$ for all possible values of x_{N-1}. We continue backwards until we finally reach $V_N^{\mathrm{OPT}}(x_0)$. Having reached this point, we realise that we actually do know x_0, and we can thus find u_0^*. Then, running forward in time, we can calculate x_1^* and, provided we have recorded u_1^* for all possible x_1, we can then proceed to x_2^* and so on.

The above procedure can be applied, in principle, to very general problems assuming a finite number of admissible states. This can be achieved, for example, by quantisation of the admissible region in the state space. For problems with large horizons and high state space dimension, however, the number of computations and the storage required by the dynamic programming technique may be impractical, or even prohibitive. Bellman called this problem the "curse of dimensionality." For some special problems, however, the technique can be successfully applied, even to obtain closed form expressions for the optimal constrained control law, as we will see in Chapters 6 and 7.

Dynamic programming can also be used to find the optimal solution of constrained estimation problems, see Sections 9.5 and 9.6 of Chapter 9.

3.5 An Interpretation of the Adjoint Variables

For a given optimal trajectory $\{x_k^*\}$, $\{u_k^*\}$, it follows from (3.85) that

$$V_{N-k}^{\text{OPT}}(x_k^*) = V_{N-(k+1)}^{\text{OPT}}(f(x_k^*, u_k^*)) + L(x_k^*, u_k^*). \qquad (3.86)$$

Hence,

$$\frac{\partial V_{N-k}^{\text{OPT}}(x_k^*)}{\partial x_k} = \frac{\partial V_{N-(k+1)}^{\text{OPT}}(x_{k+1}^*)}{\partial x_{k+1}} \frac{\partial f(x_k^*, u_k^*)}{\partial x_k} + \frac{\partial L(x_k^*, u_k^*)}{\partial x_k}. \qquad (3.87)$$

However, letting

$$(\lambda_k^*)^{\text{T}} = \frac{\partial V_{N-(k+1)}^{\text{OPT}}(x_{k+1}^*)}{\partial x_{k+1}},$$

we see that (3.87) is the same as (3.30) using (3.29) with $\eta^* = 1$. (Note that $(\lambda_{N-1}^*)^{\text{T}} = \partial V_0^{\text{OPT}}(x_N^*)/\partial x_N = \partial F(x_N^*)/\partial x_N$, which agrees with (3.31) in the case where there is no equality constraint on the terminal state and with $\eta^* = 1$.) Hence, we can interpret the adjoint variable $(\lambda_k^*)^{\text{T}}$ as being the "sensitivity" (partial derivative) of the partial value function with respect to changes in the current state.

3.6 Further Reading

For complete list of references cited, see References section at the end of book.

General

The minimum principle for systems described by ordinary differential equations is extensively covered in the literature. See, for example, Pontryagin (1959), Pontryagin, Boltyanskii, Gamkrelidze and Mischenko (1962), Athans and Falb (1966), Lee and Markus (1967), Bryson and Ho (1969), Kirk (1970), Bertsekas (2000).

The discrete minimum principle presented here is based on the work of Halkin (1966); see also Holtzman and Halkin (1966), Holtzman (1966a), Holtzman (1966b).

Connections Between Optimal Control and Optimisation

For more on connections between optimal control and optimisation theory (mathematical programming), see Canon, Cullum and Polak (1970).

4

Receding Horizon Optimal Control
with Constraints

4.1 Overview

The goal of this chapter is to introduce the principle of receding horizon optimal control. The idea is to start with a fixed optimisation horizon, of length N say, using the current state of the plant as the initial state. We then optimise the objective function over this fixed interval *accounting for constraints*, obtain an optimal sequence of N control moves, and apply only the first control move to the plant. Time then advances one step and the same N-step optimisation problem is considered using the new state of the plant as the initial state. Thus one continuously revises the *current control action* based on the *current state* and accounting for the constraints over an optimisation horizon of length N. This chapter will expand on this intuitively reasonable idea.

4.2 The Receding Horizon Optimisation Principle

Fixed horizon optimisation leads to a control sequence $\{u_i, \ldots, u_{i+N-1}\}$, which begins at the current time i and ends at some future time $i + N - 1$. This fixed horizon solution suffers from two potential drawbacks:

(i) Something unexpected may happen to the system at some time over the future interval $[i, i + N - 1]$ that was not predicted by (or included in) the model. This would render the fixed control choices $\{u_i, \ldots, u_{i+N-1}\}$ obsolete.

(ii) As one approaches the final time $i + N - 1$, the control law typically "gives up trying" since there is too little time to go to achieve anything useful in terms of objective function reduction. Of course, there do exist problems where time does indeed "run out" because the problem is simply such that no further time is available. This is typical of so-called, batch control problems. However, in other cases, the use of a fixed optimisation horizon is principally dictated by computational needs rather than the absolute requirement that everything must be "wrapped up" at some fixed future time $i + N - 1$.

The above two problems are addressed by the idea of *receding horizon optimisation*. As foreshadowed in Section 4.1, this idea can be summarised as follows:

(i) At time i and for the current state x_i, solve an optimal control problem over a fixed future interval, say $[i, i + N - 1]$, taking into account the *current* and *future* constraints.

(ii) Apply only the first step in the resulting optimal control sequence.

(iii) Measure the state reached at time $i + 1$.

(iv) Repeat the fixed horizon optimisation at time $i+1$ over the future interval $[i + 1, i + N]$, starting from the (now) current state x_{i+1}.

Of course, in the absence of disturbances, the state measured at step (iii) will be the same as that predicted by the model. Nonetheless, it seems prudent to use the *measured state* rather than the predicted state just to be sure. The above description assumes that the state is indeed measured at time $i + 1$. In practice, the available measurements would probably cover only a subset of the full state vector. In this case, it seems reasonable that one should use some form of observer to estimate x_{i+1} based on the available data. More will be said about the use of observers in Section 5.5 of Chapter 5, and on the general topic of *output feedback* in Chapter 12. For the moment, we will assume that the full state vector is indeed measured and we will ignore the impact of disturbances.

If the model and objective function are time invariant, then it is clear that the same input u_i will result whenever the state takes the same value. That is, the receding horizon optimisation strategy is really an "alibi" for generating a particular time-invariant feedback control law. In particular, we can set $i = 0$ in the formulation of the open loop control problem without loss of generality. Then at the current time, and for the current state x, we solve:

$$\mathcal{P}_N(x): \qquad V_N^{\mathrm{OPT}}(x) \triangleq \min V_N(\{x_k\}, \{u_k\}), \qquad (4.1)$$

subject to:

$$x_{k+1} = f(x_k, u_k) \quad \text{for } k = 0, \dots, N-1, \qquad (4.2)$$

$$x_0 = x, \qquad (4.3)$$

$$u_k \in \mathbb{U} \quad \text{for } k = 0, \dots, N-1, \qquad (4.4)$$

$$x_k \in \mathbb{X} \quad \text{for } k = 0, \dots, N, \qquad (4.5)$$

$$x_N \in \mathbb{X}_f \subset \mathbb{X}, \qquad (4.6)$$

where

$$V_N(\{x_k\}, \{u_k\}) \triangleq F(x_N) + \sum_{k=0}^{N-1} L(x_k, u_k), \tag{4.7}$$

and where $\{x_k\}$, $x_k \in \mathbb{R}^n$, $\{u_k\}$, $u_k \in \mathbb{R}^m$, denote the state and control sequences $\{x_0, \ldots, x_N\}$ and $\{u_0, \ldots, u_{N-1}\}$, respectively, and $\mathbb{U} \subset \mathbb{R}^m$, $\mathbb{X} \subset \mathbb{R}^n$, and $\mathbb{X}_f \subset \mathbb{R}^n$ are constraint sets. All sequences $\{u_0, \ldots, u_{N-1}\}$ and $\{x_0, \ldots, x_N\}$ satisfying the constraints (4.2)–(4.6) are called *feasible* sequences. A pair of feasible sequences $\{u_0, \ldots, u_{N-1}\}$ and $\{x_0, \ldots, x_N\}$ constitute a *feasible solution* of (4.1)–(4.7). The functions F and L in the objective function (4.7) are the *terminal state weighting* and the *per-stage weighting*, respectively.

In the sequel we make the following assumptions:

- f, F and L are continuous functions of their arguments;
- $\mathbb{U} \subset \mathbb{R}^m$ is a compact set, $\mathbb{X} \subset \mathbb{R}^n$ and $\mathbb{X}_f \subset \mathbb{R}^n$ are closed sets;
- there exists a feasible solution to the optimisation problem (4.1)–(4.7).

Because N is finite, these assumptions are sufficient to ensure the existence of a minimum by Weierstrass' theorem (see Theorem 2.2.2 of Chapter 2). Typical choices for the weighting functions F and L are quadratic functions of the form $F(x) = x^T P x$ and $L(x, u) = x^T Q x + u^T R u$, where $P = P^T \geq 0$, $Q = Q^T \geq 0$ and $R = R^T > 0$. More generally, one could use functions of the form $F(x) = \|Px\|_p$ and $L(x, u) = \|Qx\|_p + \|Ru\|_p$, where $\|y\|_p$ with $p = 1, 2, \ldots, \infty$, is the p-norm of the vector y.

Denote the minimising control sequence, which is a function of the current state x_i, by

$$\mathscr{U}_{x_i}^{\text{OPT}} \triangleq \{u_0^{\text{OPT}}, u_1^{\text{OPT}}, \ldots, u_{N-1}^{\text{OPT}}\}; \tag{4.8}$$

then the control applied to the plant at time i is the first element of this sequence, that is,

$$u_i = u_0^{\text{OPT}}. \tag{4.9}$$

Time is then stepped forward one instant, and the above procedure is repeated for another N-step-ahead optimisation horizon. The first element of the new N-step input sequence is then applied. The above procedure is repeated endlessly. The idea is illustrated in Figure 4.1 for a horizon $N = 5$. In this figure, each plot shows the minimising control sequence $\mathscr{U}_{x_i}^{\text{OPT}}$ given in (4.8), computed at time $i = 0, 1, 2$. Note that only the shaded inputs are actually applied to the system. We can see that we are continually looking ahead to judge the impact of current and future decisions on the future response before we "lock in" the current input by applying it to the plant.

The above receding horizon procedure *implicitly* defines a time-invariant control policy $\mathcal{K}_N : \mathbb{X} \to \mathbb{U}$ of the form

$$\mathcal{K}_N(x) = u_0^{\text{OPT}}. \tag{4.10}$$

Note that the strict definition of the function $\mathcal{K}_N(\cdot)$ requires the minimiser to be unique. Most of the problems treated in this book are convex and hence

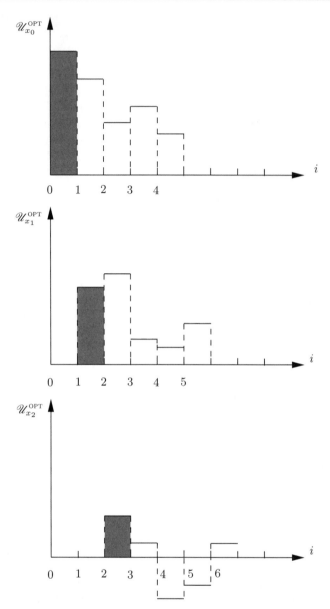

Figure 4.1. Receding horizon optimisation principle. The shaded rectangles indicate the inputs actually applied to the plant.

satisfy this condition. One exception is the "finite alphabet" optimisation case of Chapter 13, where the minimiser is not necessarily unique. However, in such cases, one can adopt a rule to select one of the minimisers (see, for example, the discussion following Definition 13.3.1 in Chapter 13).

It is common in receding horizon control applications to compute *numerically*, at time i, and for the current state $x_i = x$, the optimal control move $\mathcal{K}_N(x)$. In this case, we call it an *implicit receding horizon optimal policy*. In some cases, we can explicitly evaluate the *control law* $\mathcal{K}_N(\cdot)$. In this case, we say that we have an *explicit receding horizon optimal policy*. We will expand on the above skeleton description of receding horizon optimal constrained control as the book evolves. For example, we will treat linear constrained problems in subsequent chapters. When the system model is linear, the objective function quadratic and the constraint sets polyhedral, the fixed horizon optimal control problem $\mathcal{P}_N(\cdot)$ is a quadratic programme of the type discussed in Section 2.5.6 of Chapter 2. In Chapters 5 to 8 we will study the solution of this quadratic program in some detail. If, on the other hand, the system model is nonlinear, $\mathcal{P}_N(\cdot)$ is, in the general case, nonconvex, so that only local solutions are available.

The remainder of the present chapter is devoted to the analysis of the stability properties of receding horizon optimal control. However, before we embark on these issues, we pause to review concepts from stability theory. As for the results on optimisation presented in Chapter 2, the results on stability presented below in Section 4.3 find widespread application beyond constrained control and estimation.

4.3 Background on Stability Theory

4.3.1 Notions of Stability

We will utilise the following notions of stability:

Definition 4.3.1 (Stability Properties) *Let $\mathbb{S}$ be a set in $\mathbb{R}^n$ that contains the origin. Let $f : \mathbb{R}^n \to \mathbb{R}^n$ be such that $f(\mathbb{S}) \subset \mathbb{S}$. Suppose that the system*

$$x_{i+1} = f(x_i), \qquad (4.11)$$

with $x_i \in \mathbb{R}^n$, has an equilibrium point at the origin $x = 0$, that is, $f(0) = 0$. Let $x_0 \in \mathbb{S}$ and let $\{x_i\} \subset \mathbb{S}$, $i \geq 0$, be the resulting sequence satisfying (4.11). We say that the equilibrium point is:

(i) (Lyapunov) stable in $\mathbb{S}$: if for any $\varepsilon > 0$, there exists $\delta > 0$ such that

$$x_0 \in \mathbb{S} \text{ and } \|x_0\| < \delta \implies \|x_i\| < \varepsilon \text{ for all } i \geq 0; \qquad (4.12)$$

(ii) attractive in $\mathbb{S}$: if there exists $\eta > 0$ such that

$$x_0 \in \mathbb{S} \text{ and } \|x_0\| < \eta \implies \lim_{i \to \infty} x_i = 0;$$

(iii) globally attractive in $\mathbb{S}$: *if*

$$x_0 \in \mathbb{S} \implies \lim_{i \to \infty} x_i = 0 \,;$$

(iv) asymptotically stable in $\mathbb{S}$: *if it is both stable in $\mathbb{S}$ and attractive in $\mathbb{S}$;*
(v) exponentially stable in $\mathbb{S}$: *if there exist constants $\theta > 0$ and $\rho \in (0,1)$ such that*

$$x_0 \in \mathbb{S} \implies \|x_i\| \le \theta \|x_0\| \rho^i \text{ for all } i \ge 0\,; \qquad (4.13)$$

In cases (iii), and (v) above, we say that the set $\mathbb{S}$ is contained in the region of attraction[1] of the equilibrium point.   ∘

4.3.2 Tests for Stability

Testing for stability properties is facilitated if one can find a function $V :$ $\mathbb{S} \to [0, \infty)$ (called a Lyapunov function) satisfying certain conditions. The following results use this fact.

Theorem 4.3.1 (Attractivity in $\mathbb{S}$) *Let $\mathbb{S}$ be a nonempty set in $\mathbb{R}^n$. Let $f : \mathbb{R}^n \to \mathbb{R}^n$ be such that $f(0) = 0$ and $f(\mathbb{S}) \subset \mathbb{S}$. Assume that there exists a (Lyapunov) function $V : \mathbb{S} \to [0, \infty)$ satisfying the following properties:[2]*

(i) $V(\cdot)$ decreases along the trajectories of (4.11) that start in $\mathbb{S}$ in the following way: there exists a continuous function $\gamma : [0, \infty) \to [0, \infty)$, $\gamma(t) > 0$ for all $t > 0$, such that

$$V(f(x)) - V(x) \le -\gamma(\|x\|) \quad \text{for all } x \in \mathbb{S}. \qquad (4.14)$$

(ii) for every unbounded sequence $\{y_i\} \subset \mathbb{S}$ there is some j such that [3]

$$\limsup_{i \to \infty} V(y_i) > V(y_j)\,.$$

Then:

(a) $0 \in \mathrm{cl}\,\mathbb{S}$, and
(b) For all $x_0 \in \mathbb{S}$, the resulting sequence $\{x_i\}$, $i \ge 0$, satisfying (4.11) is such that $\lim_{i \to \infty} x_i = 0$, that is, if $0 \in \mathbb{S}$, the origin is globally attractive in $\mathbb{S}$.

---

[1] The region of attraction of an equilibrium point of (4.11) is the set of all initial states $x_0 \in \mathbb{R}^n$ that originate state trajectories $\{x_i\}$, $i \ge 0$, solution of (4.11), which converge to the equilibrium point as $i \to \infty$.
[2] Property (ii) can be omitted if $\mathbb{S}$ is bounded.
[3] We recall that, if $\{a_i\}$ is a sequence in $[-\infty, \infty]$, and $b_k = \sup\{a_k, a_{k+1}, a_{k+2}, \dots\}$, $k = 1, 2, 3, \dots$, then the *upper limit* of $\{a_i\}$, denoted by $\beta = \limsup_{i \to \infty} a_i$, is defined as $\beta \triangleq \inf\{b_1, b_2, b_3, \dots\}$.

Proof. Let $x_0 \in \mathbb{S}$ and let $\{x_i\}$, $i \geq 0$, be the resulting sequence satisfying (4.11). The associated sequence of Lyapunov function values $\{V(x_i)\} \subset [0, \infty)$ is nonincreasing, since, from (4.14),

$$V(x_{i+1}) = V(f(x_i)) \leq V(x_i) - \gamma(\|x_i\|) \leq V(x_i).$$

Hence, $c = \lim_{i \to \infty} V(x_i) \geq 0$, exists.

The sequence $\{x_i\}$ is bounded; otherwise, from property (ii) above, there would exist j such that $c > V(x_j)$, but $c \leq V(x_i)$ for all i. Thus, there exists $R > 0$ such that $\|x_i\| \leq R$ for all $i \geq 0$.

Now assume that there exists μ, $0 < \mu < R$, such that $\|x_i\| \geq \mu$ for infinitely many i. Let

$$\alpha = \min_{\mu \leq t \leq R} \gamma(t).$$

Note that α exists by Weierstrass' theorem (Theorem 2.2.2 in Chapter 2) and that $\alpha > 0$ since $\gamma(t) > 0$ for all $t > 0$. From

$$V(x_k) = V(x_0) + \sum_{j=0}^{k-1} V(x_{j+1}) - V(x_j),$$

it follows that

$$c = V(x_0) + \sum_{j=0}^{\infty} V(x_{j+1}) - V(x_j)$$

$$\leq V(x_0) - \sum_{j=0}^{\infty} \gamma(\|x_j\|)$$

$$= -\infty,$$

since $\gamma(\|x_j\|) \geq \alpha > 0$ for infinitely many j and $\gamma(t) \geq 0$ for all $t \geq 0$. The above is a contradiction since $c \geq 0$. It follows that x_i converges to 0 as i tends to infinity, showing that $0 \in \mathrm{cl}\,\mathbb{S}$ and that, if $0 \in \mathbb{S}$, the origin is attractive in $\mathbb{S}$ for (4.11). The theorem is then proved. □

Remark 4.3.1. If the Lyapunov function $V : \mathbb{S} \to [0, \infty)$ is continuous, and $f : \mathbb{S} \to \mathbb{S}$ in (4.11) is continuous, $\mathbb{S}$ is closed, and $V(0) = 0$, then inequality (4.14) in Theorem 4.3.1 can be replaced by

$$V(f(x)) - V(x) < 0 \quad \text{for all } x \in \mathbb{S}, \ x \neq 0.$$

○

Theorem 4.3.2 (Stability) *Let $\mathbb{S}$ be a set in $\mathbb{R}^n$ that contains an open neighbourhood of the origin $N_\eta(0) \triangleq \{x \in \mathbb{R}^n : \|x\| < \eta\}$. Let $f : \mathbb{R}^n \to \mathbb{R}^n$ be such that $f(0) = 0$ and $f(\mathbb{S}) \subset \mathbb{S}$. Assume that there exists a (Lyapunov) function $V : \mathbb{S} \to [0, \infty)$, $V(0) = 0$, satisfying the following properties:*

(i) $V(\cdot)$ is continuous on $N_\eta(0)$;
(ii) if $\{y_k\} \subset \mathbb{S}$ is such that $\lim_{k\to\infty} V(y_k) = 0$ then $\lim_{k\to\infty} y_k = 0$;
(iii) $V(f(x)) - V(x) \le 0$ for all $x \in N_\eta(0)$.

Then the origin is a stable equilibrium point for (4.11) in $\mathbb{S}$.

Proof. Let $\varepsilon \in (0, \eta)$ and $N_\varepsilon(0) \triangleq \{x \in \mathbb{R}^n : \|x\| < \varepsilon\}$. We first show that there exists $\beta > 0$ such that $V^{-1}[0, \beta] \triangleq \{x \in \mathbb{S} : V(x) \in [0, \beta]\} \subset N_\varepsilon(0)$. Suppose no such β exists. Then for every $k = 1, 2, \ldots$, there exists $y_k \in V^{-1}[0, \frac{1}{k}]$ such that $\|y_k\| > \varepsilon$. But, from property (ii), we have that $\lim_{k\to\infty} y_k = 0$, which is a contradiction. Thus,

$$V^{-1}[0, \beta] \subset N_\varepsilon(0). \tag{4.15}$$

Since $V(\cdot)$ is continuous on $N_\eta(0)$ and $V(0) = 0$, there exists $\delta \in (0, \varepsilon)$ such that $\|x\| < \delta \implies V(x) < \beta$. Then, combining with (4.15), we have

$$\|x\| < \delta \implies V(x) < \beta \implies x \in V^{-1}[0, \beta] \implies \|x\| < \varepsilon.$$

Now let $\|x_0\| < \delta$. We show by induction that $x_i \in V^{-1}[0, \beta]$ for all $i \ge 0$. It clearly holds for $i = 0$. Suppose $x_i \in V^{-1}[0, \beta]$. Note that, from (4.15), $\|x_i\| < \varepsilon$, so that $x_i \in N_\eta(0)$. Then, using property (iii), we have

$$V(x_{i+1}) = V(f(x_i)) \le V(x_i) \le \beta \implies x_{i+1} \in V^{-1}[0, \beta].$$

Hence, $\|x_0\| < \delta \implies x_0 \in V^{-1}[0, \beta] \implies x_i \in V^{-1}[0, \beta]$ for all $i \ge 0 \implies \|x_i\| < \varepsilon$ for all $i \ge 0$. We have thus shown that given $\varepsilon \in (0, \eta)$ there exists $\delta > 0$ such that (4.12) holds. The result then follows. $\qquad\square$

The following theorem gives a sufficient condition for exponential stability.

Theorem 4.3.3 (Exponential Stability) *Let $\mathbb{S}$ be a set in $\mathbb{R}^n$ containing a nonzero element. Let $f : \mathbb{R}^n \to \mathbb{R}^n$ be such that $f(0) = 0$ and $f(\mathbb{S}) \subset \mathbb{S}$. Assume that there exists a (Lyapunov) function $V : \mathbb{S} \to \mathbb{R}$, and positive constants a, b, c and σ satisfying*

(i) $a\|x\|^\sigma \le V(x) \le b\|x\|^\sigma$ for all $x \in \mathbb{S}$,
(ii) $V(f(x)) - V(x) \le -c\|x\|^\sigma$ for all $x \in \mathbb{S}$.

Then, if $0 \in \mathbb{S}$, the origin is exponentially stable in $\mathbb{S}$ for the system (4.11).

Proof. Let $f^0(x) \triangleq x$, $f^1(x) \triangleq f(x)$, $\ldots$, $f^{i+1}(x) \triangleq f^i(f(x))$. We first show that

$$V(f^i(x)) \le \left(1 - \frac{c}{b}\right)^i V(x), \quad \text{for all } x \in \mathbb{S}, \tag{4.16}$$

for all $i \ge 0$. Clearly, (4.16) holds for $i = 0$. Moreover, from the assumptions on $V(x)$, we have

$$V(f(x)) - V(x) \le -c\|x\|^\sigma \le -\frac{c}{b} V(x).$$

Thus,

$$V(f(x)) \leq \left(1 - \frac{c}{b}\right) V(x) \text{ for all } x \in \mathbb{S}.$$

Choose $0 \neq y \in \mathbb{S}$. Then $V(y) \geq a\|y\|^{\sigma} > 0$. Thus, $1 - \frac{c}{b} \geq 0$, and therefore $0 \leq 1 - \frac{c}{b} < 1$.

Now assume that (4.16) holds for some $i \geq 1$. Then,

$$V(f^{i+1}(x)) = V(f^i(f(x))) \leq \left(1 - \frac{c}{b}\right)^i V(f(x)) \leq \left(1 - \frac{c}{b}\right)^{i+1} V(x).$$

Hence, by induction, (4.16) holds for all $i \geq 0$. Finally, for all $x \in \mathbb{S}$ and all $i \geq 0$, we have that

$$\left\|f^i(x)\right\|^{\sigma} \leq \frac{1}{a} V(f^i(x)) \leq \frac{1}{a}\left(1 - \frac{c}{b}\right)^i V(x) \leq \frac{b}{a}\left(1 - \frac{c}{b}\right)^i \|x\|^{\sigma},$$

from which (4.13) follows with $\theta = \left(\frac{b}{a}\right)^{1/\sigma} > 0$ and $\rho = \left(1 - \frac{c}{b}\right)^{1/\sigma} \in (0,1)$.

$\square$

4.4 Stability of Receding Horizon Optimal Control

4.4.1 Ingredients

We now return to receding horizon control as described in Section 4.2. Although the receding horizon control idea seems intuitively reasonable, it is important that one be able to establish concrete results about its associated properties. Here we examine the question of closed loop stability which is a minimal performance goal.

Unfortunately, proving/guaranteeing that an optimisation scheme (such as receding horizon optimal control) leads to a stable closed loop system is a nontrivial task. One may well ask what possible tool could be used. After all, the only thing we know is that the *fixed horizon control sequence* is optimal. Luckily, *optimality* can be turned into a notion of stability by utilising the value function (that is, the function $V_N^{\text{OPT}}(x)$ in (4.1), which is a function of the initial state x only) as a *Lyapunov function*.

However, another difficulty soon arises. Namely, the optimisation problems that we are solving are only defined over a *finite* future horizon, yet stability is a property that must hold over an *infinite* future horizon. A trick that is frequently utilised to resolve this conflict is to add an appropriate weighting on the terminal state in the finite horizon problem so as to account for the impact of events that lie beyond the end of the fixed horizon. This effectively turns the fixed horizon problem into an infinite horizon one.

Following this line of reasoning, we will define a terminal control law and an associated terminal state weighting in the objective function that captures the impact of using the terminal control law over *infinite time*. Usually, the chosen terminal control laws are relatively simple and only "feasible" in a restricted (local) region. This implies that one must be able to steer the system into this restricted terminal region over the finite time period available in the optimisation window. (More will be said about this crucial point later.) It is also important to ensure that the terminal region is invariant under the terminal control law, that is, once the state reaches the terminal set, it remains inside the set if the terminal control law is used. Thus, in summary, the ingredients typically employed to provide *sufficient* (though by no means necessary) conditions for stability are captured by the following *terminal triple*:

Ingredients for Stability: The Terminal Triple $(\mathbb{X}_f, \mathcal{K}_f, F)$

(i) a *terminal constraint set* $\mathbb{X}_f$ in the state space which is invariant under the terminal control law;

(ii) a feasible *terminal control law* $\mathcal{K}_f$ that holds in the terminal constraint set;

(iii) a *terminal state weighting* F on the finite horizon optimisation problem, which usually corresponds to the objective function value generated by the use of the terminal control law over infinite time.

We will show below how, based on these "ingredients," Lyapunov-like tests, such as those described in Section 4.3.2, can be used to establish stability of receding horizon control.

4.4.2 Stability Results for Receding Horizon Control

As mentioned above, we will employ the value function $V_N^{\text{OPT}}(x)$ of the fixed horizon optimal control problem (4.1)–(4.7) as a Lyapunov function to establish asymptotic stability of the receding horizon implementation. We will first establish stability under simplifying assumptions. A more general stability analysis will be given later; however, this will follow essentially the same lines as the simplified "prototype" proof given below.

Let us define the set $\mathbb{S}_N$ of *feasible initial states*.

Definition 4.4.1 *The set $\mathbb{S}_N$ of feasible initial states is the set of initial states $x \in \mathbb{X}$ for which there exist feasible state and control sequences for the fixed horizon optimal control problem $\mathcal{P}_N(x)$ in (4.1)–(4.7).*   ∘

We also require the following definition.

Definition 4.4.2 *The set $\mathbb{S} \subset \mathbb{R}^n$ is said to be* positively invariant *for the system $x_{i+1} = f(x_i, u_i)$ under the control $u_i = \mathcal{K}(x_i)$ (or positively invariant for the closed loop system $x_{i+1} = f(x_i, \mathcal{K}(x_i))$) if $f(x, \mathcal{K}(x)) \in \mathbb{S}$ for all $x \in \mathbb{S}$.*

∘

We make the following assumptions on the data of problem $\mathcal{P}_N(x)$ in (4.1)–(4.7).

A1 The terminal constraint set in (4.6) is the origin, that is, $\mathbb{X}_f = \{0\}$.

A2 The control constraint set in (4.4) contains the origin, that is, $0 \in \mathbb{U}$.

A3 $L(x, u)$ in (4.7) satisfies $L(0, 0) = 0$ and $L(x, u) \geq \gamma(\|x\|)$ for all $x \in \mathbb{S}_N, u \in \mathbb{U}$, where $\gamma : [0, \infty) \to [0, \infty)$ is continuous, $\gamma(t) > 0$ for all $t > 0$, and $\lim_{t \to \infty} \gamma(t) = \infty$.

A4 There is no terminal state weighting in the objective function, that is, $F(x) \equiv 0$ in (4.7).

Under these conditions, we have the following stability result:

Theorem 4.4.1 *Consider the system*

$$x_{i+1} = f(x_i, u_i) \quad for\ i \geq 0, \qquad f(0, 0) = 0, \tag{4.17}$$

controlled by the receding horizon algorithm (4.1)–(4.9) *and subject to Assumptions* **A1**–**A4** *above. Then:*

(i) The set $\mathbb{S}_N$ of feasible initial states is positively invariant for the closed loop system.

(ii) The origin is globally attractive in $\mathbb{S}_N$ for the closed loop system.

(iii) If, in addition to **A1**–**A4**, *$0 \in \text{int } \mathbb{S}_N$ and the value function $V_N^{\text{OPT}}(x)$ in* (4.1) *is continuous on some neighbourhood of the origin, then the origin is asymptotically stable in $\mathbb{S}_N$ for the closed loop system.*

Proof. (i) *Positive invariance of $\mathbb{S}_N$.*

Let $x_i = x \in \mathbb{S}_N$. At step i, and for the current state $x_i = x$, the receding horizon algorithm solves the optimisation problem $\mathcal{P}_N(x)$ in (4.1)–(4.7) to obtain the optimal control and state sequences

$$\mathcal{U}_x^{\text{OPT}} \triangleq \{u_0^{\text{OPT}}, u_1^{\text{OPT}}, \ldots, u_{N-1}^{\text{OPT}}\}, \tag{4.18}$$

$$\mathcal{X}_x^{\text{OPT}} \triangleq \{x_0^{\text{OPT}}, x_1^{\text{OPT}}, \ldots, x_{N-1}^{\text{OPT}}, x_N^{\text{OPT}}\}. \tag{4.19}$$

Then the actual control applied to (4.17) at time i is the first element of (4.18), that is,

$$u_i = \mathcal{K}_N(x) = u_0^{\text{OPT}}. \tag{4.20}$$

Note that, in the optimal state sequence (4.19), we have, from Assumption **A1**, that

$$x_N^{\text{OPT}} = 0. \tag{4.21}$$

Let $x^+ \triangleq x_{i+1} = f(x, \mathcal{K}_N(x)) = f(x, u_0^{\text{OPT}})$ be the successor state. A *feasible* (but not necessarily optimal) control sequence, and corresponding feasible state sequence for the next step $i + 1$ in the receding horizon computation $\mathcal{P}_N(x^+)$ are then

$$\tilde{\mathscr{U}} = \{u_1^{\mathrm{OPT}}, \dots, u_{N-1}^{\mathrm{OPT}}, 0\}, \qquad (4.22)$$

$$\tilde{\mathscr{X}} = \{x_1^{\mathrm{OPT}}, \dots, x_{N-1}^{\mathrm{OPT}}, 0, 0\}, \qquad (4.23)$$

where the last two zeros in (4.23) follow from (4.21) and $f(0,0) = 0$. Thus, there exist feasible sequences (4.22) and (4.23) for the successor state $x^+ = f(x, \mathcal{K}_N(x))$ and hence $x^+ \in \mathbb{S}_N$. This shows that $\mathbb{S}_N$ is positively invariant for the closed loop system $x^+ = f(x, \mathcal{K}_N(x))$.

(ii) *Attractivity.*

Note first that, since $L(0,0) = 0$, $F(0) = 0$, $0 \in \mathbb{U}$ and $0 \in \mathbb{X}_f$, then the optimal sequences in (4.1)–(4.7) corresponding to $x = 0$ have all their elements equal to zero. Thus, $\mathcal{K}_N(0) = 0$. Since, in addition, $f(0,0) = 0$, then the origin is an equilibrium point for the closed loop system $x^+ = f(x, \mathcal{K}_N(x))$.

We will next use the value function $V_N^{\mathrm{OPT}}(\cdot)$ in (4.1) as a Lyapunov function. We first show that $V_N^{\mathrm{OPT}}(\cdot)$ satisfies property (i) in Theorem 4.3.1. Let $x \in \mathbb{S}_N$. The increment of the Lyapunov function, upon using the true optimal input (4.20) and moving from x to $x^+ = f(x, \mathcal{K}_N(x))$, satisfies

$$V_N^{\mathrm{OPT}}(x^+) - V_N^{\mathrm{OPT}}(x) = V_N(\mathscr{X}_{x^+}^{\mathrm{OPT}}, \mathscr{U}_{x^+}^{\mathrm{OPT}}) - V_N(\mathscr{X}_x^{\mathrm{OPT}}, \mathscr{U}_x^{\mathrm{OPT}}). \qquad (4.24)$$

However, by optimality we know that

$$V_N(\mathscr{X}_{x^+}^{\mathrm{OPT}}, \mathscr{U}_{x^+}^{\mathrm{OPT}}) \leq V_N(\tilde{\mathscr{X}}, \tilde{\mathscr{U}}), \qquad (4.25)$$

where $\tilde{\mathscr{U}}$ and $\tilde{\mathscr{X}}$ are the feasible sequences defined in (4.22)–(4.23). Combining (4.24) and (4.25) yields

$$V_N^{\mathrm{OPT}}(x^+) - V_N^{\mathrm{OPT}}(x) \leq V_N(\tilde{\mathscr{X}}, \tilde{\mathscr{U}}) - V_N(\mathscr{X}_x^{\mathrm{OPT}}, \mathscr{U}_x^{\mathrm{OPT}}). \qquad (4.26)$$

Substituting (4.18), (4.19), (4.22) and (4.23) in the objective function expression (4.7), and using the fact that the optimal and feasible sequences share common terms, we obtain that the right hand side of (4.26) is equal to $-L(x, \mathcal{K}_N(x))$. It then follows that

$$V_N^{\mathrm{OPT}}(x^+) - V_N^{\mathrm{OPT}}(x) \leq -L(x, \mathcal{K}_N(x))$$
$$\leq -\gamma(\|x\|),$$

where, in the last inequality, we have used Assumption **A3**. Thus, $V_N^{\mathrm{OPT}}(\cdot)$ satisfies property (i) in Theorem 4.3.1.

In addition, from Assumptions **A3** and **A4**, $V_N^{\mathrm{OPT}}(\cdot)$ satisfies

$$V_N^{\mathrm{OPT}}(x) \geq L(x, u_0^{\mathrm{OPT}}) \geq \gamma(\|x\|) \quad \text{for all } x \in \mathbb{S}_N. \qquad (4.27)$$

Hence, from the assumption on γ, $V_N^{\text{OPT}}(x) \to \infty$ when $\|x\| \to \infty$, and therefore $V_N^{\text{OPT}}(\cdot)$ satisfies property (ii) in Theorem 4.3.1. It then follows from Theorem 4.3.1 that the origin is globally attractive in $\mathbb{S}_N$ for the closed loop system.

(iii) *Asymptotic stability.*

To show asymptotic stability of the origin, note first that $V_N^{\text{OPT}}(0) = 0$ (since, as shown before, the optimal sequences in (4.1)–(4.7) corresponding to $x = 0$ have all their elements equal to zero). Next, note from (4.27) and the properties of γ that $V_N^{\text{OPT}}(\cdot)$ satisfies property (ii) in Theorem 4.3.2 with $\mathbb{S} = \mathbb{S}_N$. If, in addition, $0 \in \text{int } \mathbb{S}_N$ and $V_N^{\text{OPT}}(\cdot)$ is continuous on some neighbourhood of the origin, then Theorem 4.3.2 shows that the origin is a stable equilibrium point for the closed loop system, and hence it is asymptotically stable in $\mathbb{S}_N$ (that is, both stable and attractive in $\mathbb{S}_N$).

□

Assumptions **A1** to **A4** were made to keep the proof of Theorem 4.4.1 simple in order to introduce the reader to the core idea of the stability proof. The assumptions can be relaxed. (For example, Assumption **A1** can be replaced by the assumption that x_N enters a terminal set in which "nice properties" hold. Similarly, Assumption **A3** can be relaxed to requiring that the system be "detectable" in the objective function.)

We next modify the assumptions given above to provide a more comprehensive result by specifying some more general terminal conditions.

Conditions for Stability:

B1 The per-stage weighting $L(x, u)$ in (4.7) satisfies $L(0, 0) = 0$ and $L(x, u) \geq \gamma(\|x\|)$ for all $x \in \mathbb{S}_N, u \in \mathbb{U}$, where $\gamma : [0, \infty) \to [0, \infty)$ is continuous, $\gamma(t) > 0$ for all $t > 0$, and $\lim_{t \to \infty} \gamma(t) = \infty$.

B2 The terminal state weighting $F(x)$ in (4.7) satisfies $F(0) = 0$, $F(x) \geq 0$ for all $x \in \mathbb{X}_f$, and the following property: there exists a terminal control law $\mathcal{K}_f : \mathbb{X}_f \to \mathbb{U}$ such that $F(f(x, \mathcal{K}_f(x))) - F(x) \leq -L(x, \mathcal{K}_f(x))$ for all $x \in \mathbb{X}_f$.

B3 The set $\mathbb{X}_f$ is positively invariant for the system (4.17) under $\mathcal{K}_f(x)$, that is, $f(x, \mathcal{K}_f(x)) \in \mathbb{X}_f$ for all $x \in \mathbb{X}_f$.

B4 The terminal control $\mathcal{K}_f(x)$ satisfies the control constraints in $\mathbb{X}_f$, that is, $\mathcal{K}_f(x) \in \mathbb{U}$ for all $x \in \mathbb{X}_f$.

B5 The sets $\mathbb{U}$ and $\mathbb{X}_f$ contain the origin of their respective spaces.

Using the above conditions, which include more general conditions on the terminal triple $(\mathbb{X}_f, \mathcal{K}_f, F)$, we obtain the following more general theorem.

Theorem 4.4.2 (Stability of Receding Horizon Control) *Consider the closed loop system formed by system (4.17), controlled by the receding* *horizon algorithm* (4.1)–(4.9), *and suppose that Conditions* **B1** *to* **B5** *are satisfied. Then:*

(i) *The set* $\mathbb{S}_N$ *of feasible initial states is positively invariant for the closed loop system.*

(ii) *The origin is globally attractive in* $\mathbb{S}_N$ *for the closed loop system.*

(iii) *If, in addition to* **B1**–**B5**, $0 \in \text{int } \mathbb{S}_N$ *and the value function* $V_N^{\text{OPT}}(\cdot)$ *in* (4.1) *is continuous on some neighbourhood of the origin, then the origin is asymptotically stable in* $\mathbb{S}_N$ *for the closed loop system.*

(iv) *If, in addition to* **B1**–**B5**, $0 \in \text{int } \mathbb{X}_f$, $\mathbb{S}_N$ *is compact,* $\gamma(t) \geq at^\sigma$ *in* **B1**, $F(x) \leq b\|x\|^\sigma$ *for all* $x \in \mathbb{X}_f$ *in* **B2**, *where* $a > 0$, $b > 0$ *and* $\sigma > 0$ *are some real constants, and the value function* $V_N^{\text{OPT}}(\cdot)$ *in* (4.1) *is continuous on* $\mathbb{S}_N$, *then the origin is exponentially stable in* $\mathbb{S}_N$ *for the closed loop system.*

Proof. (i) *Positive invariance of* $\mathbb{S}_N$.

We will use the optimal sequences (4.18), (4.19) for the initial state $x \in \mathbb{S}_N$, and the following feasible sequences for the successor state $x^+ = f(x, \mathcal{K}_N(x))$:

$$\tilde{\mathscr{U}} = \{u_1^{\text{OPT}}, \ldots, u_{N-1}^{\text{OPT}}, \mathcal{K}_f(x_N^{\text{OPT}})\}, \tag{4.28}$$

$$\tilde{\mathscr{X}} = \{x_1^{\text{OPT}}, \ldots, x_{N-1}^{\text{OPT}}, x_N^{\text{OPT}}, f(x_N^{\text{OPT}}, \mathcal{K}_f(x_N^{\text{OPT}}))\}. \tag{4.29}$$

Indeed, the first $N - 1$ elements of (4.28) lie in $\mathbb{U}$ (see the control constraint (4.4)) since they are elements of (4.18); also, by **B4**, the last element of (4.28) lies in $\mathbb{U}$ since $x_N^{\text{OPT}} \in \mathbb{X}_f$. Finally, by **B3**, the terminal state $f(x_N^{\text{OPT}}, \mathcal{K}_f(x_N^{\text{OPT}}))$ in (4.29) also lies in $\mathbb{X}_f$. Thus, there exist feasible sequences (4.28) and (4.29) for the successor state $x^+ = f(x, \mathcal{K}_N(x))$ and hence $x^+ \in \mathbb{S}_N$. This shows the result (i) that $\mathbb{S}_N$ is positively invariant for the closed loop system $x^+ = f(x, \mathcal{K}_N(x))$.

(ii) *Attractivity.*

As in Theorem 4.4.1, we can show that the origin is an equilibrium point for the closed loop system $x^+ = f(x, \mathcal{K}_N(x))$.

We next show that the value function $V_N^{\text{OPT}}(\cdot)$ satisfies property (i) in Theorem 4.3.1. The increment of $V_N^{\text{OPT}}(\cdot)$, upon using the receding horizon optimal input (4.20) and moving from $x \in \mathbb{S}_N$ to $x^+ = f(x, \mathcal{K}_N(x))$ satisfies (4.24), and, by optimality, (4.25) also holds for the feasible sequences (4.28), (4.29). We thus have, in a fashion similar to the proof of Theorem 4.4.1,

$$V_N^{\text{OPT}}(x^+) - V_N^{\text{OPT}}(x) \leq V_N(\tilde{\mathscr{X}}, \tilde{\mathscr{U}}) - V_N(\mathscr{X}_x^{\text{OPT}}, \mathscr{U}_x^{\text{OPT}})$$
$$= -L(x, \mathcal{K}_N(x)) + L(x_N^{\text{OPT}}, \mathcal{K}_f(x_N^{\text{OPT}}))$$
$$+ F(f(x_N^{\text{OPT}}, \mathcal{K}_f(x_N^{\text{OPT}}))) - F(x_N^{\text{OPT}}).$$

From **B2**, and since $x_N^{\text{OPT}} \in \mathbb{X}_f$, the sum of the last three terms on the right hand side of the above inequality is less than or equal to zero. Thus,

$$V_N^{\text{OPT}}(x^+) - V_N^{\text{OPT}}(x) \leq -L(x, \mathcal{K}_N(x)) \leq -\gamma(\|x\|) \text{ for all } x \in \mathbb{S}_N, \quad (4.30)$$

where, in the last inequality, we have used the bound in Condition **B1**. Thus $V_N^{\text{OPT}}(\cdot)$ satisfies property (i) of Theorem 4.3.1. In a fashion similar to the proof of Theorem 4.4.1, we can show that

$$V_N^{\text{OPT}}(x) \geq \gamma(\|x\|) \text{ for all } x \in \mathbb{S}_N, \quad (4.31)$$

and hence $V_N^{\text{OPT}}(\cdot)$ also satisfies property (ii) of Theorem 4.3.1. We then conclude using Theorem 4.3.1 that the origin is globally attractive in $\mathbb{S}_N$ for the closed loop system. The result (ii) is then proved.

(iii) *Asymptotic stability.*

As in Theorem 4.4.1, we can show that the value function $V_N^{\text{OPT}}(\cdot)$ satisfies property (ii) in Theorem 4.3.2 with $\mathbb{S} = \mathbb{S}_N$, and that $V_N^{\text{OPT}}(0) = 0$. If, in addition, the origin is in the interior of $\mathbb{S}_N$ and $V_N^{\text{OPT}}(\cdot)$ is continuous on a neighbourhood of the origin, then Theorem 4.3.2 shows that the origin is a stable equilibrium point for the closed loop system, and hence, combined with attractivity in $\mathbb{S}_N$, it is asymptotically stable in $\mathbb{S}_N$. This shows the result (iii).

(iv) *Exponential stability.*

By assumption, $F(x) \leq b\|x\|^\sigma$ for all $x \in \mathbb{X}_f$, for some constants $b > 0$ and $\sigma > 0$. It is easily shown that $V_N^{\text{OPT}}(x) \leq F(x)$ for all $x \in \mathbb{X}_f$. To see this, let x be an arbitrary point in $\mathbb{X}_f$ and denote by $\{x_k^f(x) : k = 0, 1, 2, \ldots\}$, $x_0^f(x) \triangleq x$, the state sequence resulting from initial state x and controller $\mathcal{K}_f(x)$ in (4.17). Then, by **B2**,

$$F(x) \geq \sum_{k=0}^{N-1} L(x_k^f(x), \mathcal{K}_f(x_k^f(x))) + F(x_N^f(x)),$$

where, by **B3**, $x_k^f(x) \in \mathbb{X}_f$ for all $k = 0, 1, \ldots, N$ and, by **B4**, $\mathcal{K}_f(x_k^f(x)) \in \mathbb{U}$ for all $k = 0, 1, \ldots, N - 1$. Note that the above state and control sequences are feasible since $x_N^f(x) \in \mathbb{X}_f$. Hence, by optimality,

$$V_N^{\text{OPT}}(x) \leq \sum_{k=0}^{N-1} L(x_k^f(x), \mathcal{K}_f(x_k^f(x))) + F(x_N^f(x)).$$

Thus, $V_N^{\text{OPT}}(x) \leq F(x) \leq b\|x\|^\sigma$ for all $x \in \mathbb{X}_f$. We now show that there exists a constant $\bar{b} > 0$ such that $V_N^{\text{OPT}}(x) \leq \bar{b}\|x\|^\sigma$ for all $x \in \mathbb{S}_N$. Consider the function $h : \mathbb{S}_N \to \mathbb{R}$ defined as

$$h(x) \triangleq \begin{cases} \frac{V_N^{\mathrm{OPT}}(x)}{\|x\|^\sigma} & \text{if } x \neq 0, \\ b & \text{if } x = 0 \,. \end{cases}$$

Then $h(x)$ is continuous on the compact set $\mathrm{cl}\,(\mathbb{S}_N \setminus \mathbb{X}_f)$, since $V_N^{\mathrm{OPT}}(x)$ is continuous on $\mathbb{S}_N$ and $\mathbb{X}_f$ contains a neighbourhood of the origin. Hence, $h(x)$ is bounded in $\mathrm{cl}\,(\mathbb{S}_N \setminus \mathbb{X}_f)$, say $h(x) \leq M$. It then follows that

$$V_N^{\mathrm{OPT}}(x) \leq \bar{b}\|x\|^\sigma \quad \text{for all } x \in \mathbb{S}_N,$$

where $\bar{b} \geq \max\{b, M\}$. Combining the above inequality with (4.30) and (4.31), and using the assumption that $\gamma(t) \geq at^\sigma$ for some constant $a > 0$, it follows from Theorem 4.3.3 that the closed loop system has an exponentially stable equilibrium point at the origin. This shows (iv) and concludes the proof of the theorem. $\qquad\square$

4.5 Terminal Conditions for Stability

In this section, we consider possible choices for the terminal triple $(\mathbb{X}_f, \mathcal{K}_f, F)$ that satisfy conditions **B1–B5** of Theorem 4.4.2.

One choice for the terminal state weighting $F(x)$ is the value function $V_\infty^{\mathrm{OPT}}(x)$ for the associated infinite horizon constrained optimal control problem, defined as follows:

$$\mathcal{P}_\infty(x): \qquad V_\infty^{\mathrm{OPT}}(x) \triangleq \min V_\infty(\{x_k\}, \{u_k\}), \qquad (4.32)$$

$$\text{subject to:}$$

$$x_{k+1} = f(x_k, u_k) \quad \text{for } k = 0, 1, \ldots,$$

$$x_0 = x,$$

$$u_k \in \mathbb{U} \quad \text{for } k = 0, 1, \ldots,$$

$$x_k \in \mathbb{X} \quad \text{for } k = 0, 1, \ldots,$$

where $\{x_k\}$ and $\{u_k\}$ are now infinite sequences, and

$$V_\infty(\{x_k\}, \{u_k\}) \triangleq \sum_{k=0}^{\infty} L(x_k, u_k) \,. \qquad (4.33)$$

Note that $\mathcal{P}_\infty(x)$ does not have either a terminal state weighting nor a terminal state constraint; both are irrelevant since, if a solution to the problem exists, the state must converge to zero as $k \to \infty$ (since L is assumed to satisfy condition **B1**). In this case, it follows from the principle of optimality (see Section 3.4 of Chapter 3) that the finite horizon value function for problem $\mathcal{P}_N(x)$ in (4.1) is $V_N^{\mathrm{OPT}}(x) = V_\infty^{\mathrm{OPT}}(x)$. With this choice, on-line optimisation is unnecessary, and the advantages of an infinite horizon problem automatically accrue. However, constraints generally render this approach impossible.

Usually, then, $\mathbb{X}_f$ is chosen to be an appropriate neighbourhood of the origin in which $V_\infty^{\mathrm{OPT}}(x)$ is exactly (or approximately) known, and $F(x)$ is set equal to $V_\infty^{\mathrm{OPT}}(x)$ or its approximation.

In the rather general framework discussed so far, it is hard to visualise how Theorem 4.4.2 might be utilised in practice. However, when we specialise to *linear constrained* control problems it turns out that it is rather easy to satisfy the required conditions. This will be taken up in the next chapter.

4.6 Further Reading

For complete list of references cited, see References section at the end of book.

General

General treatment of nonlinear receding horizon control can be found in the book Allgöwer and Zhen (2000), and in, for example, the papers Keerthi and Gilbert (1988), Mayne and Michalska (1990), Alamir and Bornard (1994), Jadbabaie, Yu and Hauser (2001). See also the recent special issue Magni (2003).

An overview of industrial applications of receding horizon control is given in Qin and Badgwell (1997).

For a more detailed treatment of stability for general discrete time systems see Vidyasagar (2002), Kalman and Bertram (1960), Scokaert, Rawlings and Meadows (1997).

Stability for continuous-time nonlinear systems is thoroughly covered in several recent books, including Khalil (1996), Sastry (1999) and Vidyasagar (2002).

Section 4.4

The idea of using terminal state weighting to turn the finite horizon optimisation problem into an infinite horizon problem can be traced back to Kleinman (1970), Thomas (1975), and Kwon and Pearson (1977). More recent work appears in Chen and Shaw (1982), Kwon, Bruckstein and Kailath (1983), Garcia, Prett and Morari (1989), Bitmead et al. (1990). Related results for input-output systems appear in Mosca, Lemos and Zhang (1990), Clarke and Scattolini (1981) and Mosca and Zhang (1992).

The three main "ingredients" for stability used in Section 4.4 are implicit (in various combinations) in early literature dealing with constrained receding horizon control, see Sznaier and Damborg (1987), (1990), Keerthi and Gilbert (1988), Mayne and Michalska (1990), Rawlings and Muske (1993), Bemporad, Chisci and Mosca (1995), Chmielewski and Manousiouthakis (1996), De Nicolao, Magni and Scattolini (1996), Scokaert and Rawlings (1998). The form in which we have presented them is based on the clear and elegant synthesis provided by Mayne, Rawlings, Rao and Scokaert (2000).

5

Constrained Linear Quadratic Optimal Control

5.1 Overview

Up to this point we have considered rather general nonlinear receding horizon optimal control problems. Whilst we have been able to establish some important properties for these algorithms (for example, conditions for asymptotic stability), the algorithms remain relatively complex. However, remarkable simplifications occur if we specialise to the particular case of linear systems subject to linear inequality constraints. This will be the topic of the current chapter.

We will show how a fixed horizon optimal control problem for linear systems with a quadratic objective function and linear constraints can be set up as a quadratic program. We then discuss some practical aspects of the controller implementation, such as the use of observers to estimate states and disturbances. In particular, we will introduce the *certainty equivalence principle* and address several associated matters including steady state disturbance rejection (that is, provision of integral action) and how one can treat time delays in multivariable plants.

Finally, we show how closed loop stability of the receding horizon control [RHC] implementation can be achieved by specialising the results of Sections 4.4 and 4.5 in Chapter 4.

5.2 Problem Formulation

We consider a system described by the following linear, time-invariant model:

$$x_{k+1} = Ax_k + Bu_k, \tag{5.1}$$
$$y_k = Cx_k + d_k, \tag{5.2}$$

where $x_k \in \mathbb{R}^n$ is the state, $u_k \in \mathbb{R}^m$ is the control input, $y_k \in \mathbb{R}^m$ is the output, and $d_k \in \mathbb{R}^m$ is a time-varying output disturbance.

We assume that (A, B, C) is stabilisable and detectable and, for the moment, that one is not an eigenvalue of A. (See Section 5.4, where we relax the latter assumption.) So as to illustrate the principles involved, we go beyond the set-up described in Chapter 4 to include reference tracking and disturbance rejection.

Thus, we consider the problem where the output y_k in (5.2) is required to track a constant reference y^* in the presence of the disturbance d_k. That is, we wish to regulate, to zero, the output error

$$e_k \triangleq y_k - y^* = Cx_k + d_k - y^*. \tag{5.3}$$

Let $\bar{d}$ denote the steady state value of the output disturbance d_k, that is,

$$\bar{d} \triangleq \lim_{k \to \infty} d_k, \tag{5.4}$$

and denote by u_s, x_s, y_s and e_s, the *setpoints*, or desired steady state values for u_k, x_k, y_k and e_k, respectively. We then have that

$$y_s = y^* = Cx_s + \bar{d}, \tag{5.5}$$

$$e_s = 0, \tag{5.6}$$

and hence

$$u_s = [C(I - A)^{-1}B]^{-1}(y^* - \bar{d}), \tag{5.7}$$

$$x_s = (I - A)^{-1}Bu_s. \tag{5.8}$$

Without loss of generality, we take the current time as zero.

Here we assume knowledge of the disturbance d_k for all $k = 0, \ldots, N - 1$, and the current state measurement $x_0 = x$. (In practice, these signals will be obtained from an observer/predictor of some form; see Section 5.5.)

Our aim is to find, for the system (5.1)–(5.3), the M-move control sequence $\{u_0, \ldots, u_{M-1}\}$, and corresponding state sequence $\{x_0, \ldots, x_N\}$ and error sequence $\{e_0, \ldots, e_{N-1}\}$, that minimise the finite horizon objective function:

$$V_{N,M}(\{x_k\}, \{u_k\}\{e_k\}) \triangleq \frac{1}{2}(x_N - x_s)^{\mathrm{T}}P(x_N - x_s) + \frac{1}{2}\sum_{k=0}^{N-1} e_k^{\mathrm{T}}Qe_k$$

$$+ \frac{1}{2}\sum_{k=0}^{M-1}(u_k - u_s)^{\mathrm{T}}R(u_k - u_s), \tag{5.9}$$

where $P \geq 0$, $Q \geq 0$, $R > 0$. In (5.9), N is the *prediction horizon*, $M \leq N$ is the *control horizon*, and u_s, x_s are the input and state setpoints given by (5.7) and (5.8), respectively. The control is set equal to its steady state setpoint after M steps, that is, $u_k = u_s$ for all $k \geq M$.

In the following section, we will show how the minimisation of (5.9) is performed under constraints on the input and output.

The above fixed horizon minimisation problem is solved at each time step for the current state and disturbance values. Then, the first move of the resulting control sequence is used as the current control, and the procedure is repeated at the next time step in a RHC fashion, as described in Chapter 4.

5.3 Quadratic Programming

In the presence of linear constraints on the input and output, the fixed horizon optimisation problem described in Section 5.2 can be transformed into a *quadratic program* [QP] (see Section 2.5.6 in Chapter 2). We show below how this is accomplished.

5.3.1 Objective Function Handling

We will begin by showing how (5.9) can be transformed into an objective function of the form used in QP. We start by writing, from (5.1) with $x_0 = x$, and using the constraint that $u_k = u_s$ for all $k \geq M$, the following set of equations:

$$
\begin{aligned}
x_1 &= Ax + Bu_0, \\
x_2 &= A^2 x + ABu_0 + Bu_1, \\
&\;\;\vdots \\
x_M &= A^M x + A^{M-1} Bu_0 + \cdots + Bu_{M-1}, \\
x_{M+1} &= A^{M+1} x + A^M Bu_0 + \cdots + ABu_{M-1} + Bu_s, \\
&\;\;\vdots \\
x_N &= A^N x + A^{N-1} Bu_0 + \cdots + A^{N-M} Bu_{M-1} + \sum_{i=0}^{N-M-1} A^i Bu_s.
\end{aligned}
\tag{5.10}
$$

Using $x_s = Ax_s + Bu_s$ (from (5.8)) recursively, we can write a similar set of equations for x_s as follows:

$$
\begin{aligned}
x_s &= Ax_s + Bu_s, \\
x_s &= A^2 x_s + ABu_s + Bu_s, \\
&\;\;\vdots \\
x_s &= A^M x_s + A^{M-1} Bu_s + \cdots + Bu_s, \\
x_s &= A^{M+1} x_s + A^M Bu_s + \cdots + ABu_s + Bu_s, \\
&\;\;\vdots \\
x_s &= A^N x_s + A^{N-1} Bu_s + \cdots + A^{N-M} Bu_s + \sum_{i=0}^{N-M-1} A^i Bu_s.
\end{aligned}
\tag{5.11}
$$

We now subtract the set of equations (5.11) from the set (5.10), and rewrite the resulting difference in vector form to obtain

$$\mathbf{x} - \mathbf{x}_s = \Gamma(\mathbf{u} - \mathbf{u}_s) + \Omega(x - x_s), \tag{5.12}$$

where

$$\mathbf{x} \triangleq \begin{bmatrix} x_1 \\ x_2 \\ \vdots \\ x_N \end{bmatrix}, \quad \mathbf{x}_s \triangleq \begin{bmatrix} x_s \\ x_s \\ \vdots \\ x_s \end{bmatrix}, \quad \mathbf{u} \triangleq \begin{bmatrix} u_0 \\ u_1 \\ \vdots \\ u_{M-1} \end{bmatrix}, \quad \mathbf{u}_s \triangleq \begin{bmatrix} u_s \\ u_s \\ \vdots \\ u_s \end{bmatrix}, \tag{5.13}$$

($\mathbf{x}_s$ is an $nN \times 1$ vector, and $\mathbf{u}_s$ is an $mM \times 1$ vector), and where

$$\Gamma \triangleq \begin{bmatrix} B & 0 & \cdots & 0 & 0 \\ AB & B & \cdots & 0 & 0 \\ \vdots & \vdots & \ddots & \vdots & \vdots \\ A^{M-1}B & A^{M-2}B & \cdots & AB & B \\ A^M B & A^{M-1}B & \cdots & A^2 B & AB \\ \vdots & \vdots & \ddots & \vdots & \vdots \\ A^{N-1}B & A^{N-2}B & \cdots & \cdots & A^{N-M}B \end{bmatrix}, \quad \Omega \triangleq \begin{bmatrix} A \\ A^2 \\ \vdots \\ A^N \end{bmatrix}. \tag{5.14}$$

We also define the disturbance vector

$$\mathbf{d} \triangleq \left[(d_1 - \bar{d})^{\mathrm{T}} \; (d_2 - \bar{d})^{\mathrm{T}} \; \cdots \; (d_{N-1} - \bar{d})^{\mathrm{T}} \; 0_{1 \times m}\right]^{\mathrm{T}}, \tag{5.15}$$

and the matrices

$$\begin{aligned} \mathbf{Q} &\triangleq \mathrm{diag}\{C^{\mathrm{T}}QC, \ldots, C^{\mathrm{T}}QC, P\}, \\ \mathbf{R} &\triangleq \mathrm{diag}\{R, \ldots, R\}, \\ \mathbf{Z} &\triangleq \mathrm{diag}\{C^{\mathrm{T}}Q, C^{\mathrm{T}}Q, \ldots, C^{\mathrm{T}}Q\}, \end{aligned} \tag{5.16}$$

where $\mathrm{diag}\{A_1, A_2, \ldots, A_p\}$ denotes a block diagonal matrix having the matrices A_i as its diagonal blocks. Next, adding $-Cx_s - \bar{d} + y^* = 0$ (from (5.5)) to (5.3), we can express the error as

$$e_k = C(x_k - x_s) + (d_k - \bar{d}). \tag{5.17}$$

We now substitute (5.17) into the objective function (5.9), and rewrite it using the vector notation (5.13), (5.16) and (5.15), as follows:

$$\begin{aligned} V_{N,M} = \frac{1}{2}e_0^{\mathrm{T}}Qe_0 &+ \frac{1}{2}(\mathbf{x} - \mathbf{x}_s)^{\mathrm{T}}\mathbf{Q}(\mathbf{x} - \mathbf{x}_s) + \frac{1}{2}(\mathbf{u} - \mathbf{u}_s)^{\mathrm{T}}\mathbf{R}(\mathbf{u} - \mathbf{u}_s) \\ &+ (\mathbf{x} - \mathbf{x}_s)^{\mathrm{T}}\mathbf{Z}\mathbf{d} + \frac{1}{2}\mathbf{d}^{\mathrm{T}}\mathrm{diag}\{Q, Q, \ldots, Q\}\mathbf{d}. \end{aligned} \tag{5.18}$$

Next, we substitute (5.12) into (5.18) to yield

$$
\begin{aligned}
V_{N,M} &= \bar{V} + \frac{1}{2}\mathbf{u}^\mathsf{T}(\mathbf{\Gamma}^\mathsf{T}\mathbf{Q}\mathbf{\Gamma} + \mathbf{R})\mathbf{u} + \mathbf{u}^\mathsf{T}\mathbf{\Gamma}^\mathsf{T}\mathbf{Q}\mathbf{\Omega}(x - x_s) \\
&\quad - \mathbf{u}^\mathsf{T}(\mathbf{\Gamma}^\mathsf{T}\mathbf{Q}\mathbf{\Gamma} + \mathbf{R})\mathbf{u}_s + \mathbf{u}^\mathsf{T}\mathbf{\Gamma}^\mathsf{T}\mathbf{Z}\mathbf{d}, \\
&\triangleq \bar{V} + \frac{1}{2}\mathbf{u}^\mathsf{T}H\mathbf{u} + \mathbf{u}^\mathsf{T}\big[F(x - x_s) - H\mathbf{u}_s + D\mathbf{d}\big], \qquad (5.19)
\end{aligned}
$$

where $\bar{V}$ is independent of $\mathbf{u}$ and

$$
H \triangleq \mathbf{\Gamma}^\mathsf{T}\mathbf{Q}\mathbf{\Gamma} + \mathbf{R}, \qquad F \triangleq \mathbf{\Gamma}^\mathsf{T}\mathbf{Q}\mathbf{\Omega}, \qquad D \triangleq \mathbf{\Gamma}^\mathsf{T}\mathbf{Z}. \qquad (5.20)
$$

The last calculation is equivalent to the elimination of the equality constraints given by the state equations (5.1)–(5.3) by substitution into the objective function.

Note that H in (5.20) is positive definite because we have assumed $R > 0$ in the objective function (5.9).

From (5.19) it is clear that, if the problem is *unconstrained*, $V_{N,M}$ is minimised by taking

$$
\mathbf{u} = \mathbf{u}_{\mathrm{UC}}^{\mathrm{OPT}} \triangleq -H^{-1}\big[F(x - x_s) - H\mathbf{u}_s + D\mathbf{d}\big]. \qquad (5.21)
$$

The vector formed by the first m components of (5.21), $u_{0,\mathrm{UC}}^{\mathrm{OPT}}$, has a linear time-invariant feedback structure of the form

$$
u_{0,\mathrm{UC}}^{\mathrm{OPT}} = -K(x - x_s) + u_s + K_d\mathbf{d}, \qquad (5.22)
$$

where K and K_d are defined as the first m rows of the matrices $H^{-1}F$ and $-H^{-1}D$, respectively. By appropriate selection of the weightings in the objective function (5.9), the resulting K is such that the matrix $(A - BK)$ is Hurwitz, that is, all its eigenvalues have moduli smaller than one (see, for example, Bitmead et al. 1990). The control law (5.22) is the control used by the RHC algorithm if the problem is unconstrained. More interestingly, even in the constrained case, the optimal RHC solution has the form (5.22) in a region of the state space that contains the steady state setpoint $x = x_s$. This point will be discussed in detail in Chapters 6 and 7.

5.3.2 Constraint Handling

We now introduce inequality constraints into the problem formulation. Magnitude and rate constraints on the plant *input* and *output* can be expressed as follows:

$$
\begin{aligned}
u_{\min} &\leq u_k \leq u_{\max}, & k &= 0, \dots, M - 1, \\
y_{\min} &\leq y_k \leq y_{\max}, & k &= 1, \dots, N - 1, \qquad (5.23) \\
\delta u_{\min} &\leq u_k - u_{k-1} \leq \delta u_{\max}, & k &= 0, \dots, M - 1,
\end{aligned}
$$

where u_{-1} is the input used in the previous step of the receding horizon implementation, which has to be stored for use in the current fixed horizon optimisation.

More generally, we may require to impose *state constraints* of the form

$$x_k \in \mathbb{X}_k \quad \text{for } k = 1, \ldots, N, \tag{5.24}$$

where $\mathbb{X}_k$ is a polyhedral set of the form

$$\mathbb{X}_k = \{x \in \mathbb{R}^n : L_k x \le W_k\}. \tag{5.25}$$

For example, the constraint $x_N \in \mathbb{X}_f$, where $\mathbb{X}_f$ is a set satisfying certain properties, is useful to establish closed loop stability, as discussed in Chapter 4 (see also Section 5.6).

When constraints are present, we require that the setpoint $y_s = y^*$ and the corresponding input and state setpoints u_s and x_s be feasible, that is, that they satisfy the required constraints. For example, in the case of the constraints given in (5.23), we assume that $u_{\min} \le u_s \le u_{\max}$, and $y_{\min} \le y^* \le y_{\max}$. When the desired setpoint is not feasible in the presence of constraints, then one has to search for a feasible setpoint that is close to the desired setpoint in some sense. A procedure to do this is described in Section 5.4.

The constraints (5.23)–(5.25) can be written as *linear* constraints on **u** of the form

$$L\mathbf{u} \le W, \tag{5.26}$$

where

$$L = \begin{bmatrix} I_{Mm} \\ \Psi \\ E \\ -I_{Mm} \\ -\Psi \\ -E \\ \tilde{L} \end{bmatrix}, \qquad W = \begin{bmatrix} \mathbf{u}_{\max} \\ \mathbf{y}_{\max} \\ \delta\mathbf{u}_{\max} \\ \mathbf{u}_{\min} \\ \mathbf{y}_{\min} \\ \delta\mathbf{u}_{\min} \\ \tilde{W} \end{bmatrix}. \tag{5.27}$$

In (5.27), I_{Mm} is the $Mm \times Mm$ identity matrix (where M is the control horizon and m is the number of inputs). Ψ is the following $(N-1)m \times Mm$ matrix:

$$\Psi = \begin{bmatrix} CB & 0 & \ldots & 0 & 0 \\ CAB & CB & \ldots & 0 & 0 \\ \vdots & \vdots & \ddots & \vdots & \vdots \\ CA^{M-1}B & CA^{M-2}B & \ldots & CAB & CB \\ CA^M B & CA^{M-1}B & \ldots & CA^2 B & CAB \\ \vdots & \vdots & \ddots & \vdots & \vdots \\ CA^{N-2}B & CA^{N-3}B & \ldots & \ldots & CA^{N-M-1}B \end{bmatrix}.$$

E is the following $Mm \times Mm$ matrix:

$$E = \begin{bmatrix} I_m & 0 & \cdots & 0 \\ -I_m & I_m & \cdots & 0 \\ & \ddots & \ddots & \\ 0 & \cdots & -I_m & I_m \end{bmatrix},$$

where I_m is the $m \times m$ identity matrix; and

$$\tilde{L} \triangleq \mathrm{diag}\{L_1, L_2, \ldots, L_N\}\Gamma,$$

where $L_1, \ldots, L_N$ are the state constraint matrices given in (5.25) and Γ is given in (5.14).

The vectors forming W in (5.27) are as follows

$$\mathbf{u}_{\max} = \begin{bmatrix} u_{\max} \\ \vdots \\ u_{\max} \end{bmatrix}, \qquad \mathbf{u}_{\min} = \begin{bmatrix} -u_{\min} \\ \vdots \\ -u_{\min} \end{bmatrix},$$

$$\delta\mathbf{u}_{\max} = \begin{bmatrix} u_{-1} + \delta u_{\max} \\ \delta u_{\max} \\ \vdots \\ \delta u_{\max} \end{bmatrix}, \qquad \delta\mathbf{u}_{\min} = \begin{bmatrix} -u_{-1} - \delta u_{\min} \\ -\delta u_{\min} \\ \vdots \\ -\delta u_{\min} \end{bmatrix},$$

$$\mathbf{y}_{\max} = \begin{bmatrix} y_{\max} - CAx - d_1 \\ \vdots \\ y_{\max} - CA^M x - d_M \\ y_{\max} - CA^{M+1}x - d_{M+1} - CBu_s \\ \vdots \\ y_{\max} - CA^{N-1}x - d_{N-1} - \sum_{i=0}^{N-M-2} CA^i Bu_s \end{bmatrix},$$

$$\mathbf{y}_{\min} = \begin{bmatrix} -y_{\min} + CAx + d_1 \\ \vdots \\ -y_{\min} + CA^M x + d_M \\ -y_{\min} + CA^{M+1}x + d_{M+1} + CBu_s \\ \vdots \\ -y_{\min} + CA^{N-1}x + d_{N-1} + \sum_{i=0}^{N-M-2} CA^i Bu_s \end{bmatrix},$$

$$\tilde{W} \triangleq -\mathrm{diag}\{L_1, L_2, \ldots, L_N\} \begin{bmatrix} Ax \\ \vdots \\ A^M x \\ A^{M+1}x + Bu_s \\ A^{M+2}x + ABu_s \\ \vdots \\ A^N x + \sum_{i=0}^{N-M-1} A^i Bu_s \end{bmatrix} + \begin{bmatrix} W_1 \\ W_2 \\ \vdots \\ W_N \end{bmatrix},$$

where x is the initial state, $u_{\max}$, $u_{\min}$, $\delta u_{\max}$, $\delta u_{\min}$, $y_{\max}$, $y_{\min}$ are the vectors of constraint limits defined in (5.23) and $L_1, \ldots, L_N$, $W_1, \ldots, W_N$ are the matrices and vectors of the state constraint polyhedra (5.25).

5.3.3 The QP Problem

Using the above formalism, we can express the problem of minimising (5.9) subject to the inequality constraints (5.23)–(5.25) as the QP problem of minimising (5.19) subject to (5.26), that is,

$$\min_{\mathbf{u}} \frac{1}{2}\mathbf{u}^{\mathrm{T}} H \mathbf{u} + \mathbf{u}^{\mathrm{T}}\big[F(x - x_s) - H\mathbf{u}_s + D\mathbf{d}\big],$$

subject to:

$$L\mathbf{u} \leq W. \tag{5.28}$$

Note that the term $\bar{V}$ in (5.19) has not been included in (5.28) since it is independent of $\mathbf{u}$.

The optimal solution $\mathbf{u}^{\mathrm{OPT}}(x)$ to (5.28) is then:

$$\mathbf{u}^{\mathrm{OPT}}(x) = \arg\min_{L\mathbf{u}\leq W} \frac{1}{2}\mathbf{u}^{\mathrm{T}} H \mathbf{u} + \mathbf{u}^{\mathrm{T}}\big[F(x - x_s) - H\mathbf{u}_s + D\mathbf{d}\big]. \tag{5.29}$$

The matrix H is called the *Hessian* of the QP. If the Hessian is positive definite, the QP is convex. This is indeed the case for H given (5.20), which, as already mentioned, is positive definite because we have assumed $R > 0$ in the objective function (5.9). In Chapter 11 we will investigate the structure of the Hessian in detail and formulate numerically stable ways to compute it from the problem data.

Standard numerical procedures (called *QP algorithms*) are available to solve the above optimisation problem. In Chapter 8 we will review some of these algorithms.

Once the QP problem (5.29) is solved, the receding horizon algorithm applies, at the current time k, only the first control move, formed by the first m components of the optimal vector $\mathbf{u}^{\mathrm{OPT}}(x)$ in (5.29). This yields a control law of the form

$$u_k = \mathcal{K}(x_k, \bar{d}, y^*, \mathbf{d}), \tag{5.30}$$

where $x_k = x$ is the current state, and where the dependency on $\bar{d}$ and y^* is via u_s, x_s and $\mathbf{d}$ (see (5.7), (5.8), and (5.15)) as data for the optimisation (5.29). Then the whole procedure is repeated at the next time instant, with the optimisation horizon kept constant.

5.4 Embellishments

(i) **Systems with integrators.** In the above development, we have assumed that one is not an eigenvalue of A. This assumption allowed us to invert the matrix $(I - A)$ in (5.7) and (5.8).

There are several ways to treat the case when one is an eigenvalue of A. For example, in the single input-single output case, when one is an eigenvalue of A, then $u_s = 0$. To calculate x_s, we can write the state space model so that the integrator is shifted to the output, that is,

$$\tilde{x}_{k+1} = \tilde{A}\tilde{x}_k + \tilde{B}u_k,$$
$$x'_{k+1} = \tilde{C}\tilde{x}_k + x'_k,$$
$$y_k = x'_k + d_k,$$

where $\tilde{A}$, $\tilde{B}$, and $\tilde{C}$ correspond to the state space model of the reduced-order plant, that is, the plant without the integrator.

With this transformation, $(\tilde{A} - I)$ is nonsingular, and hence the state setpoint is

$$\begin{bmatrix} \tilde{x} \\ x' \end{bmatrix}_s = \begin{bmatrix} 0 \\ \vdots \\ 0 \\ y^* - \bar{d} \end{bmatrix},$$

where y^* is the reference and $\bar{d}$ is the steady state value of the output disturbance.

The optimisation problem is then solved in terms of these transformed state variables.

(ii) **Setpoint Calculation for Underactuated Systems.** By "underactuation" we mean that the actuators have insufficient authority to cancel the disturbance and reach the desired setpoint in steady state, that is, the desired setpoint is not feasible in the presence of constraints. This is not an uncommon situation. For example, in shape control in rolling mills, the actuators are cooling water sprays across the strip. These sprays change the radius of the rolls and hence influence the cross-directional reduction of the strip. However, these sprays have limited control authority, and thus they are frequently incapable of cancelling certain disturbances. More will be said about this cross-directional control problem in Chapter 15.

Under these conditions, it is clear that, in steady state, we will generally not be able to bring the output to the desired setpoint. Blind application of the receding horizon algorithm will lead to a saturated control, which achieves an optimal (in the quadratic objective function sense) compromise. However, there is an unexpected difficulty that arises from the weighting on the control effort in the objective function. In particular, the term $(u_k - u_s)^\mathsf{T} R(u_k - u_s)$ in (5.9), where u_s is the unconstrained steady state input required to achieve the desired setpoint $y_s = y^*$ (see (5.7) and (5.5)) may bias the solution of the optimisation problem.

One way to address this issue is to search for a feasible setpoint that is closest to the desired setpoint in a mean-square sense. In this case, the values of u_s and x_s that are used in the objective function (5.9) may be computed from the following quadratic program (see, for example, Muske and Rawlings 1993):

$$\min_{u_s} [(y^* - \bar{d}) - Cx_s]^{\mathrm{T}}[(y^* - \bar{d}) - Cx_s]$$

subject to:

$$x_s = (I - A)^{-1}Bu_s, \tag{5.31}$$

$$u_{\min} \leq u_s \leq u_{\max},$$

$$y_{\min} \leq Cx_s + \bar{d} \leq y_{\max}.$$

The actual value for the output that will be achieved in steady state is then $Cx_s + \bar{d}$ and u_s has been automatically defined to be consistent so that no bias results.

5.5 Observers and Integral Action

The above development has assumed that the system evolves in a deterministic fashion and that the full state (including disturbances) is measured.

When the state and disturbances are not measured, it is possible to obtain combined state and disturbance estimates via an *observer*. Those estimates can then be used in the control algorithm by means of the *certainty equivalence* [CE] *principle*, which consists of designing the control law assuming knowledge of the states and disturbances (as was done in Sections 5.2 and 5.3), and then using their estimates as if they were the true ones when implementing the controller. (More will be said about the CE principle in Chapter 12.)

In practice, it is also important to ensure that the true system output reaches its desired steady state value, or setpoint, despite the presence of unaccounted constant disturbances and modelling errors. In linear control, this is typically achieved by the inclusion of integrators in the feedback loop; hence, we say that a control algorithm that achieves this property has *integral action*.

In the context of constrained control, there are several alternative ways in which integral action can be included into a control algorithm. For example, using CE, the key idea is to include a model for constant disturbances at the input or output of the system and design an observer for the composite model including system and disturbance models. Then the control is designed to reject the disturbance, assuming knowledge of states and disturbance. Finally, the control is implemented using CE. The resulting observer-based closed loop system has integral action, as we will next show.

We will consider a *model* of the system of the form (5.1)-(5.2) with a constant output disturbance. (One could equally assume a constant input disturbance.) This leads to a composite model of the form

$$x_{k+1} = Ax_k + Bu_k,$$
$$d_{k+1} = d_k = \bar{d},$$
$$y_k = Cx_k + d_k.$$
(5.32)

We do not assume that the *model* (5.32) is a correct representation of the *real* system. In fact, we do not assume knowledge of the real system at all, but we assume that we can measure its output, which we denote y_k^{REAL}.

The RHC algorithm described in Sections 5.2 and 5.3 is now applied to the model (5.32) to design a controller for rejection of the constant output disturbance (and tracking of a constant reference) assuming knowledge of the model state and disturbance measurements. At each k the algorithm consists of solving the QP problem (5.29) for the current state $x_k = x$, with u_s and x_s computed from (5.7) and (5.8), and with $\mathbf{d} = 0$ (this follows from (5.15), since $d_k = \bar{d}$ for all k).

To apply the CE principle, we use the model (5.32) and the real system output y_k^{REAL} to construct an observer of the form

$$\hat{x}_{k+1} = A\hat{x}_k + Bu_k + L_1[y_k^{\text{REAL}} - C\hat{x}_k - \hat{d}_k],$$
$$\hat{d}_{k+1} = \hat{d}_k + L_2[y_k^{\text{REAL}} - C\hat{x}_k - \hat{d}_k],$$
(5.33)

where L_1 and L_2 are determined via any observer design method (such as the Kalman filter) that ensures that the matrix

$$\begin{bmatrix} A - L_1 C & -L_1 \\ -L_2 C & I - L_2 \end{bmatrix}$$

is Hurwitz. Then we simply use the estimates $(\hat{x}_k, \hat{d}_k)$ given by (5.33) in the RHC algorithm as if they were the true states, that is, $\hat{x}_k$ replaces x (which is the current state) and $\hat{d}_k$ replaces $\bar{d}$. Specifically, the QP problem (5.29) is solved at each k for $x = \hat{x}_k$, $\mathbf{d} = 0$ and with $u_s = u_{s,k}$ and $x_s = x_{s,k}$ computed as

$$u_{s,k} \triangleq [C(I - A)^{-1}B]^{-1}(y^* - \hat{d}_k),$$
(5.34)

$$x_{s,k} \triangleq (I - A)^{-1}Bu_{s,k}.$$
(5.35)

Note that now $u_{s,k}$ and $x_{s,k}$ are time-varying variables (compare with (5.7) and (5.8)). Thus, the resulting CE control law has the form (5.30) evaluated at $x_k = \hat{x}_k$, $\bar{d} = \hat{d}_k$, $y^* = y^*$, and $\mathbf{d} = 0$, that is,

$$u_k = \mathcal{K}(\hat{x}_k, \hat{d}_k, y^*, 0).$$
(5.36)

We will next show how integral action is achieved. We make the following assumption.

Assumption 5.1 *We assume that the real system in closed loop with the (constrained) control law (5.36) reaches a steady state in which no constraints are active and where $\{y_k^{\text{REAL}}\}$ and $\{\hat{d}_k\}$ converge to the constant values $\bar{y}^{\text{REAL}}$, $\bar{\hat{d}}$.*

○

Note that the assumption of no constraints being active when steady state is achieved implies that the control law, in steady state, must satisfy equation (5.22) (with $\mathbf{d} = 0$) evaluated at the steady state values, that is,

$$\bar{u} = -K(\bar{\hat{x}} - \bar{x}_s) + \bar{u}_s,$$ (5.37)

where $\bar{u} \triangleq \lim_{k \to \infty} u_k$, $\bar{\hat{x}} \triangleq \lim_{k \to \infty} \hat{x}_k$, $\bar{x}_s \triangleq \lim_{k \to \infty} x_{s,k}$, and $\bar{u}_s \triangleq \lim_{k \to \infty} u_{s,k}$. In (5.37), $\bar{u}_s$ and $\bar{x}_s$ satisfy, from (5.34) and (5.35),

$$\bar{u}_s = [C(I - A)^{-1}B]^{-1}(y^* - \bar{\hat{d}}),$$ (5.38)

$$\bar{x}_s = (I - A)^{-1}B\bar{u}_s,$$ (5.39)

where $\bar{\hat{d}} \triangleq \lim_{k \to \infty} \hat{d}_k$.

We make the following assumption on the matrix K in (5.37)

Assumption 5.2 *The matrix* $(A - BK)$ *is Hurwitz, that is, all its eigenvalues have moduli smaller than one.*

We then have the following result.

Lemma 5.5.1 *Under Assumptions 5.1 and 5.2, the real system output converges to the desired setpoint* y^*, *that is*

$$\bar{y}^{\text{REAL}} = y^*.$$ (5.40)

Proof. From the observer equations (5.33) in steady state we have

$$(I - A)\bar{\hat{x}} = B\bar{u},$$ (5.41)

$$\bar{y}^{\text{REAL}} = C\bar{\hat{x}} + \bar{\hat{d}}.$$ (5.42)

Substituting (5.37) in (5.41) and using $(I - A)\bar{x}_s = B\bar{u}_s$ from (5.39), we obtain

$$(I - A)\bar{\hat{x}} = B\bar{u}_s - BK\bar{\hat{x}} + BK\bar{x}_s = (I - A)\bar{x}_s - BK\bar{\hat{x}} + BK\bar{x}_s.$$

Reordering terms in the above equation yields

$$(I - A + BK)\bar{\hat{x}} = (I - A + BK)\bar{x}_s,$$

or

$$\bar{\hat{x}} = \bar{x}_s,$$ (5.43)

since $(A - BK)$ is Hurwitz by Assumption 5.2. We then have, from (5.43), (5.39) and (5.38), that

$$C\bar{\hat{x}} = C\bar{x}_s = y^* - \bar{\hat{d}}.$$ (5.44)

Thus, the result (5.40) follows upon substitution of (5.44) into (5.42).   □

Note that we have not shown (and indeed it will not be true in general) that $\bar{\hat{d}}$ is equal to the true output disturbance. In fact, the disturbance could actually be at the system input. Moreover, since we have not assumed that the model is correct, there need be no connection between $\hat{x}$ and the states of the real system. Lemma 5.5.1 is then an important result since it shows that, subject to the assumption that a steady state is achieved, the *required* output setpoint can be achieved despite uncertainty of different sources.

5.5.1 Observers for Systems with Time Delays

Many systems incorporate time delays between input and output. There is, thus, an issue of how best to deal with this. Naively, one could simply add extra states corresponding to the delays on each input. For example, suppose the delays on inputs 1 to m are τ_1 to τ_m samples, respectively. Let the input vector at time k be

$$u_k \triangleq \begin{bmatrix} u_k^1 \ u_k^2 \ \cdots \ u_k^m \end{bmatrix}^{\mathrm{T}}. \tag{5.45}$$

Then we can use the model

$$x_{k+1} = Ax_k + B\xi_k, \tag{5.46}$$

where

$$\xi_k \triangleq \begin{bmatrix} \xi_k^1 \ \xi_k^2 \ \cdots \ \xi_k^m \end{bmatrix}^{\mathrm{T}}, \tag{5.47}$$

and where each component ξ_k^i, $i = 1, \ldots, m$, has a model of the form

$$\eta_{k+1}^i = \begin{bmatrix} 0 & 0 & \cdots & 0 & 0 \\ 1 & 0 & \cdots & 0 & 0 \\ 0 & 1 & \cdots & 0 & 0 \\ \vdots & \vdots & \ddots & \vdots & \vdots \\ 0 & 0 & \cdots & 1 & 0 \end{bmatrix} \eta_k^i + \begin{bmatrix} 1 \\ 0 \\ 0 \\ \vdots \\ 0 \end{bmatrix} u_k^i,$$

$$\xi_k^i = \begin{bmatrix} 0 & 0 & \cdots & 0 & 1 \end{bmatrix} \eta_k^i, \tag{5.48}$$

with η_k^i a vector having τ_i components. That is, ξ_k^i is the input u_k^i delayed τ_i samples. A block diagram illustrating the model (5.46)–(5.48) is shown in Figure 5.1.

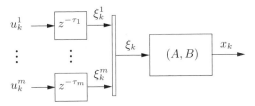

Figure 5.1. Block diagram of the input delayed system modelled by (5.48).

However, there are more parsimonious ways to proceed. As an illustration, consider the case where all input-output transfer functions contain a common delay. (This is typical in cases where measurements are made downstream from a process and all variables suffer the same transport delay. A specific example is the cross-directional control problem of the type discussed in Chapter 15.) Let the common delay be τ samples. Then, since the system model is linear, we can lump this delay at the output (whether it appears there or not). Hence, given output data $\{y_0, \ldots, y_k\}$, we can readily estimate $\hat{x}_{k-\tau}$ using a standard observer without considering any delays, other than delaying the inputs to ensure that we use the correct inputs in the model. By causality, this will utilise inputs up to $u_{k-\tau}$. The reason is that y_k is equivalent to $y'_{k-\tau}$, where y'_k denotes the undelayed output of the system.

For the purpose of the RHC calculations, we need $\hat{x}_k$. Since no measurements are available to compute this value, the best estimate of $\hat{x}_k$ is simply obtained by running the system model in open loop starting from $\hat{x}_{k-\tau}$.

Now we carry out the RHC calculations as usual to evaluate the sequence $\{u_k^{\mathrm{OPT}}, \ldots, u_{k+N}^{\mathrm{OPT}}\}$ and apply the first element u_k^{OPT} to the plant.

The reader will have observed that none of the above calculations have increased complexity resulting from the delay, save for the step of running the model forward from $\hat{x}_{k-\tau}$ to $\hat{x}_k$. Indeed, those readers who are familiar with the Smith predictor of classical control (see, for example, Goodwin et al. 2001) will recognise that the above procedure is a version of the scheme.

Of course, the problem becomes more complicated when there is not a common delay. In this case, we suggest that one should extract the delay of minimum value of all delays and treat that as a bulk delay of τ samples as described above. The residual (interaction) delays can then be dealt with as in (5.45)–(5.48), save that only the difference between the actual input delay and the bulk delay needs to be explicitly modelled.

5.6 Stability

In this section, we study closed loop stability of the receding horizon algorithm described in Sections 5.2 and 5.3. For simplicity, we assume that there are no reference or disturbance signals (that is, $d_k = 0$ for all k, $u_s = 0$ and $x_s = 0$). Also, in the objective function (5.9) we take $M = N$, $R > 0$ and we choose $Q > 0$ as the state (rather than output error) weighting matrix. We thus consider the following optimisation problem:

$$\mathcal{P}_N(x): \quad V_N^{\mathrm{OPT}}(x) \triangleq \min \left[F(x_N) + \frac{1}{2} \sum_{k=0}^{N-1} (x_k^{\mathrm{T}} Q x_k + u_k^{\mathrm{T}} R u_k) \right], \quad (5.49)$$

subject to:

$$x_{k+1} = A x_k + B u_k \quad \text{for } k = 0, \ldots, N-1, \tag{5.50}$$

$$x_0 = x, \tag{5.51}$$

$$u_k \in \mathbb{U} \quad \text{for } k = 0, \ldots, N-1, \tag{5.52}$$

$$x_k \in \mathbb{X} \quad \text{for } k = 0, \ldots, N, \tag{5.53}$$

$$x_N \in \mathbb{X}_f \subset \mathbb{X}, \tag{5.54}$$

as the underlying fixed horizon optimisation problem for the receding horizon algorithm.

Sufficient conditions for stability in the above linear constrained case can be obtained by specialising the results presented in Sections 4.4 and 4.5 of Chapter 4. Note that, with the choices $Q > 0$ and $R > 0$ in (5.49), condition **B1** of Section 4.4 is satisfied with $\gamma(t) = \lambda_{\min}(Q)\, t^2$, where $\lambda_{\min}(Q)$ is the minimum eigenvalue of the matrix Q. In the remainder of this section we will assume that the sets $\mathbb{U}$, $\mathbb{X}$ and $\mathbb{X}_f$ are convex and that the sets $\mathbb{U}$ and $\mathbb{X}_f$ contain the origin of their respective spaces (that is, condition **B5** is satisfied).

The fixed horizon optimal control problem $\mathcal{P}_N(x)$ in (5.49)–(5.54) has an associated set of feasible initial states $\mathbb{S}_N$. We recall from Definition 4.4.1 in Chapter 4 that $\mathbb{S}_N$ is the set of initial states $x \in \mathbb{X}$ for which there exist feasible state and control sequences, that is, sequences $\{x_0, x_1, \ldots, x_N\}$, $\{u_0, u_1, \ldots, u_{N-1}\}$ satisfying (5.50)–(5.54). The following lemma shows that $\mathbb{S}_N$ is convex if the sets $\mathbb{U}$, $\mathbb{X}$ and $\mathbb{X}_f$ are convex.

Lemma 5.6.1 (Convexity of the Set of Feasible Initial States)   *Let the sets $\mathbb{U}$, $\mathbb{X}$ and $\mathbb{X}_f$ in (5.52)–(5.54) be convex. Then the set $\mathbb{S}_N$ of feasible initial states for problem $\mathcal{P}_N(x)$ in (5.49)–(5.54) is convex.*

Proof. Let $x \in \mathbb{S}_N$. Hence there exist feasible state and control sequences $\{x_0, x_1, \ldots, x_N\}$, $\{u_0, u_1, \ldots, u_{N-1}\}$ satisfying (5.50)–(5.54). Similarly, let $\check{x} \in \mathbb{S}_N$, so that there exist feasible state and control sequences $\{\check{x}_0, \check{x}_1, \ldots, \check{x}_N\}$, $\{\check{u}_0, \check{u}_1, \ldots, \check{u}_{N-1}\}$ satisfying (5.50)–(5.54).

Let $x^\alpha \triangleq \alpha x + (1 - \alpha)\check{x}$, $\alpha \in [0, 1]$, and consider the sequences

$$\{x_k^\alpha\} \triangleq \{x_0^\alpha, x_1^\alpha, \ldots, x_N^\alpha\}, \tag{5.55}$$

$$\{u_k^\alpha\} \triangleq \{u_0^\alpha, u_1^\alpha, \ldots, u_{N-1}^\alpha\}, \tag{5.56}$$

where $x_k^\alpha \triangleq \alpha x_k + (1 - \alpha)\check{x}_k$, $k = 0, \ldots, N$, and $u_k^\alpha \triangleq \alpha u_k + (1 - \alpha)\check{u}_k$, $k = 0, \ldots, N - 1$. The above sequences are feasible since

$$\begin{aligned}
x_{k+1}^\alpha &= \alpha x_{k+1} + (1 - \alpha)\check{x}_{k+1} \\
&= \alpha(Ax_k + Bu_k) + (1 - \alpha)(A\check{x}_k + B\check{u}_k) \\
&= Ax_k^\alpha + Bu_k^\alpha \quad \text{for } k = 0, \ldots, N - 1, \\
x_0^\alpha &= \alpha x_0 + (1 - \alpha)\check{x}_0 \\
&= \alpha x + (1 - \alpha)\check{x} \\
&= x^\alpha,
\end{aligned}$$

and, also,

$$\begin{aligned}
u_k^\alpha &\triangleq \alpha u_k + (1 - \alpha)\check{u}_k \in \mathbb{U} \quad \text{for } k = 0, \ldots, N - 1, \\
x_k^\alpha &\triangleq \alpha x_k + (1 - \alpha)\check{x}_k \in \mathbb{X} \quad \text{for } k = 0, \ldots, N, \\
x_N^\alpha &\triangleq \alpha x_N + (1 - \alpha)\check{x}_N \in \mathbb{X}_f,
\end{aligned}$$

by the convexity of $\mathbb{U}$, $\mathbb{X}$ and $\mathbb{X}_f$. Hence, $x^\alpha \in \mathbb{S}_N$, proving that $\mathbb{S}_N$ is convex. $\square$

With the aid of Lemma 5.6.1, we can show that the value function $V_N^{\text{OPT}}(\cdot)$ in (5.49) is convex.

Lemma 5.6.2 (Convexity of the Value Function) *Let the sets* $\mathbb{U}$, $\mathbb{X}$ *and* $\mathbb{X}_f$ *in* (5.52)–(5.54) *be convex. Suppose that, in* (5.49), $Q \geq 0$, $R \geq 0$ *and the terminal state weighting is of the form* $F(x) = \frac{1}{2}x^{\mathrm{T}}Px$ *with* $P \geq 0$. *Then the value function* $V_N^{\text{OPT}}(\cdot)$ *in* (5.49) *is convex.*

Proof. Let $x \in \mathbb{S}_N$, with associated optimal (and hence, feasible) state and control sequences $\{x_k^{\text{OPT}}\} \triangleq \{x_0^{\text{OPT}}, x_1^{\text{OPT}}, \ldots, x_N^{\text{OPT}}\}$ and $\{u_k^{\text{OPT}}\} \triangleq \{u_0^{\text{OPT}}, u_1^{\text{OPT}}, \ldots, u_{N-1}^{\text{OPT}}\}$, respectively, solution of $\mathcal{P}_N(x)$ in (5.49)–(5.54). Similarly, let $\check{x} \in \mathbb{S}_N$, with associated optimal (and hence, feasible) sequences $\{\check{x}_k^{\text{OPT}}\} \triangleq \{\check{x}_0^{\text{OPT}}, \check{x}_1^{\text{OPT}}, \ldots, \check{x}_N^{\text{OPT}}\}$ and $\{\check{u}_k^{\text{OPT}}\} \triangleq \{\check{u}_0^{\text{OPT}}, \check{u}_1^{\text{OPT}}, \ldots, \check{u}_{N-1}^{\text{OPT}}\}$.
Let

$$V_N(\{x_k\}, \{u_k\}) \triangleq \frac{1}{2}x_N^{\mathrm{T}}Px_N + \frac{1}{2}\sum_{k=0}^{N-1}(x_k^{\mathrm{T}}Qx_k + u_k^{\mathrm{T}}Ru_k).$$

Then, $V_N^{\text{OPT}}(x) = V_N(\{x_k^{\text{OPT}}\}, \{u_k^{\text{OPT}}\})$ and $V_N^{\text{OPT}}(\check{x}) = V_N(\{\check{x}_k^{\text{OPT}}\}, \{\check{u}_k^{\text{OPT}}\})$.
Now consider $x^{\alpha} \triangleq \alpha x + (1-\alpha)\check{x}$, $\alpha \in [0,1]$. Similarly to the proof of Lemma 5.6.1, we can show that the sequences (5.56) and (5.55), with $u_k^{\alpha} \triangleq \alpha u_k^{\text{OPT}} + (1-\alpha)\check{u}_k^{\text{OPT}}$, $k = 0, \ldots, N-1$, and $x_k^{\alpha} \triangleq \alpha x_k^{\text{OPT}} + (1-\alpha)\check{x}_k^{\text{OPT}}$, $k = 0, \ldots, N$, are feasible. Hence, by optimality, we have that

$$V_N^{\text{OPT}}(x^{\alpha}) \leq V_N(\{x_k^{\alpha}\}, \{u_k^{\alpha}\}). \tag{5.57}$$

Also, by convexity of the quadratic functions $F(x) = \frac{1}{2}x^{\mathrm{T}}Px$ and $L(x,u) = \frac{1}{2}(x^{\mathrm{T}}Qx + u^{\mathrm{T}}Ru)$ we have

$$\begin{aligned}
V_N(\{x_k^{\alpha}\}, \{u_k^{\alpha}\}) &= F(x_N^{\alpha}) + \sum_{k=0}^{N-1} L(x_k^{\alpha}, u_k^{\alpha}) \\
&\leq \alpha F(x_N^{\text{OPT}}) + (1-\alpha)F(\check{x}_N^{\text{OPT}}) \\
&\quad + \sum_{k=0}^{N-1}\left[\alpha L(x_k^{\text{OPT}}, u_k^{\text{OPT}}) + (1-\alpha)L(\check{x}_k^{\text{OPT}}, \check{u}_k^{\text{OPT}})\right] \\
&= \alpha V_N(\{x_k^{\text{OPT}}\}, \{u_k^{\text{OPT}}\}) + (1-\alpha)V_N(\{\check{x}_k^{\text{OPT}}\}, \{\check{u}_k^{\text{OPT}}\}) \\
&= \alpha V_N^{\text{OPT}}(x) + (1-\alpha)V_N^{\text{OPT}}(\check{x}). \tag{5.58}
\end{aligned}$$

Combining the inequalities in (5.57) and (5.58), it follows that $V_N^{\text{OPT}}(\cdot)$ is convex, and the result is then proved. $\qquad\square$

Lemmas 5.6.1 and 5.6.2 show that $V_N^{\text{OPT}}(\cdot)$ is a convex function defined on the convex set $\mathbb{S}_N$. Since we have assumed that the sets $\mathbb{U}$ and $\mathbb{X}_f$ contain the origin of their respective spaces, then $0 \in \mathbb{S}_N$ and hence $\mathbb{S}_N$ is nonempty. From Theorem 2.3.8 in Chapter 2, we conclude that $V_N^{\text{OPT}}(\cdot)$ is *continuous* on int $\mathbb{S}_N$. This fact will be used below in the proof of asymptotic (exponential) stability.

We present in the following sections three instances of the application of the stability theory for receding horizon control developed in Sections 4.4 and 4.5 of Chapter 4.

5.6.1 Open Loop Stable System with Input Constraints

Assume that the matrix A in (5.50) is Hurwitz, that is, all its eigenvalues have moduli smaller than one. Suppose that there are no state constraints (that is, $\mathbb{X} = \mathbb{X}_f = \mathbb{R}^n$ in (5.53) and (5.54)) and, as a consequence, $\mathbb{S}_N = \mathbb{R}^n$. Then, to apply Theorem 4.4.2 of Chapter 4, we simply choose the terminal state weighting as

$$F(x) = \frac{1}{2}x^\mathrm{T}Px, \tag{5.59}$$

where P satisfies the discrete Lyapunov equation

$$P = A^\mathrm{T}PA + Q. \tag{5.60}$$

A feasible terminal control is

$$\mathcal{K}_f(x) = 0 \ \text{ for all } x \in \mathbb{X}_f = \mathbb{R}^n. \tag{5.61}$$

Note that, by assumption, the system is open loop stable, hence $F(x)$ is the infinite horizon objective function beginning in state x and using the terminal control (5.61).

Recall that, as discussed before, conditions **B1** and **B5** of Theorem 4.4.2 are satisfied from the assumptions on problem $\mathcal{P}_N(x)$ in (5.49)–(5.54). Clearly conditions **B3** and **B4** hold with the above choices for the terminal triple. Direct calculation yields that $F(x) = x^\mathrm{T}Px$ satisfies

$$
\begin{aligned}
F(f(x, \mathcal{K}_f(x))) - F(x) &= \frac{1}{2}(Ax + B\mathcal{K}_f(x))^\mathrm{T}P(Ax + B\mathcal{K}_f(x)) - \frac{1}{2}x^\mathrm{T}Px \\
&= \frac{1}{2}x^\mathrm{T}(A^\mathrm{T}PA - P)x \\
&= -\frac{1}{2}x^\mathrm{T}Qx \\
&= -L(x, \mathcal{K}_f(x))
\end{aligned}
$$

so that condition **B2** is also satisfied. Thus far, we have verified conditions **B1**–**B5** of Theorem 4.4.2, which establishes global attractivity of the origin. To prove exponential stability, we further need to show that the conditions in part (iv) of the theorem are also fulfilled. Note that $F(x) \leq \lambda_{\max}(P)\|x\|^2$. Also, as shown above, $V_N^{\mathrm{OPT}}(\cdot)$ is continuous. Hence, exponential stability holds in any arbitrarily large compact set of the state space.

5.6.2 General Linear System with Input Constraints

This case is a slight generalisation of the result in Section 5.6.1. Assume that the system has no eigenvalue with modulus equal to one. Suppose that $\mathbb{X} = \mathbb{R}^n$ in (5.53). We factor the system into stable and unstable subsystems as follows:

$$x_{k+1}^s = A_s x_k^s + B_s u_k,$$
$$x_{k+1}^u = A_u x_k^u + B_u u_k,$$

where the eigenvalues of A_s have moduli less than one, and the eigenvalues of A_u have moduli greater than one. Next, we choose $Q > 0$ in (5.49) of the form

$$Q = \begin{bmatrix} Q_s & 0 \\ 0 & Q_u \end{bmatrix}$$

and use as terminal state weighting

$$F(x) = \frac{1}{2}(x^s)^{\mathrm{T}} P_s x^s,$$

where P_s satisfies the discrete Lyapunov equation

$$P_s = A_s^{\mathrm{T}} P_s A_s + Q_s.$$

Finally, we choose $\mathcal{K}_f(x) = 0$ and $\mathbb{X}_f$ in (5.54) as

$$\mathbb{X}_f = \left\{ x = \begin{bmatrix} x^s \\ x^u \end{bmatrix} \in \mathbb{R}^n : x^u = 0 \right\}.$$

It can be easily verified, as done in Section 5.6.1, that the conditions **B1**–**B5** of Theorem 4.4.2 are satisfied with the above choices. Hence, if[1] $0 \in$ int $\mathbb{S}_N$, asymptotic stability of the origin follows, as proved in part (iii) of Theorem 4.4.2.

Note that the condition $x_N^u = 0$ is not very restrictive because the system $x_{k+1}^u = A_u x_k^u + B_u u_k$ is bounded input-bounded output stable *in reverse time*. Hence the set of initial states x_0^u that are taken by feasible control sequences into *any* terminal set is largely determined by the constraints on the input rather than the values of x_N^u; that is,

$$x_0^u = A_u^{-N} x_N^u - \sum_{k=0}^{N-1} A_u^{-k-1} B_u u_k,$$

and

$$A_u^{-N} \overset{\exp}{\to} 0 \quad \text{as } N \to \infty.$$

(See Sections 11.2 and 11.3 in Chapter 11 for further discussion on solving for unstable modes in reverse time.)

---

[1] This is the case if $0 \in$ int $\mathbb{U}$ and N is greater than or equal to the dimension of x^u, since the system is assumed stabilisable.

5.6.3 General Linear Systems with State and Input Constraints

In this case, $F(x)$ in (5.49) is often chosen to be the value function of the infinite horizon, *unconstrained* optimal control problem for the same system (see Scokaert and Rawlings 1998, Sznaier and Damborg 1987). This problem, defined as in (4.32)-(4.33) in Chapter 4, but with no constraints ($\mathbb{U} = \mathbb{R}^m$, $\mathbb{X} = \mathbb{R}^n$), is a standard linear quadratic regulator problem whose value function is $x^{\mathrm{T}}Px$, where P is the positive definite solution of the algebraic Riccati equation

$$P = A^{\mathrm{T}}PA + Q - K^{\mathrm{T}}\bar{R}K,$$

where

$$K \triangleq \bar{R}^{-1}B^{\mathrm{T}}PA, \qquad \bar{R} \triangleq R + B^{\mathrm{T}}PB. \tag{5.62}$$

The terminal state weighting used in this case is then

$$F(x) = \frac{1}{2}x^{\mathrm{T}}Px.$$

The local controller $\mathcal{K}_f(x)$ is chosen to be the optimal linear controller $\mathcal{K}_f(x) = -Kx$, where K is given by (5.62).

The terminal set $\mathbb{X}_f$ is usually taken to be the *maximal output admissible set* $\mathcal{O}_\infty$ (Gilbert and Tan 1991) for the closed loop system using the local controller $\mathcal{K}_f(x)$, defined as

$$\mathcal{O}_\infty \triangleq \{x : K(A-BK)^k x \in \mathbb{U} \text{ and } (A-BK)^k x \in \mathbb{X} \text{ for } k = 0, 1, \ldots\}. \tag{5.63}$$

$\mathcal{O}_\infty$ is the maximal positively invariant set for the system $x_{k+1} = (A-BK)x_k$ (see Definition 4.4.2 in Chapter 4) in which constraints are satisfied.

With the above choice for the terminal triple $(\mathbb{X}_f, \mathcal{K}_f, F)$, conditions **B1–B5** of Theorem 4.4.2 are readily established, similarly to Section 5.6.1. This proves attractivity of the origin in $\mathbb{S}_N$. To prove exponential stability, we further need to show that the conditions in part (iv) of the theorem are also fulfilled. Note that $F(x) \leq \lambda_{\max}(P)\|x\|^2$. Also, as shown above, $V_N^{\mathrm{OPT}}(\cdot)$ is continuous on int $\mathbb{S}_N$. Hence, exponential stability holds in any arbitrarily large compact subset contained in the interior of $\mathbb{S}_N$.

An interesting consequence of this choice for the terminal triple is that $V_\infty^{\mathrm{OPT}}(x) = F(x)$ for all x in $\mathbb{X}_f$ and that $V_N^{\mathrm{OPT}}(x) = V_\infty^{\mathrm{OPT}}(x)$ for all $x \in \mathbb{S}_N$. Actually, the horizon N can be chosen large enough for the predicted terminal state x_N^{OPT} (corresponding to the Nth step of the optimal state sequence for initial state x) to belong to $\mathbb{X}_f$ (see Section 5.8 for references to methods to compute lower bounds on such N). If N is so chosen, the terminal constraint may be omitted from the optimisation problem $\mathcal{P}_N(x)$.

5.7 Stability with Observers

A final question raised by the use of observers and the CE principle in RHC is whether or not closed loop stability is retained when the true states are replaced by state estimates in the control law. We will not explicitly address this issue but, instead, refer the reader, in Section 5.8 below, to recent literature dealing with this topic. Again, from a practical perspective, it seems fair to anticipate that, provided the state estimates are reasonably accurate, then stability should not be compromised; recall that we have established, in Sections 5.6.1 and 5.6.3, that exponential stability of the origin holds (under mild conditions) in the case where the states are known.

5.8 Further Reading

For complete list of references cited, see References section at the end of book.

General

The following books give detailed description of receding horizon control in the linear constrained case: Camacho and Bordons (1999), Maciejowski (2002), Borrelli (2003), Rossiter (2003). See also the early paper Muske and Rawlings (1993), as well as the survey paper Bemporad and Morari (1999).

Section 5.6

A method to compute a lower bound on the optimisation horizon N such that the predicted terminal state x_N^{OPT} in the fixed horizon optimal control problem (5.49)–(5.54) belongs to the terminal set $\mathbb{X}_f$ for all initial conditions in a given compact set is presented in Bemporad, Morari, Dua and Pistikopoulos (2002); this method, in turn, uses an algorithm proposed in Chmielewski and Manousiouthakis (1996).

There are various embellishments of the basic idea described in Section 5.6.3. For example, a new terminal triple has been provided for receding horizon control of input constrained linear systems in De Doná, Seron, Goodwin and Mayne (2002). The new triple is an improvement over those previously used in that the terminal constraint set $\mathbb{X}_f$, which we define below, is strictly larger than $\mathcal{O}_\infty$, thus facilitating the solution of the fixed horizon optimal control problem. The improved terminal conditions employ the results of Section 7.3 in Chapter 7 that show that the nonlinear controller

$$\mathcal{K}_f(x) = -\mathrm{sat}(Kx) \tag{5.64}$$

is optimal in a region $\bar{\mathbb{Z}}$, which includes the maximal output admissible set $\mathcal{O}_\infty$. The terminal constraint set $\mathbb{X}_f$ is then selected as the maximal positively invariant set for the system $x_{k+1} = Ax_k - B\mathrm{sat}(Kx)$. We refer the reader to the literature to follow up this and related ideas.

Stability of RHC has been established for neutrally stable systems (that is, systems having nonrepeated roots on the unit circle) using a nonquadratic terminal weighting (see Jadbabaie, Persis and Yoon 2002, Yoon, Kim, Jadbabaie and Persis 2003).

Section 5.7

The stability of the CE implementation of RHC has been addressed for constrained linear systems in, for example, Zheng and Morari (1995), where global asymptotic stability is shown for open loop stable systems, and in Muske, Meadows and Rawlings (1994), where a local stability result is given for general linear systems. Local results for nonlinear systems are reported in, for example, Scokaert et al. (1997), and Magni, De Nicolao and Scattolini (2001). A stability result for nonlinear systems using a moving horizon observer is given in Michalska and Mayne (1995).

See also Findeisen, Imsland, Allgöwer and Foss (2003) for a recent survey and new results on output feedback nonlinear RHC.

6

Global Characterisation of Constrained Linear Quadratic Optimal Control

6.1 Overview

As stated earlier, a common strategy in applications of receding horizon optimal control is to compute the optimal input u_k at time k by solving, on line, the associated finite horizon optimisation problem. As explained in Chapter 4, when the system and objective function are time-invariant, then this procedure implicitly defines a time-invariant control policy $\mathcal{K}_N : \mathbb{X} \to \mathbb{U}$ of the form $\mathcal{K}_N(x) = u_0^{\mathrm{OPT}}$ (see (4.10) of Chapter 4).

This leads to the obvious question: "Should we repeat the calculation of $\mathcal{K}_N(x)$ in the event that x returns to a value that has been previously visited?" Heuristically, one would immediately answer "No!" to this question. However, a little more thought reveals that to make this a feasible proposition in practice, we need to solve three problems; namely,

(i) how to efficiently calculate $\mathcal{K}_N(x)$ for all x of interest;
(ii) how to store $\mathcal{K}_N(x)$ as a function of x;
(iii) how to retrieve the value of $\mathcal{K}_N(x)$ given x.

In a general setting, these problems present substantial difficulties. However, for the case of constrained linear systems, there exists a relatively simple *finitely parameterised* characterisation of $\mathcal{K}_N(x)$, which can be computed and stored efficiently for *small* state dimensions and *short* horizons. We present this characterisation in this chapter, leading to an explicit form for the receding horizon control policy. Even if, on the balance of computational time, one decides to still solve the quadratic programming [QP] problem at each step, we believe that this finitely parameterised characterisation of $\mathcal{K}_N(x)$ gives practically valuable insights into the nature of the constrained control policy.

We first derive the characterisation using dynamic programming (introduced in Section 3.4 of Chapter 3). We consider systems with a single input constrained to lie in an interval and optimisation horizon $N = 2$. We then analyse the geometric structure of the fixed horizon optimal control problem $\mathcal{P}_N$ when seen as a QP of the form discussed in Section 5.3 of Chapter 5.

Using this geometric structure, we will re-derive the result obtained via dynamic programming. This will give insight into the solution by clarifying the interrelationship between dynamic programming and the inherent geometry of the QP solution space (the space where the QP is defined).

We then show how the same ideas can be used to characterise the solution of the fixed horizon optimal control problem $\mathcal{P}_N$ for cases with arbitrary horizons and more general linear constraints. As for the simpler case $N = 2$, the general solution is obtained by exploiting the geometry of the associated QP in particular coordinates of its solution space. Once projected onto the state space, the result is a *piecewise affine* characterisation, that is, a partition of the state space into regions in which the corresponding control law is affine.

Finally, we discuss the use of the KKT optimality conditions (Sections 2.5.4–2.5.5 in Chapter 2) in the derivation of the piecewise affine characterisation.

6.2 Global Characterisation for Horizon 2 via Dynamic Programming

For ease of exposition, here we will fix the optimisation horizon to $N = 2$.

We consider single input, linear, discrete time systems in which the magnitude of the control input is constrained to be less than or equal to a positive constant. In particular, let the system be given by

$$x_{k+1} = Ax_k + Bu_k, \qquad |u_k| \leq \Delta, \tag{6.1}$$

where $x_k \in \mathbb{R}^n$ and $\Delta > 0$ is the input constraint level. Consider the following fixed horizon optimal control problem

$$\mathcal{P}_2(x): \qquad V_2^{\text{OPT}}(x) \triangleq \min V_2(\{x_k\}, \{u_k\}), \tag{6.2}$$

subject to:

$$x_{k+1} = Ax_k + Bu_k \quad \text{for } k = 0, 1,$$

$$x_0 = x,$$

$$u_k \in \mathbb{U} \triangleq [-\Delta, \Delta] \quad \text{for } k = 0, 1,$$

where the objective function in (6.2) is

$$V_2(\{x_k\}, \{u_k\}) \triangleq \frac{1}{2} x_2^{\text{T}} P x_2 + \frac{1}{2} \sum_{k=0}^{1} (x_k^{\text{T}} Q x_k + u_k^{\text{T}} R u_k). \tag{6.3}$$

The matrices Q and R in (6.3) are positive definite and P satisfies the algebraic Riccati equation

$$P = A^{\text{T}} P A + Q - K^{\text{T}} \bar{R} K, \tag{6.4}$$

where

$$K \triangleq \bar{R}^{-1}B^{\mathrm{T}}PA, \qquad \bar{R} \triangleq R + B^{\mathrm{T}}PB. \tag{6.5}$$

Let the control sequence that minimises (6.3) be

$$\{u_0^{\mathrm{OPT}}, u_1^{\mathrm{OPT}}\}. \tag{6.6}$$

Then the RHC law is given by the first element of (6.6) (which depends on the current state $x_0 = x$), that is,

$$\mathcal{K}_2(x) = u_0^{\mathrm{OPT}}. \tag{6.7}$$

Before proceeding with the solution to the above problem, we observe that, at this stage, the choice of the matrix P as the solution to (6.4) is not necessary; however, as we showed in Chapter 5, this choice is useful to establish stability of the receding horizon implementation given by the state equation in (6.1) in closed loop with $u_k = \mathcal{K}_2(x_k)$. Also, this choice effectively gives an infinite horizon objective function with the restriction that constraints not be active after the first two steps. Similarly, the assumption that Q is positive definite is not required at this stage, but it will be used later in the stability result of Section 7.2.2 in the following chapter.

In Theorem 6.2.1 below, we will derive the solution of $\mathcal{P}_2$ defined in (6.2)–(6.5) using dynamic programming. Following Section 3.4 of Chapter 3, the partial value functions at each step of the dynamic programming algorithm, are defined by

$$V_0^{\mathrm{OPT}}(x_2) \triangleq \frac{1}{2}x_2^{\mathrm{T}}Px_2, \tag{6.8}$$

$$V_1^{\mathrm{OPT}}(x_1) \triangleq \min_{\substack{u_1 \in \mathbb{U} \\ x_2 = Ax_1 + Bu_1}} \frac{1}{2}x_2^{\mathrm{T}}Px_2 + \frac{1}{2}x_1^{\mathrm{T}}Qx_1 + \frac{1}{2}u_1^{\mathrm{T}}Ru_1,$$

and $V_2^{\mathrm{OPT}}(x)$ is the value function of $\mathcal{P}_2$ defined in (6.2)–(6.3).

The dynamic programming algorithm makes use of the principle of optimality, which states that any portion of the optimal trajectory is itself an optimal trajectory. That is, for $k = 0, 1$, (see (3.85) in Chapter 3)

$$V_k^{\mathrm{OPT}}(x) = \min_{u \in \mathbb{U}} \frac{1}{2}x^{\mathrm{T}}Qx + \frac{1}{2}u^{\mathrm{T}}Ru + V_{k-1}^{\mathrm{OPT}}(Ax + Bu), \tag{6.9}$$

where u and x denote, $u = u_k$ and $x = x_k$, respectively.

In the sequel we will use the saturation function $\mathrm{sat}_\Delta(\cdot)$ defined, for the saturation level Δ, as

$$\mathrm{sat}_\Delta(u) \triangleq \begin{cases} \Delta & \text{if } u > \Delta, \\ u & \text{if } |u| \leq \Delta, \\ -\Delta & \text{if } u < -\Delta. \end{cases} \tag{6.10}$$

The following result gives a finitely parameterised characterisation of the RHC law (6.7).

Theorem 6.2.1 (RHC Characterisation for $N = 2$) *The RHC law* (6.7) *has the form*

$$K_2(x) = \begin{cases} -\mathrm{sat}_\Delta(Gx + h) & \text{if } x \in \mathbb{Z}^-, \\ -\mathrm{sat}_\Delta(Kx) & \text{if } x \in \mathbb{Z}, \\ -\mathrm{sat}_\Delta(Gx - h) & \text{if } x \in \mathbb{Z}^+, \end{cases} \tag{6.11}$$

where K is given by (6.5), *the gain $G \in \mathbb{R}^{1 \times n}$ and the constant $h \in \mathbb{R}$ are given by*

$$G \triangleq \frac{K + KBKA}{1 + (KB)^2}, \qquad h \triangleq \frac{KB}{1 + (KB)^2}\Delta, \tag{6.12}$$

and the sets $(\mathbb{Z}^-, \mathbb{Z}, \mathbb{Z}^+)$ are defined by

$$\mathbb{Z}^- \triangleq \{x : K(A - BK)x < -\Delta\}, \tag{6.13}$$

$$\mathbb{Z} \triangleq \{x : |K(A - BK)x| \leq \Delta\}, \tag{6.14}$$

$$\mathbb{Z}^+ \triangleq \{x : K(A - BK)x > \Delta\}. \tag{6.15}$$

Proof. We start from the last partial value function (6.8), at time $k = N = 2$, and solve the problem backwards in time using (6.9).

(i) The partial value function V_0^{OPT} ($k = N = 2$):
Here $x = x_2$. By definition, the partial value function at time $k = N = 2$ is

$$V_0^{\mathrm{OPT}}(x) \triangleq \frac{1}{2}x^{\mathrm{T}}Px \quad \text{for all } x \in \mathbb{R}^n.$$

(ii) The partial value function V_1^{OPT} ($k = N - 1 = 1$):
Here $x = x_1$ and $u = u_1$. By the principle of optimality, for all $x \in \mathbb{R}^n$,

$$V_1^{\mathrm{OPT}}(x) = \min_{u \in \mathbb{U}} \left\{ \frac{1}{2}x^{\mathrm{T}}Qx + \frac{1}{2}u^{\mathrm{T}}Ru + V_0^{\mathrm{OPT}}(Ax + Bu) \right\}$$

$$= \min_{u \in \mathbb{U}} \left\{ \frac{1}{2}x^{\mathrm{T}}Qx + \frac{1}{2}u^{\mathrm{T}}Ru + \frac{1}{2}(Ax + Bu)^{\mathrm{T}}P(Ax + Bu) \right\}$$

$$= \min_{u \in \mathbb{U}} \left\{ \frac{1}{2}x^{\mathrm{T}}Px + \frac{1}{2}\bar{R}(u + Kx)^2 \right\}, \tag{6.16}$$

where $\bar{R}$ is defined in (6.5). In deriving (6.16) we have made use of (6.4). It is clear that the *unconstrained* ($u \in \mathbb{R}$) optimal control is given by $u = -Kx$. From the convexity of the function $\bar{R}(u + Kx)^2$ it then follows that the *constrained* ($u \in \mathbb{U}$) optimal control law, which corresponds to the second element of the sequence (6.6), is given by

$$u_1^{\mathrm{OPT}} = \mathrm{sat}_\Delta(-Kx) = -\mathrm{sat}_\Delta(Kx) \quad \text{for all } x \in \mathbb{R}^n, \tag{6.17}$$

and the partial value function at time $k = N - 1 = 1$ is

$$V_1^{\mathrm{OPT}}(x) = \frac{1}{2}x^{\mathrm{T}}Px + \frac{1}{2}\bar{R}\left[Kx - \mathrm{sat}_\Delta(Kx)\right]^2 \quad \text{for all } x \in \mathbb{R}^n.$$

(iii) The (partial) value function V_2^{OPT} ($k = N - 2 = 0$):
Here $x = x_0$ and $u = u_0$. By the principle of optimality, we have that, for all $x \in \mathbb{R}^n$,

$$
\begin{aligned}
V_2^{\mathrm{OPT}}(x) &= \min_{u \in \mathbb{U}} \left\{ \frac{1}{2} x^\mathsf{T} Q x + \frac{1}{2} u^\mathsf{T} R u + V_1^{\mathrm{OPT}}(Ax + Bu) \right\} \\
&= \min_{u \in \mathbb{U}} \left\{ \frac{1}{2} x^\mathsf{T} Q x + \frac{1}{2} u^\mathsf{T} R u + \frac{1}{2}(Ax + Bu)^\mathsf{T} P(Ax + Bu) \right. \\
&\qquad\left. + \frac{1}{2} \bar{R} \left[K(Ax + Bu) - \mathrm{sat}_\Delta(K(Ax + Bu)) \right]^2 \right\} \\
&= \frac{1}{2} \min_{u \in \mathbb{U}} \left\{ x^\mathsf{T} P x + \bar{R}(u + Kx)^2 \right. \\
&\qquad\left. + \bar{R} \left[KAx + KBu - \mathrm{sat}_\Delta(KAx + KBu) \right]^2 \right\}. \qquad (6.18)
\end{aligned}
$$

Denote the terms in (6.18) by $f_1(u) \triangleq (u + Kx)^2$, and $f_2(u) \triangleq [KAx + KBu - \mathrm{sat}_\Delta(KAx + KBu)]^2$. Notice that the function $f_2(u)$ has a "cup" shape formed by three zones: (a) Half-parabola corresponding to the case $KAx + KBu < -\Delta$; (b) a flat zone corresponding to the case $|KAx + KBu| \leq \Delta$, and; (c) half-parabola corresponding to the case $KAx + KBu > \Delta$. Note also that $f_1 + f_2$ is convex. With this information, we can derive the result (6.11) as follows:

Case (a). $x \in \mathbb{Z}^-$: In this case, the minimiser of $f_1(u)$ (that is, $u = -Kx$) is such that $KAx + KB(-Kx) = K(A - BK)x < -\Delta$ (see (6.13)), that is, $u = -Kx$ is in zone (a) of function $f_2(u)$. Then, the minimum of $f_1(u) + f_2(u)$ (situated between the minimum of $f_1(u)$, at $u = -Kx$, and the minimum of $f_2(u)$) will also fall in zone (a). We conclude that the value function is

$$
V_2^{\mathrm{OPT}}(x) = \frac{1}{2} \min_{u \in \mathbb{U}} \left\{ x^\mathsf{T} P x + \bar{R}(u + Kx)^2 + \bar{R} \left[KAx + KBu + \Delta \right]^2 \right\},
$$

whose unconstrained minimum is easily found to be at $u = -(Gx + h)$, with G and h as given in (6.12). From the convexity of $f_1(u) + f_2(u)$ it then follows that the *constrained* ($u \in \mathbb{U}$) optimal control law, which corresponds to the first element of the sequence (6.6), is given by

$$
u_0^{\mathrm{OPT}} = -\mathrm{sat}_\Delta(Gx + h) \quad \text{for all } x \in \mathbb{Z}^-.
$$

This shows the result in (6.11) for this case.

Case (b). $x \in \mathbb{Z}$: This case corresponds to the situation where $u = -Kx$ is in zone (b) of $f_2(u)$ and hence, the unconstrained minimum of $f_1(u) + f_2(u)$ occurs at $u = -Kx$. Again, using the convexity of $f_1(u) + f_2(u)$, it follows that the *constrained* optimal control law, which corresponds to the first element of the sequence (6.6), is

$$u_0^{\text{OPT}} = -\text{sat}_\Delta(Kx) \quad \text{for all } x \in \mathbb{Z}. \tag{6.19}$$

Case (c). $x \in \mathbb{Z}^+$: The result follows from a similar analysis to the case (a).

□

We illustrate the above result by a simple example.

Example 6.2.1. Consider again the double integrator of Example 1.2.1 in Chapter 1. For this system, we consider an input saturation level $\Delta = 1$.

In the fixed horizon objective function (6.3) we take $Q = \begin{bmatrix} 1 & 0 \\ 0 & 0 \end{bmatrix}$, and $R = 0.1$. The gain K is computed from (6.4)–(6.5). Equation (6.12) gives $G = [-0.6154 \quad -1.2870]$ and $h = 0.4156$.

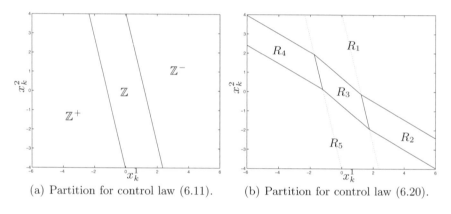

(a) Partition for control law (6.11).   (b) Partition for control law (6.20).

Figure 6.1. State space partitions for Example 6.2.1.

In Figure 6.1 (a) we show the sets $\mathbb{Z}^-$, $\mathbb{Z}$ and $\mathbb{Z}^+$ that define the controller (6.11). In this figure, x_k^1 and x_k^2 denote the two components of the state vector x_k.

Actually, we can be even more explicit regarding the form of the control law. In particular, if we discriminate between the regions where each component of the controller (6.11) is saturated from the ones where it is not, we can parameterise the controller (6.11) in the following equivalent, but more explicit, form:

$$\mathcal{K}_2(x) = \begin{cases} -\Delta & \text{if } x \in R_1, \\ -Gx - h & \text{if } x \in R_2, \\ -Kx & \text{if } x \in R_3, \\ -Gx + h & \text{if } x \in R_4, \\ \Delta & \text{if } x \in R_5. \end{cases} \tag{6.20}$$

In Figure 6.1 (b) we show the state space partition that corresponds to the parameterisation (6.20) for this example.   ∘

Theorem 6.2.1 shows that the solution to the simple RHC problem (6.2)–(6.7) has the form of a piecewise affine feedback control law defined on a partition of the state space. We will see below that a similar characterisation can be obtained for more general cases. Rather than extending the above procedure, we will utilise alternative geometric arguments.

6.3 The Geometry of Quadratic Programming

As mentioned in Chapter 5, when the system model is linear and the objective function quadratic, the fixed horizon optimal control problem $\mathcal{P}_N$ can be transformed into a QP of the form (5.28). We will start by re-examining the optimal control problem for horizon $N = 2$ defined in (6.2)–(6.5). In this case, the corresponding QP optimal solution is (see (5.29) in Chapter 5 and (6.4)–(6.5))

$$\text{QP:} \qquad \mathbf{u}^{\text{OPT}}(x) = \arg\min_{\mathbf{u} \in R_{\text{UC}}} \frac{1}{2}\mathbf{u}^{\mathsf{T}} H \mathbf{u} + \mathbf{u}^{\mathsf{T}} F x, \qquad (6.21)$$

where

$$\mathbf{u} = \begin{bmatrix} u_0 \\ u_1 \end{bmatrix}, \qquad H = \bar{R} \begin{bmatrix} 1 + (KB)^2 & KB \\ KB & 1 \end{bmatrix}, \qquad F = \bar{R} \begin{bmatrix} K + KBKA \\ KA \end{bmatrix}, \qquad (6.22)$$

and R_{UC} is the square $[-\Delta, \Delta] \times [-\Delta, \Delta] \subset \mathbb{R}^2$. Note that the Hessian H is positive definite since $\bar{R} = R + B^{\mathsf{T}} P B$ is positive because we have assumed that $R > 0$ in (6.3).

The QP in (6.21) has a nice geometric interpretation in the $\mathbf{u}$-space. Consider the equation

$$\frac{1}{2}\mathbf{u}^{\mathsf{T}} H \mathbf{u} + \mathbf{u}^{\mathsf{T}} F x = c, \qquad (6.23)$$

where c is a constant. This defines ellipsoids in $\mathbb{R}^2$ centred at $\mathbf{u}^{\text{OPT}}_{\text{UC}}(x) = -H^{-1}Fx$. Then (6.21) can be regarded as the problem of finding the smallest ellipsoid that intersects the boundary of R_{UC}, and $\mathbf{u}^{\text{OPT}}(x)$ is the point of intersection. This is illustrated in Figure 6.2.

The problem can be significantly simplified if we make a coordinate transformation via the square root of the Hessian, that is,

$$\tilde{\mathbf{u}} = H^{1/2}\mathbf{u}. \qquad (6.24)$$

In the new coordinates defined by (6.24), the constraint set R_{UC} is mapped into another set, denoted also by R_{UC} for simplicity of notation. The ellipsoids (6.23) take the form of spheres centred at $\tilde{\mathbf{u}}^{\text{OPT}}_{\text{UC}}(x) = -H^{-1/2}Fx$. Thus

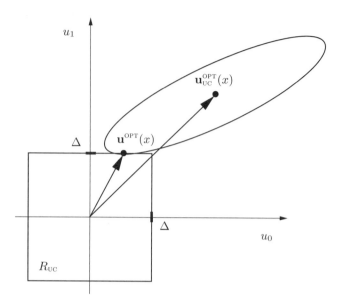

Figure 6.2. Geometric interpretation of QP.

(6.21) is transformed into the problem of finding the point in R_{UC} that is closest to $\tilde{\mathbf{u}}_{\text{UC}}^{\text{OPT}}(x)$ in the Euclidean distance. This is qualitatively illustrated in Figure 6.3.

This transformed geometric picture allows us to immediately write down the solution to the fixed horizon optimal control problem for this special case. In particular, the solution of the QP is obtained by partitioning $\mathbb{R}^2$ into nine regions; the first region is the parallelogram R_{UC}. The remaining regions, denoted by R_1 to R_8, are delimited by lines that are normal to the faces of the parallelogram and pass through its vertices, as shown in Figure 6.3. The optimal constrained solution $\tilde{\mathbf{u}}^{\text{OPT}}(x)$ is determined by the region in which the optimal unconstrained solution $\tilde{\mathbf{u}}_{\text{UC}}^{\text{OPT}}(x)$ lies, in the following way: First, it is clear that $\tilde{\mathbf{u}}^{\text{OPT}}(x) = \tilde{\mathbf{u}}_{\text{UC}}^{\text{OPT}}(x)$ if $\tilde{\mathbf{u}}_{\text{UC}}^{\text{OPT}}(x) \in R_{\text{UC}}$; that is, the optimal constrained solution coincides with the optimal unconstrained solution in R_{UC}. Next, the optimal constrained solution in each of the regions R_1, R_3, R_5 and R_7 is simply equal to the vertex that is contained in the region. Finally, the optimal constrained solution in the regions R_2, R_4, R_6 and R_8 is defined by the orthogonal projection of $\tilde{\mathbf{u}}_{\text{UC}}^{\text{OPT}}(x)$ onto the faces of the parallelogram. This can be seen from Figure 6.3, where a case in which the solution falls in R_8 is illustrated.

Whilst we have concentrated on the simple case $N = 2$, it is easy to see that this methodology can be applied also to more complex cases. Indeed, in Section 6.5 we will apply these geometric arguments to arbitrary horizons and multiple input systems.

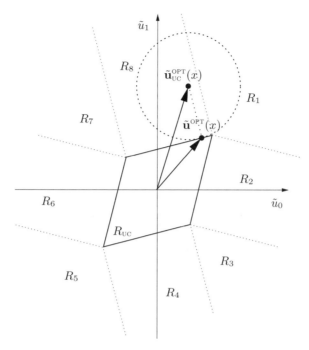

Figure 6.3. Geometry of QP as a minimum Euclidean distance problem.

We thus see that the geometry of the QP problem gives insight into its solution. We will expand on these ideas in the following sections.

6.4 Geometric Characterisation for Horizon 2

In this section, we will use the geometric ideas outlined in Section 6.3 to recover the finitely parameterised characterisation derived via dynamic programming in Theorem 6.2.1.

To solve (6.21), we use the transformation (6.24), which maps the square R_{UC} into the parallelogram R_{UC} shown in Figure 6.3. We next note that the unconstrained solution in the $\tilde{\mathbf{u}}$–coordinates is given by

$$\tilde{\mathbf{u}}_{\mathrm{UC}}^{\mathrm{OPT}}(x) = -H^{-1/2}Fx. \qquad (6.25)$$

We then derive the constrained solution in the $\tilde{\mathbf{u}}$–coordinates using geometric arguments, and finally use the transformation

$$\tilde{\mathbf{u}} = -H^{-1/2}Fx \qquad (6.26)$$

to retrieve the solution in the state space.

Following Section 6.3, we partition $\mathbb{R}^2$ into nine regions; the first region is the parallelogram R_{UC}, and the remaining regions, denoted by R_1 to R_8, are delimited by lines that are normal to the faces of R_{UC} and pass through its vertices, as shown in Figure 6.3.

In R_{UC} we have

$$\tilde{\mathbf{u}}^{\mathrm{OPT}}(x) = \tilde{\mathbf{u}}_{\mathrm{UC}}^{\mathrm{OPT}}(x) = -H^{-1/2}Fx \quad \text{for all } \mathbf{u} \in R_{\mathrm{UC}}.$$

That is, in R_{UC} the optimal constrained solution coincides with the optimal unconstrained solution.

To describe the solution in regions R_1 to R_8 in more detail, we introduce the following notation, which will be used in the remainder of the chapter.

Notation 6.4.1 *Given any matrix (column vector) M, and a set of indices $\bar{\ell}$ (with, at most, as many elements as the number of rows of M), the notation $M_{\bar{\ell}}$ identifies the submatrix (subvector) of M formed by selecting the rows with indices given by the elements of $\bar{\ell}$ and all of its columns.* ○

For example, given H defined in (6.22), and the set $\bar{\ell} = \{2\}$, $H_{\bar{\ell}} = H_2$ denotes its second row.

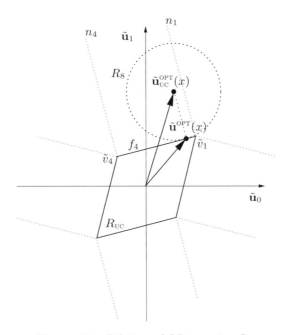

Figure 6.4. Solution of QP in region R_8.

Consider now, for example, region R_8 in Figure 6.4. It is delimited by face f_4 and its normals n_1 and n_4 passing through the vertices $\tilde{v}_1$ and $\tilde{v}_4$, respectively. The line that contains face f_4 (the line is also denoted by f_4, for simplicity) corresponds to the second control u_1 equal to the saturation level Δ, that is $u_1 = [0\ 1]\mathbf{u} = \Delta$ in the original $\mathbf{u}$-coordinates. In the new $\tilde{\mathbf{u}}$-coordinates (6.24), it is then defined by the equation

$$f_4: \qquad [0\ 1]H^{-1/2}\tilde{\mathbf{u}} = \Delta. \qquad (6.27)$$

Lines normal to f_4 are defined by the equation

$$[1\ 0]H^{1/2}\tilde{\mathbf{u}} = c, \qquad c \in \mathbb{R}. \qquad (6.28)$$

Hence, the equation defining n_1 is obtained by setting

$$\tilde{\mathbf{u}} = \tilde{v}_1 = H^{1/2}\Delta[1\ 1]^{\mathrm{T}}$$

in (6.28). This yields

$$n_1: \qquad [1\ 0]H^{1/2}\tilde{\mathbf{u}} = H_1\Delta[1\ 1]^{\mathrm{T}}. \qquad (6.29)$$

In a similar way, the equation defining n_4 is given by

$$n_4: \qquad [1\ 0]H^{1/2}\tilde{\mathbf{u}} = H_1\Delta[-1\ 1]^{\mathrm{T}}. \qquad (6.30)$$

Combining (6.27), (6.29) and (6.30), region R_8 is defined in the $\tilde{\mathbf{u}}$-coordinates by

$$R_8: \qquad \begin{cases} [0\ 1]H^{-1/2}\tilde{\mathbf{u}} \geq \Delta \\ H_1\Delta[-1\ 1]^{\mathrm{T}} \leq [1\ 0]H^{1/2}\tilde{\mathbf{u}} \leq H_1\Delta[1\ 1]^{\mathrm{T}}, \end{cases} \qquad (6.31)$$

and, using the transformation (6.26), it is defined in the state space coordinates $x \in \mathbb{R}^n$ by

$$R_8: \qquad \begin{cases} -K(A - BK)x \geq \Delta \\ H_1\Delta[-1\ 1]^{\mathrm{T}} \leq -F_1 x \leq H_1\Delta[1\ 1]^{\mathrm{T}}. \end{cases} \qquad (6.32)$$

The optimal constrained solution in R_8 is given by the normal projection of the unconstrained solution $\tilde{\mathbf{u}}_{\mathrm{UC}}^{\mathrm{OPT}}(x)$ onto face f_4; that is, the solution is obtained by intersecting face f_4 with the normal to it passing through $\tilde{\mathbf{u}}_{\mathrm{UC}}^{\mathrm{OPT}}(x)$. From (6.27) and (6.28), $\tilde{\mathbf{u}}^{\mathrm{OPT}}(x)$ satisfies the equations

$$[0\ 1]H^{-1/2}\tilde{\mathbf{u}}^{\mathrm{OPT}}(x) = \Delta,$$
$$[1\ 0]H^{1/2}\tilde{\mathbf{u}}^{\mathrm{OPT}}(x) = [1\ 0]H^{1/2}\tilde{\mathbf{u}}_{\mathrm{UC}}^{\mathrm{OPT}}(x).$$

Using (6.24) and $\tilde{\mathbf{u}}_{\mathrm{UC}}^{\mathrm{OPT}}(x)$ from (6.25) in the above equations yields

$$H_1 \begin{bmatrix} u_0^{\mathrm{OPT}}(x) \\ \Delta \end{bmatrix} = -F_1 x.$$

Further substitution of $H_1 = \bar{R}[1 + (KB)^2\ KB]$ and $F_1 = \bar{R}(K + KBKA)$ from (6.22) gives

$$u_0^{\mathrm{OPT}}(x) = -\frac{(K + KBKA)x + KB\Delta}{1 + (KB)^2},$$

and using the definitions (6.12), we obtain

$$u_0^{\mathrm{OPT}}(x) = -Gx - h. \tag{6.33}$$

If we proceed in a similar way with the remaining regions, we obtain a characterisation of the RHC problem (6.2)–(6.7) in the form

$$\mathcal{K}_2(x) = \begin{cases} -Kx & \text{if } x \in R_{\mathrm{UC}}, \\ \Delta & \text{if } x \in R_1 \cup R_2 \cup R_3, \\ -Gx + h & \text{if } x \in R_4, \\ -\Delta & \text{if } x \in R_5 \cup R_6 \cup R_7, \\ -Gx - h & \text{if } x \in R_8, \end{cases} \tag{6.34}$$

where

$$R_{\mathrm{UC}} : \quad \begin{cases} |Kx| \leq \Delta \\ |K(A - BK)x| \leq \Delta \end{cases}$$

$$R_1 : \quad \begin{cases} -F_1 x \geq H_1 \Delta [1 \ 1]^{\mathrm{T}} \\ -F_2 x \geq H_2 \Delta [1 \ 1]^{\mathrm{T}} \end{cases}$$

$$R_2 : \quad \begin{cases} -Kx \geq \Delta \\ H_2 \Delta [1 \ -1]^{\mathrm{T}} \leq -F_2 x \leq H_2 \Delta [1 \ 1]^{\mathrm{T}} \end{cases}$$

$$R_3 : \quad \begin{cases} -F_2 x \leq H_2 \Delta [1 \ -1]^{\mathrm{T}} \\ -F_1 x \geq H_1 \Delta [1 \ -1]^{\mathrm{T}} \end{cases}$$

$$R_4 : \quad \begin{cases} -K(A - BK)x \leq -\Delta \\ H_1 \Delta [-1 \ -1]^{\mathrm{T}} \leq -F_1 x \leq H_1 \Delta [1 \ -1]^{\mathrm{T}} \end{cases} \tag{6.35}$$

$$R_5 : \quad \begin{cases} -F_1 x \leq H_1 \Delta [-1 \ -1]^{\mathrm{T}} \\ -F_2 x \leq H_2 \Delta [-1 \ -1]^{\mathrm{T}} \end{cases}$$

$$R_6 : \quad \begin{cases} -Kx \leq -\Delta \\ H_2 \Delta [-1 \ -1]^{\mathrm{T}} \leq -F_2 x \leq H_2 \Delta [-1 \ 1]^{\mathrm{T}} \end{cases}$$

$$R_7 : \quad \begin{cases} -F_2 x \geq H_2 \Delta [-1 \ 1]^{\mathrm{T}} \\ -F_1 x \leq H_1 \Delta [-1 \ 1]^{\mathrm{T}} \end{cases}$$

and R_8 is given in (6.32).

Example 6.4.1. Consider again the system and data of Example 6.2.1. In Figure 6.5 we show the state space partition that corresponds to the controller (6.34)–(6.35) for this example. As can be seen, the RHC law derived using geometric arguments coincides with the one obtained in Example 6.2.1 by dynamic programming. (Compare with Figure 6.1 (b).) In the following section, we will further discuss this relationship.   ○

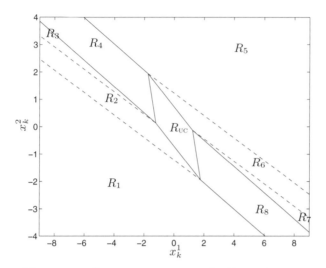

Figure 6.5. State space partitions for Example 6.4.1.

6.4.1 Relationship with the Characterisation Obtained via Dynamic Programming

We will show here that the characterisation (6.11), obtained using dynamic programming, and the characterisation (6.34), obtained using geometric arguments, are equivalent. Indeed, faces f_2 and f_4 of the parallelogram R_{UC} (see Figure 6.6) correspond to the second control u_1 being equal to the saturation limits $-\Delta$ and Δ, respectively. Hence, using the transformation $\mathbf{u} = -H^{-1}Fx$, the definitions of H and F from (6.22), and equating the second component of $\mathbf{u}$ to $-\Delta$ and Δ, we find that these faces are given in the state space by

$$f_2 : \quad -K(A-BK)x = -\Delta,$$
$$f_4 : \quad -K(A-BK)x = \Delta.$$

Comparing the above equations with (6.14), we conclude that region $\mathbb{Z}$ corresponds, in the $\tilde{\mathbf{u}}$-coordinates, to the shaded region in Figure 6.6. Similarly, the half-planes above and below the shaded region in Figure 6.6 correspond to $\mathbb{Z}^-$ and $\mathbb{Z}^+$, defined in (6.13) and (6.15), respectively. Moreover, since faces f_1 and f_3 of the parallelogram R_{UC} (which correspond to the first control u_0 equal to the saturation limits Δ and $-\Delta$, respectively) are given in the state space by

$$f_1: \qquad -Kx = \Delta,$$
$$f_3: \qquad -Kx = -\Delta,$$

it is not difficult to see, using (6.34), (6.35), that $\mathcal{K}_2(x)$ in $\mathbb{Z}$ is given by $-\mathrm{sat}_\Delta(Kx)$, as stated in (6.11).

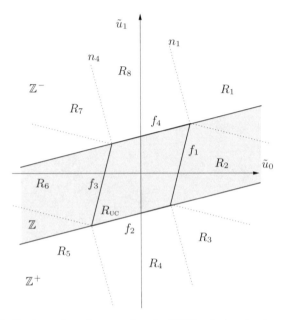

Figure 6.6. Regions that characterise the RHC solution in the case $N = 2$.

On the other hand, substituting (6.26) in (6.29), (6.30), we find that the normals n_1 and n_4 are given in the state space by

$$n_1: \qquad -F_1 x = H_1 \begin{bmatrix} \Delta & \Delta \end{bmatrix}^{\mathrm{T}},$$
$$n_4: \qquad -F_1 x = H_1 \begin{bmatrix} -\Delta & \Delta \end{bmatrix}^{\mathrm{T}},$$

and, using $H_1 = \bar{R}[1+(KB)^2 \quad KB]$, $F_1 = \bar{R}(K+KBKA)$, and the definitions (6.12), we have

$$n_1: \qquad -Gx - h = \Delta$$
$$n_4: \qquad -Gx - h = -\Delta.$$

Thus, comparing with (6.33), we can see that n_1 and n_4 define the switching lines where $\mathcal{K}_2(x) = -Gx - h$ saturates to $\mathcal{K}_2(x) = \Delta$ and $\mathcal{K}_2(x) = -\Delta$, respectively. Then, using (6.34), (6.35), it is immediately seen that $\mathcal{K}_2(x)$ in $\mathbb{Z}^-$ is given by $-\mathrm{sat}_\Delta(Gx + h)$, as stated in (6.11). A similar analysis can be performed for region $\mathbb{Z}^+$.

We have thus far obtained a partition of $\mathbb{R}^2$ that gives a complete geometric picture of the solution to the RHC problem for the case of single input, $N = 2$. One may anticipate that this kind of argument can be extended to larger horizons. This is, indeed, the case, as we will show in Section 6.5, where we will generalise this procedure to higher dimensional spaces.

6.5 Geometric Characterisation for Arbitrary Horizon

In this section we explore the RHC structure for more general problems. We again focus on linear, time-invariant, discrete time models with a quadratic objective function, but we consider arbitrary horizons and more general linear constraints. For these systems, and under a particular constraint qualification, we derive a finitely parameterised characterisation of the RHC solution. In particular, let the system be given by

$$x_{k+1} = Ax_k + Bu_k, \tag{6.36}$$

where $x_k \in \mathbb{R}^n$ and $u_k \in \mathbb{R}^m$. Consider the following fixed horizon optimal control problem:

$$\mathcal{P}_N(x): \quad V_N^{\mathrm{OPT}}(x) \triangleq \min V_N(\{x_k\}, \{u_k\}), \tag{6.37}$$

subject to:
$$x_{k+1} = Ax_k + Bu_k \quad \text{for } k = 0, \ldots, N-1,$$
$$x_0 = x,$$
$$u_k \in \mathbb{U}_k \quad \text{for } k = 0, \ldots, N-1, \tag{6.38}$$
$$x_k \in \mathbb{X}_k \quad \text{for } k = 0, \ldots, N, \tag{6.39}$$

where $\mathbb{U}_k$ and $\mathbb{X}_k$ are polyhedral constraint sets, whose description we leave unspecified at this stage. The objective function in (6.37) is

$$V_N(\{x_k\}, \{u_k\}) \triangleq \frac{1}{2} x_N^{\mathsf{T}} P x_N + \frac{1}{2} \sum_{k=0}^{N-1} \left(x_k^{\mathsf{T}} Q x_k + u_k^{\mathsf{T}} R u_k \right), \tag{6.40}$$

with $Q > 0$, $R > 0$ and $P \geq 0$.

The associated QP and optimiser are (see (5.29) in Chapter 5)

$$\mathbf{u}^{\mathrm{OPT}}(x) = \arg \min_{L\mathbf{u} \leq W} \frac{1}{2} \mathbf{u}^{\mathsf{T}} H \mathbf{u} + \mathbf{u}^{\mathsf{T}} F x, \tag{6.41}$$

where $\mathbf{u} \in \mathbb{R}^{Nm}$, and where $H = \Gamma^{\mathsf{T}} \mathbf{Q} \Gamma + \mathbf{R}$, $F = \Gamma^{\mathsf{T}} \mathbf{Q} \Omega$, with Γ, Ω, $\mathbf{Q}$, $\mathbf{R}$, defined as in (5.14) and (5.16) of Chapter 5 (with $M = N$ and $C = I$).

The constraint set

$$L\mathbf{u} \leq W \tag{6.42}$$

in (6.41) is a polyhedron in $\mathbb{R}^{Nm}$ obtained from the sets $\mathbb{U}_k$ and $\mathbb{X}_k$ in (6.38) and (6.39), respectively. We will assume that the matrix L and the vector W have the form

$$L = \begin{bmatrix} \Phi \\ -\Phi \end{bmatrix}, \qquad W = \begin{bmatrix} \bar{\Delta} \\ \underline{\Delta} \end{bmatrix} + \begin{bmatrix} -\Lambda \\ \Lambda \end{bmatrix} x. \qquad (6.43)$$

In (6.43), Φ is a $q \times Nm$ matrix, $\bar{\Delta}$ and $\underline{\Delta}$ are $q \times 1$ vectors such that $\bar{\Delta} + \underline{\Delta} > 0$ (componentwise), and Λ is a $q \times n$ matrix. Note that (6.42)–(6.43) can also be written as the interval-type constraint $-\underline{\Delta} \leq \Phi\mathbf{u} + \Lambda x \leq \bar{\Delta}$. As was shown in Section 5.3.2 of Chapter 5, the structure of L and W in (6.43) easily accommodates typical constraint requirements, such as, for example, magnitude or rate constraints on the inputs or outputs.

As in Section 6.4, we will study the solution to the above RHC problem using a special coordinate basis for the QP solution space $\mathbb{R}^{Nm}$. To this end, we use the transformation (6.24) to take the original QP coordinates into the new $\tilde{\mathbf{u}}$-coordinates. The convex constraint polyhedron in the $\mathbf{u}$-coordinates defined by (6.42)–(6.43), is mapped into a convex constraint polyhedron in the $\tilde{\mathbf{u}}$-coordinates, given by

$$\tilde{\Phi}\tilde{\mathbf{u}} \leq \bar{\Delta} - \Lambda x, \qquad (6.44)$$

$$-\tilde{\Phi}\tilde{\mathbf{u}} \leq \underline{\Delta} + \Lambda x, \qquad (6.45)$$

where

$$\tilde{\Phi} \triangleq \Phi H^{-1/2}.$$

Notice that the dimension of the constraint polyhedron is the constraint horizon $q = \text{rank } \Phi$. A *face* of the constraint polyhedron is defined by the intersection, with the constraint polyhedron, of the hyperplane defined by a subset of equalities (or active constraints) within (6.44)–(6.45). To each face of the constraint polyhedron, we will associate an *active pair* (ℓ, Δ), whose elements are defined below. We will use the notation 6.4.1, and introduce the set $J \triangleq \{1, 2, \ldots, q\}$ of the first q natural numbers. We then define, for each face with $\bar{N} \in J$ active constraints:

- The *active set* $\ell \triangleq \{\ell_1, \ell_2, \ldots, \ell_{\bar{N}} : \ell_k \in J\}$, which identifies the indices of the constraints that are active; that is, the indices of the rows within (6.44)–(6.45) that hold as equalities for the face. Note that the gradient of the active constraints is $\tilde{\Phi}_\ell$.

- The *active value vector* $\Delta \in \mathbb{R}^{\bar{N}}$, which identifies whether the active constraint whose index is ℓ_k corresponds to either row ℓ_k of (6.44) or row ℓ_k of (6.45). More precisely, the kth element of Δ is given by

$$\begin{cases} \Delta_k = \bar{\Delta}_{\ell_k} & \text{if } \tilde{\Phi}_{\ell_k}\tilde{\mathbf{u}} = \bar{\Delta}_{\ell_k} - \Lambda_{\ell_k} x, \\ \Delta_k = -\underline{\Delta}_{\ell_k} & \text{if } \tilde{\Phi}_{\ell_k}\tilde{\mathbf{u}} = -\underline{\Delta}_{\ell_k} - \Lambda_{\ell_k} x. \end{cases} \qquad (6.46)$$

- The *inactive set* $s \triangleq J-\ell = \{s_1, s_2, \ldots, s_{q-\bar{N}} : s_k \in J \text{ and } s_k \notin \ell\}$, which identifies the indices of the constraints that are not active in each face. Note that the gradient of the inactive constraints is $\tilde{\Phi}_s$.

In addition, we will impose the *constraint qualification* that the gradient of active constraints $\tilde{\Phi}_\ell$ has full row rank.

Each active pair (ℓ, Δ) fully characterises an *active face* of the constraint polyhedron, which is given by the intersection with the constraint polyhedron of the hyperplane defined by the equality constraint

$$\tilde{\Phi}_\ell \tilde{\mathbf{u}} = \Delta - \Lambda_\ell x. \tag{6.47}$$

For example, in Figure 6.7, (6.47) may represent the plane that contains face f_1, in which case one constraint is active, or the line that contains e_1, in which case two constraints are active.

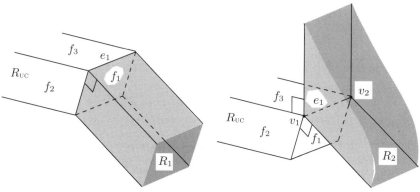

(a) Active region associated with a face. (b) Active region associated with an edge.

Figure 6.7. Illustration of active regions associated with a face and an edge of the constraint polyhedron.

To each active face, we will then associate an *active region* of $\mathbb{R}^{Nm}$ defined as the set of all points $v \in \mathbb{R}^{Nm}$ for which the point of the constraint polyhedron that is closest to v, in Euclidean distance, belongs to the corresponding active face. For example, region R_1 in Figure 6.7 (a) is an active region associated with face f_1 and region R_2 in Figure 6.7 (b) is an active region associated with edge e_1. In these regions, the optimal solution $\tilde{\mathbf{u}}^{\text{OPT}}(x)$ is simply given by the point on the corresponding active face of the constraint polyhedron that is closest, in the Euclidean distance, to the unconstrained solution $\tilde{\mathbf{u}}_{\text{UC}}^{\text{OPT}}(x) = -H^{-1/2}Fx$ (see, for example, Figure 6.4).

The following lemma characterises, for an arbitrary active pair (ℓ, Δ), the corresponding active region in $\mathbb{R}^{Nm}$.

Lemma 6.5.1 (Active Regions) *Suppose that the gradient of active constraints $\tilde{\Phi}_\ell$ has full row rank. Then the active region corresponding to the face characterised by the equality constraint (6.47) is given by*

$$S\left\{[\tilde{\Phi}_\ell\tilde{\Phi}_\ell^{\mathrm{T}}]^{-1}[\tilde{\Phi}_\ell\tilde{\mathbf{u}} + \Lambda_\ell x - \Delta]\right\} \leq 0, \qquad (6.48)$$

$$-\underline{\Delta}_s \leq \tilde{\Phi}_s\tilde{\mathbf{u}} + \Lambda_s x - \tilde{\Phi}_s\tilde{\Phi}_\ell^{\mathrm{T}}[\tilde{\Phi}_\ell\tilde{\Phi}_\ell^{\mathrm{T}}]^{-1}[\tilde{\Phi}_\ell\tilde{\mathbf{u}} + \Lambda_\ell x - \Delta] \leq \bar{\Delta}_s, \qquad (6.49)$$

where Δ is defined in (6.46) and S is a sign diagonal matrix such that its (k,k)-entry is $S_{kk} = 1$ if $\Delta_k = -\underline{\Delta}_{\ell_k}$ and $S_{kk} = -1$ if $\Delta_k = \bar{\Delta}_{\ell_k}$.

Proof. Geometrically, the active region corresponding to the face characterised by the active constraints (6.47) is delimited by[1]:

- *Each hyperplane that contains the corresponding active face (6.47) and is normal to one of the faces that share with the active face all but one of its active constraints.* (For example, in Figure 6.7 (b), these hyperplanes are the two planes delimiting R_2 that are normal to faces f_1 and f_3 and contain edge e_1.)

  We will show that these hyperplanes are given by the equality in each row of (6.48). First, it is easy to see that these equalities contain (6.47) simply by substitution of (6.47) in (6.48). Next, note that each face that shares with (6.47) all the active constraints except constraint ℓ_k is characterised by the equality constraint

$$\tilde{\Phi}_{\ell-\ell_k}\tilde{\mathbf{u}} = \Delta' - \Lambda_{\ell-\ell_k}x, \qquad (6.50)$$

  where Δ' is formed from Δ by eliminating its kth element. We now rewrite the equality in (6.48) as

$$\Psi\tilde{\mathbf{u}} = [\tilde{\Phi}_\ell\tilde{\Phi}_\ell^{\mathrm{T}}]^{-1}[-\Lambda_\ell x + \Delta] \qquad (6.51)$$

  where $\Psi \triangleq [\tilde{\Phi}_\ell\tilde{\Phi}_\ell^{\mathrm{T}}]^{-1}\tilde{\Phi}_\ell$. Since $\Psi\tilde{\Phi}_\ell^{\mathrm{T}} = I$, it follows that $\Psi_k\tilde{\Phi}_{\ell-\ell_k} = 0_{1\times(\bar{N}-1)}$. Hence, the hyperplane defined by the kth row of (6.51) is normal to the face (6.50), as claimed. Note that this holds for $k = 1, \ldots, \bar{N}$, which covers all rows of (6.48).

- *Each hyperplane that is normal to the corresponding active face (6.47) and contains one of the faces that share with the active face all its active constraints and has one more active constraint.* (For example, in Figure 6.7 (b), these hyperplanes are the two parallel planes delimiting R_2 that are normal to e_1 and contain vertices v_1 and v_2.)

  We will show that these hyperplanes are given by the equalities in each row of (6.49). The matrix multiplying $\tilde{\mathbf{u}}$ in (6.49), $\tilde{\Phi}_s[I - \tilde{\Phi}_\ell^{\mathrm{T}}[\tilde{\Phi}_\ell\tilde{\Phi}_\ell^{\mathrm{T}}]^{-1}\tilde{\Phi}_\ell]$, satisfies

$$\tilde{\Phi}_s[I - \tilde{\Phi}_\ell^{\mathrm{T}}[\tilde{\Phi}_\ell\tilde{\Phi}_\ell^{\mathrm{T}}]^{-1}\tilde{\Phi}_\ell]\tilde{\Phi}_\ell^{\mathrm{T}} = \tilde{\Phi}_s[\tilde{\Phi}_\ell^{\mathrm{T}} - \tilde{\Phi}_\ell^{\mathrm{T}}I] = 0, \qquad (6.52)$$

  and hence each equality in (6.49) is normal to the active face (6.47). Next, note that each face that shares with the active face (6.47) all its active

---

[1] See, for example, region R_2 in Figure 6.7 (b), corresponding to the edge e_1.

constraints and has one more active constraint, s_k, say, satisfies both (6.47) and an additional constraint of the form

$$\tilde{\Phi}_{s_k}\tilde{\mathbf{u}} = \bar{\Delta}_{s_k} - \Lambda_{s_k}x, \tag{6.53}$$

or of the same form with $-\underline{\Delta}_{s_k}$ replacing $\bar{\Delta}_{s_k}$. Then, substituting (6.47) and (6.53) in (6.49) it is easy to see that the kth row satisfies the right equality constraint. Thus, the hyperplane defined by the kth row of the right equality in (6.49) contains the face that is characterised by the equality constraints (6.47) and (6.53). A similar analysis can be done for the left equality constraint. Note that this holds for $k = 1, \ldots, q - \bar{N}$, which covers all rows of (6.49). The result then follows.

$\square$

For each active region characterised in Lemma 6.5.1 we can then compute the optimal solution $\tilde{\mathbf{u}}^{\mathrm{OPT}}(x)$ of the QP (6.41) as the point on the corresponding active face of the constraint polyhedron that is closest, in the Euclidean distance, to the unconstrained solution $\tilde{\mathbf{u}}_{\mathrm{UC}}^{\mathrm{OPT}}(x) = -H^{-1/2}Fx$. Also, to each active region characterised in Lemma 6.5.1 the transformation (6.26) assigns a corresponding region in the state space. The following theorem summarises this procedure by characterising for an arbitrary active pair (ℓ, Δ) the corresponding active region in the state space and the corresponding optimal solution of the QP (6.41) with constraints given by (6.42)–(6.43).

Theorem 6.5.2 (QP Solution in an Active Region) *Under the conditions of Lemma 6.5.1, the projection X_ℓ onto the state space of the active region defined by (6.48)–(6.49) is given by*

$$S\left\{[\tilde{\Phi}_\ell\tilde{\Phi}_\ell^{\mathrm{T}}]^{-1}[(-\tilde{\Phi}_\ell H^{-1/2}F + \Lambda_\ell)x - \Delta]\right\} \leq 0,$$
$$-\tilde{\Phi}_s H^{-1/2}Fx + \Lambda_s x - \tilde{\Phi}_s\tilde{\Phi}_\ell^{\mathrm{T}}[\tilde{\Phi}_\ell\tilde{\Phi}_\ell^{\mathrm{T}}]^{-1}[-\tilde{\Phi}_\ell H^{-1/2}Fx + \Lambda_\ell x - \Delta] \leq \bar{\Delta}_s,$$
$$\tilde{\Phi}_s H^{-1/2}Fx - \Lambda_s x + \tilde{\Phi}_s\tilde{\Phi}_\ell^{\mathrm{T}}[\tilde{\Phi}_\ell\tilde{\Phi}_\ell^{\mathrm{T}}]^{-1}[-\tilde{\Phi}_\ell H^{-1/2}Fx + \Lambda_\ell x - \Delta] \leq \underline{\Delta}_s. \tag{6.54}$$

Moreover, if $x \in X_\ell$, the optimal constrained control $\mathbf{u}^{\mathrm{OPT}}(x)$ in (6.41) is given by

$$\mathbf{u}^{\mathrm{OPT}}(x) = H^{-1/2}\tilde{\Phi}_\ell^{\mathrm{T}}[\tilde{\Phi}_\ell\tilde{\Phi}_\ell^{\mathrm{T}}]^{-1}(\Delta - \Lambda_\ell x) - H^{-1/2}[I - \tilde{\Phi}_\ell^{\mathrm{T}}[\tilde{\Phi}_\ell\tilde{\Phi}_\ell^{\mathrm{T}}]^{-1}\tilde{\Phi}_\ell]H^{-1/2}Fx. \tag{6.55}$$

Proof. Equations (6.54) follow immediately upon substitution of (6.26) into (6.48) and (6.49). We now show that the optimal control inside each region (6.54) has the form (6.55). Indeed, the optimal constrained control in each of the active regions is obtained by intersecting the active face (6.47) with the hyperplane normal to it and passing through the unconstrained solution $\tilde{\mathbf{u}}_{\mathrm{UC}}^{\mathrm{OPT}}(x)$. That is, $\tilde{\mathbf{u}}^{\mathrm{OPT}}(x)$ satisfies both (6.47) and the following equation (see (6.52)):

$$[I - \tilde{\Phi}_\ell^{\mathrm{T}}[\tilde{\Phi}_\ell \tilde{\Phi}_\ell^{\mathrm{T}}]^{-1}\tilde{\Phi}_\ell]\tilde{\mathbf{u}}^{\mathrm{OPT}}(x) = [I - \tilde{\Phi}_\ell^{\mathrm{T}}[\tilde{\Phi}_\ell \tilde{\Phi}_\ell^{\mathrm{T}}]^{-1}\tilde{\Phi}_\ell]\tilde{\mathbf{u}}_{\mathrm{UC}}^{\mathrm{OPT}}(x).$$

Substituting (6.47) and $\tilde{\mathbf{u}}_{\mathrm{UC}}^{\mathrm{OPT}}(x) = -H^{-1/2}Fx$ from (6.25) into the above equation, and solving for $\tilde{\mathbf{u}}^{\mathrm{OPT}}(x)$, yields

$$\tilde{\mathbf{u}}^{\mathrm{OPT}}(x) = \tilde{\Phi}_\ell^{\mathrm{T}}[\tilde{\Phi}_\ell \tilde{\Phi}_\ell^{\mathrm{T}}]^{-1}(\Delta - \Lambda_\ell x) - [I - \tilde{\Phi}_\ell^{\mathrm{T}}[\tilde{\Phi}_\ell \tilde{\Phi}_\ell^{\mathrm{T}}]^{-1}\tilde{\Phi}_\ell]H^{-1/2}Fx. \quad (6.56)$$

Equation (6.55) follows using the transformation (6.24). The theorem is then proved.   □

Theorem 6.5.2 gives the solution of the QP (6.41) if x belongs to the region X_ℓ defined by (6.54). The RHC law $\mathcal{K}_N(x) = u_0^{\mathrm{OPT}}(x)$ is then simply obtained by selecting the first m elements of $\mathbf{u}^{\mathrm{OPT}}(x)$ in (6.55), that is,

$$\mathcal{K}_N(x) = u_0^{\mathrm{OPT}}(x) = \begin{bmatrix} I & 0 & \cdots & 0 \end{bmatrix} \mathbf{u}^{\mathrm{OPT}}(x). \quad (6.57)$$

To obtain the complete solution in the state space, one would require a procedure to enumerate all possible combinations of active constraints and compute the corresponding region and optimal control for each combination using Theorem 6.5.2. An algorithm that implements such a procedure is described in Seron, Goodwin and De Doná (2003). We observe that, if the state space dimension n is $n < Nm$, then the image of the transformation $\mathbf{u}_{\mathrm{UC}}^{\mathrm{OPT}}(x) = -H^{-1}Fx$ is a lower dimensional subspace of $\mathbb{R}^{Nm}$, and so some of the regions X_ℓ in (6.54) will be empty. Hence, the partition has to be post-processed to eliminate redundant inequalities and empty regions. If $n \geq Nm$, then the computation of the region partition using Theorem 6.5.2 combined with the enumeration of all possible combinations of active constraints directly gives the complete state space partition with no need for further processing.

We illustrate the procedure with a numerical example.

Example 6.5.1. Consider a system of the form (6.36) with matrices

$$A = \begin{bmatrix} 0.8955 & -0.1897 \\ 0.0948 & 0.9903 \end{bmatrix}, \qquad B = \begin{bmatrix} 0.0948 \\ 0.0048 \end{bmatrix}.$$

In the objective function (6.40) we take $N = 4$, $Q = \begin{bmatrix} 0 & 0 \\ 0 & 2 \end{bmatrix}$ and $R = 0.01$. The terminal state weighting matrix P is chosen as the solution of the algebraic Riccati equation $P = A^{\mathrm{T}}PA + Q - K^{\mathrm{T}}\bar{R}K$, where $K \triangleq \bar{R}^{-1}B^{\mathrm{T}}PA$ and $\bar{R} \triangleq R + B^{\mathrm{T}}PB$. We consider constraints of the form (6.42), with $m = 1$, $N = 4$, $\Phi = I$, $\Lambda = 0_{Nm \times n}$, and $\Delta = \begin{bmatrix} 2 & 2 & 2 & 2 \end{bmatrix}^{\mathrm{T}}$.

The state space partition for this case, computed using Theorem 6.5.2 and the enumeration algorithm of Seron et al. (2003), is shown in Figure 6.8 (a). A "zoom" of this partition is shown in Figure 6.8 (b) to display the smaller regions in more detail. The region denoted by X_0 is the projection onto the state space of the constraint polyhedron; in regions X_2, X_3 and X_4 only one constraint is active; in regions X_5 and X_6 two constraints are active; in region

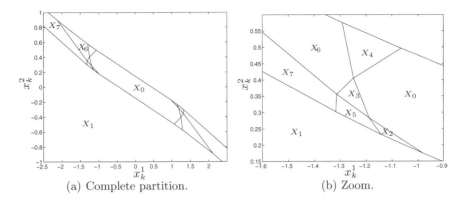

Figure 6.8. State space partition for Example 6.5.1 for $N = 4$.

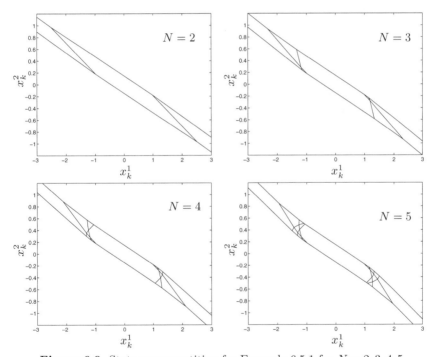

Figure 6.9. State space partition for Example 6.5.1 for $N = 2, 3, 4, 5$.

X_7 three constraints are active; finally, X_1 is the union of all regions where the control is saturated to the value -2.

The resulting RHC law (6.57) is

$$\mathcal{K}_4(x) = G_i x + h_i, \quad \text{if } x \in X_i, \quad i = 0, \dots, 7, \tag{6.58}$$

where

$$G_0 = -[4.4650 \quad 13.5974], \qquad h_0 = 0,$$
$$G_1 = [\quad 0 \qquad 0 \quad], \qquad h_1 = -2,$$
$$G_2 = -[5.6901 \quad 15.9529], \qquad h_2 = -0.7894,$$
$$G_3 = -[4.9226 \quad 13.8202], \qquad h_3 = -0.4811,$$
$$G_4 = -[4.5946 \quad 13.3346], \qquad h_4 = -0.2684,$$
$$G_5 = -[6.6778 \quad 16.8644], \qquad h_5 = -1.7057,$$
$$G_6 = -[5.1778 \quad 13.4855], \qquad h_6 = -0.9355,$$
$$G_7 = -[7.4034 \quad 16.8111], \qquad h_7 = -2.6783.$$

Similar expressions hold in the remaining unlabelled regions. These can be obtained by symmetry.

To see how the partitions are affected by the constraint horizon, we take, successively, $N = 2$, $N = 3$, $N = 4$ and $N = 5$ in the objective function (6.40). The state space partitions corresponding to each value of N are shown in Figure 6.9.

We next consider an initial condition $x_0 = [-1.2 \quad 0.53]^{\mathrm{T}}$ and simulate the system under the RHC (6.58). Figure 6.10 shows the resulting state space trajectory. The trajectory starts in region X_4 and moves, successively, into regions X_6, X_5, X_1, X_1, X_0, and stays in X_0 thereafter. Table 6.1 shows the trajectory points x_k for $k = 0, \ldots, 6$, the regions X_i such that $x_k \in X_i$, and the corresponding RHC controls computed using (6.58).   ∘

k	x_k	Region X_i	RHC control u_k
0	$[-1.2000 \quad 0.5300]^{\mathrm{T}}$	X_4	$-[4.5946 \quad 13.3346]x_k - 0.2684$
1	$[-1.3480 \quad 0.4023]^{\mathrm{T}}$	X_6	$-[5.1778 \quad 13.4855]x_k - 0.9355$
2	$[-1.2247 \quad 0.2735]^{\mathrm{T}}$	X_5	$-[6.6778 \quad 16.8644]x_k - 1.7057$
3	$[-0.9722 \quad 0.1637]^{\mathrm{T}}$	X_1	$[0 \quad 0]x_k - 2.0000$
4	$[-0.7120 \quad 0.0796]^{\mathrm{T}}$	X_1	$[0 \quad 0]x_k - 2.0000$
5	$[-0.4630 \quad 0.0209]^{\mathrm{T}}$	X_0	$-[4.4650 \quad 13.5974]x_k$
6	$-[0.2495 \quad 0.0146]^{\mathrm{T}}$	X_0	$-[4.4650 \quad 13.5974]x_k$
⋮	⋮	⋮	⋮

Table 6.1. Example 6.5.1: Trajectory x_k, $k = 0, \ldots, 6$, regions X_i such that $x_k \in X_i$, and corresponding RHC controls (6.58).

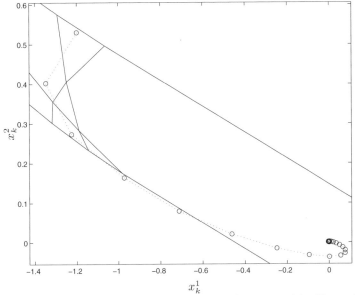

Figure 6.10. State space trajectory for Example 6.5.1 with $N = 4$, $x_0 = [-1.2 \ 0.53]^{\mathrm{T}}$.

6.6 Characterisation Using the KKT Conditions

The parameterisation given in Theorem 6.5.2 can be related to the KKT optimality conditions studied in Chapter 2. Consider again the QP (6.41) with constraints given by (6.42)–(6.43). For convenience, we will work in the $\tilde{u}$-coordinates given by the transformation (6.24), that is, we consider the following QP with inequality constraints:

$$\text{minimise } \frac{1}{2}\tilde{\mathbf{u}}^{\mathrm{T}}\tilde{\mathbf{u}} + \tilde{\mathbf{u}}^{\mathrm{T}}H^{-1/2}Fx,$$

subject to: (6.59)

$$\tilde{\Phi}\tilde{\mathbf{u}} \leq \bar{\Delta} - \Lambda x,$$

$$-\tilde{\Phi}\tilde{\mathbf{u}} \leq \underline{\Delta} + \Lambda x.$$

When solved for different values of x, the above problem is sometimes referred to as a *multiparametric quadratic program* [mp-QP], that is, a QP in which the linear term in the objective function and the right hand side of the constraints depend linearly on a vector of parameters (the state vector x in this case).

Let $\tilde{\mathbf{u}}^{\mathrm{OPT}}(x)$ be the optimal solution of (6.59). As discussed in Section 2.5.6 of Chapter 2, the KKT optimality conditions are both necessary and sufficient conditions for this problem. Using (2.41) of Chapter 2, we can write the KKT conditions for the above problem as

$$-\underline{\Delta} \le \tilde{\Phi}\tilde{\mathbf{u}}^{\text{OPT}}(x) + \Lambda x \le \bar{\Delta}, \tag{6.60}$$

$$\tilde{\mathbf{u}}^{\text{OPT}}(x) + H^{-1/2}Fx + \tilde{\Phi}^{\text{T}}(\mu - \sigma) = 0, \tag{6.61}$$

$$\mu \ge 0, \qquad \sigma \ge 0, \tag{6.62}$$

$$\mu^{\text{T}}[\tilde{\Phi}\tilde{\mathbf{u}}^{\text{OPT}}(x) - \bar{\Delta} + \Lambda x] = 0, \tag{6.63}$$

$$\sigma^{\text{T}}[\tilde{\Phi}\tilde{\mathbf{u}}^{\text{OPT}}(x) + \underline{\Delta} + \Lambda x] = 0, \tag{6.64}$$

where $\mu \in \mathbb{R}^q$ and $\sigma \in \mathbb{R}^q$ are vectors of Lagrange multipliers corresponding to the two sets of inequality constraints in (6.59).

Consider now an active pair (ℓ, Δ) for $\tilde{\mathbf{u}}^{\text{OPT}}(x)$, where Δ satisfies (6.46), and let s be the corresponding inactive set. We then have that $\tilde{\mathbf{u}}^{\text{OPT}}(x)$ satisfies the equality constraints (6.47), that is,

$$\tilde{\Phi}_\ell \tilde{\mathbf{u}}^{\text{OPT}}(x) = \Delta - \Lambda_\ell x. \tag{6.65}$$

From the above equation, the complementary slackness conditions (6.63) and (6.64), the (second) dual feasibility condition (6.62), and equation (6.46), we have that

$$\mu_s = 0, \qquad \sigma_s = 0, \tag{6.66}$$

$$\mu_{\ell_k} - \sigma_{\ell_k} \begin{cases} \ge 0 & \text{if } \Delta_k = \bar{\Delta}_{\ell_k}, \\ \le 0 & \text{if } \Delta_k = -\underline{\Delta}_{\ell_k}. \end{cases} \tag{6.67}$$

Using (6.66) in the (first) dual feasibility condition (6.61) and solving for $\tilde{\mathbf{u}}^{\text{OPT}}(x)$ yields

$$\tilde{\mathbf{u}}^{\text{OPT}}(x) = -H^{-1/2}Fx - \tilde{\Phi}_\ell^{\text{T}}[\mu_\ell - \sigma_\ell]. \tag{6.68}$$

Using (6.68) in the active constraint equality (6.65) and solving for $[\mu_\ell - \sigma_\ell]$ gives

$$[\mu_\ell - \sigma_\ell] = [\tilde{\Phi}_\ell \tilde{\Phi}_\ell^{\text{T}}]^{-1}[-\tilde{\Phi}_\ell H^{-1/2}Fx + \Lambda_\ell x - \Delta]. \tag{6.69}$$

Substituting the above equation into (6.68) we obtain

$$\tilde{\mathbf{u}}^{\text{OPT}}(x) = \tilde{\Phi}_\ell^{\text{T}}[\tilde{\Phi}_\ell \tilde{\Phi}_\ell^{\text{T}}]^{-1}(\Delta - \Lambda_\ell x) - [I - \tilde{\Phi}_\ell^{\text{T}}[\tilde{\Phi}_\ell \tilde{\Phi}_\ell^{\text{T}}]^{-1}\tilde{\Phi}_\ell]H^{-1/2}Fx, \tag{6.70}$$

which is identical to (6.56). We then recover the expression (6.55) for the optimal solution using the transformation (6.24).

The inequalities (6.54) that define the region in the state space where the optimal solution (6.55) is valid can be recovered in the following way: Combining the expression for the difference of Lagrange multipliers (6.69) and the sign condition (6.67) yields the first set of inequalities in (6.54) (recall that S is a sign diagonal matrix such that $S_{kk} = 1$ if $\Delta_k = -\underline{\Delta}_{\ell_k}$ and $S_{kk} = -1$ if $\Delta_k = \bar{\Delta}_{\ell_k}$). Finally, the second and third sets of inequalities in (6.54) follow from primal feasibility (see (6.60)) of the inactive constraints, that is,

$$-\underline{\Delta}_s \le \tilde{\Phi}_s \tilde{\mathbf{u}}^{\text{OPT}}(x) + \Lambda_s x \le \bar{\Delta}_s,$$

upon substitution of the expression (6.70) for the optimal solution.

In summary, we can see that the characterisation of Theorem 6.5.2 is a particular arrangement of the KKT optimality conditions.

6.7 Further Reading

For complete list of references cited, see References section at the end of book.

General

The global characterisation presented here is based on the work of Bemporad et al. (2002) and Seron et al. (2003).

Section 6.5

A related development to the one presented in Section 6.5 is described in Seron et al. (2003).

Section 6.6

The approach described in Section 6.6 using KKT conditions was used in Bemporad et al. (2002) to obtain a local characterisation of the QP solution in an active region of the state space around a specific solution $\tilde{\mathbf{u}}^{\mathrm{OPT}}(x)$. In that work, once a nonempty active region has been defined, the rest of the state space is explored in the search for new active regions.

In the above context, an algorithm to explore and partition the rest of the state space in polyhedral regions was originally proposed in Dua and Pistikopoulos (2000). This algorithm recursively reverses the inequalities that define each active region to obtain a new region partition. In each new region of the resulting partition a feasible point x is found (possibly by solving a linear program). Then, a QP of the form (6.59) is solved for that value of x to determine the active constraints that characterise the active region using the KKT conditions as described above. One drawback of this algorithm is that it can split active regions since the regions to be explored are not necessarily related to the active regions.

A more efficient algorithm to accomplish the state space partitioning was proposed by Tøndel, Johansen and Bemporad (2002). The main contribution of this algorithm is that the active constraints in the region of interest can be determined from the active constraints in a neighbouring region by examining the separating hyperplane between these regions. Thus, QPs do not need to be solved to determine the active set of constraints in each region and, moreover, unnecessary partitioning is avoided.

7

Regional Characterisation of Constrained Linear Quadratic Optimal Control

7.1 Overview

In Chapter 6 we provided a global characterisation of receding horizon constrained optimal control. This gives practically valuable insights into the form of the control law. Indeed, for many problems it is feasible to compute, store and retrieve the function $u = \mathcal{K}_N(x)$, thus eliminating the need to solve the associated QP on line.

In other cases, this approach may be too complex. Thus, in Chapter 8 we will explore various numerical algorithms aimed at solving the QP on line. The current chapter addresses a question with similar motivation but a different end result. Here we ask the following question: Given that we only ever apply the *first* move from the optimal control sequence of length N, is it possible that the first element of the control law might not change as the horizon increases beyond some modest size at least locally in the state space? We will show, via dynamic programming arguments, that this is indeed the case, at least for special classes of problems. To illustrate the ideas we will consider single input systems with an amplitude input constraint. This class of problems is simple and is intended to motivate the idea of local solutions. In particular, we will consider the simple control law

$$u_k = -\text{sat}_\Delta(Kx_k), \qquad (7.1)$$

where $\text{sat}_\Delta(\,\cdot\,)$ is the saturation function defined in (6.10) of Chapter 6, and K is the feedback gain resulting from the *infinite horizon unconstrained* optimal control problem. We will show that there exists a nontrivial region of the state space (which we denote by $\bar{\mathbb{Z}}$) such that (7.1) is the *constrained* optimal control law with arbitrary large horizon. This is a very interesting result which has important practical implications. For example, (7.1) can be thought of as a simple type of anti-windup control law (see Section 7.4) when used in a certainty equivalence form with an appropriate observer for the system state and disturbances. Thus, the result establishes a link between anti-windup and RHC. Also, the result explains the (local) success of this control law in Example 1.2.1 of Chapter 1.

We will see that the characterisation of the region $\bar{\mathbb{Z}}$ is relatively complicated for large horizons. Hence, we will first, in Section 7.2, present the result for horizon $N = 2$. We will then establish the result for general horizons.

7.2 Regional Characterisation for Horizon 2

We consider again single input, linear, discrete time systems in which the magnitude of the control input is constrained to be less than or equal to a positive constant. In particular, let the system be given by

$$x_{k+1} = Ax_k + Bu_k, \tag{7.2}$$

where $x_k \in \mathbb{R}^n$ and $u_k \in \mathbb{R}$. As in Section 6.2 of Chapter 6, we consider the following fixed horizon optimal control problem with horizon 2:

$$\mathcal{P}_2(x) : \quad V_2^{\text{OPT}}(x) \triangleq \min V_2(\{x_k\}, \{u_k\}), \tag{7.3}$$

subject to:

$$x_{k+1} = Ax_k + Bu_k \quad \text{for } k = 0, 1,$$

$$x_0 = x,$$

$$u_k \in \mathbb{U} \triangleq [-\Delta, \Delta] \quad \text{for } k = 0, 1,$$

where $\Delta > 0$ is the input constraint level, and the objective function in (7.3) is

$$V_2(\{x_k\}, \{u_k\}) \triangleq \frac{1}{2} x_2^{\text{T}} P x_2 + \frac{1}{2} \sum_{k=0}^{1} (x_k^{\text{T}} Q x_k + u_k^{\text{T}} R u_k). \tag{7.4}$$

The matrices Q and R in (7.4) are positive definite and P satisfies the algebraic Riccati equation

$$P = A^{\text{T}} P A + Q - K^{\text{T}} \bar{R} K, \tag{7.5}$$

where

$$K \triangleq \bar{R}^{-1} B^{\text{T}} P A, \qquad \bar{R} \triangleq R + B^{\text{T}} P B. \tag{7.6}$$

Let the control sequence that achieves the minimum in (7.3) be

$$\{u_0^{\text{OPT}}, u_1^{\text{OPT}}\}. \tag{7.7}$$

Then the RHC law is given by the first element of (7.7) (which depends on the current state $x = x_0$); that is,

$$\mathcal{K}_2(x) = u_0^{\text{OPT}}. \tag{7.8}$$

For the above special case, we have the following regional characterisation of the fixed horizon optimal control (7.7).

Lemma 7.2.1 (Regional Characterisation of the Fixed Horizon Control (7.7)) *Consider the fixed horizon optimal control problem $\mathcal{P}_2$ defined in (7.3)–(7.6). Then for all $x \in \mathbb{Z}$, where*

$$\mathbb{Z} \triangleq \{x : |K(A - BK)x| \leq \Delta\}, \tag{7.9}$$

the optimal control sequence (7.7) that attains the minimum is

$$u_k^{\mathrm{OPT}} = -\mathrm{sat}_\Delta(Kx_k), \quad k = 0, 1. \tag{7.10}$$

Proof. The result follows from the proof of Theorem 6.2.1 of Chapter 6, in particular, from equations (6.17) and (6.19).   □

The above result is quite remarkable in that it shows that the simple policy (7.10) provides a solution to the constrained linear quadratic fixed horizon optimal control problem (7.3)–(7.6) in a region of the state space. Using induction and similar dynamic programming arguments to the proof of Theorem 6.2.1, it is possible to show that a characterisation of the form (7.10) holds for horizon N of *arbitrary length*. We will establish this result in Section 7.3. As we will see, in the case of arbitrary horizon the characterisation is valid in a region $\mathbb{Z}$ of the state space having a more complex description than the one used in Lemma 7.2.1.

In the sequel, we will explore various aspects of the solution provided by Lemma 7.2.1, including a refinement of the set in which the result holds.

7.2.1 Regional Characterisation of RHC

We have seen above that the simple control law (7.10) solves the fixed horizon constrained linear quadratic problem in a special region of the state space. However, before we can use this control law as the solution to the associated RHC problem (see (7.8)), we need to extend the results to the receding horizon formulation of the problem. In particular, in order to guarantee that the RHC mapping (7.8) is regionally given by (7.10), it is essential to know if future states remain in the region in which the result holds or whether they are driven outside this region. Clearly, in the former case, we can implement the RHC algorithm as in (7.10) without further consideration. We thus proceed to examine the conditions under which the state remains in the region $\mathbb{Z}$ where (7.10) applies. We first define the mapping $\phi_{nl} : \mathbb{R}^n \to \mathbb{R}^n$ as

$$\phi_{nl}(x) \triangleq Ax - B\mathrm{sat}_\Delta(Kx), \tag{7.11}$$

so that when the controller (7.10) is employed, the closed loop system satisfies $x_{k+1} = \phi_{nl}(x_k)$. In the sequel we denote by ϕ_{nl}^k the concatenation of ϕ_{nl} with itself k times; for example, $\phi_{nl}^0(x) \triangleq x$, $\phi_{nl}^1(x) = \phi_{nl}(x)$, $\phi_{nl}^2(x) = \phi_{nl}(\phi_{nl}(x))$, and so on.

We also require the following definition.

Definition 7.2.1 *Define the set* $\bar{\mathbb{Z}}$ *as*

$$\bar{\mathbb{Z}} \triangleq \left\{ x : \phi_{nl}^k(x) \in \mathbb{Z}, \ k = 0, 1, 2, \ldots \right\}. \tag{7.12}$$

○

From its definition it is clear that $\bar{\mathbb{Z}}$ is the *maximal positively invari-ant set* contained in $\mathbb{Z}$ for the closed loop system $x_{k+1} = \phi_{nl}(x_k) = Ax_k - B\mathrm{sat}_\Delta(Kx_k)$ (see Definition 4.4.2 in Chapter 4).

We have from Lemma 7.2.1 that, if the initial state $x_0 = x \in \mathbb{Z}$, then $\mathcal{K}_2(x_0) = -\mathrm{sat}_\Delta(Kx_0)$. But, in general, since $\mathbb{Z}$ is not necessarily positively invariant under $\phi_{nl}(\cdot)$, after this control is applied there is no guarantee that the successor state $x_1 = \phi_{nl}(x_0)$ will stay in $\mathbb{Z}$. Hence, in general, $\mathcal{K}_2(x_1) \neq -\mathrm{sat}_\Delta(Kx_1)$. So, in order that the solution (7.10) can be applied to the RHC problem we must ensure that all successor states belong to $\mathbb{Z}$. We can then state:

Theorem 7.2.2 *For all* $x \in \bar{\mathbb{Z}}$ *defined in* (7.12), *the RHC law* $\mathcal{K}_2$ *in* (7.8) *is given by*

$$\mathcal{K}_2(x) = -\mathrm{sat}_\Delta(Kx). \tag{7.13}$$

Proof. The proof of the theorem follows from the fact that $\bar{\mathbb{Z}} \subseteq \mathbb{Z}$ and that $\bar{\mathbb{Z}}$ is positively invariant for the system (7.2) under the control $\mathcal{K}_2(x) = -\mathrm{sat}_\Delta(Kx)$. Then, for all states in $\bar{\mathbb{Z}}$ the future trajectories of the system will be such that $x \in \bar{\mathbb{Z}} \subseteq \mathbb{Z}$, and from Lemma 7.2.1 we conclude that $\mathcal{K}_2(x) = u_0^{\mathrm{OPT}} = -\mathrm{sat}_\Delta(Kx)$. $\qquad\Box$

Notice that, if the set $\bar{\mathbb{Z}}$, in which the theorem is valid, were small enough such that the control sequence $\{u_k\} = \{-\mathrm{sat}_\Delta(Kx_k)\}$ stayed unsaturated along the system trajectories, then the result of Theorem 7.2.2 would be triv-ial, since it would readily follow from the result for the unconstrained case (see, for example, Anderson and Moore 1989). We will show next that $\bar{\mathbb{Z}}$ is not smaller than this trivial case.

Consider the *maximal output admissible set* $\mathcal{O}_\infty$ (introduced in (5.63) of Chapter 5), which in this case is defined as

$$\mathcal{O}_\infty \triangleq \left\{ x : |K(A - BK)^i x| \leq \Delta \text{ for } i = 0, 1, \ldots \right\}. \tag{7.14}$$

The following proposition shows that the set $\bar{\mathbb{Z}}$ contains $\mathcal{O}_\infty$.

Proposition 7.2.3 $\mathcal{O}_\infty \subseteq \bar{\mathbb{Z}}$.

Proof. $\bar{\mathbb{Z}}$ is the maximal positively invariant set in $\mathbb{Z}$ for the closed loop system $x_{k+1} = \phi_{nl}(x_k)$. The set $\mathcal{O}_\infty$ is also a positively invariant set for $x_{k+1} = \phi_{nl}(x_k)$ (since $\phi_{nl}(x) = (A - BK)x$ in $\mathcal{O}_\infty$). It suffices, therefore, to establish that $\mathcal{O}_\infty \subseteq \mathbb{Z}$. This is indeed true since, from (7.14), we can write

$$\mathcal{O}_\infty = \{x : |Kx| \leq \Delta\} \cap \{x : |K(A - BK)x| \leq \Delta\} \cap$$
$$\{x : |K(A - BK)^2 x| \leq \Delta\} \cap \dots$$
$$= \{x : |Kx| \leq \Delta\} \cap \mathbb{Z} \cap \{x : |K(A - BK)^2 x| \leq \Delta\} \cap \dots$$

Hence $\mathcal{O}_\infty \subseteq \mathbb{Z}$ and the result then follows.   □

We have proved that the set $\bar{\mathbb{Z}}$ contains the maximal positively invariant set in which the control constraints are avoided. Although a complete characterisation of the set $\bar{\mathbb{Z}}$ is not currently known, examples (see Example 7.3.2 at the end of the chapter) show that, in general, the set $\bar{\mathbb{Z}}$ is considerably larger than the set $\mathcal{O}_\infty$. In other words, that the motions of the system $x_{k+1} = \phi_{nl}(x_k)$ involve control sequences $\{u_k\} = \{-\text{sat}_\Delta(Kx_k)\}$ which remain saturated for several steps and, in accordance with Theorem 7.2.2, coincide with the solution provided by the RHC strategy (see the simulation of Example 7.3.2).

7.2.2 An Ellipsoidal Approximation to the Set $\bar{\mathbb{Z}}$

The set $\bar{\mathbb{Z}}$ is, in general, very difficult to characterise explicitly since it involves nonlinear inequalities. Notice however that, for any positively invariant set contained in $\bar{\mathbb{Z}}$, the result of Theorem 7.2.2 is also valid. In principle, a positively invariant inner approximation of the set $\bar{\mathbb{Z}}$ could be obtained by considering a *family* of positively invariant sets, which can be represented with reasonable complexity, and finding the biggest member within this family which is contained in $\mathbb{Z}$. The set $\mathbb{Z}$ is a polyhedral set, which suggests that polyhedral sets could be good candidates for this approximation, having also the advantage of flexibility. However, these sets could be arbitrarily complex (see, for example, Blanchini 1999).

In this section we will consider an alternative mechanism for obtaining positively invariant sets under the control $u = -\text{sat}_\Delta(Kx)$, based on the use of ellipsoidal sets. We will show how to construct an ellipsoidal invariant set $\mathcal{E} \subseteq \bar{\mathbb{Z}}$ based on a quadratic Lyapunov function constructed from the solution P of (7.5)–(7.6). In view of the discussion following Theorem 7.2.2, we will be interested in positively invariant ellipsoidal sets which extend to regions wherein the controls are saturated (that is, such that $|Kx| > \Delta$ for some x in the set). This result is related to the fact that a linear system, with optimal LQR controller, remains closed loop stable when sector-bounded input nonlinearities are introduced (see, for example, Anderson and Moore 1989). This, in turn, translates into bigger positively invariant ellipsoidal sets, under the control $u = -\text{sat}_\Delta(Kx)$, than the case $|Kx| \leq \Delta$ for all x.

In the sequel we will need the following result.

Lemma 7.2.4 *Let K be a nonzero row vector, P a symmetric positive definite matrix and ρ a positive constant. Then,*

$$\min\{Kx : x^T Px \leq \rho\} = -\sqrt{\rho K P^{-1} K^T}, \tag{7.15}$$

and

$$\max\{Kx : x^T Px \le \rho\} = +\sqrt{\rho K P^{-1} K^T}. \tag{7.16}$$

Proof. We first prove (7.15). The KKT conditions (2.32) in Chapter 2 are, in this case (convex objective function and convex constraint), sufficient for optimality. Now, let $\bar{x}$ be a (the) KKT point and denote by $\mu \ge 0$ its associated Lagrange multiplier. From the dual feasibility condition $K^T + \mu P\bar{x} = 0$ we can see that $\mu > 0$ (since $K^T \ne 0$) and $\bar{x} = -P^{-1}K^T/\mu$. Moreover, from the complementary slackness condition $\mu(\bar{x}^T P\bar{x} - \rho) = 0$ we have that the constraint is active at $\bar{x}$, that is $\bar{x}^T P\bar{x} = \rho$, from where we obtain that $\mu = \sqrt{KP^{-1}K^T/\rho}$. Thus, the minimum value is $K\bar{x} = -KP^{-1}K^T/\mu = -\sqrt{\rho K P^{-1} K^T}$, which proves (7.15). Finally, (7.16) follows from the fact that $\max\{Kx : x^T Px \le \rho\} = -\min\{-Kx : x^T Px \le \rho\}$. □

We define the ellipsoidal set

$$\mathcal{E} \triangleq \{x : x^T Px \le \rho\}.$$

From Lemma 7.2.4 we can see that, if the ellipsoidal radius ρ is computed from $\rho = (1 + \bar{\beta})^2 \Delta^2 / (KP^{-1}K^T)$, $\bar{\beta} \ge 0$, then the ellipsoid has the property: $|Kx| \le (1 + \bar{\beta})\Delta$ for all $x \in \mathcal{E}$.

Notice that whenever $\bar{\beta}$ is bigger than zero the ellipsoid extends to regions where saturation levels are reached. For this reason, $\bar{\beta}$ is called the *over-saturation index*. We compute the *maximum over-saturation index* $\bar{\beta}_{\max}$ from:

$$\bar{\beta}_{\max} \triangleq \begin{cases} \dfrac{\sqrt{KK^T\left[R(KK^T\bar{R} - q_\varepsilon) + \bar{R}q_\varepsilon\right]} + q_\varepsilon}{KK^T\bar{R} - q_\varepsilon} & \text{if} \quad \dfrac{q_\varepsilon}{KK^T\bar{R}} < 1, \\ M^+ & \text{otherwise,} \end{cases} \tag{7.17}$$

where $q_\varepsilon = (1 - \varepsilon)\lambda_{\min}(Q)$, $\varepsilon \in [0,1)$ is an arbitrarily small nonnegative number (introduced to ensure exponential stability; see the result in (7.23) below), $\lambda_{\min}(Q)$ is the minimum eigenvalue of the matrix Q (strictly positive, since Q is assumed positive definite), and M^+ is an arbitrarily large positive number.

Then, the *maximum radius* $\bar{\rho}_{\max}$ is computed from

$$\bar{\rho}_{\max} = \frac{(1 + \bar{\beta}_{\max})^2 \Delta^2}{KP^{-1}K^T}. \tag{7.18}$$

Theorem 7.2.5 *The ellipsoid* $\mathcal{E} = \{x : x^T Px \le \rho\}$, *with radius* $\rho < \bar{\rho}_{\max}$, *has the following properties:*

(i) $\mathcal{E}$ *is a positively invariant set for system* (7.2) *under the control* $u = -\text{sat}_\Delta(Kx)$.

(ii) *The origin is exponentially stable in* $\mathcal{E}$ *for system* (7.2) *with control* $u = -\text{sat}_\Delta(Kx)$ *(and, in particular,* $\mathcal{E}$ *is contained in the region of attraction of* (7.2) *for all admissible controls* $u \in \mathbb{U} = [-\Delta, \Delta]$*).*

Proof. Consider the quadratic Lyapunov function: $V(x) = x^{\mathrm{T}}Px$. Let $x = x_k$ and $x^+ = x_{k+1}$. Then, by using the system equation (7.2) with control $u = -\mathrm{sat}_\Delta(Kx)$, and the Riccati equation (7.5), we can express the increment of $V(\cdot)$ along the system trajectory as

$$\Delta V(x) \triangleq V(x^+) - V(x) = [Ax - B\mathrm{sat}_\Delta(Kx)]^{\mathrm{T}}P[Ax - B\mathrm{sat}_\Delta(Kx)] - x^{\mathrm{T}}Px$$
$$= -\varepsilon x^{\mathrm{T}}Qx - (1-\varepsilon)x^{\mathrm{T}}Qx$$
$$+ \bar{R}\left[|Kx|^2 - 2|Kx|\mathrm{sat}_\Delta(|Kx|) + \frac{B^{\mathrm{T}}PB}{\bar{R}}\mathrm{sat}_\Delta(|Kx|)^2\right].$$
$$(7.19)$$

Next, define the sequence $\{\bar{\beta}_i\}_{i=1}^\infty$ as

$$\bar{\beta}_1 = 0, \quad \ldots, \quad \bar{\beta}_{i+1} = \sqrt{\frac{R}{\bar{R}} + \frac{q_\varepsilon(1+\bar{\beta}_i)^2}{KK^{\mathrm{T}}\bar{R}}}, \quad \ldots, \quad (7.20)$$

where $q_\varepsilon = (1-\varepsilon)\lambda_{\min}(Q)$ and $\varepsilon \in [0,1)$ is an arbitrarily small nonnegative number. It can be shown that the sequence $\{\bar{\beta}_i\}_{i=1}^\infty$ grows monotonically, and converges to $\bar{\beta}_{\max}$ defined by (7.17) in the case when $q_\varepsilon/KK^{\mathrm{T}}\bar{R} < 1$, or diverges to $+\infty$ if $q_\varepsilon/KK^{\mathrm{T}}\bar{R} \geq 1$, in which case, for any arbitrarily large positive number M^+, there exists i^+ such that $\bar{\beta}_i > M^+$ for all $i > i^+$.

Consider now the following cases:

Case (a). $|Kx| \leq (1+\bar{\beta}_1)\Delta$: Suppose first that $x \in \mathcal{E}$, $x \neq 0$, is such that

$$|Kx| \leq \Delta = (1+\bar{\beta}_1)\Delta,$$

then $\Delta V(x)$ in (7.19) is equal to

$$\Delta V(x) = -\varepsilon x^{\mathrm{T}}Qx - x^{\mathrm{T}}\left((1-\varepsilon)Q + K^{\mathrm{T}}RK\right)x < -\varepsilon x^{\mathrm{T}}Qx,$$

(from the positive definiteness of Q and R).

Case (b). $(1+\bar{\beta}_1)\Delta < |Kx| \leq (1+\bar{\beta}_2)\Delta$: Suppose next that $x \in \mathcal{E}$, $x \neq 0$, is such that

$$\Delta = (1+\bar{\beta}_1)\Delta < |Kx| \leq (1+\bar{\beta}_2)\Delta,$$

then $\mathrm{sat}_\Delta(|Kx|) = \Delta$, and, by the Cauchy-Schwarz inequality we obtain:

$$|Kx|^2 > (1+\bar{\beta}_1)^2\Delta^2 \quad \Rightarrow \quad -x^{\mathrm{T}}x < -(1+\bar{\beta}_1)^2\frac{\Delta^2}{KK^{\mathrm{T}}}. \quad (7.21)$$

Therefore, an upper bound for $\Delta V(\cdot)$ in (7.19) is

$$\Delta V(x) < -\varepsilon x^{\mathrm{T}}Qx + \bar{R}\left[|Kx|^2 - 2|Kx|\Delta + \left(\frac{B^{\mathrm{T}}PB}{\bar{R}} - \frac{q_\varepsilon(1+\bar{\beta}_1)^2}{KK^{\mathrm{T}}\bar{R}}\right)\Delta^2\right].$$
$$(7.22)$$

It is easy to see that the quadratic term in (7.22) is nonpositive if $\Delta < |Kx| \le (1 + \bar{\beta}_2)\Delta$, in which case

$$\Delta V(x) < -\varepsilon x^\mathrm{T} Q x.$$

Case (c). $(1 + \bar{\beta}_i)\Delta < |Kx| \le (1 + \bar{\beta}_{i+1})\Delta$, $i = 2, 3, \ldots$: Repeating the above argument for $x \in \mathcal{E}$, $x \ne 0$, such that

$$\Delta < (1 + \bar{\beta}_i)\Delta < |Kx| \le (1 + \bar{\beta}_{i+1})\Delta,$$

for $i = 2, 3, \ldots$, and, since $\bar{\beta}_i \to \bar{\beta}_{\max}$ (or diverges to $+\infty$, in which case $\bar{\beta}_i$ eventually becomes bigger than $\bar{\beta}_{\max} = M^+$), we can see that

$$\Delta V(x) < -\varepsilon x^\mathrm{T} Q x$$

if $|Kx| < (1 + \bar{\beta}_{\max})\Delta$, which (from the construction of $\mathcal{E} = \{x : x^\mathrm{T} P x \le \rho\}$ with $\rho < \bar{\rho}_{\max}$) is true for all $x \in \mathcal{E}$.

It follows that,

$$\Delta V(x) < -\varepsilon x^\mathrm{T} Q x \quad \text{for all } x \in \mathcal{E}, \tag{7.23}$$

and hence:

(i) The trajectories that start in the ellipsoid $\mathcal{E} \triangleq \{x : x^\mathrm{T} P x \le \rho\}$ will never leave it since $\Delta V(x)$, along the trajectories, is negative definite on the ellipsoid. Therefore the ellipsoid $\mathcal{E}$ is a positively invariant set under the control $u = -\mathrm{sat}_\Delta(Kx)$.

(ii) From Theorem 4.3.3 in Chapter 4, the origin is exponentially stable in $\mathcal{E}$ for system (7.2) with control $u = -\mathrm{sat}_\Delta(Kx)$, with a region of attraction that includes the ellipsoid $\mathcal{E}$. Notice that if we choose $\varepsilon = 0$ we only guarantee asymptotic stability.

□

We have thus found an ellipsoidal set $\mathcal{E}$ that is positively invariant for the system (7.2) with control $u = -\mathrm{sat}_\Delta(Kx)$. Moreover, this control exponentially stabilises (7.2) with a region of attraction that contains $\mathcal{E}$. However, to guarantee that $u = -\mathrm{sat}_\Delta(Kx)$ is also the receding horizon optimal control law in $\mathcal{E}$, we need to further restrict the radius of the ellipsoid so that the trajectories inside $\mathcal{E}$ also remain within the set $\bar{\mathbb{Z}}$ defined in (7.12).

Recall that $\bar{\mathbb{Z}}$ is the maximal positively invariant set for the closed loop system $x_{k+1} = \phi_{nl}(x_k)$ contained in the set $\mathbb{Z}$ given by

$$\mathbb{Z} \triangleq \{x : |K(A - BK)x| \le \Delta\}. \tag{7.24}$$

We then compute the ellipsoidal radius equal to

$$\bar{\rho} \triangleq \min\left\{ \frac{(1 + \bar{\beta})^2 \Delta^2}{K P^{-1} K^\mathrm{T}}, \frac{\Delta^2}{(K(A - BK)) P^{-1} (K(A - BK))^\mathrm{T}} \right\}, \tag{7.25}$$

where $\bar{\beta} < \bar{\beta}_{\max}$, and $\bar{\beta}_{\max}$ is computed from (7.17) (in practice, one can choose $\bar{\beta}$ arbitrarily close to $\bar{\beta}_{\max}$). Then we have the following corollary of Theorems 7.2.2 and 7.2.5.

Corollary 7.2.6 *Consider the ellipsoidal set $\mathcal{E} = \{x : x^{\mathrm{T}} P x \leq \bar{\rho}\}$ where $\bar{\rho}$ is computed from (7.25). Then:*

(i) The set $\mathcal{E}$ is a positively invariant set for system (7.2) under the control $u = -\mathrm{sat}_\Delta(Kx)$.

(ii) The set $\mathcal{E}$ is a subset of $\bar{\mathbb{Z}}$.

(iii) The RHC law (7.8) is

$$\mathcal{K}_2(x) = -\mathrm{sat}_\Delta(Kx) \quad \text{for all } x \in \mathcal{E}. \tag{7.26}$$

(iv) System (7.2), with the RHC sequence (7.26), is exponentially stable in $\mathcal{E}$.

Proof.

(i) Notice from (7.18), (7.25), and the fact that $\bar{\beta} < \bar{\beta}_{\max}$, that $\bar{\rho} < \bar{\rho}_{\max}$. Then it follows from Theorem 7.2.5 (i) that $\mathcal{E} = \{x : x^{\mathrm{T}} P x \leq \bar{\rho}\}$ is positively invariant.

(ii) From Lemma 7.2.4 and the definitions (7.24) and (7.25) it follows that $x \in \mathcal{E} \Rightarrow x \in \mathbb{Z}$, and, since $\mathcal{E}$ is positively invariant, this implies that $\phi_{nl}^k(x) \in \mathbb{Z}$, $k = 0, 1, 2, \ldots$. Clearly then, from the definition of $\bar{\mathbb{Z}}$, $x \in \bar{\mathbb{Z}}$.

(iii) This result follows immediately from (ii) above and Theorem 7.2.2.

(iv) This follows from $\bar{\rho} < \bar{\rho}_{\max}$ and Theorem 7.2.5 (ii).

$\square$

7.3 Regional Characterisation for Arbitrary Horizon

Here we extend the result presented in Section 7.2 to arbitrary horizons. We will build on the special case presented above. This development is somewhat involved and the reader might prefer to postpone reading the remainder of this chapter until a second reading of the book. For clarity of exposition, we present first in Section 7.3.1 some notation and preliminary results. In particular various sets in $\mathbb{R}^n$ are defined. These sets are used in the characterisation of the state space regions in which the solution of the form (7.1) holds.

7.3.1 Preliminaries

We consider the discrete time system

$$x_{k+1} = Ax_k + Bu_k, \tag{7.27}$$

where $x_k \in \mathbb{R}^n$ and $u_k \in \mathbb{R}$. The pair (A, B) is assumed to be stabilisable. We consider the following fixed horizon optimal control problem:

$$\mathcal{P}_N(x): \qquad V_N^{\mathrm{OPT}}(x) \triangleq \min V_N(\{x_k\}, \{u_k\}), \qquad (7.28)$$

subject to:

$$x_{k+1} = Ax_k + Bu_k \quad \text{for } k = 0, 1, \ldots, N-1,$$

$$x_0 = x,$$

$$u_k \in \mathbb{U} \triangleq [-\Delta, \Delta] \quad \text{for } k = 0, 1, \ldots, N-1,$$

where $\Delta > 0$ is the input saturation level, and the objective function in (7.28) is

$$V_N(\{x_k\}, \{u_k\}) \triangleq \frac{1}{2} x_N^{\mathrm{T}} P x_N + \frac{1}{2} \sum_{k=0}^{N-1} (x_k^{\mathrm{T}} Q x_k + u_k^{\mathrm{T}} R u_k). \qquad (7.29)$$

We assume that Q and R are positive definite, and P satisfies the algebraic Riccati equation (7.5)–(7.6).

Let the control sequence that achieves the minimum in (7.28) be $\{u_0^{\mathrm{OPT}}, \ldots, u_{N-1}^{\mathrm{OPT}}\}$. The associated RHC law, which depends on the current state $x = x_0$, is

$$\mathcal{K}_N(x) = u_0^{\mathrm{OPT}}. \qquad (7.30)$$

For each $i = 0, 1, 2, \ldots, N-1$, the partial value function, is defined by (see similar definitions in (6.8) of Chapter 6)

$$V_{N-i}^{\mathrm{OPT}}(x) \triangleq \min_{u_k \in \mathbb{U}} V_{N-i}(\{x_k\}, \{u_k\}), \qquad (7.31)$$

where V_{N-i} is the partial objective function

$$V_{N-i}(\{x_k\}, \{u_k\}) \triangleq \frac{1}{2} x_N^{\mathrm{T}} P x_N + \frac{1}{2} \sum_{k=i}^{N-1} (x_k^{\mathrm{T}} Q x_k + u_k^{\mathrm{T}} R u_k),$$

with x_k, $k = i, \ldots, N$ satisfying (7.27) starting from $x_i = x$. We refer to V_{N-i}^{OPT} as the *partial value function* (or, just the *value function*) "at time i," meaning that the (partial) value function "starts at time i." The partial value function at time N is defined as

$$V_0^{\mathrm{OPT}}(x) \triangleq \frac{1}{2} x^{\mathrm{T}} P x.$$

We also define the functions $\delta_i : \mathbb{R}^n \to \mathbb{R}$ as

$$\delta_i(x) \triangleq Kx - \mathrm{sat}_{\Delta_i}(Kx), \quad i = 1, 2, \ldots, N, \qquad (7.32)$$

where the saturation bounds Δ_i are defined as

$$\Delta_i \triangleq \left(1 + \sum_{k=0}^{i-2} |KA^k B| \right) \Delta, \quad i = 1, 2, \ldots, N. \qquad (7.33)$$

In summations, it is to be understood that $\sum_{k=k_1}^{k_2} (\cdot) = 0$ whenever $k_2 < k_1$, so that, in (7.33) we have

$$\Delta_1 = \Delta, \quad \Delta_2 = \Delta_1 + |KB|\Delta, \quad \ldots, \quad \Delta_{i+1} = \Delta_i + |KA^{i-1}B|\Delta, \ldots.$$

We define, for future use, the sets $X_i \subseteq \mathbb{R}^n$:

$$X_i \triangleq \left\{ x : \delta_i \left(A^{i-1}(A - BK)x \right) = 0 \right\}, \qquad i = 1, 2, \ldots, N-1. \tag{7.34}$$

Denote

$$\bar{K}_i \triangleq KA^{i-1}(A - BK), \qquad i = 1, 2, \ldots, N-1.$$

Then, the sets X_i are given by the set of linear inequalities:

$$X_i = \left\{ x : \bar{K}_i x \leq \Delta_i, \ -\bar{K}_i x \leq \Delta_i \right\}, \quad i = 1, 2, \ldots, N-1. \tag{7.35}$$

Recall the definition of the nonlinear mapping ϕ_{nl} in (7.11), which we repeat here for convenience:

$$\phi_{nl}(x) \triangleq Ax - B\mathrm{sat}_\Delta(Kx). \tag{7.36}$$

Also, $\phi_{nl}^0(x) = x$ and ϕ_{nl}^k, $k \geq 1$, denotes the concatenation of ϕ_{nl} with itself k times.

We define, for future use, the sets Y_i, $Z_i \subseteq \mathbb{R}^n$:

$$\begin{aligned}
&Y_0 \triangleq Y_1 \triangleq \mathbb{R}^n, \\
&Y_i = \bigcap_{j=1}^{i-1} X_j, \quad i = 2, 3, \ldots, N, \\
&Z_0 \triangleq Z_1 \triangleq \mathbb{R}^n, \\
&Z_i \triangleq \left\{ x : \phi_{nl}^k(x) \in Y_{i-k}, \ k = 0, 1, \ldots, i-2 \right\}, \ i = 2, 3, \ldots, N,
\end{aligned} \tag{7.37}$$
$$\tag{7.38}$$

so that

$$\begin{aligned}
Z_2 &= Y_2, \\
Z_3 &= \{ x : x \in Y_3, \ \phi_{nl}(x) \in Y_2 \}, \\
Z_4 &= \{ x : x \in Y_4, \ \phi_{nl}(x) \in Y_3, \ \phi_{nl}^2(x) \in Y_2 \},
\end{aligned}$$

and so on.

We have the following properties of these sets:

Proposition 7.3.1

(i) $Y_{i+1} = Y_i \cap X_i$, $i = 1, 2, \ldots, N-1$.
(ii) The set sequence $\{Z_i : i = 0, 1, \ldots, N\}$ is monotonically nonincreasing (with respect to inclusion), that is, $Z_{i+1} \subseteq Z_i$, $i = 0, 1, \ldots, N-1$.
(iii) $Z_{i+1} = Y_{i+1} \cap \{x : \phi_{nl}(x) \in Z_i\}$, $i = 0, 1, \ldots, N-1$.

Proof.

(i) This follows trivially from (7.37).

(ii) Certainly $Z_{i+1} \subseteq Z_i$ for $i = 0$ and 1. For $i \geq 2$:

$$Z_{i+1} = \left\{ x : \phi_{nl}^k(x) \in \bigcap_{j=1}^{i-k} X_j, \ k = 0, 1, \ldots, i - 1 \right\}$$

$$= \left\{ x : \phi_{nl}^k(x) \in \bigcap_{j=1}^{i-k} X_j, \ k = 0, 1, \ldots, i - 2 \right\} \cap \left\{ x : \phi_{nl}^{i-1}(x) \in X_1 \right\}$$

$$= \left\{ x : \phi_{nl}^k(x) \in \bigcap_{j=1}^{i-k-1} X_j, \ k = 0, 1, \ldots, i - 2 \right\}$$

$$\cap \left\{ x : \phi_{nl}^k(x) \in X_{i-k}, \ k = 0, 1, \ldots, i - 2 \right\} \cap \left\{ x : \phi_{nl}^{i-1}(x) \in X_1 \right\}$$

$$= Z_i \cap \left\{ x : \phi_{nl}^k(x) \in X_{i-k}, \ k = 0, 1, \ldots, i - 1 \right\}.$$

(iii) This is trivial for $i = 0$. For $i \geq 1$:

$$Z_{i+1} = \left\{ x : \phi_{nl}^k(x) \in \bigcap_{j=1}^{i-k} X_j, \ k = 0, 1, \ldots, i - 1 \right\}$$

$$= \left\{ x : \phi_{nl}^{k+1}(x) \in \bigcap_{j=1}^{i-k-1} X_j, \ k = -1, 0, \ldots, i - 2 \right\}$$

$$= \left\{ x : x \in \bigcap_{j=1}^{i} X_j \right\}$$

$$\cap \left\{ x : \phi_{nl}^k(\phi_{nl}(x)) \in \bigcap_{j=1}^{i-k-1} X_j, \ k = 0, 1, \ldots, i - 2 \right\}$$

$$= Y_{i+1} \cap \left\{ x : \phi_{nl}(x) \in Z_i \right\}.$$

$\square$

Finally, we require the following key result.

Lemma 7.3.2 *For any $i \in \{1, 2, \ldots, N - 1\}$ define the functions $\phi_{nl}(\cdot)$ and $\delta_i(\cdot)$, $\delta_{i+1}(\cdot)$ as in (7.36) and (7.32), respectively, and the set X_i as in (7.34). Define, for $i \in \{1, 2, \ldots, N - 1\}$ the functions $\mu_1, \mu_2 : \mathbb{R}^n \to [0, +\infty)$ as*

$$\mu_1(x) \triangleq \delta_i \left(A^{i-1} \phi_{nl}(x) \right)^2,$$

$$\mu_2(x) \triangleq \delta_{i+1} \left(A^i x \right)^2.$$

Then, $\mu_1(x) = \mu_2(x)$ for all $x \in X_i$.

Proof. The functions $\mu_1, \mu_2 : \mathbb{R}^n \to [0, +\infty)$ can be written as

$$\mu_1(x) \triangleq \delta_i \left(A^{i-1} \phi_{nl}(x) \right)^2$$

$$= \Big[K A^i x - K A^{i-1} B \operatorname{sat}_\Delta(Kx)$$

$$- \operatorname{sat}_{\Delta_i} \left(K A^i x - K A^{i-1} B \operatorname{sat}_\Delta(Kx) \right) \Big]^2,$$

$$\mu_2(x) \triangleq \delta_{i+1} \left(A^i x \right)^2 = \left[K A^i x - \operatorname{sat}_{\Delta_{i+1}} \left(K A^i x \right) \right]^2,$$

for $i \in \{1, 2, \ldots, N - 1\}$. Notice, from (7.35), that:

$$x \in X_i \Leftrightarrow \left| K A^i x - K A^{i-1} B K x \right| \leq \Delta_i = \left(1 + \sum_{k=0}^{i-2} |K A^k B| \right) \Delta. \qquad (7.39)$$

We will prove that $\mu_1(x) = \mu_2(x)$ for all $x \in X_i$ by considering two separate cases, case (a) where $x \in X_i$ and $|Kx| \leq \Delta$, and case (b) where $x \in X_i$ and $|Kx| > \Delta$.

Case (a). $x \in X_i$ and $|Kx| \leq \Delta$:
  Suppose

$$|Kx| \leq \Delta. \tag{7.40}$$

It follows from (7.39) and (7.40) that

$$\begin{aligned}
\mu_1(x) &= \left[KA^i x - KA^{i-1} B \, \text{sat}_\Delta(Kx) \right. \\
&\quad \left. - \text{sat}_{\Delta_i} \left(KA^i x - KA^{i-1} B \, \text{sat}_\Delta(Kx) \right) \right]^2 \\
&= \left[KA^i x - KA^{i-1} BKx - \text{sat}_{\Delta_i} \left(KA^i x - KA^{i-1} BKx \right) \right]^2 \\
&= 0.
\end{aligned} \tag{7.41}$$

Also, notice from (7.39) and (7.40) that

$$\begin{aligned}
\Delta_i &\geq |KA^i x - KA^{i-1} BKx| \geq |KA^i x| - |KA^{i-1} B||Kx| \\
&\geq |KA^i x| - |KA^{i-1} B|\Delta
\end{aligned} \tag{7.42}$$

$$\Rightarrow$$

$$|KA^i x| \leq \Delta_i + |KA^{i-1} B|\Delta = \Delta_{i+1}, \tag{7.43}$$

then it follows that

$$\mu_2(x) = \left[KA^i x - \text{sat}_{\Delta_{i+1}} \left(KA^i x \right) \right]^2 = 0, \tag{7.44}$$

and we conclude that, for case (a):

$$\mu_1(x) = \mu_2(x) = 0.$$

Case (b). $x \in X_i$ and $|Kx| > \Delta$:
  Suppose:

$$|Kx| > \Delta. \tag{7.45}$$

We will consider two cases for case (b): case (b1) where $x \in X_i$ satisfies (7.45) and $|KA^i x| \leq \Delta_{i+1}$ and case (b2) where $x \in X_i$ satisfies (7.45) and $|KA^i x| > \Delta_{i+1}$.

Case (b1). $x \in X_i$, $|Kx| > \Delta$ and $|KA^i x| \leq \Delta_{i+1}$:
  Suppose

$$|KA^i x| \leq \Delta_{i+1} = \Delta_i + |KA^{i-1} B|\Delta. \tag{7.46}$$

Now, suppose also that $KA^{i-1} BKx \leq 0$. Then from (7.39) we have

$$-KA^{i-1} B\text{sat}_\Delta(Kx) \leq -KA^{i-1} BKx \leq -KA^i x + \Delta_i, \tag{7.47}$$

and from (7.45) and (7.46):

$$-KA^{i-1}B\mathrm{sat}_\Delta(Kx) = |KA^{i-1}B|\Delta$$
$$\geq |KA^ix| - \Delta_i \geq -KA^ix - \Delta_i. \tag{7.48}$$

Suppose next $KA^{i-1}BKx > 0$, then it follows from (7.45) and (7.46) that

$$-KA^{i-1}B\mathrm{sat}_\Delta(Kx) = -|KA^{i-1}B|\Delta$$
$$\leq -|KA^ix| + \Delta_i \leq -KA^ix + \Delta_i, \tag{7.47'}$$

and from (7.39)

$$-KA^{i-1}B\mathrm{sat}_\Delta(Kx) \geq -KA^{i-1}BKx \geq -KA^ix - \Delta_i. \tag{7.48'}$$

We conclude from (7.47) and (7.48) (or, (7.47)$'$ and (7.48)$'$) that

$$|KA^ix - KA^{i-1}B\mathrm{sat}_\Delta(Kx)| \leq \Delta_i, \tag{7.49}$$

and, hence:

$$\mu_1(x) = [KA^ix - KA^{i-1}B\mathrm{sat}_\Delta(Kx)$$
$$- \mathrm{sat}_{\Delta_i}\left(KA^ix - KA^{i-1}B\mathrm{sat}_\Delta(Kx)\right)]^2 = 0. \tag{7.50}$$

Also, it follows immediately from (7.46) that

$$\mu_2(x) = \left[KA^ix - \mathrm{sat}_{\Delta_{i+1}}\left(KA^ix\right)\right]^2 = 0, \tag{7.51}$$

and we conclude that, for case (b1),

$$\mu_1(x) = \mu_2(x) = 0.$$

Case (b2). $x \in X_i$, $|Kx| > \Delta$ and $|KA^ix| > \Delta_{i+1}$:
Suppose

$$|KA^ix| > \Delta_{i+1} = \Delta_i + |KA^{i-1}B|\Delta. \tag{7.52}$$

We will next show that case (b2) is not compatible with

$$\mathrm{sign}(KA^ix) = -\mathrm{sign}(KA^{i-1}BKx). \tag{7.53}$$

To see this, notice that (7.39), (7.45) and (7.53) imply

$$\Delta_i \geq |KA^ix - KA^{i-1}BKx| = |KA^ix| + |KA^{i-1}BKx|$$
$$> |KA^ix| + |KA^{i-1}B|\Delta$$
$$\Rightarrow$$
$$|KA^ix| < \Delta_i - |KA^{i-1}B|\Delta \leq \Delta_i + |KA^{i-1}B|\Delta,$$

which, clearly, contradicts (7.52). We conclude, then, that for case (b2),

$$\text{sign}(KA^i x) = \text{sign}(KA^{i-1}BKx). \tag{7.54}$$

We then have from (7.45) and (7.54), that

$$\begin{aligned}
\mu_1(x) &= \big[KA^i x - KA^{i-1}B\text{sat}_\Delta(Kx) \\
&\quad - \text{sat}_{\Delta_i}\left(KA^i x - KA^{i-1}B\text{sat}_\Delta(Kx)\right)\big]^2, \\
&= [\text{sign}(KA^i x)(|KA^i x| - |KA^{i-1}B|\Delta \\
&\quad - \text{sat}_{\Delta_i}\left(|KA^i x| - |KA^{i-1}B|\Delta\right))]^2, \\
&= \big[|KA^i x| - |KA^{i-1}B|\Delta - \text{sat}_{\Delta_i}\left(|KA^i x| - |KA^{i-1}B|\Delta\right)\big]^2.
\end{aligned} \tag{7.55}$$

Notice, finally, that (7.52) implies

$$|KA^i x| - |KA^{i-1}B|\Delta > \Delta_i, \tag{7.56}$$

which, in turn, implies in (7.55) that

$$\mu_1(x) = \big[|KA^i x| - |KA^{i-1}B|\Delta - \Delta_i\big]^2 = \big[|KA^i x| - \Delta_{i+1}\big]^2. \tag{7.57}$$

It also follows from (7.52) that

$$\mu_2(x) = \big[KA^i x - \text{sat}_{\Delta_{i+1}}\left(KA^i x\right)\big]^2 \tag{7.58}$$

$$= \big[\text{sign}(KA^i x)|KA^i x| - \text{sign}(KA^i x)\Delta_{i+1}\big]^2 \tag{7.59}$$

$$= \big[|KA^i x| - \Delta_{i+1}\big]^2, \tag{7.60}$$

and we conclude that, for case (b2),

$$\mu_1(x) = \mu_2(x) = \big[|KA^i x| - \Delta_{i+1}\big]^2.$$

We can see that for all the cases considered (which cover all the possibilities for $x \in X_i$) the equality $\mu_1(x) = \mu_2(x)$ is satisfied. □

7.3.2 Main Result

The following theorem gives a characterisation of the partial value function (7.31). The proof extends to the case of arbitrary horizon the dynamic programming arguments used in Theorem 6.2.1 of Chapter 6 for horizon $N = 2$.

Theorem 7.3.3 *For all $i \in \{0, 1, \ldots, N\}$, provided $x \in Z_{N-i}$ (see (7.38)), the partial value function (7.31) is given by*

$$V_{N-i}^{\text{OPT}}(x) = \frac{1}{2}x^\mathsf{T} P x + \frac{1}{2}\bar{R}\sum_{k=1}^{N-i} \delta_k(A^{k-1}x)^2, \tag{7.61}$$

where δ_k is the function defined in (7.32).

Proof. We prove the theorem by induction. We start from the last value function at $i = N$, and solve the problem backwards in time by using the principle of optimality:

$$V_{N-i}^{\text{OPT}}(x) = \min_{u \in \mathbb{U}} \left\{ \frac{1}{2} x^\mathsf{T} Q x + \frac{1}{2} u^\mathsf{T} R u + V_{N-(i+1)}^{\text{OPT}}(Ax + Bu) \right\},$$

where u and x denote, $u = u_i$ and $x = x_i$, respectively.

(i) The value function V_0^{OPT} $(i = N)$:

By definition, the optimal value function at time N is

$$V_0^{\text{OPT}}(x) \triangleq \frac{1}{2} x^\mathsf{T} P x \quad \text{for all } x \in Z_0 \equiv \mathbb{R}^n.$$

(ii) The value function V_1^{OPT} $(i = N - 1)$:

By the principle of optimality, for all $x \in \mathbb{R}^n$,

$$\begin{aligned}
V_1^{\text{OPT}}(x) &= \min_{u \in \mathbb{U}} \left\{ \frac{1}{2} x^\mathsf{T} Q x + \frac{1}{2} u^\mathsf{T} R u + V_0^{\text{OPT}}(Ax + Bu) \right\} \\
&= \min_{u \in \mathbb{U}} \left\{ \frac{1}{2} x^\mathsf{T} Q x + \frac{1}{2} u^\mathsf{T} R u + \frac{1}{2}(Ax + Bu)^\mathsf{T} P (Ax + Bu) \right\} \\
&= \min_{u \in \mathbb{U}} \left\{ \frac{1}{2} x^\mathsf{T} P x + \frac{1}{2} \bar{R}(u + Kx)^2 \right\}.
\end{aligned} \tag{7.62}$$

In deriving the last line we have made use of the algebraic Riccati equation (7.5)–(7.6). It is clear that the *unconstrained* optimal control is given by $u = -Kx$. From the convexity of the function $\bar{R}(u + Kx)^2$ it then follows that the *constrained* optimal control law is given by

$$u_{N-1}^{\text{OPT}} = \text{sat}_\Delta(-Kx) = -\text{sat}_\Delta(Kx) \quad \text{for all } x \in Z_1 \equiv \mathbb{R}^n, \tag{7.63}$$

and the optimal value function at time $N - 1$ is

$$V_1^{\text{OPT}}(x) = \frac{1}{2} x^\mathsf{T} P x + \frac{1}{2} \bar{R} \delta_1(x)^2 \quad \text{for all } x \in Z_1 \equiv \mathbb{R}^n.$$

(iii) The value function V_2^{OPT} $(i = N - 2)$:

By the principle of optimality, for all $x \in \mathbb{R}^n$,

$$\begin{aligned}
V_2^{\text{OPT}}(x) &= \min_{u \in \mathbb{U}} \left\{ \frac{1}{2} x^\mathsf{T} Q x + \frac{1}{2} u^\mathsf{T} R u + V_1^{\text{OPT}}(Ax + Bu) \right\} \\
&= \min_{u \in \mathbb{U}} \left\{ \frac{1}{2} x^\mathsf{T} Q x + \frac{1}{2} u^\mathsf{T} R u + \frac{1}{2}(Ax + Bu)^\mathsf{T} P(Ax + Bu) \right. \\
&\qquad \left. + \frac{1}{2} \bar{R} \delta_1 (Ax + Bu)^2 \right\} \\
&= \min_{u \in \mathbb{U}} \left\{ \frac{1}{2} x^\mathsf{T} P x + \frac{1}{2} \bar{R}(u + Kx)^2 + \frac{1}{2} \bar{R} \delta_1 (Ax + Bu)^2 \right\}.
\end{aligned}$$

Since $\delta_1(Ax - BKx) = 0$ for $x \in X_1$ (see (7.34)) the *unconstrained* minimum of the right hand side of the above equation occurs at $u = -Kx$ if $x \in X_1$. Because the right hand side is convex in u, the *constrained* minimum occurs at

$$u_{N-2}^{\text{OPT}} = \text{sat}_\Delta(-Kx) = -\text{sat}_\Delta(Kx) \quad \text{for all } x \in Z_2 \equiv X_1,$$

and the optimal partial value function at time $N - 2$ is

$$V_2^{\text{OPT}}(x) = \frac{1}{2}x^{\mathsf{T}}Px + \frac{1}{2}\bar{R}\delta_1(x)^2 + \frac{1}{2}\bar{R}\delta_1(\phi_{nl}(x))^2 \quad \text{for all } x \in Z_2 \equiv X_1.$$

Now we can use the result of Lemma 7.3.2 to express $V_2^{\text{OPT}}(x)$ as

$$V_2^{\text{OPT}}(x) = \frac{1}{2}x^{\mathsf{T}}Px + \frac{1}{2}\bar{R}\delta_1(x)^2 + \frac{1}{2}\bar{R}\delta_2(Ax)^2$$

$$= \frac{1}{2}x^{\mathsf{T}}Px + \frac{1}{2}\bar{R}\sum_{k=1}^{2}\delta_k(A^{k-1}x)^2 \quad \text{for all } x \in Z_2 \equiv X_1.$$

(iv) The value functions V_{N-i}^{OPT} and $V_{N-(i-1)}^{\text{OPT}}$ $(i \in \{1, 2, \ldots, N-1\})$:

We have established above the theorem for $N - i$, $i = N$, $N - 1$ and $N - 2$. We will now introduce the *induction hypothesis*. Assume that the value function V_{N-i}^{OPT}, for some $i \in \{1, 2, \ldots, N-1\}$, is given by the general expression (7.61). Based on this assumption, we will now derive the partial value function at time $i - 1$.

By the principle of optimality,

$$V_{N-(i-1)}^{\text{OPT}} = \min_{u \in \mathbb{U}}\left\{\frac{1}{2}x^{\mathsf{T}}Qx + \frac{1}{2}u^{\mathsf{T}}Ru + V_{N-i}^{\text{OPT}}(Ax + Bu)\right\}$$

$$= \min_{u \in \mathbb{U}}\left\{\frac{1}{2}x^{\mathsf{T}}Qx + \frac{1}{2}u^{\mathsf{T}}Ru + \frac{1}{2}(Ax + Bu)^{\mathsf{T}}P(Ax + Bu)\right.$$

$$\left. + \frac{1}{2}\bar{R}\sum_{k=1}^{N-i}\delta_k(A^{k-1}(Ax + Bu))^2\right\}$$

$$= \min_{u \in \mathbb{U}}\left\{\frac{1}{2}x^{\mathsf{T}}Px + \frac{1}{2}\bar{R}(u + Kx)^2\right.$$

$$\left. + \frac{1}{2}\bar{R}\sum_{k=1}^{N-i}\delta_k(A^{k-1}(Ax + Bu))^2\right\}, \tag{7.64}$$

for all x such that

$$Ax + Bu_{i-1}^{\text{OPT}} \in Z_{N-i}, \tag{7.65}$$

(since the expression used above for $V_{N-i}^{\text{OPT}}(\cdot)$ is only valid in Z_{N-i}).

Since $\delta_k(A^{k-1}(Ax - BKx)) = 0$ for $k = 1, 2, \ldots, N-i$ if $x \in Y_{N-(i-1)} = X_1 \cap X_2 \cap \cdots \cap X_{N-i}$ (see (7.34)) the *unconstrained* minimum of the right hand side of (7.64) occurs at $u = -Kx$ if $x \in Y_{N-(i-1)}$. Because the right hand side of (7.64) is convex in u, the *constrained* minimum occurs at:

$$u_{i-1}^{\text{OPT}} = \text{sat}_\Delta(-Kx) = -\text{sat}_\Delta(Kx),$$

for all $x \in Y_{N-(i-1)} = \bigcap_{j=1}^{N-i} X_j$ and, such that $Ax - B\text{sat}_\Delta(Kx) = \phi_{nl}(x) \in Z_{N-i}$ (see (7.65), that is, for all $x \in Z_{N-(i-1)}$ (Proposition 7.3.1 (iii)).

Therefore the optimal partial value function at time $i - 1$ is

$$V_{N-(i-1)}^{\text{OPT}}(x) = \frac{1}{2}x^{\mathrm{T}}Px + \frac{1}{2}\bar{R}\delta_1(x)^2$$

$$+ \frac{1}{2}\bar{R}\sum_{k=1}^{N-i}\delta_k(A^{k-1}\phi_{nl}(x))^2 \quad \text{for all } x \in Z_{N-(i-1)},$$

and, using the result of Lemma 7.3.2, we can express $V_{N-(i-1)}^{\text{OPT}}(\cdot)$ as

$$V_{N-(i-1)}^{\text{OPT}}(x) = \frac{1}{2}x^{\mathrm{T}}Px + \frac{1}{2}\bar{R}\delta_1(x)^2 + \frac{1}{2}\bar{R}\sum_{k=1}^{N-i}\delta_{k+1}(A^k x)^2$$

$$= \frac{1}{2}x^{\mathrm{T}}Px + \frac{1}{2}\bar{R}\sum_{k=1}^{N-(i-1)}\delta_k(A^{k-1}x)^2 \quad \text{for all } x \in Z_{N-(i-1)}.$$

This expression for $V_{N-(i-1)}^{\text{OPT}}(\cdot)$ is of the same form as that of (7.61) for $V_{N-i}^{\text{OPT}}(\cdot)$. The result then follows by induction.

$\square$

The optimal solution of the fixed horizon control problem $\mathcal{P}_N$ easily follows as a corollary of the above result. For a horizon $N \geq 1$, consider the set

$$\mathbb{Z} \triangleq Z_N = \mathbb{R}^n, \quad \text{if } N = 1,$$

$$\mathbb{Z} \triangleq Z_N = \left\{x : \phi_{nl}^k(x) \in Y_{N-k}, \ k = 0, 1, \ldots, N-2\right\}, \quad \text{if } N \geq 2. \quad (7.66)$$

Note that, for $N = 2$, (7.66) coincides with (7.9) since $\phi_{nl}^0(x) = x$ and, hence, $\mathbb{Z} = Z_2 = Y_2 = X_1 = \{x : |K(A - BK)x| \leq \Delta\}$ (see (7.37) and (7.35)) in this case.

We then have:

Corollary 7.3.4 *Consider the fixed horizon optimal control problem $\mathcal{P}_N$ defined in (7.28)–(7.29), where x denotes the initial state $x = x_0$ of system (7.27). Then for all $x \in \mathbb{Z}$ the minimum value is*

$$V_N^{\text{OPT}}(x) = \frac{1}{2}x^{\mathrm{T}}Px + \frac{1}{2}\bar{R}\sum_{k=1}^{N}\delta_k(A^{k-1}x)^2, \quad (7.67)$$

and, for all $x \in \mathbb{Z}$ the minimising sequence $\{u_0^{\text{OPT}}, \ldots, u_{N-1}^{\text{OPT}}\}$ is

$$u_k^{\text{OPT}} = \text{sat}_\Delta(-Kx_k) = -\text{sat}_\Delta(Kx_k), \tag{7.68}$$

for $k = 0, 1, \ldots, N-1$, where $x_k = \phi_{nl}^k(x)$.

Proof. Equation (7.67) follows from (7.61) for $i = 0$. From Proposition 7.3.1 (iii) it follows that $x = x_0 \in \mathbb{Z} = Z_N \Rightarrow x_k = \phi_{nl}^k(x) \in Z_{N-k}$, $k = 0, 1, \ldots N - 1$. Then (7.68) follows from the proof by induction of Theorem 7.3.3. □

The above result extends Lemma 7.2.1 to arbitrary horizons. We next present a simple example for which the solution (7.68) holds globally.

Example 7.3.1 (Scalar System with Cheap Control). Consider a scalar system $x_{k+1} = ax_k + bu_k$, $x_0 = x$, with $b \neq 0$ and weights $Q = 1$, $R = 0$, in the objective function (7.29). Such a design, with no weight on the control input, is a limiting case of the controller considered known as *cheap control*.

For this case, the *unconstrained* optimal control is $u = -Kx$, where K computed from (7.6) is $K = a/b$. Now, notice that, with $K = a/b$, the gain $A - BK$ is zero and, hence, the sets in (7.34)–(7.38) are: $X_i \equiv \mathbb{R}^n$, $Y_i \equiv \mathbb{R}^n$, $Z_i \equiv \mathbb{R}^n$, for all i. It then follows from Corollary 7.3.4 that the optimal control sequence for all $x \in \mathbb{R}$ in this case is

$$u_k^{\text{OPT}} = \text{sat}_\Delta\left(\frac{-ax_k}{b}\right) = -\text{sat}_\Delta\left(\frac{ax_k}{b}\right), \tag{7.69}$$

for $k = 0, 1, \ldots, N-1$, where $x_k = \phi_{nl}^k(x)$. Note that here the result (7.69) holds *globally* in the state space. ○

7.3.3 Regional Characterisation of RHC

As in Section 7.2.1, we turn here to the regional characterisation of the RHC law. That is, we will extend the regional characterisation given in Corollary 7.3.4 for the fixed horizon optimal control problem to its receding horizon formulation. To this end, we define the set

$$\bar{\mathbb{Z}} \triangleq \left\{x : \phi_{nl}^k(x) \in Y_N, \ k = 0, 1, 2, \ldots\right\}, \qquad N \geq 2. \tag{7.70}$$

Notice that, from the definitions, $\bar{\mathbb{Z}} \subset \mathbb{Z} \subset Y_N$. The set $\bar{\mathbb{Z}}$ is the maximal positively invariant set contained in $\mathbb{Z}$ and Y_N for the closed loop system $x_{k+1} = \phi_{nl}(x_k) = Ax_k - B\text{sat}_\Delta(Kx_k)$. It is easy to see that (7.70) coincides with the set (7.12) introduced in Definition 7.2.1 for horizon $N = 2$ (since $\mathbb{Z} = Z_2 = Y_2$ in this case, see the discussion after (7.66)).

We then have the following result.

Theorem 7.3.5 *For all $x \in \bar{\mathbb{Z}}$ the RHC law $\mathcal{K}_N$ in (7.30) is given by*

$$\mathcal{K}_N(x) = \text{sat}_\Delta(-Kx) = -\text{sat}_\Delta(Kx). \tag{7.71}$$

Proof. The proof of the theorem follows from the fact that $\bar{\mathbb{Z}} \subseteq \mathbb{Z}$ and that $\bar{\mathbb{Z}}$ is positively invariant under the control $\mathcal{K}_N(x) = -\text{sat}_\Delta(Kx)$. Then, for all states in $\bar{\mathbb{Z}}$ the future trajectories of the system will be such that $x \in \bar{\mathbb{Z}} \subseteq \mathbb{Z}$, and from Corollary 7.3.4 we conclude that $\mathcal{K}_N(x) = u_0^{\text{OPT}} = \text{sat}_\Delta(-Kx) = -\text{sat}_\Delta(Kx)$. □

As we discussed before, if the set $\bar{\mathbb{Z}}$, in which the theorem is valid, were such that the control sequence $\{u_k\} = \{-\text{sat}_\Delta(Kx_k)\}$ stayed unsaturated along the system trajectories, then the result of Theorem 7.3.5 would be trivial. Also, if this set were such that only the first control in the sequence $\{u_k\} = \{-\text{sat}_\Delta(Kx_k)\}$ stayed saturated, then the result would also be trivial (although this is not as evident). This fact can be seen from the proof of Theorem 7.3.3. Assume, for this purpose, that the horizon is $N \geq 2$. Notice that the step $i = N - 1$ of the dynamic programming procedure involves the minimisation of the quadratic function in (7.62), whose *constrained minimum* is simply given by $u_{N-1}^{\text{OPT}} = -\text{sat}_\Delta(Kx_{N-1}) \equiv -Kx_{N-1}$ (since we are assuming that *only the first control* saturates; see (7.63)). Following the same argument backwards in time, and assuming that the controls $u_i^{\text{OPT}} = -Kx_i$, $i = N-1, N-2, \ldots, 1$ are not saturated, it can be easily seen—since P satisfies (7.5)—that the same quadratic equation (7.62) will propagate until the initial step $i = 0$, in which case no assumption would be needed for the optimal control to be $u_0^{\text{OPT}} = -\text{sat}_\Delta(Kx_0)$. In fact, $\bar{\mathbb{Z}}$ can be considerably bigger than both of these trivial cases, as we will see later in Example 7.3.2.

Proposition 7.2.3 also extends to the case of horizons of arbitrary length, that is, the set $\bar{\mathbb{Z}}$ defined in (7.70) contains the maximal output admissible set $\mathcal{O}_\infty$, defined in (7.14). We show this below.

Proposition 7.3.6 $\mathcal{O}_\infty \subseteq \bar{\mathbb{Z}}$.

Proof. As in the proof of Proposition 7.2.3, since $\bar{\mathbb{Z}}$ is the maximal positively invariant set in Y_N, it suffices to show that $\mathcal{O}_\infty \subseteq Y_N \triangleq \bigcap_{i=1}^{N-1} X_i$ (see (7.37)). Assume, therefore, that $x \in \mathcal{O}_\infty$, so that (see (7.14))

$$|KA_K^j x| \leq \Delta, \quad j = 0, 1, \ldots, \tag{7.72}$$

where $A_K \triangleq A - BK$. For any $i \in \{1, 2, \ldots, N-1\}$,

$$
\begin{aligned}
A_K^i &= (A - BK)A_K^{i-1} = AA_K^{i-1} - BKA_K^{i-1} \\
&= A(A - BK)A_K^{i-2} - BKA_K^{i-1} = A^2 A_K^{i-2} - ABKA_K^{i-2} - BKA_K^{i-1} \\
&= A^2(A - BK)A_K^{i-3} - ABKA_K^{i-2} - BKA_K^{i-1} \\
&= A^3 A_K^{i-3} - A^2 BKA_K^{i-3} - ABKA_K^{i-2} - BKA_K^{i-1} \\
&\vdots \\
&= A^{i-1}A_K - \sum_{j=0}^{i-2} A^j BKA_K^{i-1-j},
\end{aligned}
$$

which implies

$$KA^{i-1}A_K x = KA^i_K x + \sum_{j=0}^{i-2} KA^j BK A^{i-1-j}_K x. \qquad (7.73)$$

From (7.72) and (7.73), we obtain the inequality

$$|KA^{i-1}A_K x| \le |KA^i_K x| + \sum_{j=0}^{i-2} |KA^j B||K A^{i-1-j}_K x| \qquad (7.74)$$

$$\le \left(1 + \sum_{j=0}^{i-2} |KA^j B|\right)\Delta = \Delta_i \quad \text{(see (7.33))}. \qquad (7.75)$$

This implies $x \in X_i$ for all $i \in \{1, 2, \dots, N-1\}$ (see (7.35)), yielding the desired result. □

7.3.4 An Ellipsoidal Approximation to the Set $\bar{\mathbb{Z}}$

We can also construct an ellipsoidal inner approximation to the set $\bar{\mathbb{Z}}$, as was done in Section 7.2.2. To this end, recall that $\bar{\mathbb{Z}}$ is the largest positively invariant set, under the mapping $\phi_{nl}(\cdot)$, contained in the set $Y_N \triangleq \bigcap_{i=1}^{N-1} X_i$. Also, recall from (7.35) that the sets X_i are given by

$$X_i = \{x : |\bar{K}_i x| \le \Delta_i\}, \quad i = 1, 2, \dots, N-1.$$

We then compute the ellipsoidal radius from

$$\bar{\rho} = \min\left\{\frac{(1+\bar{\beta})^2\Delta^2}{KP^{-1}K^{\mathrm{T}}}, \frac{\Delta_1^2}{\bar{K}_1 P^{-1}\bar{K}_1^{\mathrm{T}}}, \frac{\Delta_2^2}{\bar{K}_2 P^{-1}\bar{K}_2^{\mathrm{T}}}, \dots, \frac{\Delta_{N-1}^2}{\bar{K}_{N-1}P^{-1}\bar{K}_{N-1}^{\mathrm{T}}}\right\}, \qquad (7.76)$$

where $\bar{\beta} < \bar{\beta}_{\max}$, and $\bar{\beta}_{\max}$ is computed from (7.17) (in practice, one can choose $\bar{\beta}$ arbitrarily close to $\bar{\beta}_{\max}$).

Then we have that Corollary 7.2.6 holds for the ellipsoidal set

$$\mathcal{E} = \{x : x^{\mathrm{T}} P x \le \bar{\rho}\},$$

that is,

(i) $\mathcal{E}$ is positively invariant for system (7.27) under the control $u_k = -\mathrm{sat}_\Delta(Kx_k)$.
(ii) $\mathcal{E} \subseteq \bar{\mathbb{Z}}$.
(iii) The RHC law (7.71) holds, and it is optimal, for all $x \in \mathcal{E}$.
(iv) System (7.27), with the RHC sequence (7.71), is exponentially stable in $\mathcal{E}$.

The following example illustrates the regional characterisation of RHC and the different sets used to describe it.

Example 7.3.2. Consider the system $x_{k+1} = Ax_k + Bu_k$ with

$$A = \begin{bmatrix} 1 & 0 \\ 0.4 & 1 \end{bmatrix}, \qquad B = \begin{bmatrix} 0.4 \\ 0.08 \end{bmatrix},$$

which is the zero-order hold discretisation with a sampling period of 0.4 sec of the double integrator

$$\dot{x}^1(t) = u(t), \quad \dot{x}^2(t) = x^1(t).$$

The input constraint level is taken as $\Delta = 1$. The fixed horizon objective function is of the form (7.29) using $N = 10$, $Q = I$ and $R = 0.25$. The matrix P and the gain K were computed from (7.5) and (7.6). The maximum over-saturation index was computed from (7.17) with $\varepsilon = 0$ and is equal to $\bar{\beta}_{\max} = 1.3397$. We then take $\bar{\beta} = 1.3396 < \bar{\beta}_{\max}$ and compute the ellipsoid radius $\bar{\rho}$ from (7.76).

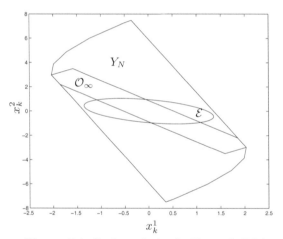

Figure 7.1. Set boundaries for Example 7.3.2.

In Figure 7.1 we show the following sets: $Y_N = \bigcap_{i=1}^{N-1} X_i$ (from (7.37)); the maximal output admissible set $\mathcal{O}_\infty$; and the ellipsoid $\mathcal{E} = \{x : x^{\mathrm{T}} P x \leq \bar{\rho}\}$. In this figure, x_k^1 and x_k^2 denote the components of the state vector x_k in the discrete time model. The sets $\mathcal{O}_\infty$ and $\mathcal{E}$ are positively invariant and are contained in $\bar{\mathbb{Z}}$ (Proposition 7.3.6 and (ii) above), and hence we have that $\mathcal{O}_\infty \cup \mathcal{E} \subseteq \bar{\mathbb{Z}} \subseteq Y_N$, which gives an estimate of the size of $\bar{\mathbb{Z}}$.

In Figure 7.2 we show the boundaries of the sets discussed above, together with the result of simulating the system with control $u = -\mathrm{sat}_\Delta(Kx)$, and with RHC performed numerically via quadratic programming, for an initial condition contained in the invariant ellipsoid $\mathcal{E} = \{x : x^{\mathrm{T}} P x \leq \bar{\rho}\}$. Notice

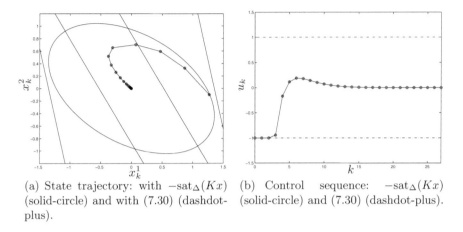

(a) State trajectory: with $-\mathrm{sat}_\Delta(Kx)$ (solid-circle) and with (7.30) (dashdot-plus).

(b) Control sequence: $-\mathrm{sat}_\Delta(Kx)$ (solid-circle) and (7.30) (dashdot-plus).

Figure 7.2. State trajectories and control sequence for the initial condition $x_0 = [1.27 \ -0.1]^{\mathrm{T}}$. Also shown in the left figure are the set boundaries for Y_N, $\mathcal{O}_\infty$, $\mathcal{E}$.

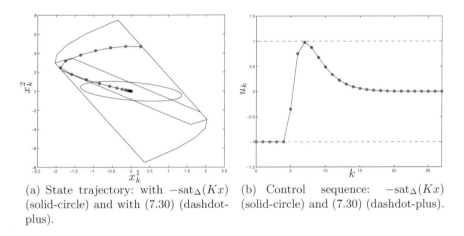

(a) State trajectory: with $-\mathrm{sat}_\Delta(Kx)$ (solid-circle) and with (7.30) (dashdot-plus).

(b) Control sequence: $-\mathrm{sat}_\Delta(Kx)$ (solid-circle) and (7.30) (dashdot-plus).

Figure 7.3. State trajectories and control sequence for the initial condition $x_0 = [0.25 \ 4.7]^{\mathrm{T}}$. Also shown in the left figure are the set boundaries for Y_N, $\mathcal{O}_\infty$, $\mathcal{E}$.

that both strategies coincide, and that the control remains saturated during the initial three steps.

Figure 7.3 shows a case where the initial condition is not contained in the invariant ellipsoid $\mathcal{E} = \{x : x^{\mathrm{T}}Px \leq \bar{\rho}\}$ but is contained in the set $\bar{\mathbb{Z}}$ (since the trajectory does not leave the set Y_N). Therefore, as established in Theorem 7.3.5, both control sequences coincide and, as Figure 7.3 shows, they stay saturated during the initial five steps.

As can be seen from the simulations, the region in which both strategies coincide is such that the control remains saturated during several steps. Hence, we conclude that this region—the set $\bar{\mathbb{Z}}$—is, in fact, nontrivial (see the discussion following Theorem 7.3.5).   ∘

7.4 Further Reading

For complete list of references cited, see References section at the end of book.

General

Further discussion on the regional solution may be found in De Doná (2000), De Doná and Goodwin (2000) and De Doná, Goodwin and Seron (2000).

Link to Anti-Windup Strategies

Another line of attack on the problem of input constraints was developed, beginning from a different perspective to the optimisation approach taken here, grouped under the name of *anti-windup techniques*. The first versions of these techniques can be traced back to PID and integral control, where limitations on the controller's ability to quickly regulate errors to zero, imposed by input saturation, led to unnecessarily high values of the state of the controller integrator. The term "anti-windup" is then used to describe the capability of the technique to prevent the state of the integrator from "winding up" to an excessively high value. Distinctive characteristics of the anti-windup technique are (see, for example, Teel 1999): (i) The original controller is used locally as long as it does not encounter input saturation, and (ii) saturation effects are minimised by modifying the controller structure when the plant input reaches its saturation level. In essence, all the algorithms achieve these goals by letting the controller states "know" about saturation being reached. A unified framework that encompasses many of these algorithms is given in Kothare, Campo, Morari and Nett (1994). They are prime examples of "evolutionary" strategies (see Section 1.2 of Chapter 1).

Consider now that the original controller is the static state feedback $u = -Kx$, that is, that the controller has no dynamics. Particularising the unified framework of Kothare et al. (1994) to this case, one easily obtains that the corresponding anti-windup strategy is equivalent to saturating the control signal, that is, $u = -\text{sat}_\Delta(Kx)$. Comparing with the RHC law (7.71), we conclude that for all $x \in \bar{\mathbb{Z}}$, the RHC and anti-windup strategies have the identical characterisation $u = -\text{sat}_\Delta(Kx)$. That is, the RHC and anti-windup control laws coincide in the region where RHC has the simple finite parameterisation (7.71).

8

Computational Issues in Constrained Optimal Control

8.1 Overview

In this chapter, we will consider the implementation of RHC and the computation of the fixed horizon optimal control problem that underlies it. We will consider linear discrete time systems with quadratic objective function and linear constraints. We will first discuss possible implementations of the piecewise affine characterisation developed in Chapter 6. We will then present two popular algorithms for the computation of the optimal solution of a QP, namely, active set and interior point methods. Finally, we will briefly turn to suboptimal solutions.

8.2 Implementation of the Piecewise Affine RHC Characterisation

In Chapter 6, we showed that the RHC law can be expressed as a "look-up table" consisting of a partition of the state space into polyhedral regions in which the corresponding control law has an explicit piecewise affine form. If the look-up table is precomputed off-line and stored, the RHC control computation is then reduced to the problem of (i) identifying the "current region," that is, deciding in which region the current state belongs, and (ii) computing the control input using the affine control law corresponding to that region. The control action is then computed using a simple function evaluation. However, the problem of identifying the current region may still require a significant computational load if the number of polyhedral regions is large. It is thus relevant to find efficient implementations of this current region identification step.

A direct way of deciding in which region the current state belongs is to perform a sequential search through the regions of the polyhedral partition. For each region, the algorithm would check if the state belongs to it by evaluating the linear inequalities corresponding to the hyperplanes that delimit the region. However, this leads to potentially very long computation times since, in the worst case, every region and every hyperplane in the partition is checked. It would therefore be desirable to organise the structure of the data stored in the look-up table so that the identification of the current region could be performed by evaluating as few hyperplanes as possible. An efficient way to organise the data, one which exploits the convexity of polyhedral sets, is to use a binary search tree where each level, or "node," is associated with one hyperplane inequality. When the tree is used on-line for the current state, at each node one inequality is evaluated, its sign checked, and the left or right subtree selected based on the sign. By traversing the tree from the "root" to a "leaf" node, one finds the region corresponding to the current state. To reduce computation time, it is then important to design a tree of minimum depth so that the number of hyperplanes to be evaluated in the determination of the solution is minimised. In Tøndel et al. (2002), this goal is achieved by constructing a binary search tree with the criterion to reduce the number of remaining regions as much as possible from one level of the tree to the next. These authors show that such a tree achieves a computation time that is logarithmic in the number of polyhedral regions (for details see Tøndel et al. 2002).

For RHC problems of large dimension, for which the implementation of the explicit solution would lead to a binary search tree of high complexity, one could resort to approximations, or suboptimal solutions, of smaller complexity. Several algorithms have been recently proposed to achieve this. The reader is referred to Sections 8.4 and 8.5 for a brief description of some of these novel algorithms.

Although the availability of an explicit characterisation of the RHC law has the potential to extend the applicability of RHC to processes that require faster computation times, more research is required to assess its comparative performance with respect to traditional RHC implementations that employ on-line QP solvers. In fact, the QP problem is an optimisation paradigm that has reached a mature level of development, and efficient algorithms exist to solve it. The following section discusses some of these algorithms.

8.3 Algorithms for Quadratic Programming

There are several approaches for solving a QP problem. One could use, for example, methods of *feasible directions*, which solve the problem by moving from a feasible point to an improved feasible point. These algorithms typically proceed as follows: Given a feasible point x_k, a direction d_k is determined such that for sufficiently small $\alpha > 0$, $x_k + \alpha d_k$ is feasible and the objective value at $x_k + \alpha d_k$ is better than the objective value at x_k. Once such a direction is determined, a one-dimensional optimisation problem is solved to determine how far one needs to move along d_k. This leads to a new point x_{k+1} and the procedure is repeated. Examples of these methods are Beale's method for quadratic programming (Beale 1955), Rosen's gradient projection method (Rosen 1960), and the reduced gradient method developed by Wolfe (Wolfe 1963).

Another popular algorithm is the *active set method*, in which at each step certain constraints, indexed by the *active set*, are regarded as equalities whilst the rest are temporarily disregarded. The method then sequentially solves an equality-constrained QP and adjusts the active set in order to identify the correct active constraints at the optimal solution. In Section 8.3.1, we describe the main ideas used by the active set method; the reader is referred to, for example, Fletcher (1981) for more details.

Primal-dual interior point algorithms have also been proposed for QP. One of such methods, the *infeasible interior point method* (Wright 1997b), is described in Section 8.3.2.

8.3.1 The Active Set Method

Consider the QP problem

$$\text{minimise} \left\{ f(x) = \frac{1}{2} x^\mathrm{T} H x + x^\mathrm{T} c \right\},$$

subject to: $\qquad\qquad\qquad\qquad\qquad\qquad\qquad\qquad$ (8.1)

$$A_i^\mathrm{T} x = b_i \quad \text{for } i \in E,$$
$$A_i^\mathrm{T} x \leq b_i \quad \text{for } i \in I,$$

where H is positive definite, and where the index sets E and I correspond to equality and inequality constraints, respectively. Let x_k be a feasible solution for (8.1) and let $I_k = \{i \in I : A_i^\mathrm{T} x_k = b_i\}$ be the set that records the active constraints at x_k. We recall from Section 2.5.6 in Chapter 2 that, because the constraints are linear, and since the optimisation problem is strictly convex, a necessary and sufficient condition for a feasible solution x_k to be the unique optimal solution of (8.1) is that there exist Lagrange multipliers λ_i for $i \in I_k$, and ν_i for $i \in E$, such that the following KKT conditions hold (see equations (2.41) in Chapter 2)

$$H x_k + c + \sum_{i \in I_k} \lambda_i A_i + \sum_{i \in E} \nu_i A_i = 0,$$
$$\lambda_i \geq 0, \qquad i \in I_k. \qquad\qquad\qquad (8.2)$$

The active set method iterates to find the optimal solution of (8.1) in the following way. At the kth iteration we assume that a feasible solution x_k is available. We define the working active set

$$W_k = E \cup I_k, \qquad I_k = \{i \in I : A_i^\mathrm{T} x_k = b_i\}, \qquad (8.3)$$

and assume that the gradients of the constraints indexed in W_k are linearly independent.[1] The method now seeks to minimise the objective function $f(x)$ in (8.1) subject only to the equality constraints recorded in W_k and ignoring the remaining inequality constraints. That is, we seek to solve the problem

$$\text{minimise } \left\{ f(x) = \frac{1}{2} x^{\mathrm{T}} H x + x^{\mathrm{T}} c \right\},$$
$$\text{subject to:}$$
$$A_i^{\mathrm{T}} x = b_i \quad \text{for } i \in W_k. \tag{8.4}$$

Note that, if we parameterise x as $x = x_k + d$, we have that

$$f(x) = f(x_k + d) = f(x_k) + \frac{1}{2} d^{\mathrm{T}} H d + d^{\mathrm{T}} (c + H x_k).$$

Also, $A_i^{\mathrm{T}} x_k = b_i$ for all $i \in W_k$ (see (8.3)). Hence, instead of solving (8.4), we can solve the following equivalent direction-finding problem:

$$\text{minimise } \frac{1}{2} d^{\mathrm{T}} H d + d^{\mathrm{T}} (c + H x_k),$$
$$\text{subject to:}$$
$$A_i^{\mathrm{T}} d = 0 \quad \text{for } i \in W_k. \tag{8.5}$$

Let d_k be the minimiser of (8.5) and let v_i^* for $i \in W_k$ be the corresponding optimal Lagrange multipliers in the KKT conditions for (8.5), that is,

$$H d_k + c + H x_k + \sum_{i \in W_k} v_i^* A_i = 0, \tag{8.6}$$
$$A_i^{\mathrm{T}} d_k = 0 \quad \text{for } i \in W_k.$$

We can distinguish between two cases.

Case $d_k = 0$. Suppose $d_k = 0$.

- If $v_i^* \geq 0$ for $i \in I_k$, then x_k is the optimal solution of (8.1). To see this, recall that $W_k = I_k \cup E$ and rewrite (8.6) at $d_k = 0$ as

$$c + H x_k + \sum_{i \in I_k} v_i^* A_i + \sum_{i \in E} v_i^* A_i = 0. \tag{8.7}$$

  Hence, if $v_i^* \geq 0$ for $i \in I_k$, the KKT conditions (8.2) hold for x_k and $\lambda_i = v_i^*$ for $i \in I_k$ and $\nu_i = v_i^*$ for $i \in E$. Since x_k is feasible for the original problem, it is thus the optimal solution and the algorithm stops.

---

[1] This is to obtain well-defined QP problems at each iteration (see, for example, Fletcher 1981).

- On the other hand, suppose that there exists an index, $p \in I_k$, say, such that $v_p^* < 0$. Then the objective function $f(x)$ can be reduced by allowing the pth constraint to become inactive. To see this, first note that, using (8.7), we can write

$$\nabla f(x_k)^{\mathrm{T}} = Hx_k + c = -v_p^* A_p - \sum_{i \in W_k : i \neq p} v_i^* A_i. \qquad (8.8)$$

In addition, it is always possible to find a direction s such that $s^{\mathrm{T}} A_i = 0$ for $i \in W_k$, $i \neq p$, and $s^{\mathrm{T}} A_p = -1$ (for example, selecting s^{T} to be minus the pth row of $A^\dagger$, where A is the matrix whose columns are the constraint gradients A_i, $i \in W_k$, and $\dagger$ denotes pseudoinverse). Premultiplying (8.8) by such a direction s^{T} we then have that

$$s^{\mathrm{T}} \nabla f(x_k)^{\mathrm{T}} = v_p^* < 0,$$

and hence s is a descent direction (see Theorem 2.4.1 in Chapter 2). It is also a feasible direction, since an incremental step $x_k + \delta s$, with $\delta > 0$ sufficiently small, satisfies $A_i^{\mathrm{T}}(x_k + \delta s) = A_i^{\mathrm{T}} x_k = b_i$ for $i \in W_k$, $i \neq p$, $A_p^{\mathrm{T}}(x_k + \delta s) = A_p^{\mathrm{T}} x_k - \delta < b_i$ and $A_i^{\mathrm{T}}(x_k + \delta s) < b_i$ for $i \in I \setminus I_k$. Moreover, the strict inequality $A_p^{\mathrm{T}}(x_k + \delta s) < b_i$, together with the fact that s is a feasible descent direction, imply that the objective function $f(x)$ can be reduced by moving away from the boundary of constraint p.

Hence, the algorithm removes the pth constraint from the working active set, that is, we let $I_{k+1} = I_k - \{p\}$, $W_{k+1} = E \cup I_{k+1}$, and $x_{k+1} = x_k$, and solve problem (8.5) again.

If there is more than one index for which the corresponding Lagrange multiplier is negative, then it is usual to select p to solve $v_p^* = \min\{v_i^* : i \in I_k\}$. This selection works quite well, although it has the slight disadvantage that it is not invariant to scaling of the constraints. An invariant but more complex test can be used based on the expected objective function reduction (Fletcher and Jackson 1974).

Case $d_k \neq 0$. Suppose $d_k \neq 0$.

- If $x_k + d_k$ is feasible for (8.1), then we have found a feasible descent direction since $d_k \neq 0$ implies that the optimal value in (8.5) is strictly negative, which yields $f(x_k + d_k) < f(x_k)$. Hence, the algorithm lets $W_{k+1} = W_k$, $x_{k+1} = x_k + d_k$, and solves problem (8.5) again.

- If $x_k + d_k$ is not feasible for (8.1), then a line search is performed in the direction of d_k to find the best feasible point. This can be done in the following way. Note that $x_k + d_k$ is feasible for all constraints indexed in W_k since $A_i^T d_k = 0$ for $i \in W_k$. In fact, infeasibility can only occur amongst constraints with indexes $i \notin I_k$ such that $A_i^T d_k > 0$. Thus, to maintain feasibility, we need to choose a step length $\alpha_k \in [0, 1]$ such that

$$A_i^T(x_k + \alpha_k d_k) \leq b_i \quad \text{for } i \notin I_k \text{ such that } A_i^T d_k > 0,$$

or, equivalently, $\alpha_k \leq (b_i - A_i^T x_k)/(A_i^T d_k)$ for $i \notin I_k$ such that $A_i^T d_k > 0$. We therefore select α_k to be

$$\alpha_k = \min_{i \notin I_k : A_i^T d_k > 0} \frac{b_i - A_i^T x_k}{A_i^T d_k} \triangleq \frac{b_p - A_p^T x_k}{A_p^T d_k}. \tag{8.9}$$

Note that the pth constraint becomes active at $x_k + \alpha_k d_k$, and hence this index is added to the active set. The algorithm then lets $I_{k+1} = I_k \cup \{p\}$, $W_{k+1} = E \cup I_{k+1}$, and $x_{k+1} = x_k + \alpha_k d_k$, and solves problem (8.5) again.

Except in degenerate cases where $\alpha_k = 0$ or $\alpha_k = 1$ in (8.9), termination of the algorithm can be proved easily. First, although we will not prove it, it is a consequence of (8.9) that if the first active working set has the property that the set of gradients that it indexes are linearly independent, then any vector A_p added to the set is not dependent on the other vectors of the set; hence it is possible to prove by induction that the linear independence condition is retained. This means that the equality-constrained problem (8.5) (equivalently, (8.4)) is always well-defined. The termination proof uses the fact that there is a subsequence $\{x_k\}$ of iterates that solve the current problem (8.4). Note that only when $\alpha_k < 1$ in (8.9) is x_{k+1} *not* a solution to (8.4); in this case an index p is added to the working active set. This can happen, at most, n times (where n is the dimension of x), in which case x_k is then a vertex and so it solves the corresponding problem (8.4). Since the number of possible problems (8.4) is finite, since each x_k in the subsequence is the unique global minimum of (8.4), and since the objective function $f(x_k)$ is monotonically decreasing, it follows that termination must occur. The argument fails in cases of degeneracy, where "ties" or "cycles" can occur, and hence some modifications have to be done to the algorithm to ensure convergence (see, for example, Fletcher 1981).

Important practical features of an active set method, which we will not discuss here, include the computation of an initial feasible point, the efficient solution of the equality constrained QP (8.5) by factorisation methods, and the use of factor updates instead of re-factorising when changes are made to the active set. In the case of the particular form of QP that arises in RHC, a more efficient formulation is possible by adding the system dynamic equations as equality constraints and then rearranging the variables so that the constraint matrices become "banded" (Wright 1997b). This can speed up factorisation, albeit at the expense of increasing the dimension of the problem, so a careful analysis has to be performed for each particular application to assess the convenience of the "banded" formulation (Maciejowski 2002).

8.3.2 Interior Point Methods

Consider again the QP problem (8.1), which we write in a more convenient form below:

$$\text{minimise } \frac{1}{2}x^{\mathrm{T}}Hx + x^{\mathrm{T}}c,$$

$$\text{subject to:} \tag{8.10}$$

$$A_E^{\mathrm{T}}x = b_E,$$

$$A_I^{\mathrm{T}}x \leq b_I.$$

The KKT optimality conditions for the above problem, including the primal feasibility conditions, are

$$Hx + c + A_E\nu + A_I\lambda = 0,$$
$$A_E^{\mathrm{T}}x = b_E,$$
$$A_I^{\mathrm{T}}x + \xi = b_I, \tag{8.11}$$
$$\lambda \geq 0, \quad \xi \geq 0, \quad \lambda^{\mathrm{T}}\xi = 0,$$

where ν and λ are vectors of Lagrange multipliers corresponding to the equality and inequality constraints, respectively, and where we have introduced a vector of nonnegative slack variables ξ to transform the inequality constraints into equalities. We can express (8.11) in the following form:

$$\begin{bmatrix} H & A_E & A_I \\ -A_E^{\mathrm{T}} & 0 & 0 \\ -A_I^{\mathrm{T}} & 0 & 0 \end{bmatrix} \begin{bmatrix} x \\ \nu \\ \lambda \end{bmatrix} + \begin{bmatrix} c \\ b_E \\ b_I \end{bmatrix} = \begin{bmatrix} 0 \\ 0 \\ \xi \end{bmatrix}, \tag{8.12}$$

$$\lambda \geq 0, \quad \xi \geq 0, \quad \lambda^{\mathrm{T}}\xi = 0.$$

The above equations have the form of the *mixed linear complementarity problem* [mLCP], a standard paradigm in optimisation that generalises the optimality conditions for linear and quadratic programming. The mLCP is a convenient platform for interior point methods (Wright 1997b). We will next describe one such method, the infeasible interior point algorithm of Wright (1996), (1997b). We will first use a generic form of the mLCP and then apply the algorithm to problem (8.12).

The mLCP seeks to find vectors z, λ and ξ that satisfy

$$\begin{bmatrix} M_{11} & M_{12} \\ M_{21} & M_{22} \end{bmatrix} \begin{bmatrix} z \\ \lambda \end{bmatrix} + \begin{bmatrix} q_1 \\ q_2 \end{bmatrix} = \begin{bmatrix} 0 \\ \xi \end{bmatrix}, \tag{8.13}$$

$$\lambda^{\mathrm{T}}\xi = 0, \tag{8.14}$$

$$\lambda \geq 0, \quad \xi \geq 0, \tag{8.15}$$

where the matrix on the left of (8.13) is positive semidefinite and $M_{11} \in \mathbb{R}^{n_1 \times n_1}$, $M_{22} \in \mathbb{R}^{n_2 \times n_2}$, $q_1 \in \mathbb{R}^{n_1}$, $q_2 \in \mathbb{R}^{n_2}$.

Note that we can write the system of nonlinear equations formed by the "feasibility conditions" (8.13) and the "complementarity condition" (8.14), in compact form as

$$
F(z, \lambda, \xi) = \begin{bmatrix} f_1(z, \lambda, \xi) \\ \vdots \\ f_{n_1 + 2n_2}(z, \lambda, \xi) \end{bmatrix} \triangleq \begin{bmatrix} M_{11}z + M_{12}\lambda + q_1 \\ M_{21}z + M_{22}\lambda - \xi + q_2 \\ \Lambda \Xi e \end{bmatrix} = 0, \qquad (8.16)
$$

where

$$
\begin{aligned}
\Lambda &= \mathrm{diag}\{\lambda_1, \ldots, \lambda_{n_2}\}, \\
\Xi &= \mathrm{diag}\{\xi_1, \ldots, \xi_{n_2}\}, \\
e &= \begin{bmatrix} 1 \ldots 1 \end{bmatrix}^{\mathrm{T}},
\end{aligned} \qquad (8.17)
$$

and where λ_i, ξ_i, for $i = 1, \ldots, n_2$, are the components of the vectors λ and ξ, respectively. The system of nonlinear equations (8.16) can be solved, for example, using Newton's method (see, for example, Fletcher 1981). In Newton's method, given an estimate of the solution $w^k \triangleq (z^k, \lambda^k, \xi^k)$, the function F is approximated by the linear function consisting of the first two terms of its Taylor series at the point w^k. The resulting linear system is then solved to obtain a new estimate of the solution $w^{k+1} \triangleq (z^{k+1}, \lambda^{k+1}, \xi^{k+1})$. That is, a standard Newton step solves

$$
\frac{\partial F(w^k)}{\partial w^k} \Delta w^k = -F(w^k) \qquad (8.18)
$$

for $\Delta w^k \triangleq (\Delta z^k, \Delta \lambda^k, \Delta \xi^k)$ and then sets $w^{k+1} = w^k + \Delta w^k$. In (8.18), $\partial F(\cdot)/\partial w^k$ is the Jacobian matrix of F, whose (i, j)th entry is $\partial f_i(\cdot)/\partial w_j^k$, for $i, j = 1, \ldots, n_1 + 2n_2$. We can write (8.18) in terms of the matrices in (8.16) as

$$
\begin{bmatrix} M_{11} & M_{12} & 0 \\ M_{21} & M_{22} & -I \\ 0 & \Xi^k & \Lambda^k \end{bmatrix} \begin{bmatrix} \Delta z^k \\ \Delta \lambda^k \\ \Delta \xi^k \end{bmatrix} = \begin{bmatrix} -r_1^k \\ -r_2^k \\ -r_3^k \end{bmatrix}, \qquad (8.19)
$$

where

$$
\begin{aligned}
r_1^k &\triangleq M_{11}z^k + M_{12}\lambda^k + q_1, \\
r_2^k &\triangleq M_{21}z^k + M_{22}\lambda^k - \xi^k + q_2, \\
r_3^k &\triangleq \Lambda^k \Xi^k e.
\end{aligned} \qquad (8.20)
$$

The *infeasible interior point* [IIP] algorithm of Wright (1997b) uses a modified Newton method to solve (8.16), which ensures convergence. The algorithm is initialised with a point (z^0, λ^0, ξ^0) that satisfies $(\lambda^0, \xi^0) > 0$, that is, *interior* to the positive orthant $(\lambda, \xi) > 0$, but possibly *infeasible* with respect to the constraints (8.13). Then all iterates (z^k, λ^k, ξ^k) maintain $(\lambda^k, \xi^k) > 0$, but the infeasibilities and the *complementarity gap*, defined by

$$\mu_k \triangleq \frac{(\lambda^k)^{\mathsf{T}}\xi^k}{n_2}, \tag{8.21}$$

are reduced to zero as $k \to \infty$.

In particular, the kth step of the IIP algorithm proceeds as follows. Given the current estimate of the solution (z^k, λ^k, ξ^k) with $(\lambda^k, \xi^k) > 0$, and given $\sigma_k \in (0,1)$, μ_k defined in (8.21), and r_i^k, $i = 1,2,3$ defined in (8.20), the algorithm solves

$$\begin{bmatrix} M_{11} & M_{12} & 0 \\ M_{21} & M_{22} & -I \\ 0 & \Xi^k & \Lambda^k \end{bmatrix} \begin{bmatrix} \Delta z^k \\ \Delta \lambda^k \\ \Delta \xi^k \end{bmatrix} = \begin{bmatrix} -r_1^k \\ -r_2^k \\ -r_3^k + \sigma_k \mu_k e \end{bmatrix} \tag{8.22}$$

for $(\Delta z^k, \Delta \lambda^k, \Delta \xi^k)$. Then it sets

$$(z^{k+1}, \lambda^{k+1}, \xi^{k+1}) = (z^k, \lambda^k, \xi^k) + \alpha_k(\Delta z^k, \Delta \lambda^k, \Delta \xi^k), \tag{8.23}$$

for some $\alpha_k \in (0,1]$ that preserves the positivity condition $(\lambda^{k+1}, \xi^{k+1}) > 0$.

Comparing (8.22) with (8.19), we see that the only modification to the standard Newton method is the term $\sigma_k \mu_k e$ on the right hand side of (8.22). This term ensures that the algorithm converges to the solution of (8.13)–(8.15), whilst remaining inside the positive orthant $(\lambda, \xi) > 0$. The convergence analysis requires the values of σ_k and α_k to satisfy certain conditions (see Wright (1996) for details). A heuristic rule often used in practice is to select $\sigma_k \in [0.001, 0.8]$ and $\alpha_k = \min\{1, 0.995\bar{\alpha}_k\}$, where

$$\bar{\alpha}_k = \sup\{\alpha \in (0,1] : (z^k, \lambda^k, \xi^k) + \alpha_k(\Delta z^k, \Delta \lambda^k, \Delta \xi^k) > 0\}.$$

The main step of the IIP algorithm is the solution of the linear system (8.22). Note that (8.22) can be simplified by eliminating $\Delta \xi^k$, which is possible since the matrix Λ^k has positive diagonal entries. From the last block row of (8.22) we have

$$\begin{aligned} \Delta \xi^k &= (\Lambda^k)^{-1}(-r_3^k + \sigma_k \mu_k e) - (\Lambda^k)^{-1}\Xi^k \Delta \lambda^k, \\ &= -\xi^k + (\Lambda^k)^{-1}\sigma_k \mu_k e - (\Lambda^k)^{-1}\Xi^k \Delta \lambda^k, \end{aligned}$$

where we have used $(\Lambda^k)^{-1}r_3^k = \Xi^k e = \xi^k$, which follows from the definitions (8.17) and (8.20). Substituting the above in the first two rows of (8.22), we obtain

$$\begin{bmatrix} M_{11} & M_{12} \\ M_{21} & M_{22} + (\Lambda^k)^{-1}\Xi^k \end{bmatrix} \begin{bmatrix} \Delta z^k \\ \Delta \lambda^k \end{bmatrix} = \begin{bmatrix} -r_1^k \\ -r_2^k - \xi^k + \sigma_k \mu_k (\Lambda^k)^{-1}e \end{bmatrix}. \tag{8.24}$$

We can now go back to the original problem (8.12) and, comparing with the mLCP (8.13)–(8.15), we can make the following identifications of matrices and variables:

$$M_{11} = \begin{bmatrix} H & A_E \\ -A_E^{\mathrm{T}} & 0 \end{bmatrix}, \quad M_{12} = \begin{bmatrix} A_I \\ 0 \end{bmatrix}, \quad q_1 = \begin{bmatrix} c \\ b_E \end{bmatrix},$$

$$M_{21} = \begin{bmatrix} -A_I^{\mathrm{T}} & 0 \end{bmatrix}, \quad M_{22} = 0, \quad q_2 = b_I, \tag{8.25}$$

$$z = \begin{bmatrix} x \\ \nu \end{bmatrix}, \quad \lambda = \lambda, \quad \xi = \xi.$$

Substituting (8.25) into (8.24), we obtain the main linear equation to be solved at the kth iteration by the IIP algorithm for problem (8.12):

$$\begin{bmatrix} H & A_E & A_I \\ -A_E^{\mathrm{T}} & 0 & 0 \\ -A_I^{\mathrm{T}} & 0 & (\Lambda^k)^{-1}\Xi^k \end{bmatrix} \begin{bmatrix} \Delta x^k \\ \Delta \nu^k \\ \Delta \lambda^k \end{bmatrix} = \begin{bmatrix} -r_1^k \\ -r_2^k - \xi^k + \sigma_k \mu_k (\Lambda^k)^{-1} e \end{bmatrix}. \tag{8.26}$$

In (8.26), the corresponding values of r_1^k and r_2^k are obtained by substitution of (8.25) into the first two lines of (8.20). Also, the complementarity gap μ used in (8.26) has the form (8.21) with n_2 equal to the number of inequality constraints in the QP problem (8.10). Usually, the last two rows of (8.26) are multiplied by -1 so that the matrix on the left becomes symmetric indefinite, for which a substantial amount of factorisation software is available. Further simplifications are possible by eliminating $\Delta\lambda^k$ from the third row.

As was the case for the active set methods of Section 8.3.1, in the application of the IIP algorithm to the particular QP problem that arises in RHC, it is possible to include the system dynamic equations as equality constraints and then rearrange the variables so that the matrices in the problem become block-banded. This has significant computational advantages when performing matrix factorisations and can lead to dramatic computational savings. The interested reader is referred to Wright (1997b) for details.

8.4 Suboptimal RHC Strategies

In the previous sections, we discussed computational aspects of the on-line implementation of RHC, for both the case where the explicit solution is to be used in the form of a look-up table, and the case where the optimisation is to be performed on-line by solving the underlying QP. For RHC problems of large dimension (large state and input dimensions and long constraint horizons), however, it is probably unrealistic to compute the exact explicit solution, and even if that were possible, its implementation would be impractical due to the large amount of memory that would be required to store a complex region partition. On the other hand, on-line optimisation to solve the associated QP may also be impractical from the computation time perspective.

These problems motivated the development of suboptimal, or simplified, solutions aimed at reducing computational burden and controller complexity. In Section 8.5 below, we briefly mention some novel schemes that implement suboptimal simplified versions of the explicit solution. Alternative suboptimal approaches that avoid, or simplify, on-line optimisation but do not necessarily exploit the explicit RHC solution have also been recently proposed (see, for example, Kouvaritakis, Cannon and Rossiter 2002).

Other suboptimal strategies will be explored in Chapter 11 in connection with the singular value decomposition structure of the Hessian of the associated QP.

8.5 Further Reading

For complete list of references cited, see References section at the end of book.

General

There is a growing literature in computational issues in constrained control. The reader is referred to contemporary journals and conference proceedings for recent applications.

Section 8.2

Several new algorithms that implement approximate explicit solutions to RHC have been proposed. For example, Tøndel et al. (2002) suggest a method for generating an approximate state feedback law based on a binary search tree that only allows orthogonal hyperplanes in the tree nodes. This has the advantage that the on-line evaluation at each node involves one comparison only. In Johansen, Petersen and Slupphaug (2002), a suboptimal strategy is considered where an approximation to the optimal objective function is utilised, imposing restrictions on the allowed switching between the active constraint sets during the prediction horizon. Input trajectory parameterisation is studied in Tøndel and Johansen (2002) in order to reduce the degrees of freedom in the optimisation problem. In Bemporad and Filippi (2001), approximate (suboptimal) solutions to the multiparametric QP problem are found by relaxing the Karush–Kuhn–Tucker optimality conditions. In Johansen (2002), it is shown how the system structure can be exploited to derive reduced dimension multiparametric QPs that lead to suboptimal explicit feedback solutions to the state and input constrained RHC problems. In Johansen and Grancharova (2002), approximate explicit solutions to the RHC problem are built in correspondence with given bounds for suboptimality and constraint violation. In Grancharova and Johansen (2002), approximate multiparametric quadratic programming is used, structuring the partition as a binary search tree.

Section 8.3

More material on optimisation algorithms can be found in the following books: Nocedal and Wright (1999), Fletcher (1981), Gill et al. (1981), Polak (1971).

In particular, interior point algorithms are analysed in detail in the books Fiacco and McCormick (1990), Wright (1997a), Nesterov and Nemirovskii (1994), den Hertog (1994), Ye (1997).

9

Constrained Estimation

9.1 Overview

This chapter introduces the reader to the issues involved in constrained estimation. We adopt a stochastic framework and model the underlying system via a set of stochastic difference equations in which the noise has a known probability density function. This leads to a stochastic interpretation of the resulting estimators. Alternatively, one can interpret the resulting optimisation problems in a purely deterministic framework.

We begin with fixed horizon *constrained* linear estimation problems. We will see that the resulting optimisation problems are *similar* to the problems that arise in constrained control. Indeed, they only differ by virtue of the boundary conditions imposed. In the next chapter we will show that the connection is actually deeper than similarity. Indeed, we will show that, for the linear constrained case, the problems are formally dual to each other. We then consider rather general nonlinear estimation problems. Finally, the moving horizon implementation of these estimators is discussed and illustrated by examples.

Potential applications of the ideas presented here include any estimation problem where the variables are known, a priori, to satisfy various constraints. Examples are:

(i) State estimation problems in physical systems where constraints are known to apply, for example, in a distillation column where the liquid levels in the trays are known to lie between two levels (empty and full).

(ii) More general state estimation problems in process control where key variables (for example, disturbances) are known to lie in certain regions.

(iii) Channel equalisation problems in digital communication systems where the transmitted signal is known to belong to a finite alphabet (say ± 1).

(iv) Estimation problems with general distributions where the distribution can be approximated in different regions by different Gaussian distributions.

9.2 Simple Linear Regression

To motivate the more general results to follow, let us first consider a simple linear regression problem:

$$x_{k+1} = x_k = x_0 \quad \text{for } k = 0, \ldots, N-1,$$
$$y_k = Cx_k + v_k \quad \text{for } k = 1, \ldots, N, \tag{9.1}$$

where $x_k \in \mathbb{R}^n$ and where $\{y_k\}$ is a given sequence of scalar observations. Say that $\{v_k\}$ is an i.i.d. sequence having a distribution $p_v(v_k)$ obtained by truncating on the interval $[-b, b]$ a Gaussian distribution with zero mean and variance σ^2, that is,

$$p_v(v_k) = \begin{cases} \dfrac{\frac{1}{\sigma\sqrt{2\pi}} \exp\left\{-\frac{v_k^2}{2\sigma^2}\right\}}{\int_{-b}^{b} \frac{1}{\sigma\sqrt{2\pi}} \exp\left\{-\frac{\alpha^2}{2\sigma^2}\right\} d\alpha} & \text{if } |v_k| \le b, \\[4mm] 0 & \text{otherwise.} \end{cases} \tag{9.2}$$

Also, assume that x_0 has a Gaussian distribution: $N(\mu_0, P_0)$ with $P_0 > 0$.

In the sequel, we will need to refer frequently to *conditional probability density functions*. These take the general form of the probability density for a random variable a evaluated at $\hat{a}$ (say), given that another random variable b takes the specific value $\hat{b}$. We will express this density as $p_{a|b}(a = \hat{a}|b = \hat{b})$. Often we will simplify the notation to $p_{a|b}(\hat{a}|\hat{b})$.

Let $\mathbf{y}_N = [y_1 \ \ldots \ y_N]^{\mathrm{T}}$ and let $\mathbf{y}_N^d = [y_1^d \ \ldots \ y_N^d]^{\mathrm{T}}$ denote the given observations. Then, using Bayes' rule and the independence assumption, the joint probability density function for the data $\mathbf{y}_N^d$ and initial state estimate $\hat{x}_0$ can be obtained as follows:

$$p_{y_1, x_0}(y_1^d, \hat{x}_0) = p_{y_1|x_0}(y_1^d|\hat{x}_0)\, p_{x_0}(\hat{x}_0),$$

$$p_{y_2, y_1, x_0}(y_2^d, y_1^d, \hat{x}_0) = p_{y_2|y_1, x_0}(y_2^d|y_1^d, \hat{x}_0)\, p_{y_1, x_0}(y_1^d, \hat{x}_0)$$
$$= p_{y_2|y_1, x_0}(y_2^d|y_1^d, \hat{x}_0)\, p_{y_1|x_0}(y_1^d|\hat{x}_0)\, p_{x_0}(\hat{x}_0)$$
$$= p_{y_2|x_0}(y_2^d|\hat{x}_0)\, p_{y_1|x_0}(y_1^d|\hat{x}_0)\, p_{x_0}(\hat{x}_0),$$

$$\vdots$$

$$p_{\mathbf{y}_N, x_0}(\mathbf{y}_N^d, \hat{x}_0) = p_{x_0}(\hat{x}_0) \prod_{k=1}^{N} p_{y_k|x_0}(y_k^d|\hat{x}_0). \tag{9.3}$$

Also note from (9.1) and (9.2) that

$$p_{y_k|x_0}(y_k^d|\hat{x}_0) = p_v(y_k^d - C\hat{x}_0)$$

$$= \begin{cases} \dfrac{\frac{1}{\sigma\sqrt{2\pi}} \exp\left\{-\frac{(y_k^d - C\hat{x}_0)^2}{2\sigma^2}\right\}}{\int_{-b}^{b} \frac{1}{\sigma\sqrt{2\pi}} \exp\left\{-\frac{\alpha^2}{2\sigma^2}\right\} d\alpha} & \text{if } |y_k^d - C\hat{x}_0| \le b, \\[4mm] 0 & \text{otherwise.} \end{cases}$$

Then, using the above in (9.3), we finally obtain

$$p_{\mathbf{y}_N, x_0}(\mathbf{y}_N^d, \hat{x}_0) = \begin{cases} f_1(\mathbf{y}_N^d, \hat{x}_0) & \text{if } |y_k^d - C\hat{x}_0| \le b, \quad k = 1, \ldots, N, \\ 0 & \text{otherwise,} \end{cases}$$

$$(9.4)$$

where

$$f_1(\mathbf{y}_N^d, \hat{x}_0) \triangleq \beta \exp\left\{ -(\hat{x}_0 - \mu_0)^{\mathsf{T}} \frac{P_0^{-1}}{2}(\hat{x}_0 - \mu_0) \right\}$$

$$\times \prod_{k=1}^{N} \frac{\frac{1}{\sigma\sqrt{2\pi}} \exp\left\{ \frac{-(y_k^d - C\hat{x}_0)^2}{2\sigma^2} \right\}}{\int_{-b}^{b} \frac{1}{\sigma\sqrt{2\pi}} \exp\left\{ \frac{-\alpha^2}{2\sigma^2} \right\} d\alpha}, \quad (9.5)$$

where $\beta \triangleq (2\pi)^{-\frac{n}{2}} (\det P_0)^{-\frac{1}{2}}$.

The estimation problem is as follows: Given $\mathbf{y}_N^d$, make some statement about the value of x_0. Based on $p_{\mathbf{y}_N, x_0}(\mathbf{y}_N^d, \hat{x}_0)$ we can express the *a posteriori* distribution of x_0 given $\mathbf{y}_N$ as follows:

$$p_{x_0|\mathbf{y}_N}(\hat{x}_0|\mathbf{y}_N^d) = \frac{p_{\mathbf{y}_N, x_0}(\mathbf{y}_N^d, \hat{x}_0)}{p_{\mathbf{y}_N}(\mathbf{y}_N^d)}, \quad (9.6)$$

where $p_{\mathbf{y}_N}(\mathbf{y}_N^d)$ is independent of x_0 and satisfies

$$p_{\mathbf{y}_N}(\mathbf{y}_N^d) = \int_{\mathbb{R}^n} p_{\mathbf{y}_N, x_0}(\mathbf{y}_N^d, \alpha) d\alpha. \quad (9.7)$$

The a posteriori distribution $p_{x_0|\mathbf{y}_N}(\hat{x}_0|\mathbf{y}_N^d)$ summarises "what we know about x_0 given the observations $\mathbf{y}_N^d$." If we require a specific estimate, then we can obtain this from $p_{x_0|\mathbf{y}_N}(\hat{x}_0|\mathbf{y}_N^d)$. Possible estimates are:

(i) Conditional mean

$$\hat{x}_0^{[1]} \triangleq \mathbf{E}\{x_0|\mathbf{y}_N^d\} = \int_{\mathbb{R}^n} \alpha \, p_{x_0|\mathbf{y}_N}(\alpha|\mathbf{y}_N^d) \, d\alpha. \quad (9.8)$$

(ii) A posteriori most probable

$$\hat{x}_0^{[2]} \triangleq \arg\max_{\hat{x}_0} \; p_{x_0|\mathbf{y}_N}(\hat{x}_0|\mathbf{y}_N^d) = \arg\max_{\hat{x}_0} \; p_{\mathbf{y}_N, x_0}(\mathbf{y}_N^d, \hat{x}_0). \quad (9.9)$$

Note that, in general, $\hat{x}_0^{[1]} \ne \hat{x}_0^{[2]}$. A simple two-state case is illustrated in Figure 9.1. In the *unconstrained* Gaussian case we have that the conditional mean coincides with the a posteriori most probable estimate (denoted $\hat{x}_0$ in the figure). However, in the presence of constraints, the a posteriori probability density is nonzero only in a restricted region illustrated by the shaded area[1] in

---

[1] For simplicity, all the truncated distributions are illustrated in this chapter without scaling.

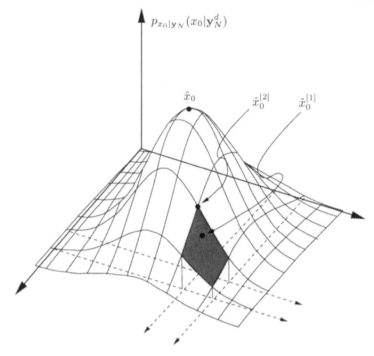

Figure 9.1. Illustration of the conditional mean and the a posteriori most probable estimate. (The points shown should actually be on the x_0-plane.)

Figure 9.1. In this case, we see that the conditional mean $\hat{x}_0^{[1]}$ will, in general, differ from the a posteriori most probable $\hat{x}_0^{[2]}$.

In the sequel, we will mainly focus on the a posteriori most probable estimate since this is found via a *constrained optimisation procedure* which is similar to the optimal control problems addressed earlier.

Returning to our special case of simple linear regression, we see from (9.9), that the a posteriori most probable estimate is obtained by maximising (9.4)–(9.5). In turn, this is equivalent to minimising $-\ln p_{\mathbf{y}_N, x_0}(\mathbf{y}_N^d, \hat{x}_0)$ where

$$-\ln p_{\mathbf{y}_N, x_0}(\mathbf{y}_N^d, \hat{x}_0) = \sum_{k=1}^{N} \frac{1}{2\sigma^2} \hat{v}_k^2 + \frac{1}{2}(\hat{x}_0 - \mu_0)^{\mathsf{T}} P_0^{-1}(\hat{x}_0 - \mu_0) + \text{ constant},$$

subject to the constraints

$$\hat{v}_k = y_k^d - C\hat{x}_0 \quad \text{for } k = 1, \ldots, N,$$
$$\hat{v}_k \in [-b, b] \quad \text{for } k = 1, \ldots, N.$$

We recognise this as a standard constrained quadratic optimisation problem in the variable $\hat{x}_0$.

9.3 Linear State Estimation with Constraints

Here we generalise the ideas presented in Section 9.2 to the following linear Markov model:

$$x_{k+1} = Ax_k + Bw_k,$$
$$y_k = Cx_k + v_k,$$

(9.10)

where $x_k \in \mathbb{R}^n$, $w_k \in \mathbb{R}^m$, $y_k \in \mathbb{R}^r$ and $v_k \in \mathbb{R}^r$. Suppose that $\{w_k\}$, $\{v_k\}$, x_0 are i.i.d. sequences having truncated Gaussian distributions, that is,

$$p_w(w_k) = \begin{cases} \dfrac{\beta_w \exp\left\{-\frac{1}{2}w_k^{\mathsf{T}}Q^{-1}w_k\right\}}{\beta_w \int_{\Omega_1} \exp\left\{-\frac{1}{2}\nu^{\mathsf{T}}Q^{-1}\nu\right\} d\nu} & \text{for } w_k \in \Omega_1, \\[4mm] 0 & \text{otherwise}, \end{cases}$$

(9.11)

$$p_v(v_k) = \begin{cases} \dfrac{\beta_v \exp\left\{-\frac{1}{2}v_k^{\mathsf{T}}R^{-1}v_k\right\}}{\beta_v \int_{\Omega_2} \exp\left\{-\frac{1}{2}\nu^{\mathsf{T}}R^{-1}\nu\right\} d\nu} & \text{for } v_k \in \Omega_2, \\[4mm] 0 & \text{otherwise}, \end{cases}$$

(9.12)

$$p_{x_0}(x_0) = \begin{cases} \dfrac{\beta_{x_0} \exp\left\{-\frac{1}{2}(x_0 - \mu_0)^{\mathsf{T}}P_0^{-1}(x_0 - \mu_0)\right\}}{\beta_{x_0} \int_{\Omega_3} \exp\left\{-\frac{1}{2}(\nu - \mu_0)^{\mathsf{T}}P_0^{-1}(\nu - \mu_0)\right\} d\nu} & \text{for } x_0 \in \Omega_3, \\[4mm] 0 & \text{otherwise}, \end{cases}$$

(9.13)

where $Q > 0$, $R > 0$, $P_0 > 0$, $\beta_w \triangleq (2\pi)^{-\frac{m}{2}}(\det Q)^{-\frac{1}{2}}$, $\beta_v \triangleq (2\pi)^{-\frac{r}{2}}(\det R)^{-\frac{1}{2}}$, $\beta_{x_0} \triangleq (2\pi)^{-\frac{n}{2}}(\det P_0)^{-\frac{1}{2}}$, $\Omega_1 \subset \mathbb{R}^m$, $\Omega_2 \subset \mathbb{R}^r$ and $\Omega_3 \subset \mathbb{R}^n$.

We define

$$\mathbf{y}_N = \begin{bmatrix} y_1^{\mathsf{T}} & \cdots & y_N^{\mathsf{T}} \end{bmatrix}^{\mathsf{T}},$$

(9.14)

$$\mathbf{y}_N^d = \begin{bmatrix} y_1^{d\,\mathsf{T}} & \cdots & y_N^{d\,\mathsf{T}} \end{bmatrix}^{\mathsf{T}},$$

(9.15)

$$\mathbf{x}_N = \begin{bmatrix} x_0^{\mathsf{T}} & \cdots & x_N^{\mathsf{T}} \end{bmatrix}^{\mathsf{T}},$$

(9.16)

$$\hat{\mathbf{x}}_N = \begin{bmatrix} \hat{x}_0^{\mathsf{T}} & \cdots & \hat{x}_N^{\mathsf{T}} \end{bmatrix}^{\mathsf{T}}.$$

(9.17)

From Bayes' rule and the Markovian structure of (9.10) we have that

$$p_{x_{k+1},\ldots,x_0}(\hat{x}_{k+1}, \hat{x}_k, \hat{x}_{k-1}, \ldots, \hat{x}_0) = p_{x_{k+1}|x_k,\ldots,x_0}(\hat{x}_{k+1}|\hat{x}_k, \hat{x}_{k-1}, \ldots, \hat{x}_0)$$
$$\times p_{x_k,\ldots,x_0}(\hat{x}_k, \hat{x}_{k-1}, \ldots, \hat{x}_0)$$
$$= p_{x_{k+1}|x_k}(\hat{x}_{k+1}|\hat{x}_k)$$
$$\times p_{x_k,\ldots,x_0}(\hat{x}_k, \hat{x}_{k-1}, \ldots, \hat{x}_0),$$

and also

$$
\begin{aligned}
p_{y_k,x_k,\ldots,x_0}(\hat{y}_k^d,\hat{x}_k,\hat{x}_{k-1},\ldots,\hat{x}_0) &= p_{y_k|x_k,\ldots,x_0}(y_k^d|\hat{x}_k,\hat{x}_{k-1},\ldots,\hat{x}_0) \\
&\quad \times p_{x_k,\ldots,x_0}(\hat{x}_k,\hat{x}_{k-1},\ldots,\hat{x}_0) \\
&= p_{y_k|x_k}(y_k^d|\hat{x}_k) \\
&\quad \times p_{x_k,\ldots,x_0}(\hat{x}_k,\hat{x}_{k-1},\ldots,\hat{x}_0).
\end{aligned}
$$

It then follows that the joint probability density function for $\mathbf{y}_N$ and $\mathbf{x}_N$ defined in (9.14) and (9.16), respectively, is given by

$$
\begin{aligned}
p_{\mathbf{y}_N,\mathbf{x}_N}(\mathbf{y}_N = \mathbf{y}_N^d, \mathbf{x}_N = \hat{\mathbf{x}}_N) = p_{x_0}(x_0 = \hat{x}_0) \prod_{k=1}^{N} \Big[& p_{y_k|x_k}(y_k = y_k^d|x_k = \hat{x}_k) \\
& \times p_{x_k|x_{k-1}}(x_k = \hat{x}_k|x_{k-1} = \hat{x}_{k-1}) \Big].
\end{aligned}
$$

$$(9.18)$$

We next develop an explicit expression for the joint density function in (9.18). We begin with the nonsingular case when $w_k \in \mathbb{R}^n$ ($m = n$) and B is nonsingular in (9.10).

Lemma 9.3.1 *For the model described in (9.10) to (9.17), and subject to $w_k \in \mathbb{R}^n$ ($m = n$) and B nonsingular, the joint probability density function (9.18) for $\mathbf{y}_N$ and $\mathbf{x}_N$ satisfies*

$$
\begin{aligned}
p_{\mathbf{y}_N,\mathbf{x}_N}(\mathbf{y}_N = \mathbf{y}_N^d, \mathbf{x}_N = \hat{\mathbf{x}}_N) &= \text{constant} \times \exp\left\{ -\frac{1}{2} \sum_{k=0}^{N-1} \hat{w}_k^{\mathsf{T}} Q^{-1} \hat{w}_k \right\} \\
&\quad \times \exp\left\{ -\frac{1}{2} \sum_{k=1}^{N} \hat{v}_k^{\mathsf{T}} R^{-1} \hat{v}_k \right\} \\
&\quad \times \exp\left\{ -\frac{1}{2}(\hat{x}_0 - \mu_0)^{\mathsf{T}} P_0^{-1}(\hat{x}_0 - \mu_0) \right\},
\end{aligned}
$$

$$(9.19)$$

whenever

$$
\begin{aligned}
\hat{w}_k &\in \Omega_1 \quad \textit{for } k = 0,\ldots,N-1, \\
\hat{v}_k &\in \Omega_2 \quad \textit{for } k = 1,\ldots,N, \\
\hat{x}_0 &\in \Omega_3,
\end{aligned}
$$

where

$$
\begin{aligned}
\hat{x}_{k+1} &= A\hat{x}_k + B\hat{w}_k \quad \textit{for } k = 0,\ldots,N-1, \\
\hat{v}_k &= y_k^d - C\hat{x}_k \quad \textit{for } k = 1,\ldots,N.
\end{aligned}
$$

Proof. From (9.10), (9.11) and (9.12), we have, using the rule of transformation of probability density functions:

$$p_{x_{k+1}|x_k}(x_{k+1} = \hat{x}_{k+1}|x_k = \hat{x}_k) = \text{constant} \times p_w(\hat{w}_k)$$

$$= \text{constant} \times \exp\left\{-\frac{1}{2}\hat{w}_k^{\mathrm{T}}Q^{-1}\hat{w}_k\right\},$$

whenever $\hat{w}_k \in \Omega_1$ and satisfies $\hat{x}_{k+1} = A\hat{x}_k + B\hat{w}_k$. Also,

$$p_{y_k|x_k}(y_k = y_k^d|x_k = \hat{x}_k) = \text{constant} \times p_v(\hat{v}_k)$$

$$= \text{constant} \times \exp\left\{-\frac{1}{2}\hat{v}_k^{\mathrm{T}}R^{-1}\hat{v}_k\right\},$$

whenever $\hat{v}_k \in \Omega_2$ and satisfies $y_k^d = C\hat{x}_k + \hat{v}_k$. Finally, using (9.13), and substituting all expressions into (9.18), the result follows.   □

Remark 9.3.1. In the general case, when $w_k \in \mathbb{R}^m$ with $m < n$ in (9.10), the linear equality $\hat{x}_{k+1} - A\hat{x}_k = B\hat{w}_k$, implies that $\hat{x}_{k+1} - A\hat{x}_k$ can only take values in the range space of B. Hence, we need to account for the fact that $\hat{x}_{k+1} - A\hat{x}_k$ has a singular distribution[2] in $\mathbb{R}^n$. We can easily deal with this situation by introducing a linear transformation in the state space as follows.

Assume that B has full column rank. Let T_1 be a basis for the range space of B (which, in particular, could be chosen equal to B) and choose any T_2 such that $T = [T_1 \ T_2]$ is nonsingular. We partition T^{-1} as follows:

$$T^{-1} = \begin{bmatrix} S_1 \\ S_2 \end{bmatrix},$$

where S_1 is an $m \times n$ matrix. Then $T^{-1}T = I_n$ implies

$$S_1 T_1 = I_m, \quad S_2 T_1 = 0_{(n-m)\times m}.$$

Hence, since $B = T_1 \bar{B}_1$ for some nonsingular $m \times m$ matrix $\bar{B}_1$, we have, using the above equations, that

$$T^{-1}B = \begin{bmatrix} S_1 \\ S_2 \end{bmatrix} T_1 \bar{B}_1 = \begin{bmatrix} \bar{B}_1 \\ 0 \end{bmatrix}. \tag{9.20}$$

Partition $\bar{x}_{k+1}$ as

$$\bar{x}_{k+1} \triangleq T^{-1}x_{k+1}. \tag{9.21}$$

Then, from (9.10), $\bar{x}_{k+1}$ satisfies

$$\bar{x}_{k+1} = \bar{A}x_k + \bar{B}w_k, \tag{9.22}$$

---

[2] A singular distribution is a distribution in $\mathbb{R}^n$ which is concentrated in a lower dimensional subspace, that is, the probability associated with any set not intersecting the subspace is zero (Anderson 1958).

where

$$\bar{A} \triangleq T^{-1} A \triangleq \begin{bmatrix} \bar{A}_1 \\ \bar{A}_2 \end{bmatrix}, \qquad \bar{B} \triangleq T^{-1} B = \begin{bmatrix} \bar{B}_1 \\ 0 \end{bmatrix}, \tag{9.23}$$

using (9.20). Let

$$\bar{x}_{k+1} \triangleq \begin{bmatrix} \bar{x}'_{k+1} \\ \bar{x}''_{k+1} \end{bmatrix},$$

where $\bar{x}'_{k+1} \in \mathbb{R}^m$. Then, from (9.22)–(9.23), we can write

$$\begin{bmatrix} \bar{x}'_{k+1} \\ \bar{x}''_{k+1} \end{bmatrix} = \begin{bmatrix} \bar{A}_1 \\ \bar{A}_2 \end{bmatrix} x_k + \begin{bmatrix} \bar{B}_1 \\ 0 \end{bmatrix} w_k. \tag{9.24}$$

Hence, using the rule of transformation of probability density functions, we have, from (9.21) and (9.24), that

$$\begin{aligned}
p_{x_{k+1}|x_k}(\hat{x}_{k+1}|\hat{x}_k) &= \text{constant} \times p_{\bar{x}_{k+1}|x_k}(\hat{\bar{x}}_{k+1}|\hat{x}_k) \\
&= \text{constant} \times p_{\bar{x}'_{k+1}|x_k}(\hat{\bar{x}}'_{k+1}|\hat{x}_k) \times \delta_{n-m}[\hat{\bar{x}}''_{k+1} - \bar{A}_2 \hat{x}_k] \\
&= \text{constant} \times p_w(\hat{w}_k) \times \delta_{n-m}[\hat{\bar{x}}''_{k+1} - \bar{A}_2 \hat{x}_k],
\end{aligned}$$

whenever $\hat{w}_k \in \Omega_1$ and satisfies $\hat{\bar{x}}'_{k+1} = \bar{A}_1 \hat{x}_k + \bar{B}_1 \hat{w}_k$. In the above equations, $\delta_{n-m}[\cdot]$ is the Dirac delta function defined on $\mathbb{R}^{n-m}$, that is, $\delta_{n-m}[\eta] = \delta(\eta_1) \times \cdots \times \delta(\eta_{n-m})$, where $\eta = \begin{bmatrix} \eta_1 & \cdots & \eta_{n-m} \end{bmatrix}^T \in \mathbb{R}^{n-m}$.

We can thus write

$$p_{x_{k+1}|x_k}(\hat{x}_{k+1}|\hat{x}_k) = \text{constant} \times p_w(\hat{w}_k) \times \delta_{n-m}[\hat{\bar{x}}''_{k+1} - \bar{A}_2 \hat{x}_k], \tag{9.25}$$

where $\hat{x}_{k+1}$ is restricted to those values reachable from $\hat{w}_k$, that is, such that $\hat{x}_{k+1} = A\hat{x}_k + B\hat{w}_k$ for some $\hat{w}_k \in \Omega_1$. We thus see that $p_{x_{k+1}|x_k}(\cdot \mid \cdot)$ has a density function in $\mathbb{R}^n$ corresponding to those values of x_{k+1} that are reachable from w_k.

When defining the joint a posteriori most probable [JAPMP] estimator below, we will maximise the envelope of the delta function in (9.25). For notational convenience, we define this envelope as

$$p'_{x_{k+1}|x_k}(\hat{x}_{k+1}|\hat{x}_k) \triangleq \text{constant} \times p_w(\hat{w}_k)$$

whenever $\hat{w}_k \in \Omega_1$ and satisfies $\hat{x}_{k+1} = A\hat{x}_k + B\hat{w}_k$. Hence, in the sequel, probability densities p corresponding to singular distributions should be interpreted as the envelope p' defined above.   ∘

The general estimation problem is: Given the observations $\mathbf{y}_N^d = [y_1^{d^T} \ldots y_N^{d^T}]^T$, make some statement about the states $\mathbf{x}_N = [x_0^T \ldots x_N^T]^T$. From the joint probability density function (9.19), we can express the a posteriori distribution of $\mathbf{x}_N$ given $\mathbf{y}_N$ as follows:

$$p_{\mathbf{x}_N|\mathbf{y}_N}(\hat{\mathbf{x}}_N|\mathbf{y}_N^d) = \frac{p_{\mathbf{y}_N, \mathbf{x}_N}(\mathbf{y}_N^d, \hat{\mathbf{x}}_N)}{p_{\mathbf{y}_N}(\mathbf{y}_N^d)}, \tag{9.26}$$

where $p_{\mathbf{y}_N}(\mathbf{y}_N^d)$ is a data dependent term which does not depend on $\mathbf{x}_N$.

The a posteriori distribution $p_{\mathbf{x}_N|\mathbf{y}_N}(\hat{\mathbf{x}}_N|\mathbf{y}_N^d)$ summarises "what we know about $\mathbf{x}_N$ given the observations $\mathbf{y}_N^d$." As foreshadowed in Remark 9.3.1, our aim is to find the *joint a posteriori most probable* [JAPMP] state estimates $\hat{\mathbf{x}}_N = [\hat{x}_0^T \dots \hat{x}_N^T]^T$ given the observations $\hat{\mathbf{y}}_N^d$; that is,

$$\hat{\mathbf{x}}_N^* \triangleq \arg\max_{\hat{\mathbf{x}}_N} \; p_{\mathbf{x}_N|\mathbf{y}_N}(\hat{\mathbf{x}}_N|\mathbf{y}_N^d). \tag{9.27}$$

Note that (9.27) is equivalent to maximising the joint probability density function, since, as noticed in (9.26), both functions are related by a term that does not depend on $\mathbf{x}_N$. Thus, the joint maximum a posteriori estimate is given by

$$\begin{aligned} \hat{\mathbf{x}}_N^* &\triangleq \arg\max_{\hat{\mathbf{x}}_N} \; p_{\mathbf{x}_N|\mathbf{y}_N}(\hat{\mathbf{x}}_N|\mathbf{y}_N^d) \\ &= \arg\max_{\hat{\mathbf{x}}_N} \; p_{\mathbf{y}_N,\mathbf{x}_N}(\mathbf{y}_N^d, \hat{\mathbf{x}}_N) \\ &= \arg\min_{\hat{\mathbf{x}}_N} \; -\ln p_{\mathbf{y}_N,\mathbf{x}_N}(\mathbf{y}_N^d, \hat{\mathbf{x}}_N). \end{aligned} \tag{9.28}$$

The preceding discussion leads, upon substitution of (9.19) into (9.28), to the following optimisation problem.

Estimation Problem

Given the observations $\{y_1^d, \dots, y_N^d\}$ and the knowledge of μ_0 (the mean value of x_0), solve:

$$\mathcal{P}_e: \qquad V_N^{\text{OPT}}(\mu_0, \{y_k^d\}) \triangleq \min V_N(\{\hat{x}_k\}, \{\hat{v}_k\}, \{\hat{w}_k\}), \tag{9.29}$$

subject to:

$$\hat{x}_{k+1} = A\hat{x}_k + B\hat{w}_k \quad \text{for } k = 0, \dots, N-1, \tag{9.30}$$

$$\hat{v}_k = y_k^d - C\hat{x}_k \quad \text{for } k = 1, \dots, N, \tag{9.31}$$

$$\hat{w}_k \in \Omega_1 \quad \text{for } k = 0, \dots, N-1, \tag{9.32}$$

$$\hat{v}_k \in \Omega_2 \quad \text{for } k = 1, \dots, N, \tag{9.33}$$

$$\hat{x}_0 \in \Omega_3, \tag{9.34}$$

where

$$\begin{aligned} V_N(\{\hat{x}_k\}, \{\hat{v}_k\}, \{\hat{w}_k\}) &\triangleq \frac{1}{2}(\hat{x}_0 - \mu_0)^T P_0^{-1}(\hat{x}_0 - \mu_0) \\ &+ \frac{1}{2}\sum_{k=0}^{N-1} \hat{w}_k^T Q^{-1} \hat{w}_k + \frac{1}{2}\sum_{k=1}^{N} \hat{v}_k^T R^{-1} \hat{v}_k. \end{aligned} \tag{9.35}$$

We see that the above problem is very *similar* to the constrained linear quadratic optimal control problems discussed earlier (see, for example, (5.49)–(5.54) in Chapter 5) save that they have different boundary conditions and initial and terminal state weightings. The two problems are compared in Table 9.1.

	Constrained control	Constrained estimation
Model	$x_{k+1} = Ax_k + Bu_k$	$\hat{x}_{k+1} = A\hat{x}_k + B\hat{w}_k$
Initial condition	x_0 (given)	$\hat{x}_0 \in \Omega_3$
Initial state weighting	$\frac{1}{2}x_0^{\mathrm{T}}Qx_0$ (given)	$\frac{1}{2}(\hat{x}_0 - \mu_0)^{\mathrm{T}}P_0^{-1}(\hat{x}_0 - \mu_0)$, μ_0 given
Terminal state weighting	$\frac{1}{2}x_N^{\mathrm{T}}Px_N$	$\frac{1}{2}(y_N^d - C\hat{x}_N)^{\mathrm{T}}R^{-1}(y_N^d - C\hat{x}_N)$, y_N^d given

Table 9.1. Comparison between the optimisation problems corresponding to constrained control and constrained estimation.

9.4 Extensions to Other Constraints and Distributions

The development in Section 9.3 was based on an assumption of truncated Gaussian noise. This result is interesting in its raw form but becomes a powerful tool when utilised as a basic building block to solve more general problems. Several alternatives are discussed below indicating how the core ideas of Section 9.3 can be used in more general problems.

9.4.1 Nonzero-mean Truncated Gaussian Noise

It is very straightforward to add a nonzero mean assumption to the truncated Gaussian noise assumption. The appropriate changes to (9.11) and (9.12) are

$$
p_w(w_k) = \begin{cases} \dfrac{\beta_w \exp\left\{-\frac{1}{2}(w_k - \mu_w)^{\mathrm{T}}Q^{-1}(w_k - \mu_w)\right\}}{\beta_w \int_{\Omega_1} \exp\left\{-\frac{1}{2}(\nu - \mu_w)^{\mathrm{T}}Q^{-1}(\nu - \mu_w)\right\} d\nu} & \text{for } w_k \in \Omega_1, \\[4mm] 0 & \text{otherwise,} \end{cases}
$$

$$
p_v(v_k) = \begin{cases} \dfrac{\beta_v \exp\left\{-\frac{1}{2}(v_k - \mu_v)^{\mathrm{T}}R^{-1}(v_k - \mu_v)\right\}}{\beta_v \int_{\Omega_2} \exp\left\{-\frac{1}{2}(\nu - \mu_v)^{\mathrm{T}}R^{-1}(\nu - \mu_v)\right\} d\nu} & \text{for } v_k \in \Omega_2, \\[4mm] 0 & \text{otherwise,} \end{cases}
$$

where μ_w and μ_v are the "prior" means, that is, the means of the Gaussian distributions before truncation.

The corresponding change in the objective function (9.35) is

$$V_N(\{\hat{x}_k\},\{\hat{v}_k\},\{\hat{w}_k\}) \triangleq \frac{1}{2}(\hat{x}_0 - \mu_0)^{\mathrm{T}} P_0^{-1}(\hat{x}_0 - \mu_0)$$

$$+ \frac{1}{2}\sum_{k=0}^{N-1}(\hat{w}_k - \mu_w)^{\mathrm{T}} Q^{-1}(\hat{w}_k - \mu_w)$$

$$+ \frac{1}{2}\sum_{k=1}^{N}(\hat{v}_k - \mu_v)^{\mathrm{T}} R^{-1}(\hat{v}_k - \mu_v).$$

The use of a nonzero mean for the underlying distribution allows one, for example, to build new zero-mean distributions such as the one illustrated in Figure 9.2.

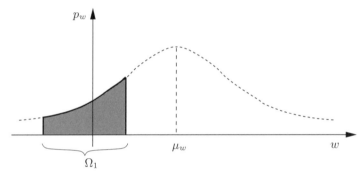

Figure 9.2. Zero-mean distribution formed by truncating a nonzero-mean Gaussian distribution.

9.4.2 Combinations of Truncated Gaussian Noise

A further embellishment is to have different truncated Gaussian distributions in different regions. For example, we could have

$$p_w(w_k) = \frac{\beta_{w_i} \exp\left\{-\frac{1}{2}(w_k - \mu_i)^{\mathrm{T}} Q_i^{-1}(w_k - \mu_i)\right\}}{\sum_{i=1}^{L} \beta_{w_i} \int_{\Omega_i} \exp\left\{-\frac{1}{2}(\nu - \mu_i)^{\mathrm{T}} Q_i^{-1}(\nu - \mu_i)\right\} d\nu},$$

for $w_k \in \Omega_i$, $i = 1,\ldots,L$, and zero otherwise, where $\Omega_i \subset \mathbb{R}^m$ are convex sets that have an empty intersection pairwise. A simple example is shown in Figure 9.3.

The associated optimisation problem can be solved by partitioning the problem into constrained sub-problems, each of which is convex in a convex

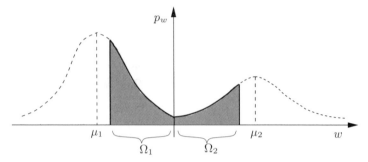

Figure 9.3. Combination of two nonzero-mean truncated Gaussian distributions.

region. One then simply chooses the global optimum as the minimum of the individual sub-problems. This idea was described in general terms in Section 2.7 of Chapter 2.

Thus, say that we have a scalar disturbance $\{w_k\}$ and an N-step optimisation horizon. Also, say that the distribution of $\{w_k\}$ is divided into L nonoverlapping regions, each containing a different truncated Gaussian distribution. Then one needs to solve L^N separate QP problems. As an illustration, with $L = 2$ (as in Figure 9.3) and $N = 5$, then one needs to solve $2^5 = 32$ QP problems.

Remark 9.4.1. Actually, the above idea is an interesting precursor to ideas that will be presented in Chapter 13 when we treat finite alphabet estimation problems. The latter case can be thought of as the limiting version of the idea presented above in which each region contains a *point mass distribution*. In this case, the optimisation problem requires L^N objective function evaluations rather than L^N QP problems.   ∘

9.4.3 Multiconvex Approximations of Arbitrary Distributions

A further generalisation of these ideas is to use a staircase approximation to an arbitrary distribution. Thus, consider the smooth, but otherwise arbitrary, distribution in Figure 9.4, together with a staircase approximation.

In each region, the probability density function is approximated by a uniform distribution, that is,

$$p_w(w_k) \approx c_i \quad \text{for} \quad w_k \in \Omega_i,$$

where $c_i > 0$ is a constant. In this case, we have

$$\ln p_w(w_k) \approx \ln c_i \quad \text{for} \quad w_k \in \Omega_i.$$

The objective function (9.35) splits into L^N (where L is the number of regions Ω_i in the staircase approximation) convex functions V_N^i, each of them

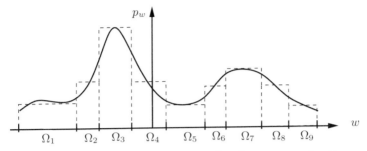

Figure 9.4. Arbitrary distribution and staircase approximation.

having the form:[3]

$$V_N^i(\{\hat{x}_k\}, \{\hat{v}_k\}, \{\hat{w}_k\}) \triangleq \frac{1}{2}(\hat{x}_0 - \mu_0)^\mathsf{T} P_0^{-1}(\hat{x}_0 - \mu_0)$$
$$- \sum_{k=0}^{N-1} \ln \ell_k + \frac{1}{2}\sum_{k=1}^{N} \hat{v}_k^\mathsf{T} R^{-1} \hat{v}_k,$$

where $\ell_k \in \{c_1, \ldots, c_L\}$, $k = 0, \ldots, N-1$.

The global solution is computed as the minimum of the L^N convex optimisation sub-problems. (Note that the term $-\sum_{k=0}^{N-1} \ln \ell_k$ is constant for each sub-problem and, hence, it does not affect each minimiser. However, these terms must be included in the evaluation of each sub-problem when computing the global optimum.)

9.4.4 Discussion

We have seen above that one can treat very general estimation problems by combining convex optimisation with constraints. Note that the juxtaposition of *constraints* and regional *convexity* is the key idea to solving these problems.

9.5 Dynamic Programming

As for constrained control problems, we can utilise dynamic programming to solve the constrained estimation problem. Here it is most convenient to use *forward dynamic programming* whereas previously we used *reverse dynamic programming* (see Section 3.4 in Chapter 3).

We return to the problem of constrained estimation described in (9.29)–(9.35). We note that the objective function from time 0 to k (that is, (9.35) for $N = k$) is a function of the initial state estimate $\hat{x}_0$, the choice of the input

---

[3] Notice that we assume that v_k has a Gaussian distribution, but the idea is readily extended to arbitrary distributions for v_k, also.

noise sequence $\hat{w}_0, \ldots, \hat{w}_{k-1}$, and the given data $\mu_0, y_1^d, \ldots y_k^d$. If, for given $\hat{x}_0$, we optimise with respect to $\hat{w}_0, \ldots, \hat{w}_{k-1}$, then the resulting partial value function (at time k) is a function of $\hat{x}_0$ and $\mu_0, y_1^d, \ldots y_k^d$. For the purposes of the dynamic programming argument it is actually more convenient to make the partial value function a function of $\hat{x}_k$ and $\mu_0, y_1^d, \ldots y_k^d$. This is possible since (9.30) allows us to express $\hat{x}_0$ as a function of $\hat{x}_k$ (together with the given sequence $\hat{w}_0, \ldots, \hat{w}_{k-1}$) provided A is nonsingular. Thus, assuming that A is nonsingular, the partial value function at time k is

$$V_k^{\text{OPT}}(\hat{x}_k, \mu_0, y_1^d, \ldots, y_k^d) \triangleq \min_{\hat{w}_0, \ldots, \hat{w}_{k-1}} \left\{ \frac{1}{2}(\hat{x}_0 - \mu_0)^{\text{T}} P_0^{-1}(\hat{x}_0 - \mu_0) \right.$$

$$\left. + \frac{1}{2}\sum_{j=0}^{k-1} \hat{w}_j^{\text{T}} Q^{-1} \hat{w}_j + \frac{1}{2}\sum_{j=1}^{k} (y_j^d - C\hat{x}_j)^{\text{T}} R^{-1}(y_j^d - C\hat{x}_j) \right\},$$

subject to:

$$\hat{x}_j = A^{-1}(\hat{x}_{j+1} - B\hat{w}_j) \quad \text{for } j = 0, \ldots, k-1, \tag{9.36}$$

$$\hat{w}_j \in \Omega_1 \quad \text{for } j = 0, \ldots, k-1, \tag{9.37}$$

$$y_j^d - C\hat{x}_j \in \Omega_2 \quad \text{for } j = 1, \ldots, k, \tag{9.38}$$

$$\hat{x}_0 \in \Omega_3. \tag{9.39}$$

Then, the forward dynamic programming algorithm proceeds as follows. We start with the partial value function at time 0, which, for $\hat{x}_0 \in \Omega_3$, is defined as

$$V_0^{\text{OPT}}(\hat{x}_0, \mu_0) \triangleq \frac{1}{2}(\hat{x}_0 - \mu_0)^{\text{T}} P_0^{-1}(\hat{x}_0 - \mu_0). \tag{9.40}$$

Next, for $\hat{x}_1 \in \mathbb{R}^n$ such that $y_1^d - C\hat{x}_1 \in \Omega_2$, the partial value function at time 1 is computed as

$$V_1^{\text{OPT}}(\hat{x}_1, \mu_0, y_1^d) = \min_{\hat{w}_0} \left\{ V_0^{\text{OPT}}(A^{-1}\hat{x}_1 - A^{-1}B\hat{w}_0, \mu_0) + \frac{1}{2}\hat{w}_0^{\text{T}} Q^{-1} \hat{w}_0 \right.$$

$$\left. + \frac{1}{2}(y_1^d - C\hat{x}_1)^{\text{T}} R^{-1}(y_1^d - C\hat{x}_1) \right\}, \tag{9.41}$$

subject to:

$$\hat{w}_0 \in \Omega_1, \tag{9.42}$$

$$A^{-1}\hat{x}_1 - A^{-1}B\hat{w}_0 \in \Omega_3. \tag{9.43}$$

Finally, for $k \geq 1$, and $\hat{x}_{k+1} \in \mathbb{R}^n$ such that $y_{k+1}^d - C\hat{x}_{k+1} \in \Omega_2$,

$$V_{k+1}^{\mathrm{OPT}}(\hat{x}_{k+1}, \mu_0, y_1^d, \ldots, y_{k+1}^d)$$

$$= \min_{\hat{w}_k} \left\{ V_k^{\mathrm{OPT}}(A^{-1}\hat{x}_{k+1} - A^{-1}B\hat{w}_k, \mu_0, y_1^d, \ldots, y_k^d) + \frac{1}{2}\hat{w}_k^{\mathrm{T}}Q^{-1}\hat{w}_k \right.$$

$$\left. + \frac{1}{2}(y_{k+1}^d - C\hat{x}_{k+1})^{\mathrm{T}}R^{-1}(y_{k+1}^d - C\hat{x}_{k+1}) \right\}, \tag{9.44}$$

subject to:

$$\hat{w}_k \in \Omega_1, \tag{9.45}$$

$$y_k^d - C(A^{-1}\hat{x}_{k+1} - A^{-1}B\hat{w}_k) \in \Omega_2. \tag{9.46}$$

In the absence of constraints, the above dynamic programming algorithm leads to the well-known Kalman filter. This is explained in the next section.

9.6 Linear Gaussian Unconstrained Problems

For the case of linear Gaussian unconstrained problems, the dynamic programming algorithm of Section 9.5 can be solved explicitly. As expected, the optimal estimator in this case is the Kalman filter, as we show in the following results.

Lemma 9.6.1 (Dynamic Programming for Linear Gaussian Estimation) *Assume that A is nonsingular.[4] In the absence of constraints (that is, $\Omega_1 = \mathbb{R}^m$ in (9.37), $\Omega_2 = \mathbb{R}^r$ in (9.38) and $\Omega_3 = \mathbb{R}^n$ in (9.39)), the dynamic programming problem specified in (9.40)–(9.46) has the solution*

$$V_k^{\mathrm{OPT}}(\hat{x}_k, \mu_0, y_1^d, \ldots, y_k^d) = \frac{1}{2}(\hat{x}_k - \hat{x}_{k|k})^{\mathrm{T}}P_{k|k}^{-1}(\hat{x}_k - \hat{x}_{k|k}) + \text{ constant}, \tag{9.47}$$

where $\hat{x}_{k|k}$ is a function of $\mu_0, y_1^d, \ldots, y_k^d$ defined via the following recursion:

$$\hat{x}_{0|0} = \mu_0, \tag{9.48}$$

$$P_{0|0} = P_0, \tag{9.49}$$

and, for $j = 0, \ldots, k-1$,

$$\hat{x}_{j+1|j} = A\hat{x}_{j|j}, \tag{9.50}$$

$$\hat{x}_{j+1|j+1} = \hat{x}_{j+1|j} + P_{j+1|j}C^{\mathrm{T}}(R + CP_{j+1|j}C^{\mathrm{T}})^{-1}(y_{j+1}^d - C\hat{x}_{j+1|j}), \tag{9.51}$$

$$P_{j+1|j} = AP_{j|j}A^{\mathrm{T}} + BQB^{\mathrm{T}}, \tag{9.52}$$

$$P_{j+1|j+1} = P_{j+1|j} - P_{j+1|j}C^{\mathrm{T}}(R + CP_{j+1|j}C^{\mathrm{T}})^{-1}CP_{j+1|j}. \tag{9.53}$$

Proof. We use induction, and assume that $V_k^{\mathrm{OPT}}(\hat{x}_k, \mu_0, y_1^d, \ldots, y_k^d)$ is a quadratic function of $\hat{x}_k$ of the form

---

[4] Here we assume A nonsingular, but the result holds for any matrix A.

$$V_k^{\text{OPT}}(\hat{x}_k, \mu_0, y_1^d, \ldots, y_k^d) = \frac{1}{2}(\hat{x}_k - \hat{x}_{k|k})^{\mathrm{T}} P_{k|k}^{-1}(\hat{x}_k - \hat{x}_{k|k}) + \text{constant}, \quad (9.54)$$

where $\hat{x}_{k|k}$ is the function of $\mu_0, y_1^d, \ldots, y_k^d$ defined via (9.48)–(9.53) for $j = 0, \ldots, k-1$. We note from (9.40) and (9.48)–(9.49) that the induction hypothesis holds for $k = 0$.

We next assume that (9.54) holds for k and show, by performing the minimisation (9.44), that the results holds for $k+1$. (Note that, in this case, we should obtain that equations (9.50)–(9.53) apply for $j = k$.)

Step 1: As the first step towards performing the minimisation in (9.44), we begin by adding the term $\frac{1}{2}\hat{w}_k^{\mathrm{T}} Q^{-1} \hat{w}_k$ to (9.54) and substituting $\hat{x}_k = A^{-1}\hat{x}_{k+1} - A^{-1}B\hat{w}_k$, and then minimise with respect to $\hat{w}_k$. We denote the resulting value function by $W_{k+1}^{\text{OPT}}(\hat{x}_{k+1}, \mu_0, y_1^d, \ldots, y_k^d)$, that is:

$$W_{k+1}^{\text{OPT}}(\hat{x}_{k+1}, \mu_0, y_1^d, \ldots, y_k^d) = \min_{\hat{w}_k} \left\{ \frac{1}{2}(A^{-1}\hat{x}_{k+1} - A^{-1}B\hat{w}_k - \hat{x}_{k|k})^{\mathrm{T}} P_{k|k}^{-1} \right.$$
$$(A^{-1}\hat{x}_{k+1} - A^{-1}B\hat{w}_k - \hat{x}_{k|k})$$
$$\left. + \frac{1}{2}\hat{w}_k^{\mathrm{T}} Q^{-1}\hat{w}_k \right\} + \text{constant}. \quad (9.55)$$

Differentiating the argument of the min in (9.55) with respect to $\hat{w}_k$ and equating to zero gives

$$B^{\mathrm{T}} A^{-T} P_{k|k}^{-1} \left(\hat{x}_{k|k} - A^{-1}\hat{x}_{k+1} + A^{-1}B\hat{w}_k\right) + Q^{-1}\hat{w}_k = 0,$$

or

$$\hat{w}_k = -\left(B^{\mathrm{T}} A^{-T} P_{k|k}^{-1} A^{-1} B + Q^{-1}\right)^{-1} B^{\mathrm{T}} A^{-T} P_{k|k}^{-1} \left(\hat{x}_{k|k} - A^{-1}\hat{x}_{k+1}\right)$$
$$\triangleq -(\Gamma + \Theta)^{-1} B^{\mathrm{T}} A^{-T} P_{k|k}^{-1} \alpha, \quad (9.56)$$

where we have used the definitions

$$\Gamma \triangleq B^{\mathrm{T}} A^{-T} P_{k|k}^{-1} A^{-1} B, \quad (9.57)$$

$$\Theta \triangleq Q^{-1}, \quad (9.58)$$

$$\alpha \triangleq \left(\hat{x}_{k|k} - A^{-1}\hat{x}_{k+1}\right). \quad (9.59)$$

Back-substituting (9.56) into (9.55), we obtain

$$
\begin{aligned}
W_{k+1}^{\mathrm{OPT}}(\hat{x}_{k+1}, \mu_0, y_1^d, \ldots, y_k^d) = {} & \frac{1}{2}\left[\alpha - A^{-1}B(\Gamma + \Theta)^{-1}B^{\mathrm{T}}A^{-T}P_{k|k}^{-1}\alpha\right]^{\mathrm{T}}P_{k|k}^{-1} \\
& \left[\alpha - A^{-1}B(\Gamma + \Theta)^{-1}B^{\mathrm{T}}A^{-T}P_{k|k}^{-1}\alpha\right] \\
& + \frac{1}{2}\alpha^{\mathrm{T}}P_{k|k}^{-1}A^{-1}B(\Gamma + \Theta)^{-1}\Theta \qquad (9.60) \\
& (\Gamma + \Theta)^{-1}B^{\mathrm{T}}A^{-T}P_{k|k}^{-1}\alpha + \text{ constant}
\end{aligned}
$$

$$
\triangleq \frac{1}{2}\alpha^{\mathrm{T}}S\alpha + \text{ constant}, \qquad (9.61)
$$

where

$$
\begin{aligned}
S = {} & \left[I - A^{-1}B(\Gamma + \Theta)^{-1}B^{\mathrm{T}}A^{-T}P_{k|k}^{-1}\right]^{\mathrm{T}}P_{k|k}^{-1} \\
& \left[I - A^{-1}B(\Gamma + \Theta)^{-1}B^{\mathrm{T}}A^{-T}P_{k|k}^{-1}\right] \\
& + P_{k|k}^{-1}A^{-1}B(\Gamma + \Theta)^{-1}\Theta(\Gamma + \Theta)^{-1}B^{\mathrm{T}}A^{-T}P_{k|k}^{-1} \\
= {} & P_{k|k}^{-1} - 2P_{k|k}^{-1}A^{-1}B(\Gamma + \Theta)^{-1}B^{\mathrm{T}}A^{-T}P_{k|k}^{-1} \\
& + P_{k|k}^{-1}A^{-1}B(\Gamma + \Theta)^{-1}\left\{B^{\mathrm{T}}A^{-T}P_{k|k}^{-1}A^{-1}B\right\}(\Gamma + \Theta)^{-1}B^{\mathrm{T}}A^{-T}P_{k|k}^{-1} \\
& \qquad\qquad\qquad\qquad\qquad\qquad\qquad\qquad\qquad\qquad (9.62) \\
& + P_{k|k}^{-1}A^{-1}B(\Gamma + \Theta)^{-1}\Theta(\Gamma + \Theta)^{-1}B^{\mathrm{T}}A^{-T}P_{k|k}^{-1}.
\end{aligned}
$$

We note that the term in the $\{\ \}$ in (9.62) is equal to Γ defined in (9.57). Hence, the last three terms above can be combined to give

$$
S = P_{k|k}^{-1} - P_{k|k}^{-1}A^{-1}B(\Gamma + \Theta)^{-1}B^{\mathrm{T}}A^{-T}P_{k|k}^{-1}. \qquad (9.63)
$$

Substituting (9.63) and (9.57)–(9.59) in (9.61), we have

$$
\begin{aligned}
W_{k+1}^{\mathrm{OPT}}(\hat{x}_{k+1}, \mu_0, y_1^d, \ldots, y_k^d) = {} & \frac{1}{2}\left(\hat{x}_{k+1} - A\hat{x}_{k|k}\right)^{\mathrm{T}}A^{-T}\left\{P_{k|k}^{-1} - P_{k|k}^{-1}A^{-1}B\right. \\
& \left[B^{\mathrm{T}}A^{-T}P_{k|k}^{-1}A^{-1}B + Q^{-1}\right]^{-1}B^{\mathrm{T}}A^{-T}P_{k|k}^{-1}\right\} \\
& A^{-1}(\hat{x}_{k+1} - A\hat{x}_{k|k}) + \text{ constant}, \\
\triangleq {} & \frac{1}{2}\left(\hat{x}_{k+1} - \hat{x}_{k+1|k}\right)^{\mathrm{T}}\left(P_{k+1|k}\right)^{-1}\left(\hat{x}_{k+1} - \hat{x}_{k+1|k}\right) \\
& + \text{ constant},
\end{aligned}
$$

where we have used

$$
\hat{x}_{k+1|k} \triangleq A\hat{x}_{k|k}, \qquad \text{(which gives (9.50) for } j = k),
$$

$$
\begin{aligned}
P_{k+1|k} \triangleq {} & A\left\{P_{k|k}^{-1} - P_{k|k}^{-1}A^{-1}B\left[B^{\mathrm{T}}A^{-T}P_{k|k}^{-1}A^{-1}B + Q^{-1}\right]^{-1}\right. \\
& \left. B^{\mathrm{T}}A^{-T}P_{k|k}^{-1}\right\}^{-1}A^{\mathrm{T}}. \qquad (9.64)
\end{aligned}
$$

We also we note from (9.64) that

$$P_{k+1|k} = \left\{ A^{-T} P_{k|k}^{-1} A^{-1} - A^{-T} P_{k|k}^{-1} A^{-1} B \right.$$
$$\left. \left[B^{T} A^{-T} P_{k|k}^{-1} A^{-1} B + Q^{-1} \right]^{-1} B^{T} A^{-T} P_{k|k}^{-1} A^{-1} \right\}^{-1}.$$

Using the matrix inversion lemma, we have

$$P_{k+1|k} = A P_{k|k} A^{T} + B Q B^{T},$$

as in (9.52) for $j = k$. Thus, summarising step 1, we have shown that

$$W_{k+1}^{\mathrm{OPT}}(\hat{x}_{k+1}, \mu_0, y_1^d, \ldots, y_k^d) = \frac{1}{2} \left(\hat{x}_{k+1} - \hat{x}_{k+1|k} \right)^{T} P_{k+1|k}^{-1} \left(\hat{x}_{k+1} - \hat{x}_{k+1|k} \right)$$
$$+ \text{ constant}, \tag{9.65}$$

where $\hat{x}_{k+1|k}$ and $P_{k+1|k}$ satisfy (9.50) and (9.52), respectively, for $j = k$.

Step 2: We next add the term $\frac{1}{2}(y_{k+1}^d - C\hat{x}_{k+1})^{T} R^{-1} (y_{k+1}^d - C\hat{x}_{k+1})$ to (9.65) to obtain

$$V_{k+1}^{\mathrm{OPT}}(\hat{x}_{k+1}, \mu_0, y_1^d, \ldots, y_{k+1}^d) = \frac{1}{2}(\hat{x}_{k+1} - \hat{x}_{k+1|k})^{T} P_{k+1|k}^{-1}(\hat{x}_{k+1} - \hat{x}_{k+1|k})$$
$$+ \frac{1}{2}(y_{k+1}^d - C\hat{x}_{k+1})^{T} R^{-1}(y_{k+1}^d - C\hat{x}_{k+1})$$
$$+ \text{ constant}. \tag{9.66}$$

We want to write (9.66) as a perfect square, that is,

$$V_{k+1}^{\mathrm{OPT}}(\hat{x}_{k+1}, \mu_0, y_1^d, \ldots, y_{k+1}^d) = \frac{1}{2}(\hat{x}_{k+1} - \hat{x}_{k+1|k+1})^{T} P_{k+1|k+1}^{-1}$$
$$(\hat{x}_{k+1} - \hat{x}_{k+1|k+1}) + \text{ constant}. \tag{9.67}$$

To find the expression for $\hat{x}_{k+1|k+1}$ used in (9.67), we note that $\hat{x}_{k+1|k+1}$ is the minimum of $V_{k+1}^{\mathrm{OPT}}(\hat{x}_{k+1}, \mu_0, y_1^d, \ldots, y_{k+1}^d)$. Hence, to obtain $\hat{x}_{k+1|k+1}$, we differentiate (9.66) with respect to $\hat{x}_{k+1}$, evaluate at $\hat{x}_{k+1} = \hat{x}_{k+1|k+1}$ and set the result to zero, that is,

$$P_{k+1|k}^{-1}(\hat{x}_{k+1|k+1} - \hat{x}_{k+1|k}) - C^{T} R^{-1}(y_{k+1}^d - C\hat{x}_{k+1|k+1}) = 0.$$

Adding and subtracting $C^{T} R^{-1} C \hat{x}_{k+1|k}$, and rearranging, we have

$$(P_{k+1|k}^{-1} + C^{T} R^{-1} C)\hat{x}_{k+1|k+1} = (P_{k+1|k}^{-1} + C^{T} R^{-1} C)\hat{x}_{k+1|k}$$
$$+ C^{T} R^{-1}(y_{k+1}^d - C\hat{x}_{k+1|k}).$$

From the above expression we obtain

$$\begin{aligned}
\hat{x}_{k+1|k+1} &= \hat{x}_{k+1|k} + (P_{k+1|k}^{-1} + C^\mathsf{T} R^{-1} C)^{-1} C^\mathsf{T} R^{-1}(y_{k+1}^d - C\hat{x}_{k+1|k}) \\
&= \hat{x}_{k+1|k} + P_{k+1|k}(I + C^\mathsf{T} R^{-1} C P_{k+1|k})^{-1} C^\mathsf{T} R^{-1}(y_{k+1}^d - C\hat{x}_{k+1|k}) \\
&= \hat{x}_{k+1|k} + P_{k+1|k} C^\mathsf{T}(I + R^{-1} C P_{k+1|k} C^\mathsf{T})^{-1} R^{-1}(y_{k+1}^d - C\hat{x}_{k+1|k}) \\
&= \hat{x}_{k+1|k} + P_{k+1|k} C^\mathsf{T}(R + C P_{k+1|k} C^\mathsf{T})^{-1}(y_{k+1}^d - C\hat{x}_{k+1|k}),
\end{aligned}$$

which gives (9.51) for $j = k$.

Similarly, to find the expression for $P_{k+1|k+1}$ used in (9.67), we differentiate (9.66) twice with respect to $\hat{x}_{k+1}$. This gives

$$P_{k+1|k+1}^{-1} \triangleq P_{k+1|k}^{-1} + C^\mathsf{T} R^{-1} C,$$

which, using the matrix inversion lemma, gives (9.53) for $j = k$.

Thus, we have established (9.67) and induction completes the proof. □

We can use the characterisation of the partial value functions given in Lemma 9.6.1 to derive the optimal estimator where we optimise with respect to both $\{\hat{w}_0, \ldots, \hat{w}_{k-1}\}$ and $\hat{x}_0$ (or, equivalently, $\hat{x}_k$). In particular, we have the following important result.

Theorem 9.6.2 (Kalman Filter) *The optimal estimate $\hat{x}_k$ for x_k given the data $\mu_0, y_1^d, \ldots y_k^d$, satisfies*

$$\hat{x}_k = \hat{x}_{k|k},$$

where $\hat{x}_{k|k}$ satisfies the recursions (9.48) to (9.53).

Proof. The optimal choice $\hat{x}_k = \hat{x}_{k|k}$ follows immediately by minimising (9.47) with respect to $\hat{x}_k$ since $\hat{x}_k$ is unconstrained here. □

Remark 9.6.1 (Optimal Smoother). Actually, the minimisation of (9.47) with respect to $\hat{x}_k$ yields optimal estimates of all states $x_0, \ldots, x_k$ given data up to time k. These are called optimal *smoothed* estimates, and will be denoted by $\hat{x}_{j|k}$ for $j = 0, \ldots, k$. They can be computed simply by running $\hat{x}_{k-1} = A^{-1}\hat{x}_k - A^{-1}B\hat{w}_{k-1}$ backwards starting from $\hat{x}_k = \hat{x}_{k|k}$ and using $\hat{w}_{k-1}, \hat{w}_{k-2}, \ldots, \hat{w}_0$ as in (9.56). Defining $\hat{w}_{j|k} \triangleq \hat{w}_j$, for $j = 0, \ldots, k-1$, the *optimal smoother* is then given by the recursion

$$\hat{x}_{j|k} = A^{-1}\hat{x}_{j+1|k} - A^{-1}B\hat{w}_{j|k} \quad \text{for } j = 0, \ldots, k-1,$$

where

$$\hat{w}_{j|k} = -\left(B^\mathsf{T} A^{-T} P_{j|j}^{-1} A^{-1} B + Q^{-1}\right)^{-1} B^\mathsf{T} A^{-T} P_{j|j}^{-1}\left(\hat{x}_{j|j} - A^{-1}\hat{x}_{j+1|k}\right),$$

and $\hat{x}_{j|j}$ and $P_{j|j}$ are given by (9.50)–(9.53). ○

9.7 Nonlinear Problems

The above circle of ideas can be extended to nonlinear and/or non-Gaussian problems. Consider the following nonlinear Markov model:

$$x_{k+1} = f(x_k, w_k), \tag{9.68}$$

$$y_k = h(x_k) + v_k, \tag{9.69}$$

where f and h are continuously differentiable functions of their arguments, and $\partial f / \partial w_k$ is nonsingular. In (9.68)–(9.69), $\{w_k\}$ and $\{v_k\}$ are i.i.d. sequences having probability density functions that satisfy

$$p_w(w_k) = \begin{cases} p_1(w_k) & \text{for } w_k \in \Omega_1, \\ 0 & \text{otherwise}, \end{cases}$$

and such that $-\ln p_1(w_k) = \ell_1(w_k)$; and

$$p_v(v_k) = \begin{cases} p_2(v_k) & \text{for } v_k \in \Omega_2, \\ 0 & \text{otherwise}, \end{cases}$$

and such that $-\ln p_2(v_k) = \ell_2(v_k)$. Also, we assume

$$p_{x_0}(x_0) = \begin{cases} p_3(x_0) & \text{for } x_0 \in \Omega_3, \\ 0 & \text{otherwise}, \end{cases}$$

and $-\ln p_3(x_0) = \ell_3(x_0)$.

Using the rule of transformation of probability density functions for the model (9.68)–(9.69) we have:

$$p_{y_k | x_k}(y_k = y_k^d | x_k = \hat{x}_k) = p_v(v_k = y_k^d - h(\hat{x}_k)),$$

$$p_{x_{k+1} | x_k}(x_{k+1} = \hat{x}_{k+1} | x_k = \hat{x}_k) = p_w(w_k = \hat{w}_k) \left| \det \frac{\partial x_{k+1}}{\partial w_k} \Big|_{\hat{x}_k, \hat{w}_k} \right|^{-1}$$

$$= p_w(w_k = \hat{w}_k) \left| \det \frac{\partial f(\hat{x}_k, \hat{w}_k)}{\partial w_k} \right|^{-1},$$

for all $\hat{w}_k \in \Omega_1$ such that $\hat{x}_{k+1} = f(\hat{x}_k, \hat{w}_k)$.

Then, using the vector definitions in (9.14)–(9.17), the negative logarithm of the joint probability density function for states and outputs satisfies

$$-\ln p_{\mathbf{y}_N, \mathbf{x}_N}(\mathbf{y}_N = \mathbf{y}_N^d, \mathbf{x}_N = \hat{\mathbf{x}}_N) = \ell_3(\hat{x}_0) + \sum_{k=1}^{N} \ell_2(y_k^d - h(\hat{x}_k))$$

$$+ \sum_{k=0}^{N-1} \left[\ell_1(\hat{w}_k) + \ln \left| \det \frac{\partial f(\hat{x}_k, \hat{w}_k)}{\partial w_k} \right| \right], \tag{9.70}$$

subject to the constraints

$$\hat{x}_{k+1} = f(\hat{x}_k, \hat{w}_k) \quad \text{for } k = 0, \ldots, N-1, \tag{9.71}$$

$$\hat{w}_k \in \Omega_1 \quad \text{for } k = 0, \ldots, N-1, \tag{9.72}$$

$$y_k^d - h(\hat{x}_k) \in \Omega_2 \quad \text{for } k = 1, \ldots, N, \tag{9.73}$$

$$\hat{x}_0 \in \Omega_3. \tag{9.74}$$

Hence, we can find the JAPMP estimate (9.27) by minimising (9.70) subject to (9.71)–(9.74) (see (9.28)).

9.8 Relationship to Chapman–Kolmogorov Equation

We next relate the above ideas to the Chapman–Kolmogorov[5] equation for recursive nonlinear filtering. The latter equation allows one to *recursively* compute $p_{x_k|y_k,\ldots,y_1}(x_k|y_k, y_{k-1}, \ldots, y_1)$. Specifically, using the Markovian structure of (9.68), (9.69), we have, from Bayes' rule:

Time Update[6] (Chapman-Kolmogorov Equation)

$$p_{x_k|y_{k-1},\ldots,y_1}(x_k|y_{k-1}, \ldots, y_1)$$

$$= \int_{\mathbb{R}^n} p_{x_k, x_{k-1}|y_{k-1},\ldots,y_1}(x_k, x_{k-1}|y_{k-1}, \ldots, y_1) dx_{k-1} \tag{9.75}$$

$$= \int_{\mathbb{R}^n} p_{x_k|x_{k-1}, y_{k-1},\ldots,y_1}(x_k|x_{k-1}, y_{k-1}, \ldots, y_1)$$

$$\times p_{x_{k-1}|y_{k-1},\ldots,y_1}(x_{k-1}|y_{k-1}, \ldots, y_1) dx_{k-1} \tag{9.76}$$

$$= \int_{\mathbb{R}^n} p_{x_k|x_{k-1}}(x_k|x_{k-1}) p_{x_{k-1}|y_{k-1},\ldots,y_1}(x_{k-1}|y_{k-1}, \ldots, y_1) dx_{k-1}, \quad k \geq 1. \tag{9.77}$$

Observation Update[7]

$$p_{x_k|y_k,\ldots,y_1}(x_k|y_k, \ldots, y_1)$$

$$= \frac{p_{y_k|x_k, y_{k-1},\ldots,y_1}(y_k|x_k, y_{k-1}, \ldots, y_1) \, p_{x_k|y_{k-1},\ldots,y_1}(x_k|y_{k-1}, \ldots, y_1)}{p_{y_k|y_{k-1},\ldots,y_1}(y_k|y_{k-1}, \ldots, y_1)} \tag{9.78}$$

$$= \frac{p_{y_k|x_k}(y_k|x_k) \, p_{x_k|y_{k-1},\ldots,y_1}(x_k|y_{k-1}, \ldots, y_1)}{p_{y_k|y_{k-1},\ldots,y_1}(y_k|y_{k-1}, \ldots, y_1)}, \quad k \geq 0, \tag{9.79}$$

---

[5] Sometimes misspelled in Australia as Kolmogoroo.

[6] In passing from (9.75) to (9.76) we use Bayes' rule, and from (9.76) to (9.77) we use the Markovian property of (9.68).

[7] Equality (9.78) follows from Bayes' rule, and, in passing from (9.78) to (9.79) we use the Markovian property of (9.69).

where

$$p_{y_k|y_{k-1},\ldots,y_1}(y_k|y_{k-1},\ldots,y_1)$$
$$= \int_{\mathbb{R}^n} p_{y_k|x_k}(y_k|x_k)p_{x_k|y_{k-1},\ldots,y_1}(x_k|y_{k-1},\ldots,y_1)dx_k. \quad (9.80)$$

Notice that $p_{x_k|x_{k-1}}$ and $p_{y_k|x_k}$, needed in the evaluation of equations (9.77) and (9.79)–(9.80) are given in Section 9.7 above.

Given $p_{x_k|y_k,\ldots,y_1}(x_k|y_k,\ldots,y_1)$, one can then compute various estimates, for example:

(i) Conditional mean

$$\hat{x}_k^{[1]} = \int_{\mathbb{R}^n} x_k \, p_{x_k|y_k,\ldots,y_1}(x_k|y_k,\ldots,y_1)dx_k. \quad (9.81)$$

(ii) A posteriori most probable

$$\hat{x}_k^{[3]} = \arg\max_{x_k} \, p_{x_k|y_k,\ldots,y_1}(x_k|y_k,\ldots,y_1). \quad (9.82)$$

Thus the Chapman–Kolmogorov equation (9.77) and the observation update equation (9.80) offer more flexibility than the optimisation approach presented in Section 9.3 (for the linear constrained case) and Section 9.7 (for the nonlinear constrained case) since they describe the entire conditional distribution of x_k given the (past) data $y_1,\ldots,y_k$. Given this distribution, one can then compute various estimates, for example, those given in (9.81) and (9.82). On the other hand, the Chapman–Kolmogorov equation is, in general, difficult to solve and require various approximations to be used, for example, those used in particle filtering (see, for example, Doucet, de Freitas and Gordon 2001). By way of contrast, the optimisation approach of Sections 9.3 and 9.7 can be solved via optimal control methods.

Finally, we note that the following two estimates are not, in general, equal:

(i) Joint a posteriori most probable [JAPMP]

$$\left[\hat{x}_0^{[2]},\ldots,\hat{x}_N^{[2]}\right] \triangleq \arg\max_{x_0,\ldots,x_N} \, p_{x_0,\ldots,x_N|y_1,\ldots,y_N}(x_0,\ldots,x_N|y_1,\ldots,y_N).$$
$$(9.83)$$

(ii) A posteriori most probable [APMP]

$$\hat{x}_N^{[3]} \triangleq \arg\max_{x_N} \, p_{x_N|y_1,\ldots,y_N}(x_N|y_1,\ldots,y_N) \quad (9.84)$$
$$= \arg\max_{x_N} \int_{\mathbb{R}^n\times\cdots\times\mathbb{R}^n} p_{x_0,\ldots,x_N|y_1,\ldots,y_N}(x_0,\ldots,x_N|y_1,\ldots,y_N)dx_0\ldots dx_{N-1}.$$
$$(9.85)$$

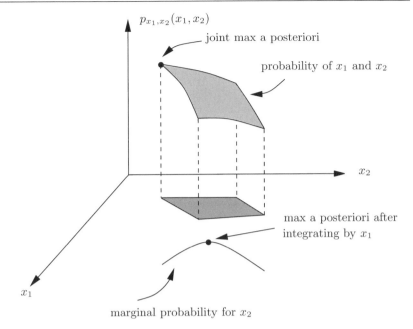

Figure 9.5. Difference between joint a posteriori maximum probability and a posteriori maximum probability.

This is illustrated in Figure 9.5.

However, if we use the conditional mean (9.81) as an estimate then we get the same answer whether we use the joint distribution for $\{x_0, \ldots, x_N\}$ or the marginal distribution for x_N. This follows because

$$
\begin{aligned}
\hat{x}_N^{[1]} &= \int_{\mathbb{R}^n} x_N \; p_{x_N|y_N,\ldots,y_1}(x_N|y_N,\ldots,y_1)dx_N \\
&= \int_{\mathbb{R}^n} x_N \left[\int_{\mathbb{R}^n \times \cdots \times \mathbb{R}^n} p_{x_N,\ldots,x_0|y_N,\ldots,y_1}(x_N,\ldots,x_0|y_N,\ldots,y_1) \; dx_{N-1}\ldots dx_0 \right] dx_N \\
&= \int_{\mathbb{R}^n \times \cdots \times \mathbb{R}^n} x_N \; p_{x_N,\ldots,x_0|y_N,\ldots,y_1}(x_N,\ldots,x_0|y_N,\ldots,y_1)dx_N \ldots dx_0,
\end{aligned}
$$

since

$$
\begin{aligned}
&p_{x_N|y_N,\ldots,y_1}(x_N|y_N,\ldots,y_1) \\
&\qquad = \int_{\mathbb{R}^n \times \cdots \times \mathbb{R}^n} p_{x_N,\ldots,x_0|y_N,\ldots,y_1}(x_N,\ldots,x_0|y_N,\ldots,y_1)dx_{N-1}\ldots dx_0.
\end{aligned}
$$

9.9 Moving Horizon Estimation

As with control, we can readily convert the fixed horizon estimators discussed above into *moving horizon estimators* [MHE]. An issue to be addressed in this context is whether or not the situation allows *data-smoothing*; that is, whether one can collect data beyond the time at which the state estimate is required.

In some applications, for example, control, one requires that the estimate apply to the most recent state; that is, it is not possible to collect data beyond the point where the state estimate is defined. In other applications, for example, telecommunications, one can tolerate a delay between the last time at which the data are collected and the time at which the estimate is defined. In the latter situation we say that a *smoothed* state estimate is required.

To cover both of the above scenarios, we let i denote the "time" at which the estimate is required. We also fix integers $L_1 \geq 0$ and $L_2 \geq 0$ and suppose for the moment that

$$x_{i-L_1} \sim N(z_{i-L_1}, P_{i-L_1}), \tag{9.86}$$

where z_{i-L_1} is a given a priori estimate for x_{i-L_1} having a Gaussian distribution. The matrix $P_{i-L_1}^{-1}$ reflects the degree of belief in this a priori estimate. We will treat the data in blocks of length $N = L_1 + L_2$. We assume that the estimate of x_i can be based on data collected between $i - L_1$ and $i + L_2 - 1$. We then formulate the fixed horizon optimisation problem as in (9.29)–(9.35) over the interval $[i - L_1, i + L_2 - 1]$. That is, the corresponding sequences are indexed by $k = i - L_1, i - L_1 + 1, \ldots, i + L_2 - 1$. This yields the required estimate (or smoother for $L_2 > 1$, see Remark 9.6.1) of x_i.

The next question is how to turn this into a moving horizon procedure. The idea is to store the final state estimate $\hat{x}_{i+L_2-1}$ obtained from the above fixed horizon optimisation together with some measure of our degree of belief in this estimate, which we denote $P_{i+L_2-1}^{-1}$. The pair $(\hat{x}_{i+L_2-1}, P_{i+L_2-1})$ will be used to initialise a fixed horizon optimisation problem $L_1 + L_2$ steps ahead (that is, they will take the role of (z_{i-L_1}, P_{i-L_1}) in (9.86)).

We use again a Gaussian approximation when we return to this estimate. Of course, due to the constraints, we appreciate that the a posteriori distribution of the state will not be Gaussian. However, a Gaussian approach is justified on the following grounds:

(i) The "initial state information" is of diminishing importance as the block length N increases.
(ii) Making a Gaussian approximation greatly simplifies the problems.
(iii) We can, at least, be compatible with the unconstrained case by determining P_{i-L_1} from ordinary linear estimation theory.

Finally, the MHE is organised as illustrated in Figure 9.6. (Note that we need storage for $L_1 + L_2$ past state estimates to initialise subsequent blocks.)

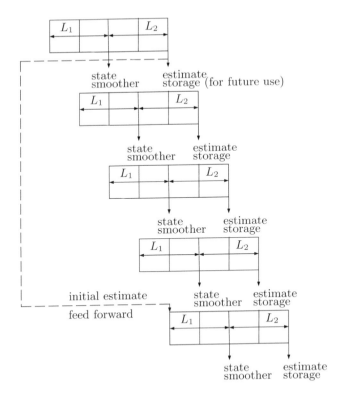

Figure 9.6. Graphical representation of MHE.

We next illustrate the idea of constrained estimation by three simple examples.

Example 9.9.1. Consider the same model as used in Example 1.3.1 of Chapter 1, which we repeat here for convenience:

$$y_k = w_k - 1.7w_{k-1} + 0.72w_{k-2} + v_k. \tag{9.87}$$

Rather than a binary signal, we here consider that the input noise w_k has a truncated Gaussian distribution. We assume that the measurement noise v_k has a Gaussian distribution. The details are:

- input noise variance prior to truncation: $Q = 1$;
- input noise mean prior to truncation: $\mu_w = 0$;
- measurement noise variance: $R = 0.2$;
- truncation interval: $w_k \in [-1, 1]$;
- input noise variance after truncation: ≈ 0.293;
- input noise mean after truncation: 0.

Two estimators were compared, namely the MHE using $N = L_1 + 1 = 2$, $L_2 = 1$, incorporating the constraint $|w_k| \leq 1$, and a standard linear Kalman filter based on $R = 0.2$ and the true input variance of 0.293. The initial estimates as in (9.86) were selected as follows: z_{i-N} is stored and propagated as in Figure 9.6; P_{i-N} is set equal to the corresponding value for the Kalman filter. The results are shown in Figure 9.7. Some observations from this figure are:

(i) The linear Kalman filter performs quite well in this example. (This is not surprising since it is, after all, the best linear unbiased estimator.)

(ii) The estimates provided by the linear Kalman filter occasionally lie outside the range ± 1. (Again, this is not surprising since this estimator is unconstrained.)

(iii) The MHE is slightly better but the result is marginal. (Again, this is not surprising in view of observation (i).)   ○

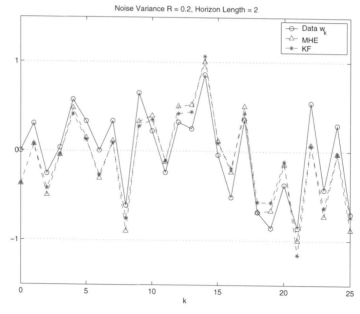

Figure 9.7. Comparison of MHE and Kalman filter with correct variance: data (circle-solid line), estimate provided by the MHE (triangle-dashed line) and estimate provided by the Kalman filter (star-dashed line).

Example 9.9.2. Here we consider the same model (9.87) as in Example 9.9.1, save that we change the input to a nonzero-mean truncated Gaussian distribution as illustrated in Figure 9.2. The details are:

- input noise variance prior to truncation: $Q = 1$;
- input noise mean prior to truncation: $\mu_w = 1.5$;
- truncation interval: $w_k \in [-1.5, 0.5]$;
- input noise variance after truncation: ≈ 0.175;
- input noise mean after truncation: ≈ 0.

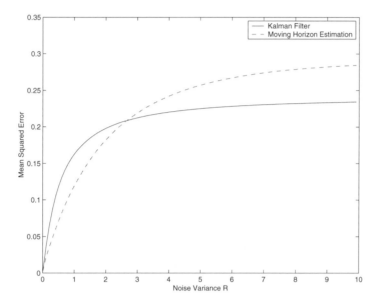

Figure 9.8. Comparison of mean square estimation error achieved by the MHE (dashed line) and the Kalman filter with correct variance (solid line).

Two estimators were compared, namely, MHE with $N = L_1 + 1 = 5$, $L_2 = 1$, and using the given constraints; and a standard linear Kalman filter based on the true variance. Figure 9.8 compares the mean square estimation errors for a range of measurement noise variances R.

It can be seen from Figure 9.8 that the MHE outperforms the Kalman filter save in the presence of large measurement noise. This result is in good accord with intuition since, for large measurement noise, the observations are basically ignored. This means that the Kalman filter gives the a priori mean, which is zero, whereas the MHE gives $w_k = 0.5$ since this corresponds to the point where the a priori probability is maximal.

○

Example 9.9.3. Here we consider the same channel model (9.87) as in Examples 9.9.1 and 9.9.2, save that now the input w_k is distributed as the combination of two nonoverlapping, nonzero-mean truncated Gaussian distributions as in Figure 9.9. The distribution can be described by two regions: the "left region" is a Gaussian distribution $N(-1.5, 0.1)$ truncated between $[-1, 0]$, and the "right region" is a Gaussian distribution $N(1.5, 0.1)$ truncated between $[0, 1]$. The resulting distribution has mean ≈ 0 and variance ≈ 0.872.

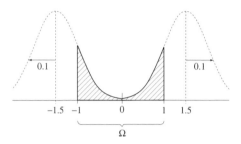

Figure 9.9. Combining the tails of two truncated Gaussian distributions.

We will compare the performance of the Kalman filter and the MHE for the above problem. The Kalman filter assumes a Gaussian approximation of the distribution, with zero mean and variance 0.872. For the MHE, we consider no smoothing, that is, $L_2 = 1$. The initial weighting P is set equal to the value of the steady state error covariance of the Kalman filter, and the initial estimate is forwarded as in Figure 9.6. To find the optimal input sequence $\{\hat{w}_0, \ldots, \hat{w}_{N-1}\}$ the estimator solves, at each step, 2^N separate QP problems (see Section 9.4.2). The global optimum is the minimum of the individual sub-problems.

In Figure 9.10, we compare the Kalman filter estimates with those of the MHE for different measurement noise variances and different horizon lengths. In Figure 9.10 (a), incorporating mixed distributions with the MHE method and horizon 1 gives estimates that are closer to the boundary. On the other hand, the unconstrained Kalman filter exceeds the limits and tends to estimate near the zero mean. In Figure 9.10 (b) we see that the MHE performs more poorly as more measurement noise is introduced, since, in this case, the MHE tends to give the point where the a priori probability is maximal. By increasing the horizon length to 2 (see Figure 9.10 (c)), the estimator uses more data, resulting in better estimates. However, the number of sub-problems also increases. In Figure 9.10 (d), the horizon was increased to 4, showing a slight improvement in performance.

It should be observed that, since the distribution of the data points w_k is close to the boundary, and with additive measurement noise, the MHE will give estimates that are close to the boundary. In the limiting case, when the distribution approaches a point mass distribution, the estimation problem will resemble that of the finite alphabet estimation problem, which is discussed in Chapter 13.

o

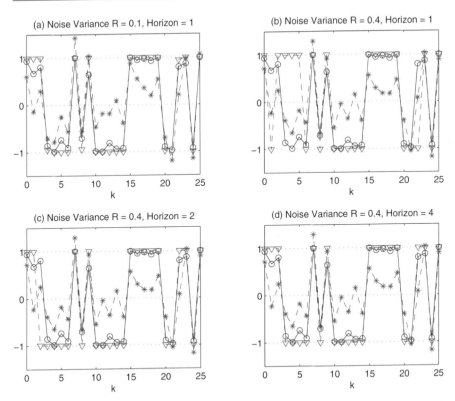

Figure 9.10. Data w_k (circle-solid line), Kalman filter estimates (star-dashed line), MHE estimates (triangle-dashed line) for different measurement noise variances R and different horizons N.

9.10 Further Reading

For complete list of references cited, see References section at the end of book.

General

A useful introduction to estimation is given in Jazwinski (1970). The original derivation of the discrete Kalman filter used the concept of orthogonal projection (Kalman 1960a). The variational approach to estimation was first taken by Bryson and Frazier (1963). The solution of the continuous least square problem via dynamic programming was first given by Cox (1964).

Section 9.4.3

The idea of utilising constraints in the context of approximating arbitrary distributions appears in Robertson and Lee (2002).

Section 9.9

Early work on moving horizon estimation appears in Michalska and Mayne (1995). See also Rao, Rawlings and Lee (2001) and Rao, Rawlings and Mayne (2003).

10

Duality Between Constrained Estimation and Control

10.1 Overview

The previous chapter showed that the problem of constrained estimation can be formulated as a constrained optimisation problem. Indeed, this problem is remarkably similar to the constrained control problem—differing only with respect to the boundary conditions. (In control, the initial condition is fixed, whereas in estimation, the initial condition can also be adjusted.) In the current chapter we show that the similarity between the two problems of constrained estimation and constrained control has deeper implications.

In particular, we derive the Lagrangian dual (see Section 2.6 of Chapter 2) of a constrained estimation problem and show that it leads to a particular unconstrained nonlinear optimal control problem. We then show that the original (primal) constrained estimation problem has an equivalent formulation as an unconstrained nonlinear optimisation problem, exposing a clear symmetry with its dual.

10.2 Lagrangian Duality of Constrained Estimation and Control

Consider the following system

$$
\begin{aligned}
x_{k+1} &= Ax_k + Bw_k \quad \text{for } k = 0, \cdots, N-1, \\
y_k &= Cx_k + v_k \quad \text{for } k = 1, \cdots, N,
\end{aligned}
\tag{10.1}
$$

where $x_k \in \mathbb{R}^n$, $w_k \in \mathbb{R}^m$, $y_k \in \mathbb{R}^p$. For clarity of exposition, we begin with the case where only the process noise sequence $\{w_k\}$ is constrained.[1] We thus assume that $\{w_k\}$ is an i.i.d. sequence having truncated Gaussian distribution

---

[1] The case of general constraints on w_k, v_k and x_0 will be treated in Sections 10.6 and 10.7.

of the form given in (9.11) of Chapter 9, with $\Omega_1 = \Omega$. We further assume that $\{v_k\}$ is an i.i.d. sequence having a Gaussian distribution $N(0, R)$, and x_0 has a Gaussian distribution $N(\mu_0, P_0)$.

For (10.1) we consider the optimisation problem defined in (9.29)–(9.35) of Chapter 9, which yields the joint a posteriori most probable state estimates. According to the assumptions, we set $\Omega_1 = \Omega$, $\Omega_2 = \mathbb{R}^p$, and $\Omega_3 = \mathbb{R}^n$ in (9.29)–(9.35). Thus, we consider:

$$\mathcal{P}_e : \quad V_N^{\text{OPT}}(\mu_0, \{y_k^d\}) \triangleq \min_{\hat{x}_k, \hat{v}_k, \hat{w}_k} V_N(\{\hat{x}_k\}, \{\hat{v}_k\}, \{\hat{w}_k\}), \tag{10.2}$$

subject to:

$$\hat{x}_{k+1} = A\hat{x}_k + B\hat{w}_k \quad \text{for } k = 0, \ldots, N-1, \tag{10.3}$$

$$\hat{v}_k = y_k^d - C\hat{x}_k \quad \text{for } k = 1, \ldots, N, \tag{10.4}$$

$$\{\hat{x}_0, \ldots, \hat{x}_N, \hat{v}_1, \ldots, \hat{v}_N, \hat{w}_0, \ldots, \hat{w}_{N-1}\} \in X, \tag{10.5}$$

where, in (10.5),

$$X = \underbrace{\mathbb{R}^n \times \cdots \times \mathbb{R}^n}_{N+1} \times \underbrace{\mathbb{R}^p \times \cdots \times \mathbb{R}^p}_{N} \times \underbrace{\Omega \times \cdots \times \Omega}_{N}. \tag{10.6}$$

In (10.2), the objective function is

$$V_N(\{\hat{x}_k\}, \{\hat{v}_k\}, \{\hat{w}_k\}) \triangleq \frac{1}{2}(\hat{x}_0 - \mu_0)^{\text{T}} P_0^{-1}(\hat{x}_0 - \mu_0)$$

$$+ \frac{1}{2} \sum_{k=0}^{N-1} \hat{w}_k^{\text{T}} Q^{-1} \hat{w}_k + \frac{1}{2} \sum_{k=1}^{N} \hat{v}_k^{\text{T}} R^{-1} \hat{v}_k, \tag{10.7}$$

where $P_0 > 0$, $Q > 0$, $R > 0$ are the covariance matrices in (9.11)–(9.13).

The following result establishes duality between the constrained *estimation* problem $\mathcal{P}_e$ and a particular unconstrained nonlinear optimal *control* problem.

Theorem 10.2.1 (Dual Problem) *Assume* Ω *in* (10.6) *is a nonempty closed convex set. Given the primal constrained fixed horizon estimation problem* $\mathcal{P}_e$ *defined by equations* (10.2)–(10.7), *the Lagrangian dual problem is*

$$\mathcal{D}_e : \quad \phi^{\text{OPT}}(\mu_0, \{y_k^d\}) \triangleq \min_{\lambda_k, u_k} \phi(\{\lambda_k\}, \{u_k\}), \tag{10.8}$$

subject to:

$$\lambda_{k-1} = A^{\text{T}} \lambda_k + C^{\text{T}} u_k \quad \text{for } k = 1, \cdots, N, \tag{10.9}$$

$$\lambda_N = 0, \tag{10.10}$$

$$\zeta_k = B^{\text{T}} \lambda_k \quad \text{for } k = 0, \cdots, N-1, \tag{10.11}$$

$$\bar{\zeta}_k = Q^{-1/2} \Pi_{\tilde{\Omega}} Q^{1/2} \zeta_k \quad \text{for } k = 0, \cdots, N-1. \tag{10.12}$$

In (10.8), *the objective function is*

$$\phi(\{\lambda_k\}, \{u_k\}) \triangleq \frac{1}{2}(A^{\mathrm{T}}\lambda_0 + P_0^{-1}\mu_0)^{\mathrm{T}}P_0(A^{\mathrm{T}}\lambda_0 + P_0^{-1}\mu_0)$$

$$+ \frac{1}{2}\sum_{k=1}^{N}(u_k - R^{-1}y_k^d)^{\mathrm{T}}R(u_k - R^{-1}y_k^d)$$

$$+ \sum_{k=0}^{N-1}\left[\frac{1}{2}\bar{\zeta}_k^{\mathrm{T}}Q\bar{\zeta}_k + (\zeta_k - \bar{\zeta}_k)^{\mathrm{T}}Q\bar{\zeta}_k\right] + \gamma \tag{10.13}$$

where γ is the constant term given by

$$\gamma \triangleq -\frac{1}{2}\mu_0^{\mathrm{T}}P_0^{-1}\mu_0 - \frac{1}{2}\sum_{k=1}^{N}(y_k^d)^{\mathrm{T}}R^{-1}y_k^d. \tag{10.14}$$

In (10.12), $\Pi_{\tilde{\Omega}}$ denotes the minimum Euclidean distance projection onto $\tilde{\Omega} \triangleq \{z : Q^{1/2}z \in \Omega\}$, that is,

$$\Pi_{\tilde{\Omega}} : \mathbb{R}^m \longrightarrow \tilde{\Omega}$$
$$s \longmapsto \bar{s} = \Pi_{\tilde{\Omega}}s \triangleq \arg\min_{z\in\tilde{\Omega}} \|z - s\|. \tag{10.15}$$

Moreover, there is no duality gap, that is, the minimum achieved in (10.2) is equal to minus the minimum achieved in (10.8).

Proof. Consider the primal constrained fixed horizon estimation problem $\mathcal{P}_e$, defined by equations (10.2)–(10.7). From (2.44) in Chapter 2, the Lagrangian dual function θ is given by:

$$\theta(\{\lambda_k\}, \{u_k\}) = \inf_{\hat{w}_k \in \Omega, \hat{x}_k, \hat{v}_k} L(\{\hat{x}_k\}, \{\hat{v}_k\}, \{\hat{w}_k\}, \{\lambda_k\}, \{u_k\}), \tag{10.16}$$

where the function L is defined as,

$$L(\{\hat{x}_k\}, \{\hat{v}_k\}, \{\hat{w}_k\}, \{\lambda_k\}, \{u_k\}) = V_N(\{\hat{x}_k\}, \{\hat{v}_k\}, \{\hat{w}_k\})$$

$$+ \sum_{k=0}^{N-1}\lambda_k^{\mathrm{T}}\left[\hat{x}_{k+1} - A\hat{x}_k - B\hat{w}_k\right]$$

$$+ \sum_{k=1}^{N}u_k^{\mathrm{T}}\left[y_k^d - C\hat{x}_k - \hat{v}_k\right]. \tag{10.17}$$

In (10.17), V_N is the primal objective function defined in (10.7), and $\{\lambda_k\}$ and $\{u_k\}$ are the Lagrange multipliers corresponding, respectively, to the linear equalities (10.3) and (10.4). Using (10.7) in (10.17), and combining terms, the function L can be rewritten as

$$L\big(\{\hat{x}_k\}, \{\hat{v}_k\}, \{\hat{w}_k\}, \{\lambda_k\}, \{u_k\}\big) = \frac{1}{2}(\hat{x}_0 - \mu_0)^\mathrm{T} P_0^{-1}(\hat{x}_0 - \mu_0) - \lambda_0^\mathrm{T} A \hat{x}_0$$

$$+ \sum_{k=1}^{N} \left\{ \frac{1}{2} \hat{v}_k^\mathrm{T} R^{-1} \hat{v}_k - u_k^\mathrm{T} \hat{v}_k + u_k^\mathrm{T} y_k^d \right\}$$

$$+ \sum_{k=0}^{N-1} \left\{ \frac{1}{2} \hat{w}_k^\mathrm{T} Q^{-1} \hat{w}_k - \lambda_k^\mathrm{T} B \hat{w}_k \right\}$$

$$+ \sum_{k=1}^{N-1} \left\{ (\lambda_{k-1} - A^\mathrm{T} \lambda_k - C^\mathrm{T} u_k)^\mathrm{T} \hat{x}_k \right\}$$

$$+ (\lambda_{N-1} - C^\mathrm{T} u_N)^\mathrm{T} \hat{x}_N. \qquad (10.18)$$

Notice that the terms that depend on the constrained variables $\hat{w}_k$ are independent of the other variables, $\hat{x}_k$ and $\hat{v}_k$, with respect to which the minimisation (10.16) is carried out. The values that achieve the infimum in (10.16), denoted $\hat{w}_k^*$, $\hat{x}_k^*$ and $\hat{v}_k^*$, can be computed from

$$\hat{w}_k^* = \arg\min_{\hat{w}_k \in \Omega} \left\{ \frac{1}{2} \hat{w}_k^\mathrm{T} Q^{-1} \hat{w}_k - \lambda_k^\mathrm{T} B \hat{w}_k \right\} \quad \text{for } k = 0, \cdots, N-1, \quad (10.19)$$

$$\frac{\partial L(\cdot)}{\partial \hat{x}_0} = P_0^{-1} (\hat{x}_0^* - \mu_0) - A^\mathrm{T} \lambda_0 = 0, \qquad (10.20)$$

$$\frac{\partial L(\cdot)}{\partial \hat{v}_k} = R^{-1} \hat{v}_k^* - u_k = 0 \quad \text{for } k = 1, \cdots, N, \qquad (10.21)$$

provided that the following two conditions are satisfied

$$\lambda_{k-1} - A^\mathrm{T} \lambda_k - C^\mathrm{T} u_k = 0 \quad \text{for } k = 1, \cdots, N-1, \qquad (10.22)$$

$$\lambda_{N-1} - C^\mathrm{T} u_N = 0. \qquad (10.23)$$

Notice from (10.18) that the infimum in (10.16) is $-\infty$ whenever $\{\lambda_k\}$ and $\{u_k\}$ are such that (10.22) and (10.23) are not satisfied. However, since we will subsequently choose $\{\lambda_k\}$ and $\{u_k\}$ so as to maximise $\theta(\{\lambda_k\}, \{u_k\})$ in (10.16) (see (2.43) and (2.44) in Chapter 2), we are here interested only in those values of $\{\lambda_k\}$ and $\{u_k\}$ satisfying (10.22) and (10.23).

We next define the variables

$$\zeta_k \triangleq B^\mathrm{T} \lambda_k, \qquad (10.24)$$

$$s \triangleq Q^{-1/2} \hat{w}_k, \qquad (10.25)$$

$$s^* \triangleq Q^{-1/2} \hat{w}_k^*, \qquad (10.26)$$

which transform the minimisation problem (10.19) into the minimum Euclidean distance problem

$$s^* = \arg\min_{s \in \tilde{\Omega},} \left\{ \frac{1}{2} s^\mathrm{T} s - (\zeta_k^\mathrm{T} Q^{1/2}) s \right\}, \qquad (10.27)$$

where $\tilde{\Omega} \triangleq \{z : Q^{1/2}z \in \Omega\}$. The solution to (10.27) can be expressed as

$$s^* = \bar{s} \triangleq \Pi_{\tilde{\Omega}} Q^{1/2} \zeta_k, \tag{10.28}$$

where $\Pi_{\tilde{\Omega}}$ is the Euclidean projection (10.15). Using (10.26) and (10.28), the solution to (10.19) is then

$$\hat{w}_k^* = Q^{1/2} \Pi_{\tilde{\Omega}} Q^{1/2} \zeta_k. \tag{10.29}$$

Finally, we define

$$\bar{\zeta}_k \triangleq Q^{-1} \hat{w}_k^* = Q^{-1/2} \Pi_{\tilde{\Omega}} Q^{1/2} \zeta_k, \tag{10.30}$$

and introduce an extra variable, $\lambda_N \triangleq 0$, for ease of notation. Thus, from (10.19)–(10.24) and (10.30), we obtain:

$$\hat{w}_k^* = Q\bar{\zeta}_k \quad \text{for } k = 0, \cdots, N-1, \tag{10.31}$$

$$\bar{\zeta}_k \triangleq Q^{-1/2} \Pi_{\tilde{\Omega}} Q^{1/2} \zeta_k \quad \text{for } k = 0, \cdots, N-1, \tag{10.32}$$

$$\zeta_k \triangleq B^{\mathrm{T}} \lambda_k \quad \text{for } k = 0, \cdots, N-1, \tag{10.33}$$

$$\lambda_N \triangleq 0, \tag{10.34}$$

$$\lambda_{k-1} = A^{\mathrm{T}} \lambda_k + C^{\mathrm{T}} u_k \quad \text{for } k = 1, \cdots, N, \tag{10.35}$$

$$\hat{x}_0^* = P_0 A^{\mathrm{T}} \lambda_0 + \mu_0, \tag{10.36}$$

$$\hat{v}_k^* = R u_k \quad \text{for } k = 1, \cdots, N. \tag{10.37}$$

Substituting (10.31)–(10.37) into (10.18) we obtain, after some algebraic manipulations, the Lagrangian dual function:

$$\begin{aligned}
\theta\big(\{\lambda_k\}, \{u_k\}\big) &= L\big(\{\hat{x}_k^*\}, \{\hat{v}_k^*\}, \{\hat{w}_k^*\}, \{\lambda_k\}, \{u_k\}\big) \\
&= -\frac{1}{2}\{\lambda_0^{\mathrm{T}} A P_0 A^{\mathrm{T}} \lambda_0 + 2\lambda_0^{\mathrm{T}} A \mu_0\} \\
&\quad - \frac{1}{2} \sum_{k=1}^{N} \{u_k^{\mathrm{T}} R u_k - 2u_k^{\mathrm{T}} y_k^d\} \\
&\quad + \sum_{k=0}^{N-1} \Big\{\frac{1}{2} \bar{\zeta}_k^{\mathrm{T}} Q \bar{\zeta}_k - \zeta_k^{\mathrm{T}} Q \bar{\zeta}_k\Big\}.
\end{aligned} \tag{10.38}$$

Finally, completing the squares in (10.38), and after further algebraic manipulations, we obtain:

$$\begin{aligned}
\theta\big(\{\lambda_k\}, \{u_k\}\big) &= -\frac{1}{2}(A^{\mathrm{T}} \lambda_0 + P_0^{-1} \mu_0)^{\mathrm{T}} P_0 (A^{\mathrm{T}} \lambda_0 + P_0^{-1} \mu_0) \\
&\quad - \frac{1}{2} \sum_{k=1}^{N} (u_k - R^{-1} y_k^d)^{\mathrm{T}} R (u_k - R^{-1} y_k^d) \\
&\quad - \sum_{k=0}^{N-1} \Big[\frac{1}{2} \bar{\zeta}_k^{\mathrm{T}} Q \bar{\zeta}_k + (\zeta_k - \bar{\zeta}_k)^{\mathrm{T}} Q \bar{\zeta}_k\Big] - \gamma,
\end{aligned}$$

where γ is the constant defined in (10.14). Defining $\phi \triangleq -\theta$, the formulation of the dual problem $\mathcal{D}_e$ in (10.8)–(10.15) follows from (2.43)–(2.44) in Chapter 2, and the fact that $\max \theta = -\min(-\theta) = -\min \phi$ and the optimisers are the same. Also, from Theorem 2.6.4 in Chapter 2, we conclude that there is no duality gap, that is, the minimum achieved in (10.2) is equal to minus the minimum achieved in (10.8). □

We can think of (10.9)–(10.11) as the state equations of a system (running in reverse time) with input u_k and output ζ_k. Theorem 10.2.1 then shows that the dual of the primal estimation problem of minimisation with *constraints on the system inputs* (the process noise w_k) is an *unconstrained* optimisation problem using *projected outputs* $\bar{\zeta}_k$ in the objective function.

A particular case of Theorem 10.2.1 is the following result for the *unconstrained* case.

Corollary 10.2.2 *In the case in which the variables $\hat{w}_k$ in the primal problem $\mathcal{P}_e$ are unconstrained (that is, $\Omega = R^m$), the dual problem becomes:*

$$\mathcal{D}_e : \quad \min_{\lambda_k, u_k} \frac{1}{2} \left\{ (A^{\mathrm{T}} \lambda_0 + P_0^{-1} \mu_0)^{\mathrm{T}} P_0 (A^{\mathrm{T}} \lambda_0 + P_0^{-1} \mu_0) \right.$$

$$\left. + \sum_{k=1}^{N} (u_k - R^{-1} y_k^d)^{\mathrm{T}} R (u_k - R^{-1} y_k^d) + \sum_{k=0}^{N-1} \lambda_k^{\mathrm{T}} B Q B^{\mathrm{T}} \lambda_k \right\} + \gamma,$$

subject to:

$$\lambda_{k-1} = A^{\mathrm{T}} \lambda_k + C^{\mathrm{T}} u_k \quad for \ k = 1, \cdots, N,$$

$$\lambda_N = 0,$$

where γ is the constant defined in (10.14).

Proof. Note that $\bar{\zeta}_k = \zeta_k$ in (10.12) since the projection (10.15) reduces to the identity mapping in the unconstrained case. The result then follows upon substituting $\bar{\zeta}_k = \zeta_k = B^{\mathrm{T}} \lambda_k$ in expression (10.13). □

10.3 An Equivalent Formulation of the Primal Problem

In the previous section we have shown that problem $\mathcal{D}_e$ is dual to problem $\mathcal{P}_e$ in (10.2)–(10.7). We can gain further insight by expressing $\mathcal{P}_e$ in a different way. This is facilitated by the following results.

Lemma 10.3.1 *Let $\tilde{\Omega} \subset R^m$ be a closed convex set with a nonempty interior. Let $s \in R^m$ such that $s \notin \tilde{\Omega}$. Then there exists a unique point $\bar{s} \in \tilde{\Omega}$ with minimum Euclidean distance from s. Furthermore, s and $\bar{s}$ satisfy the inequality*

$$(s - \bar{s})^{\mathrm{T}} (\bar{s} - \xi) > 0 \tag{10.39}$$

for any point ξ in the interior of $\tilde{\Omega}$.

Proof. By assumption, $\tilde{\Omega}$ is a nonempty closed convex set. From Theorem 2.3.1 of Chapter 2, we have that there exists a unique $\bar{s} \in \tilde{\Omega}$ with minimum Euclidean distance from s, and $\bar{s}$ is the minimiser if and only if

$$(s - \bar{s})^{\mathrm{T}}(z - \bar{s}) \leq 0 \text{ for all } z \in \tilde{\Omega}. \tag{10.40}$$

Now, let $\xi \in \operatorname{int} \tilde{\Omega}$. We will show that (10.39) holds. Since $\xi \in \tilde{\Omega}$, (10.40) holds for $z = \xi$. Thus we only need to show that (10.40) for $z = \xi \in \operatorname{int} \tilde{\Omega}$ can never be an equality. Suppose, by contradiction, that

$$(s - \bar{s})^{\mathrm{T}}(\xi - \bar{s}) = 0. \tag{10.41}$$

Note that $\|s - \bar{s}\| > 0$ since $\tilde{\Omega}$ is closed, and $s \notin \tilde{\Omega}$, $\bar{s} \in \tilde{\Omega}$. Since $\xi \in \operatorname{int} \tilde{\Omega}$, there exists an $\varepsilon > 0$ such that the ball $N_\varepsilon(\xi) \triangleq \{z : \|z - \xi\| < \varepsilon\}$ is contained in $\tilde{\Omega}$. Define

$$\tilde{\xi} = \xi + \alpha \frac{s - \bar{s}}{\|s - \bar{s}\|}, \qquad 0 < \alpha < \varepsilon; \tag{10.42}$$

hence, $\|\tilde{\xi} - \xi\| = \alpha < \varepsilon$ and $\tilde{\xi} \in N_\varepsilon(\xi)$. We then have, using (10.41) and (10.42), that

$$(s - \bar{s})^{\mathrm{T}}(\tilde{\xi} - \bar{s}) = (s - \bar{s})^{\mathrm{T}}(\xi - \bar{s}) + \alpha \frac{(s - \bar{s})^{\mathrm{T}}(s - \bar{s})}{\|s - \bar{s}\|} = \alpha \|s - \bar{s}\| > 0.$$

Thus, we have found a point $\tilde{\xi} \in \tilde{\Omega}$ (since $N_\varepsilon(\xi)$ is contained in $\tilde{\Omega}$) such that $(s - \bar{s})^{\mathrm{T}}(\tilde{\xi} - \bar{s}) > 0$, which contradicts (10.40). Thus, (10.39) must be true, and the result follows. □

Lemma 10.3.2 *Let $f : \mathbb{R}^m \to \mathbb{R}$ be any function and let $\Omega \subset \mathbb{R}^m$ be a closed convex set that contains an interior point c. Consider the optimisation problem*

$$\mathcal{P}_1' : \quad \min_w V(w), \tag{10.43}$$

with

$$V(w) \triangleq f(\bar{w}) + (w - \bar{w})^{\mathrm{T}} Q^{-1}(\bar{w} - c), \tag{10.44}$$

$$\bar{w} \triangleq Q^{1/2} \Pi_{\tilde{\Omega}} Q^{-1/2} w, \tag{10.45}$$

where $\Pi_{\tilde{\Omega}}$ is the mapping that assigns to any vector s in $\mathbb{R}^m$ the vector $\bar{s}$ in $\tilde{\Omega}$ that is closest to s in Euclidean distance, that is,

$$
\begin{aligned}
\Pi_{\tilde{\Omega}} &: \mathbb{R}^m \longrightarrow \tilde{\Omega} \\
s &\longmapsto \bar{s} = \Pi_{\tilde{\Omega}} s \triangleq \arg \min_{z \in \tilde{\Omega}} \|z - s\|,
\end{aligned} \tag{10.46}
$$

and set $\tilde{\Omega}$ is defined as

$$\tilde{\Omega} \triangleq \{z : Q^{1/2} z \in \Omega\}. \tag{10.47}$$

Then,

$$V(\bar{w}) < V(w) \quad \text{for all } w \in \mathbb{R}^m \backslash \Omega.$$

Proof. Suppose that $w^* \in \mathbb{R}^m \backslash \Omega$ and let

$$\bar{w}^* \triangleq Q^{1/2} \Pi_{\tilde{\Omega}} Q^{-1/2} w^*. \tag{10.48}$$

Notice that $\bar{w}^* \in \Omega$ since (10.45), with $\Pi_{\tilde{\Omega}}$ and $\tilde{\Omega}$ defined in (10.46) and (10.47), respectively, defines a projection of $\mathbb{R}^m$ onto Ω.
   Define,

$$s^* \triangleq Q^{-1/2} w^*, \qquad \bar{s}^* \triangleq Q^{-1/2} \bar{w}^*. \tag{10.49}$$

Then, by construction, s^* and $\bar{s}^*$ satisfy,

$$\bar{s}^* = \Pi_{\tilde{\Omega}} s^*, \tag{10.50}$$

and, in particular, $\bar{s}^* \in \tilde{\Omega}$. Using (10.48) and (10.49) in (10.44)–(10.45), we obtain,

$$V(w^*) = f(\bar{w}^*) + (w^* - \bar{w}^*)^{\mathsf{T}} Q^{-1} (\bar{w}^* - c) = f(Q^{1/2} \bar{s}^*) + (s^* - \bar{s}^*)^{\mathsf{T}} (\bar{s}^* - Q^{-1/2} c). \tag{10.51}$$

Also, since $\bar{w}^* \in \Omega$, we have $\overline{(\bar{w}^*)} \triangleq Q^{1/2} \Pi_{\tilde{\Omega}} Q^{-1/2} \bar{w}^* = \bar{w}^*$. Thus,

$$V(\bar{w}^*) = f(\bar{w}^*) + (\bar{w}^* - \bar{w}^*)^{\mathsf{T}} Q^{-1} (\bar{w}^* - c) = f(Q^{1/2} \bar{s}^*). \tag{10.52}$$

It is easy to see, from the assumptions on Ω, that $\tilde{\Omega}$ in (10.47) is a closed convex set and $Q^{-1/2} c \in \operatorname{int} \tilde{\Omega}$ since $Q^{1/2} > 0$. From Lemma 10.3.1, equation (10.50), the definition of $\Pi_{\tilde{\Omega}}$ in (10.46), and noticing that $s^* \triangleq Q^{-1/2} w^* \notin \tilde{\Omega}$, we conclude that

$$(s^* - \bar{s}^*)^{\mathsf{T}} (\bar{s}^* - Q^{-1/2} c) > 0.$$

Hence, from (10.51) and (10.52), we have

$$V(w^*) - V(\bar{w}^*) = (s^* - \bar{s}^*)^{\mathsf{T}} (\bar{s}^* - Q^{-1/2} c) > 0.$$

The result then follows.   □

   In the sequel, we consider two optimisation problems to be equivalent if they both achieve the same optimum and if the optimisers are the same.

Corollary 10.3.3 *Under the conditions of Lemma 10.3.2, problem $\mathcal{P}_1'$ defined by (10.43)–(10.47) is equivalent to the following problem*

$$\mathcal{P}_1 : \quad \min_{w \in \Omega} f(w). \tag{10.53}$$

Proof. It follows from Lemma 10.3.2 that for any point w in $\mathbb{R}^m \backslash \Omega$ we can find a point $\bar{w}$ in Ω that yields a strictly lower objective function value. Hence, we can perform the minimisation of (10.44) in Ω without losing global optimal solutions. Since the mapping $Q^{1/2} \Pi_{\tilde{\Omega}} Q^{-1/2}$ used in (10.45) reduces to the identity mapping in Ω, we conclude that (10.44) is equal to the objective function in (10.53) for all $w \in \Omega$, and thus the problems are equivalent.   □

Corollary 10.3.4 *Let $f : \mathbb{R}^n \times \mathbb{R}^m \times \cdots \times \mathbb{R}^m \to \mathbb{R}$ be any function and let $\Omega \subset \mathbb{R}^m$ be a closed convex set that contains zero in its interior. Consider the optimisation problem*

$$\mathcal{P}'_2 : \quad \min_{x_0, w_0, \dots, w_{N-1}} V(x_0, w_0, \dots, w_i, \dots, w_{N-1}), \qquad (10.54)$$

with

$$V(x_0, w_0, \dots, w_i, \dots, w_{N-1}) \triangleq f(x_0, \bar{w}_0, \dots, \bar{w}_i, \dots, \bar{w}_{N-1})$$
$$+ \sum_{k=0}^{N-1} (w_k - \bar{w}_k)^{\mathrm{T}} Q^{-1} \bar{w}_k, \qquad (10.55)$$

and

$$\bar{w}_i = Q^{1/2} \Pi_{\tilde{\Omega}} Q^{-1/2} w_i \quad \text{for } i = 0, \dots, N-1, \qquad (10.56)$$

where $\Pi_{\tilde{\Omega}}$ and $\tilde{\Omega}$ are defined as in (10.46) and (10.47), respectively. Then, if $w_i \in \mathbb{R}^m \backslash \Omega$ for some $i \in \{0, \dots, N-1\}$, we have

$$V(x_0, w_0, \dots, \bar{w}_i, \dots, w_{N-1}) < V(x_0, w_0, \dots, w_i, \dots, w_{N-1})$$

for all $x_0 \in \mathbb{R}^n$ and $w_0, \dots, w_{i-1}, w_{i+1}, \dots, w_{N-1} \in \mathbb{R}^m$.

Proof. Consider the sequence $\{x_0^*, w_0^*, \dots, w_i^*, \dots, w_{N-1}^*\}$ and suppose $w_i^* \in \mathbb{R}^m \backslash \Omega$ for some i. Via a similar argument to that used in the proof of Lemma 10.3.2 (with $c = 0$), we can show that the sequence $\{x_0^*, w_0^*, \dots, \bar{w}_i^*, \dots, w_{N-1}^*\}$, with $\bar{w}_i^* = Q^{1/2} \Pi_{\tilde{\Omega}} Q^{-1/2} w_i^*$, gives a lower value of the objective function (10.55). The result then follows. $\qquad \square$

Corollary 10.3.5 *Under the conditions of Corollary 10.3.4, problem $\mathcal{P}'_2$ defined by (10.54)–(10.56) is equivalent to the problem*

$$\mathcal{P}_2 : \quad \min_{w_k \in \Omega, x_0} f(x_0, w_0, \dots, w_i, \dots, w_{N-1}). \qquad (10.57)$$

Proof. Similar to the proof of Corollary 10.3.3. $\qquad \square$

We are now ready to express the primal estimation problem $\mathcal{P}_e$ defined by equations (10.2)–(10.7) in an equivalent form. This is done in the following theorem.

Theorem 10.3.6 (Equivalent Primal Formulation) *Assume that Ω is a closed convex set that contains zero in its interior. Then the primal estimation problem $\mathcal{P}_e$ defined by equations (10.2)–(10.7) is equivalent to the following unconstrained optimisation problem:*

$$\mathcal{P}'_e : \quad V_N^{\mathrm{OPT}}(\mu_0, y_k^d) \triangleq \min_{\hat{x}_k, \hat{v}_k, \hat{w}_k} V'_N(\{\hat{x}_k\}, \{\hat{v}_k\}, \{\hat{w}_k\}), \qquad (10.58)$$

subject to:

$$\hat{x}_{k+1} = A\hat{x}_k + B\bar{w}_k \quad \text{for } k = 0, \cdots, N-1, \qquad (10.59)$$

$$\hat{v}_k = y_k^d - C\hat{x}_k \quad \text{for } k = 1, \cdots, N, \qquad (10.60)$$

$$\bar{w}_k = Q^{1/2} \Pi_{\tilde{\Omega}} Q^{-1/2} \hat{w}_k \quad \text{for } k = 0, \dots, N-1, \qquad (10.61)$$

where

$$V_N'(\{\hat{x}_k\}, \{\hat{v}_k\}, \{\hat{w}_k\}) \triangleq \frac{1}{2}(\hat{x}_0 - \mu_0)^{\mathrm{T}} P_0^{-1}(\hat{x}_0 - \mu_0) + \frac{1}{2}\sum_{k=1}^{N} \hat{v}_k^{\mathrm{T}} R^{-1} \hat{v}_k$$

$$+ \sum_{k=0}^{N-1} \left[\frac{1}{2}\bar{w}_k^{\mathrm{T}} Q^{-1} \bar{w}_k + (\hat{w}_k - \bar{w}_k)^{\mathrm{T}} Q^{-1} \bar{w}_k \right], \quad (10.62)$$

where $\Pi_{\tilde{\Omega}}$ *and* $\tilde{\Omega}$ *are defined in* (10.46) *and* (10.47), *respectively.*

Proof. First note that, using the equations (10.3) and (10.4), the objective function (10.7) can be written in the form

$$V_N(\{\hat{x}_k\}, \{\hat{v}_k\}, \{\hat{w}_k\}) = f(\hat{x}_0, \hat{w}_0, \dots, \hat{w}_i, \dots, \hat{w}_{N-1}).$$

Since the minimisation of the above objective function is performed for $\hat{x}_0 \in \mathbb{R}^n$ and for $\hat{w}_k \in \Omega$, we conclude that problem $\mathcal{P}_e$ can be written in the form (10.57). Using Corollary 10.3.5 we can then express $\mathcal{P}_e$ in the form of problem $\mathcal{P}_2'$ defined by (10.54)–(10.56). However, this is equivalent to (10.58)–(10.62) (note the presence of $\bar{w}_k$ in (10.59)), and the result then follows.   □

Theorem 10.3.6 shows that the primal estimation problem of minimisation with *constraints on the system inputs* (the process noise w_k) can be transformed into an equivalent *unconstrained* minimisation problem *using projected inputs* $\bar{w}_k$ both in the objective function and in the state equations (10.59).

Comparing the primal problem in its equivalent formulation (10.58)–(10.62) with the dual problem (10.8)–(10.13) we observe an interesting *symmetry* between them. This is discussed in the following section.

10.4 Symmetry of Constrained Estimation and Control

In summary, we have shown that the two following problems are dual in the Lagrangian sense.

Primal Constrained Problem (Equivalent Unconstrained Form)

$$\mathcal{P}_e' : \quad \min_{\hat{x}_k, \hat{v}_k, \hat{w}_k} \left\{ \frac{1}{2}(\hat{x}_0 - \mu_0)^{\mathrm{T}} P_0^{-1}(\hat{x}_0 - \mu_0) + \frac{1}{2}\sum_{k=1}^{N} \hat{v}_k^{\mathrm{T}} R^{-1} \hat{v}_k \right.$$

$$\left. + \sum_{k=0}^{N-1} \left[\frac{1}{2}\bar{w}_k^{\mathrm{T}} Q^{-1} \bar{w}_k + (\hat{w}_k - \bar{w}_k)^{\mathrm{T}} Q^{-1} \bar{w}_k \right] \right\},$$

subject to:

$$\hat{x}_{k+1} = A\hat{x}_k + B\bar{w}_k \quad \text{for } k = 0, \cdots, N-1,$$

$$\hat{v}_k = y_k^d - C\hat{x}_k \quad \text{for } k = 1, \cdots, N,$$

$$\bar{w}_k = Q^{1/2}\Pi_{\tilde{\Omega}} Q^{-1/2}\hat{w}_k \quad \text{for } k = 0, \dots, N-1.$$

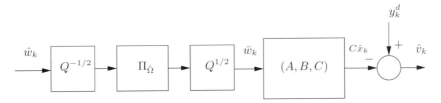

Figure 10.1. Configuration for the primal problem (equivalent formulation).

Dual Unconstrained Problem

$$\mathcal{D}_e : \quad \min_{\lambda_k, u_k} \left\{ \frac{1}{2} (\lambda_{-1} - \tilde{\mu}_0)^{\mathrm{T}} P_0 (\lambda_{-1} - \tilde{\mu}_0) + \frac{1}{2} \sum_{k=1}^{N} \hat{u}_k^{\mathrm{T}} R \hat{u}_k \right.$$

$$\left. + \sum_{k=0}^{N-1} \left[\frac{1}{2} \bar{\zeta}_k^{\mathrm{T}} Q \bar{\zeta}_k + (\zeta_k - \bar{\zeta}_k)^{\mathrm{T}} Q \bar{\zeta}_k \right] \right\} + \gamma,$$

subject to:

$$\lambda_{k-1} = A^{\mathrm{T}} \lambda_k + C^{\mathrm{T}} u_k \quad \text{for } k = 1, \cdots, N,$$

$$\lambda_N = 0, \quad \lambda_{-1} \triangleq A^{\mathrm{T}} \lambda_0,$$

$$\hat{u}_k \triangleq R^{-1} y_k^d - u_k \quad \text{for } k = 1, \cdots, N,$$

$$\zeta_k = B^{\mathrm{T}} \lambda_k \quad \text{for } k = 0, \cdots, N-1,$$

$$\bar{\zeta}_k = Q^{-1/2} \Pi_{\tilde{\Omega}} Q^{1/2} \zeta_k \quad \text{for } k = 0, \cdots, N-1,$$

where $\tilde{\mu}_0 \triangleq -P_0^{-1} \mu_0$ and γ is the constant defined in (10.14).

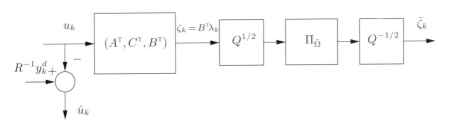

Figure 10.2. Configuration for the dual problem.

In the above two problems, $\Pi_{\tilde{\Omega}}$ is the minimum Euclidean distance projection defined in (10.46) onto the set $\tilde{\Omega}$ defined in (10.47).

Figures 10.1 and 10.2 illustrate the primal equivalent problem $\mathcal{P}'_e$ and the dual problem $\mathcal{D}_e$, respectively. Note from the figures and corresponding equations the symmetry between both problems; namely, input variables take the role of output variables in the objective function, system matrices are *swapped*: $A \longrightarrow A^T$, $B \longrightarrow C^T$, $C \longrightarrow B^T$, time is *reversed* and input projections become output projections.

10.5 Scalar Case

The above duality result takes a particularly simple form in the scalar input case, that is, when $m = 1$ in (10.1). We assume $\Omega = \{w : |w| \le \Delta\}$, where Δ is a positive constant, and take $Q = 1$ in the objective function (10.7), without loss of generality, since we can always scale by this factor. In this case, $\tilde{\Omega} = \Omega$ and the minimum Euclidean distance projection reduces to the usual saturation function defined as $\mathrm{sat}_\Delta(u) = \mathrm{sign}\,(u)\,\min(|u|, \Delta)$.

The (equivalent) primal and dual problems for the scalar case are then:

Primal Problem

$$\mathcal{P}'_e : \quad \min_{\hat{x}_k, \hat{v}_k, \hat{w}_k} \left\{ \frac{1}{2}(\hat{x}_0 - \mu_0)^{\mathrm{T}} P_0^{-1}(\hat{x}_0 - \mu_0) \right.$$

$$\left. + \frac{1}{2}\sum_{k=1}^{N} \hat{v}_k^{\mathrm{T}} R^{-1} \hat{v}_k + \frac{1}{2}\sum_{k=0}^{N-1} \left[\hat{w}_k^2 - (\hat{w}_k - \mathrm{sat}_\Delta(\hat{w}_k))^2\right] \right\},$$

subject to:

$$\hat{x}_{k+1} = A\hat{x}_k + B\mathrm{sat}_\Delta(\hat{w}_k) \quad \text{for } k = 0, \cdots, N-1,$$

$$\hat{v}_k = y_k^d - C\hat{x}_k \quad \text{for } k = 1, \cdots, N.$$

Dual Problem

$$\mathcal{D}_e : \quad \min_{\lambda_k, u_k} \left\{ \frac{1}{2}(A^{\mathrm{T}}\lambda_0 + P_0^{-1}\mu_0)^{\mathrm{T}} P_0 (A^{\mathrm{T}}\lambda_0 + P_0^{-1}\mu_0) \right.$$

$$\left. + \frac{1}{2}\sum_{k=1}^{N} \hat{u}_k^{\mathrm{T}} R \hat{u}_k + \frac{1}{2}\sum_{k=0}^{N-1} \left[\zeta_k^2 - (\zeta_k - \mathrm{sat}_\Delta(\zeta_k))^2\right] \right\} + \gamma, \quad (10.63)$$

subject to:

$$\lambda_{k-1} = A^{\mathrm{T}}\lambda_k + C^{\mathrm{T}} u_k \quad \text{for } k = 1, \cdots, N, \tag{10.64}$$

$$\lambda_N = 0, \tag{10.65}$$

$$\hat{u}_k \triangleq R^{-1} y_k^d - u_k \quad \text{for } k = 1, \cdots, N, \tag{10.66}$$

$$\zeta_k = B^{\mathrm{T}}\lambda_k \quad \text{for } k = 0, \cdots, N-1, \tag{10.67}$$

where γ is the constant defined in (10.14).

Example 10.5.1. Consider the model (10.1) with matrices

$$A = \begin{bmatrix} 0.50 & 0.01 \\ -0.70 & 0.30 \end{bmatrix}, \ B = \begin{bmatrix} 0.40 \\ 0.90 \end{bmatrix} \text{ and } C = \begin{bmatrix} 0.90 & -0.50 \end{bmatrix}.$$

The initial state x_0 has a Gaussian distribution $N(\mu_0, P_0)$, with $\mu_0 = \begin{bmatrix} 1 & 2 \end{bmatrix}^{\mathrm{T}}$. The output noise $\{v_k\}$ is an i.i.d. sequence having a Gaussian distribution $N(0, R)$, with $R = 0.1$. The process noise $\{w_k\}$ has a truncated Gaussian distribution of the form (9.11) in Chapter 9. For this example, we take $\Omega_1 = \Omega = \{w : |w| \le 1\}$ and $Q = 1$. The weighting matrix P_0 was obtained from the steady state error covariance of the Kalman filter for the system above.

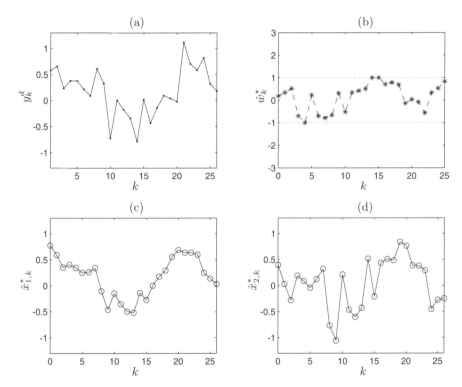

Figure 10.3. Primal problem: constrained estimation. (a) Measurement data. (b) Optimal primal input estimate. (c),(d) Optimal primal state estimates.

Given the measurement data $\{y_k^d\} \triangleq \{y_1^d, \cdots, y_N^d\}$ plotted in Figure 10.3 (a), we solve the primal problem (10.2)–(10.7). Note that by using equations (10.3)–(10.5) the minimisation is performed for $\hat{x}_0$ and for $\hat{w}_k \in \Omega$. We can then use QP to obtain the optimal initial state estimate $\hat{x}_0^*$, and the optimal input estimate $\hat{\omega}_k^*$. The latter is plotted in Figure 10.3 (b). In addition, using the state equations (10.3) with the optimal values $\hat{x}_0^*$ and $\hat{\omega}_k^*$, we obtain the optimal state estimates $\hat{x}^*_{1,k}$ and $\hat{x}^*_{2,k}$, $k = 0, \ldots, N$ (the two components of the state estimate vector $\hat{x}^*_k$), shown in Figure 10.3 (c) and (d), respectively.

The dual of the above estimation problem is the nonlinear optimal control problem (10.63)–(10.67), where the saturation value is $\Delta = 1$. Note that using equations (10.64)–(10.67), the decision variables of the minimisation problem (10.63) are $u_1, \ldots, u_N$ only. The dual problem has swapped the role of the inputs and outputs in the objective function. In the primal problem, the system outputs were the measurement data y^d_k. For the dual problem (see (10.66)), y^d_k has been scaled as the input reference $R^{-1}y^d_k$ to system (10.64)–(10.65). This scaled input reference is shown in Figure 10.4 (a). We solve the nonlinear unconstrained optimisation problem (10.63)–(10.67) to obtain the optimal input u^*_k shown in Figure 10.4 (b). Similarly, the dual system states λ^*_k, whose components are plotted in Figure 10.4 (c)–(d), respectively, can be obtained, in reverse time, via equations (10.64)–(10.65) by using the optimal values of u^*_k.

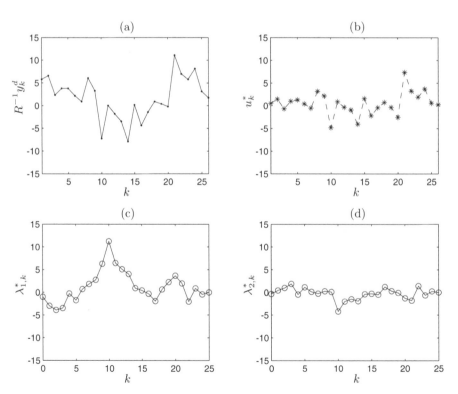

Figure 10.4. Dual problem: nonlinear optimal control. (a) Scaled data. (b) Optimal dual input. (c),(d) Optimal dual states.

   The relation of strong Lagrangian duality between constrained estimation
and control defines a relation between the optimal values in the primal and
dual problems. From equation (10.67), the "dual output" $\zeta_k^* = B^{\mathsf{T}}\lambda_k^*$ (shown
in Figure 10.5 (a)) is a combination of the states λ_k^*, and from the proof of
Theorem 10.2.1 (see (10.31)–(10.32)), we have

$$\hat{w}_k^* = Q\bar{\zeta}_k^*, \quad \text{where} \quad \bar{\zeta}_k^* = \text{sat}_\Delta(\zeta_k^*).$$

That is, the optimal input values $\hat{w}_k^*$ of the primal problem are the scaled
projections of the optimal dual outputs ζ_k^*, as can be seen by comparing
Figure 10.3 (b) with Figure 10.5 (b).   ∘

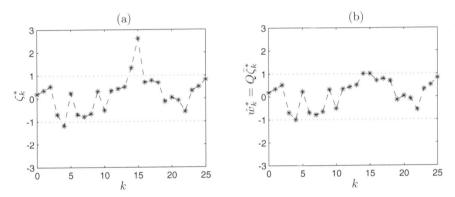

Figure 10.5. Relation between optimal values of the primal and dual problems. (a)
Optimal dual output. (b) Optimal primal input equal to scaled projected optimal
dual output.

10.6 More General Constraints

We have seen above that the dual of the estimation problem *with constraints
on the process noise sequence* $\{w_k\}$ is an unconstrained nonlinear control problem defined in terms of projected outputs. Here we generalise the estimation
problem by considering constraints on the *process noise sequence* $\{w_k\}$, the
measurement noise sequence $\{v_k\}$ and the *initial state* x_0. In this case, the
dual problem will turn out to be an unconstrained nonlinear control problem
defined in terms of projected outputs, projected inputs and projected terminal
states.
   Thus, consider the following system

$$\begin{aligned} x_{k+1} &= Ax_k + Bw_k \quad \text{for } k = 0, \cdots, N-1, \\ y_k &= Cx_k + v_k \quad \text{for } k = 1, \cdots, N, \end{aligned} \tag{10.68}$$

where $x_k \in \mathbb{R}^n$, $w_k \in \mathbb{R}^m$, $y_k \in \mathbb{R}^p$. We assume that $\{w_k\}$, $\{v_k\}$, x_0 have truncated Gaussian distributions of the forms given in (9.11), (9.12) and (9.13) of Chapter 9, respectively, where

$$w_k \in \Omega_1 \quad \text{for } k = 0, \cdots, N-1,$$
$$v_k \in \Omega_2 \quad \text{for } k = 1, \cdots, N,$$
$$x_0 \in \Omega_3.$$

For (10.68) we consider the following optimisation problem:

$$\mathcal{P}_e: \quad V_N^{\mathrm{OPT}}(\mu_0, y_k^d) \triangleq \min_{\hat{x}_k, \hat{v}_k, \hat{w}_k} V_N(\{\hat{x}_k\}, \{\hat{v}_k\}, \{\hat{w}_k\}), \tag{10.69}$$

subject to:

$$\hat{x}_{k+1} = A\hat{x}_k + B\hat{w}_k \quad \text{for } k = 0, \ldots, N-1, \tag{10.70}$$
$$\hat{v}_k = y_k^d - C\hat{x}_k \quad \text{for } k = 1, \ldots, N, \tag{10.71}$$
$$\{\hat{x}_0, \hat{x}_1, \ldots, \hat{x}_N, \hat{v}_1, \ldots, \hat{v}_N, \hat{w}_0, \ldots, \hat{w}_{N-1}\} \in X, \tag{10.72}$$

where, in (10.72),

$$X = \Omega_3 \times \underbrace{\mathbb{R}^n \cdots \times \mathbb{R}^n}_{N} \times \underbrace{\Omega_2 \times \cdots \times \Omega_2}_{N} \times \underbrace{\Omega_1 \times \cdots \times \Omega_1}_{N}, \tag{10.73}$$

and where, in (10.69), the objective function is

$$V_N(\{\hat{x}_k\}, \{\hat{v}_k\}, \{\hat{w}_k\}) \triangleq \frac{1}{2}(\hat{x}_0 - \mu_0)^{\mathrm{T}} P_0^{-1}(\hat{x}_0 - \mu_0)$$
$$+ \frac{1}{2}\sum_{k=0}^{N-1} \hat{w}_k^{\mathrm{T}} Q^{-1} \hat{w}_k + \frac{1}{2}\sum_{k=1}^{N} \hat{v}_k^{\mathrm{T}} R^{-1} \hat{v}_k. \tag{10.74}$$

The following result establishes duality between the constrained estimation problem $\mathcal{P}_e$ and an unconstrained nonlinear optimal control problem.

Theorem 10.6.1 (Dual Problem) *Assume* Ω_1, Ω_2, Ω_3 *in* (10.73) *are nonempty closed convex sets such that there exists a feasible solution* $\{\hat{x}_0, \hat{x}_1, \ldots, \hat{x}_N, \hat{v}_1, \ldots, \hat{v}_N, \hat{w}_0, \ldots, \hat{w}_{N-1}\} \in \mathrm{int}\, X$ *for the primal problem* $\mathcal{P}_e$. *Given the primal constrained fixed horizon estimation problem* $\mathcal{P}_e$ *defined by equations* (10.69)–(10.74), *the Lagrangian dual problem is*

$$\mathcal{D}_e: \quad \phi^{\mathrm{OPT}}(\mu_0, \{y_k^d\}) \triangleq \min_{\lambda_k, u_k} \phi(\{\lambda_k\}, \{u_k\}), \tag{10.75}$$

subject to:

$$\lambda_{k-1} = A^{\mathrm{T}}\lambda_k + C^{\mathrm{T}}u_k \quad \text{for } k = 1, \cdots, N, \tag{10.76}$$
$$\lambda_N = 0, \quad \lambda_{-1} = A^{\mathrm{T}}\lambda_0, \tag{10.77}$$
$$\zeta_k = B^{\mathrm{T}}\lambda_k \quad \text{for } k = 0, \cdots, N-1. \tag{10.78}$$

In (10.75), *the objective function is*

$$\phi(\{\lambda_k\}, \{u_k\}) \triangleq \frac{1}{2}(\bar{\lambda}_{-1} + P_0^{-1}\mu_0)^{\mathrm{T}} P_0 (\bar{\lambda}_{-1} + P_0^{-1}\mu_0)$$

$$+ (\lambda_{-1} - \bar{\lambda}_{-1})^{\mathrm{T}} P_0 (\bar{\lambda}_{-1} + P_0^{-1}\mu_0)$$

$$+ \sum_{k=1}^{N} \left[\frac{1}{2}(\bar{u}_k - R^{-1}y_k^d)^{\mathrm{T}} R(\bar{u}_k - R^{-1}y_k^d) \right.$$

$$\left. + (u_k - \bar{u}_k)^{\mathrm{T}} R(\bar{u}_k - R^{-1}y_k^d) \right]$$

$$+ \sum_{k=0}^{N-1} \left[\frac{1}{2}\bar{\zeta}_k^{\mathrm{T}} Q \bar{\zeta}_k + (\zeta_k - \bar{\zeta}_k)^{\mathrm{T}} Q \bar{\zeta}_k \right] + \gamma \tag{10.79}$$

where γ is the constant term given by

$$\gamma \triangleq -\frac{1}{2}\mu_0^{\mathrm{T}} P_0^{-1} \mu_0 - \frac{1}{2} \sum_{k=1}^{N} (y_k^d)^{\mathrm{T}} R^{-1} y_k^d. \tag{10.80}$$

In (10.79) *the projected variables are defined as*

$$\bar{\lambda}_{-1} \triangleq P_0^{-1/2} \Pi_{\tilde{\Omega}_3} P_0^{1/2} \lambda_{-1}, \tag{10.81}$$

$$\bar{u}_k \triangleq R^{-1/2} \Pi_{\tilde{\Omega}_2} R^{1/2} u_k \quad \textit{for } k = 1, \dots, N, \tag{10.82}$$

$$\bar{\zeta}_k \triangleq Q^{-1/2} \Pi_{\tilde{\Omega}_1} Q^{1/2} \zeta_k \quad \textit{for } k = 0, \dots, N-1, \tag{10.83}$$

where $\Pi_{\tilde{\Omega}_i}$, $\cdot i = 1, 2, 3$, denote the minimum Euclidean distance projections (defined as in (10.15)*) onto the sets*

$$\tilde{\Omega}_1 \triangleq \{z : Q^{1/2}z \in \Omega_1\}, \tag{10.84}$$

$$\tilde{\Omega}_2 \triangleq \{z : R^{1/2}z \in \Omega_2\}, \tag{10.85}$$

$$\tilde{\Omega}_3 \triangleq \{z : P_0^{1/2}z + \mu_0 \in \Omega_3\}. \tag{10.86}$$

Moreover, there is no duality gap, that is, the minimum achieved in (10.69) *is equal to minus the minimum achieved in* (10.75).

Proof. The proof follows the same lines as the proof of Theorem 10.2.1, save that we must consider the constraints on $\hat{x}_0$ and $\hat{v}_k$, as well as on $\hat{w}_k$, when optimising (10.18). Thus, instead of (10.20) and (10.21), we need to carry out a constrained optimisation as was done for $\hat{w}_k$ in (10.19). $\qquad\square$

10.7 Symmetry Revisited

We have seen in Section 10.6 that the Lagrangian dual of the general constrained estimation problem is an unconstrained nonlinear control problem involving projected variables. The symmetry in this result is revealed by transforming the primal problem into an equivalent unconstrained estimation problem using projected variables. To derive this equivalent problem, we will use the following extensions of Corollaries 10.3.4 and 10.3.5:

Corollary 10.7.1 Let $f : Z \to \mathbb{R}$ and $h : Z \to \mathbb{R}^q$, with $Z = \mathbb{R}^n \times \mathbb{R}^p \times \cdots \times \mathbb{R}^p \times \mathbb{R}^m \times \cdots \times \mathbb{R}^m$, be any functions and let $\Omega_1 \subset \mathbb{R}^m$, $\Omega_2 \subset \mathbb{R}^p$ and $\Omega_3 \subset \mathbb{R}^n$ be closed convex sets. Let $0 \in \mathrm{int}\,\Omega_1$, $0 \in \mathrm{int}\,\Omega_2$ and $\mu_0 \in \mathrm{int}\,\Omega_3$. Consider the optimisation problem

$$\mathcal{P}' : \quad \min_{x_0, v_k, w_k} V(x_0, v_1, \ldots, v_N, w_0, \ldots, w_{N-1}), \qquad (10.87)$$

subject to:

$$h(\bar{x}_0, \bar{v}_1, \ldots, \bar{v}_N, \bar{w}_0, \ldots, \bar{w}_{N-1}) = 0, \qquad (10.88)$$

with

$$
\begin{aligned}
V(x_0, v_1, \ldots, v_N, w_0, \ldots, w_{N-1}) &\triangleq f(\bar{x}_0, \bar{v}_1, \ldots, \bar{v}_N, \bar{w}_0, \ldots, \bar{w}_{N-1}) \\
&+ (x_0 - \bar{x}_0)^{\mathrm{T}} P_0^{-1} (\bar{x}_0 - \mu_0) \\
&+ \sum_{k=1}^{N} (v_k - \bar{v}_k)^{\mathrm{T}} R^{-1} \bar{v}_k \\
&+ \sum_{k=0}^{N-1} (w_k - \bar{w}_k)^{\mathrm{T}} Q^{-1} \bar{w}_k, \qquad (10.89)
\end{aligned}
$$

and

$$\bar{x}_0 = P_0^{1/2} \Pi_{\tilde{\Omega}_3} P_0^{-1/2} (x_0 - \mu_0) + \mu_0, \qquad (10.90)$$

$$\bar{v}_k = R^{1/2} \Pi_{\tilde{\Omega}_2} R^{-1/2} v_k \quad \text{for } k = 1, \ldots, N, \qquad (10.91)$$

$$\bar{w}_k = Q^{1/2} \Pi_{\tilde{\Omega}_1} Q^{-1/2} w_k \quad \text{for } k = 0, \ldots, N - 1, \qquad (10.92)$$

where $\Pi_{\tilde{\Omega}_i}$, $i = 1, 2, 3$, are the minimum Euclidean distance projections (defined as in (10.15)) onto the sets (10.84)–(10.86), respectively.

Then any global optimal solution $\{x_0^*, v_1^*, \ldots, v_i^*, \ldots, v_N^*, w_0^*, \ldots, w_i^*, \ldots, w_{N-1}^*\}$ of (10.87)–(10.92) satisfies $x_0^* \in \Omega_3$, $v_i^* \in \Omega_2$ for $i = 1, \ldots, N$ and $w_i^* \in \Omega_1$ for $i = 0, \ldots, N - 1$.

Proof. As in the proof of Corollary 10.3.4 we can show that given any feasible sequence $\{x_0^*, v_1^*, \ldots, v_i^*, \ldots, v_N^*, w_0^*, \ldots, w_i^*, \ldots, w_{N-1}^*\}$ for problem (10.87)–(10.92) and such that $x_0^* \in \mathbb{R}^n \backslash \Omega_3$, and/or $v_i^* \in \mathbb{R}^p \backslash \Omega_2$ for some i and/or $w_i^* \in \mathbb{R}^m \backslash \Omega_1$ for some i, a lower value of the objective function is achieved by replacing these variables by their projected values $\bar{x}_0^* \in \Omega_3$, $\bar{v}_i^* \in \Omega_2$ and $\bar{w}_i^* \in \Omega_1$, computed as in (10.90)–(10.92). Since the sequence so obtained satisfies the equality constraint (10.88), it is feasible and hence the result follows. $\square$

Corollary 10.7.2 *Under the conditions of Corollary 10.7.1, problem $\mathcal{P}'$ defined by (10.87)–(10.92) is equivalent to the problem*

$$\mathcal{P}: \qquad \min_{x_0 \in \Omega_3, v_k \in \Omega_2, w_k \in \Omega_1} f(x_0, v_1, \ldots, v_N, w_0, \ldots, w_{N-1}).$$

Proof. Similar to the proof of Corollary 10.3.3. □

We then have the following equivalent formulation for the primal estimation problem $\mathcal{P}_e$ defined by equations (10.69)–(10.74).

Theorem 10.7.3 (Equivalent Primal Formulation Revisited)
Suppose $\Omega_1 \subset \mathbb{R}^m$, $\Omega_2 \subset \mathbb{R}^p$ and $\Omega_3 \subset \mathbb{R}^n$ are closed convex sets such that $0 \in \mathrm{int}\,\Omega_1$, $0 \in \mathrm{int}\,\Omega_2$ and $\mu_0 \in \mathrm{int}\,\Omega_3$. Then the primal estimation problem $\mathcal{P}_e$ defined by equations (10.69)–(10.74) is equivalent to the following unconstrained optimisation problem:

$$\mathcal{P}'_e: \qquad V_N^{\mathrm{OPT}}(\mu_0, y_k^d) \triangleq \min V_N'(\hat{x}_0, \{\hat{v}_k\}, \{\hat{w}_k\}), \tag{10.93}$$

subject to:

$$\bar{x}_{k+1} = A\bar{x}_k + B\bar{w}_k \quad \text{for } k = 0, \cdots, N-1, \tag{10.94}$$

$$\bar{v}_k = y_k^d - C\bar{x}_k \quad \text{for } k = 1, \cdots, N, \tag{10.95}$$

$$\bar{x}_0 = P_0^{1/2} \Pi_{\tilde{\Omega}_3} P_0^{-1/2}(\hat{x}_0 - \mu_0) + \mu_0, \tag{10.96}$$

$$\bar{v}_k = R^{1/2} \Pi_{\tilde{\Omega}_2} R^{-1/2} \hat{v}_k \quad \text{for } k = 1, \ldots, N, \tag{10.97}$$

$$\bar{w}_k = Q^{1/2} \Pi_{\tilde{\Omega}_1} Q^{-1/2} \hat{w}_k \quad \text{for } k = 0, \ldots, N-1, \tag{10.98}$$

where $\Pi_{\tilde{\Omega}_i}$, $i = 1, 2, 3$ are the minimum Euclidean distance projections (defined as in (10.15)) onto the sets (10.84)–(10.86), respectively, and where

$$V_N'(\hat{x}_0, \{\hat{v}_k\}, \{\hat{w}_k\}) \triangleq \frac{1}{2}(\bar{x}_0 - \mu_0)^{\mathrm{T}} P_0^{-1}(\bar{x}_0 - \mu_0)$$

$$+ (\hat{x}_0 - \bar{x}_0)^{\mathrm{T}} P_0^{-1}(\bar{x}_0 - \mu_0)$$

$$+ \sum_{k=1}^{N} \left[\frac{1}{2} \bar{v}_k^{\mathrm{T}} R^{-1} \bar{v}_k + (\hat{v}_k - \bar{v}_k)^{\mathrm{T}} R^{-1} \bar{v}_k \right]$$

$$+ \sum_{k=0}^{N-1} \left[\frac{1}{2} \bar{w}_k^{\mathrm{T}} Q^{-1} \bar{w}_k + (\hat{w}_k - \bar{w}_k)^{\mathrm{T}} Q^{-1} \bar{w}_k \right]. \tag{10.99}$$

Proof. Immediate from Corollary 10.7.2 on interpreting h in (10.88) as

$$h(\bar{x}_0, \bar{v}_1, \ldots, \bar{v}_N, \bar{w}_0, \ldots, \bar{w}_{N-1}) =$$

$$\begin{bmatrix} C A \bar{x}_0 + \bar{v}_1 + C B \bar{w}_0 - y_1^d \\ C A^2 \bar{x}_0 + \bar{v}_2 + C A B \bar{w}_0 + C B \bar{w}_1 - y_2^d \\ \vdots \\ C A^N \bar{x}_0 + \bar{v}_N + \sum_{k=0}^{N-1} C A^{N-k-1} B \bar{w}_k - y_N^d \end{bmatrix}.$$

□

If we compare the equivalent form of the primal problem (10.93)–(10.99) with the dual problem (10.75)–(10.83) then aspects of the symmetry between these problems are revealed. In particular, we see that the following connections hold: time is reversed, system matrices are swapped ($A \longrightarrow A^T$, $B \longrightarrow C^T$, $C \longrightarrow B^T$), input projections become output projections and vice versa, and initial state projections become terminal state projections. These connections and other observations are summarised in Table 10.1.

	Primal	Equivalent Primal	Dual
State equations	$\hat{x}_{k+1} = A\hat{x}_k + B\hat{w}_k$	$\bar{x}_{k+1} = A\bar{x}_k + B\bar{w}_k$	$\lambda_{k-1} = A^{\mathsf{T}}\lambda_k + C^{\mathsf{T}}u_k,$ $\lambda_{-1} = A^{\mathsf{T}}\lambda_0$
Output equation	$\hat{v}_k = y_k^d - C\hat{x}_k$	$\bar{v}_k = y_k^d - C\bar{x}_k$	$\zeta_k = B^{\mathsf{T}}\lambda_k$
Input/output connection	Input constraints $\hat{w}_k \in \Omega_1$	*Unconstrained* minimisation using the *projected input* $\bar{w}_k$ in the objective function. *Projected input* used in the state equations: $\bar{x}_{k+1} = A\bar{x}_k + B\bar{w}_k$.	*Unconstrained* minimisation using *projected output* $\bar{\zeta}_k$ in the objective function.
Output/input connection	Output constraints $\hat{v}_k \in \Omega_2$	*Unconstrained* minimisation using the *projected output* $\bar{v}_k$ in the objective function. *Projected output* required to satisfy the output equation: $\bar{v}_k = y_k^d - C\bar{x}_k$.	*Unconstrained* minimisation using the *projected input* $\bar{u}_k$ in the objective function.
Initial/final state connection	Initial state constraints $\hat{x}_0 \in \Omega_3$	*Unconstrained* minimisation using the *projected initial state* $\bar{x}_0$ in the objective function. *Projected initial state* used as initial state for the state equations.	*Unconstrained* minimisation using the *projected terminal state* $\bar{\lambda}_{-1}$ in the objective function.

Table 10.1. Connections between the primal problem, its equivalent formulation and the dual problem.

10.8 Further Reading

For complete list of references cited, see References section at the end of book.

General

The relationship between linear estimation and linear quadratic control is well-known in the *unconstrained* case. Since the original work of Kalman and others (see Kalman 1960b, Kalman and Bucy 1961), many authors have contributed to further understand this relationship. For example, Kailath, Sayed and Hassibi (2000) have explored duality in the unconstrained case using the geometrical concepts of dual bases and orthogonal complements. The connection between the two unconstrained optimisation problems using Lagrangian duality has also been established in, for example, the recent work of Rao (2000).

The results in the current chapter are based on Goodwin, De Doná, Seron and Zhuo (2004).

Part II

Further Developments

The Hessian in the QP Problem: Singular Value Structure and Related Algorithms

Contributed by Osvaldo J. Rojas

11.1 Overview

We saw in earlier chapters that a core ingredient in quadratic constrained optimisation problems is the Hessian matrix H. So far we have simply given an "in principle" approach to the evaluation of this matrix. That is, for a system with state equations $x_{k+1} = Ax_k + Bu_k$, we have from (5.20) in Chapter 5, that we can compute the Hessian as $H = \Gamma^{\mathsf{T}} \mathbf{Q} \Gamma + \mathbf{R}$, where Γ, $\mathbf{Q}$, $\mathbf{R}$ are defined in (5.14) and (5.16). This will be satisfactory for simple problems. However, for more complex problems (for example, high order systems or problems having mixed stable and unstable modes) this "brute force" approach may fail. A hint as to the source of the difficulties is that the direct way of computing the Hessian depends on powers of the system matrix A, as can be readily seen from the matrix Γ in (5.14). Clearly, if the system has unstable modes, then some entries of Γ will diverge as N increases.

We will show in this chapter that this problem can be resolved by focusing attention on the stable and unstable parts of the system separately.

We note, in passing, that the problem can be addressed by several other routes. Two alternatives to the idea described here are:

(i) Introducing prestabilising feedback, for example, Rossiter, Kouvaritakis and Rice (1998); that is, putting

$$u_k = -Kx_k + \bar{u}_k,$$

where K is chosen so that $(A - BK)$ is Hurwitz; then solving the optimisation problem for the new input sequence $\{\bar{u}_k\}$.

(ii) Introducing the system equations as (a set of) linear equality constraints, that is,

$$x_{k+1} = Ax_k + Bu_k \quad \text{for } k = 0, 1, \ldots, N - 1; \qquad (11.1)$$

then treating both $\{x_1, \ldots, x_N\}$ and $\{u_0, \ldots, u_{N-1}\}$ as decision variables in the optimisation, instead of eliminating the states in the objective function using (11.1).

These alternative approaches resolve the problem at the expense of minor disadvantages. In particular, approach (i) implies that simple constraints on u_k become state-dependent constraints on the new control variable, $\bar{u}_k = u_k + Kx_k$. Approach (ii) implies that one must deal with Nn additional linear equality constraints in the optimisation where n is the state dimension and N is the optimisation horizon. By way of contrast, the methodology presented here avoids these two difficulties by considering a different formulation of the QP, one which is a function only of exponentially decaying terms. This leads to an alternative Hessian (which includes a submatrix that we will call the "regularised sub-Hessian"). An important consequence of using the regularised sub-Hessian is that it provides a direct link to the system frequency response. This latter property can be exploited to gain heuristic insights into the performance of the algorithms.

We also describe algorithms that arise from the singular value decomposition of the regularised sub-Hessian, which underlies the frequency domain viewpoint.

11.2 Separating Stable and Unstable Modes

Our eventual goal in this chapter is to gain a better understanding of the structure of the Hessian matrix particularly for large optimisation horizons. However, as mentioned in Section 11.1, the straightforward approach to evaluating the Hessian will often meet difficulties for open loop unstable plants due to exponential divergence of the system impulse response. One way of addressing this problem is to recognise that there is an intimate connection between "stability" and "causality." In particular, a system having all modes unstable becomes stable if viewed in reverse time, that is, as an anti-causal system. This line of reasoning leads to an alternative viewpoint in which unstable modes are treated differently. The new formulation of the QP problem that we propose separates the stable and unstable responses. We show below that this leads to an equivalent problem formulation with a different Hessian having different properties. In particular, the new Hessian has improved numerical properties and also provides a direct link to frequency domain insights for large horizons (as we will show in Section 11.5).

Consider a discrete time linear system and suppose that it has no eigenvalues on the unit circle. We can then partition the state vector $x_k \in \mathbb{R}^n$ as

$$x_k = \begin{bmatrix} x_k^s \\ x_k^u \end{bmatrix},$$

where the states x_k^s and x_k^u are associated with the stable and unstable modes, respectively. Correspondingly, we can factor the state equations into stable and unstable parts as follows:

$$\begin{bmatrix} x_{k+1}^s \\ x_{k+1}^u \end{bmatrix} = \begin{bmatrix} A_s & 0 \\ 0 & A_u \end{bmatrix} \begin{bmatrix} x_k^s \\ x_k^u \end{bmatrix} + \begin{bmatrix} B_s \\ B_u \end{bmatrix} u_k, \tag{11.2}$$

$$y_k = C\, x_k = \begin{bmatrix} C_s & C_u \end{bmatrix} \begin{bmatrix} x_k^s \\ x_k^u \end{bmatrix},$$

where $u_k \in \mathbb{R}^m$ and $y_k \in \mathbb{R}^p$ ($p \geq m$). The eigenvalues of A_s have moduli less than one, and the eigenvalues of A_u have moduli greater than one.

We can then express the solution of (11.2) as

$$x_k^s = A_s^k x_0^s + \sum_{j=0}^{k-1} A_s^{k-1-j} B_s u_j \quad \text{for } k = 1, \ldots, N, \tag{11.3}$$

$$x_k^u = A_u^{-(N-k)} \mu - \sum_{j=k}^{N-1} A_u^{k-1-j} B_u u_j \quad \text{for } k = 0, \ldots, N-1. \tag{11.4}$$

Equation (11.3) is the result of solving the stable states in *forward* time, whilst equation (11.4) results from solving the unstable states in *reverse* time starting from

$$x_N^u \triangleq \mu,$$

that is, the unstable state at time $k = N$.

Note that since the initial condition x_0^u is given, setting $k = 0$ in (11.4) effectively determines an equality constraint that both μ and the sequence of control signals $\{u_0, \ldots, u_{N-1}\}$ need to satisfy in order to bring the unstable states back to their correct initial values, that is,

$$A_u^{-N} \mu - \sum_{j=0}^{N-1} A_u^{-j-1} B_u u_j = x_0^u. \tag{11.5}$$

We are thus led to the following equivalent statement of the optimisation problem.

$$\mathcal{P}_N(x): \qquad V_N^{\text{OPT}}(x) \triangleq \min V_N(\{x_k\}, \{u_k\}, \mu), \qquad (11.6)$$

subject to:

$$x_k = \begin{bmatrix} x_k^s \\ x_k^u \end{bmatrix} \qquad \text{for } k = 0, \dots, N,$$

$$x_k^s = A_s^k x_0^s + \sum_{j=0}^{k-1} A_s^{k-1-j} B_s u_j \qquad \text{for } k = 1, \dots, N,$$

$$x_k^u = A_u^{-(N-k)} \mu - \sum_{j=k}^{N-1} A_u^{k-1-j} B_u u_j \qquad \text{for } k = 0, \dots, N-1,$$

$$x_0 = \begin{bmatrix} x_0^s \\ x_0^u \end{bmatrix} = x,$$

$$u_k \in \mathbb{U} \quad \text{for } k = 0, \dots, N-1,$$

$$x_k \in \mathbb{X} \quad \text{for } k = 0, \dots, N-1,$$

$$x_N = \begin{bmatrix} x_N^s \\ \mu \end{bmatrix} \in \mathbb{X}_f \subset \mathbb{X},$$

where

$$V_N(\{x_k\}, \{u_k\}, \mu) \triangleq \frac{1}{2} \begin{bmatrix} x_N^s \\ \mu \end{bmatrix}^{\text{T}} P \begin{bmatrix} x_N^s \\ \mu \end{bmatrix} + \frac{1}{2} \sum_{k=0}^{N-1} (x_k^{\text{T}} Q x_k + u_k^{\text{T}} R u_k). \qquad (11.7)$$

The above formulation of the problem, at the expense of the introduction of the additional optimisation variable μ, avoids exponentially diverging terms in the computation of the Hessian matrix. Thus, at least intuitively, it would seem to be more apposite for studying the structure of the problem for large horizons.

In the next sections we will explore the consequences of this alternative formulation of the QP problem.

11.3 Numerical Issues in the Computation of the Hessian

We first show how the split into stable and unstable modes can be a useful strategy to deal with numerical problems involved in evaluating the Hessian.

We represent the time evolution of the system output using the usual vector notation. Thus, let

$$\begin{aligned} \mathbf{y} &= \begin{bmatrix} y_1^{\text{T}} & y_2^{\text{T}} & \cdots & y_N^{\text{T}} \end{bmatrix}^{\text{T}}, \\ \mathbf{u} &= \begin{bmatrix} u_0^{\text{T}} & u_1^{\text{T}} & \cdots & u_{N-1}^{\text{T}} \end{bmatrix}^{\text{T}}. \end{aligned} \qquad (11.8)$$

We then have, using (11.3) and (11.4), that

$$\mathbf{y} = \underbrace{(\Gamma_s + \Gamma_u)}_{\bar{\Gamma}}\mathbf{u} + \Omega_s x_0^s + \Omega_u \mu, \tag{11.9}$$

where

$$\Omega_s = \begin{bmatrix} C_s A_s \\ C_s A_s^2 \\ \vdots \\ C_s A_s^N \end{bmatrix}, \quad \Gamma_s = \begin{bmatrix} C_s B_s & 0 & \cdots & 0 \\ C_s A_s B_s & C_s B_s & \cdots & 0 \\ \vdots & \vdots & \ddots & \vdots \\ C_s A_s^{N-1} B_s & C_s A_s^{N-2} B_s & \cdots & C_s B_s \end{bmatrix},$$

and

$$\Omega_u = \begin{bmatrix} C_u A_u^{-(N-1)} \\ C_u A_u^{-(N-2)} \\ \vdots \\ C_u A_u^{-1} \\ C_u \end{bmatrix}, \quad \Gamma_u = - \begin{bmatrix} 0 & C_u A_u^{-1} B_u & C_u A_u^{-2} B_u & \cdots & C_u A_u^{-(N-1)} B_u \\ 0 & 0 & C_u A_u^{-1} B_u & \cdots & C_u A_u^{-(N-2)} B_u \\ \vdots & \vdots & \ddots & \ddots & \vdots \\ 0 & 0 & \cdots & 0 & C_u A_u^{-1} B_u \\ 0 & 0 & \cdots & 0 & 0 \end{bmatrix}.$$

The matrix $\bar{\Gamma} \triangleq \Gamma_s + \Gamma_u$ has the form

$$\bar{\Gamma} \triangleq \Gamma_s + \Gamma_u = \begin{bmatrix} \bar{h}_0 & \bar{h}_{-1} & \bar{h}_{-2} & \ldots \ldots & \bar{h}_{-(N-1)} \\ \bar{h}_1 & \bar{h}_0 & \bar{h}_{-1} & \ldots \ldots & \bar{h}_{-(N-2)} \\ \vdots & \vdots & \ddots & \ddots & \vdots \\ \vdots & \vdots & & \ddots & \vdots \\ \vdots & \vdots & & \bar{h}_0 & \bar{h}_{-1} \\ \bar{h}_{N-1} & \bar{h}_{N-2} & \ldots & \ldots & \bar{h}_1 & \bar{h}_0 \end{bmatrix}, \tag{11.10}$$

where $\{\bar{h}_k : k = -(N-1), \ldots, N-1\}$, is a finite subsequence of the infinite sequence $\{\bar{h}_k : k = -\infty, \ldots, \infty\}$ defined by

$$\{\bar{h}_k : k = 0, \ldots, \infty\} \triangleq \{C_s B_s, C_s A_s B_s, C_s A_s^2 B_s, \ldots\}, \tag{11.11}$$

$$\{\bar{h}_k : k = -1, \ldots, -\infty\} \triangleq \{-C_u A_u^{-1} B_u, -C_u A_u^{-2} B_u, -C_u A_u^{-3} B_u, \ldots\}. \tag{11.12}$$

In what follows, we will set the terminal state weighting matrix $P = Q$ in (11.7) (note that this choice is not restrictive here since our main interest is in the case $N \to \infty$). Also, we consider

$$Q = C^{\mathsf{T}} C \quad \text{and} \quad R = \rho I_m > 0.$$

If we adopt the vector notation (11.8), we can express the objective function V_N as follows[1]:

---

[1] We keep the function V_N but change its arguments as appropriate.

$$V_N(x, \mathbf{u}, \mathbf{y}, \mu) = \frac{1}{2}(x^\mathsf{T}Qx + \mathbf{y}^\mathsf{T}\mathbf{y} + \rho\mathbf{u}^\mathsf{T}\mathbf{u}).$$

Replacing the expression for $\mathbf{y}$ provided in (11.9) we have

$$
\begin{aligned}
V_N(x, \mathbf{u}, \mu) &= \frac{1}{2}\left[x^\mathsf{T}Qx + \left(\bar{\Gamma}\mathbf{u} + \Omega_s x_0^s + \Omega_u \mu\right)^\mathsf{T}\left(\bar{\Gamma}\mathbf{u} + \Omega_s x_0^s + \Omega_u \mu\right) + \rho\mathbf{u}^\mathsf{T}\mathbf{u}\right] \\
&= \frac{1}{2}\mathbf{u}^\mathsf{T}[\bar{\Gamma}^\mathsf{T}\bar{\Gamma} + \rho I]\mathbf{u} + \mathbf{u}^\mathsf{T}\bar{\Gamma}^\mathsf{T}\left(\Omega_s x_0^s + \Omega_u \mu\right) \\
&\quad + \frac{1}{2}\left(\Omega_s x_0^s + \Omega_u \mu\right)^\mathsf{T}\left(\Omega_s x_0^s + \Omega_u \mu\right).
\end{aligned}
\tag{11.13}
$$

With respect to the new optimisation variables $(\mathbf{u}, \mu)$, the modified Hessian of the quadratic objective function (11.13) is

$$H' \triangleq \begin{bmatrix} \bar{\Gamma}^\mathsf{T}\bar{\Gamma} + \rho I & \bar{\Gamma}^\mathsf{T}\Omega_u \\ \Omega_u^\mathsf{T}\bar{\Gamma} & \Omega_u^\mathsf{T}\Omega_u \end{bmatrix}. \tag{11.14}$$

For future use, we extract the left-upper submatrix, which we call the "regularised sub-Hessian":

$$\bar{H}_N \triangleq \bar{\Gamma}^\mathsf{T}\bar{\Gamma} + \rho I. \tag{11.15}$$

Note that we have made use of the subindex N in (11.15) in order to make explicit the dependence on the prediction horizon.

The standard Hessian H can be recovered from (11.13) on noting from (11.5) that

$$
\begin{aligned}
\mu &= A_u^N x_0^u + \begin{bmatrix} A_u^{N-1}B_u & \cdots & A_u B_u & B_u \end{bmatrix}\mathbf{u} \\
&\triangleq A_u^N x_0^u + L_u \mathbf{u}.
\end{aligned}
\tag{11.16}
$$

Hence, substituting into (11.13), we obtain

$$
\begin{aligned}
V_N(x, \mathbf{u}) &= \frac{1}{2}\mathbf{u}^\mathsf{T}[\bar{\Gamma}^\mathsf{T}\bar{\Gamma} + \rho I]\mathbf{u} + \mathbf{u}^\mathsf{T}\bar{\Gamma}^\mathsf{T}\left[\Omega_s x_0^s + \Omega_u\left(A_u^N x_0^u + L_u\mathbf{u}\right)\right] \\
&\quad + \frac{1}{2}\left[\Omega_s x_0^s + \Omega_u\left(A_u^N x_0^u + L_u\mathbf{u}\right)\right]^\mathsf{T}\left[\Omega_s x_0^s + \Omega_u\left(A_u^N x_0^u + L_u\mathbf{u}\right)\right].
\end{aligned}
$$

Forming the Hessian H of the above equation and comparing it with the expression for the regularised sub-Hessian (11.15), it follows that

$$H = \bar{H}_N + \bar{\Gamma}\Omega_u L_u + L_u^\mathsf{T}\Omega_u^\mathsf{T}\bar{\Gamma} + L_u^\mathsf{T}\Omega_u^\mathsf{T}\Omega_u L_u. \tag{11.17}$$

The problem formulation adopted in Section 11.2 has the ability to ameliorate the numerical difficulties encountered when dealing with unstable plants. Indeed, comparing (11.14) with (11.17) we see that all terms in (11.14) depend only on exponentially decaying quantities whereas (11.17) also depends on L_u, which as seen from (11.16), contains exponentially exploding terms. The following example illustrates this point.

Example 11.3.1. Consider a single input-single output system of the form (11.2) with stable and unstable modes defined via the following matrices:

$$A_s = \begin{bmatrix} 1.442 & -0.64 \\ 1 & 0 \end{bmatrix}, \quad B_s = \begin{bmatrix} 1 \\ 0 \end{bmatrix}, \quad C_s = \begin{bmatrix} 0.721 & -0.64 \end{bmatrix},$$

and

$$A_u = 2, \quad B_u = 1, \quad C_u = -1.$$

In the objective function (11.7) we take

$$Q = \begin{bmatrix} C_s & C_u \end{bmatrix}^{\mathrm{T}} \begin{bmatrix} C_s & C_u \end{bmatrix}, \quad R = 0.1, \quad P = Q.$$

We consider input constraints of the form $|u_k| \leq 1$ and no state constraints.

We we will formulate the optimisation problem for the above system in the following two ways:

(i) solving both stable and unstable modes in forward time using the procedure described in Section 5.3.1 of Chapter 5; the Hessian of the associated QP is the "standard Hessian;"

(ii) solving the stable modes in forward time and the unstable modes in reverse time using the procedure described in Sections 11.2 and 11.3; the Hessian of the associated QP is the "modified Hessian."

We will compare the numerical conditioning of the optimisation problem for the above two formulations.

Figure 11.1 shows the condition number of the standard Hessian and the modified Hessian (11.14), plotted as a function of the prediction horizon. We see that the condition number of the modified Hessian is never excessively large even when long prediction horizons are used. On the other hand, when the Hessian is constructed with both stable and unstable modes solved in forward time, the condition number explodes dramatically fast. When $N = 5$ the condition number of H is already equal to 4484, reaching values of the order of 10^{39} when $N = 50$.

The poor conditioning of the standard Hessian will inevitably generate difficulties with the solution to the QP problem. If the condition number of H is large, the solution to the QP problem can become unreliable. This is illustrated in Figure 11.2 where the computed objective function values of the optimisation problem for the two cases analysed here are compared for different prediction horizon. We observe that the solution to the QP problem based on the modified Hessian quickly settles to its infinite horizon value of 2.2546. However, the solution to the QP problem based on the standard Hessian, produces a nonoptimal objective function value from $N = 30$ onwards. This behaviour suggests that the optimisation algorithm has not been able to find the optimal solution to the problem due to the poor conditioning of the Hessian.

○

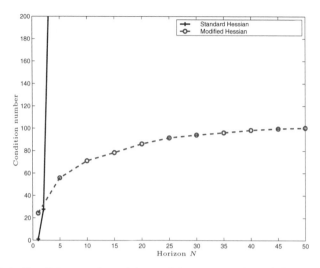

Figure 11.1. Condition number of the modified Hessian (11.14) (circle-dashed line) and that of the standard Hessian (plus-solid line).

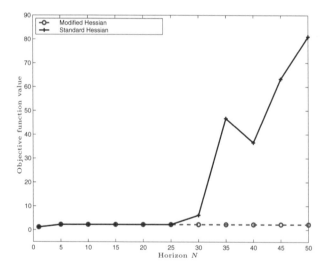

Figure 11.2. Comparison of the objective function value for different prediction horizons: values achieved using the modified Hessian (circle-dashed line) and using the standard Hessian (plus-solid line).

Our goal in the next two sections will be to examine the properties of the regularised sub-Hessian (11.15) in the limit as N increases. Using a large value of N together with a time domain ℓ_2 norm in the objective function suggests, via the application of Parseval's theorem, that there should be a connection to the system frequency response. Before formally establishing this connection we pause to review aspects of frequency response.

11.4 The System Frequency Response

Consider again the system split into stable and unstable parts as in (11.2). The system transfer function is clearly

$$G(z) = G_s(z) + G_u(z), \tag{11.18}$$

where

$$G_s(z) = C_s(zI - A_s)^{-1}B_s, \tag{11.19}$$
$$G_u(z) = C_u(zI - A_u)^{-1}B_u, \tag{11.20}$$

are the transfer functions of the stable and unstable parts of (11.2), respectively.

In Section 11.5, we will study properties of the frequency response of (11.2), $G(e^{j\omega})$, via its *singular value decomposition* [SVD], which has the form

$$G(e^{j\omega}) = U(e^{j\omega})\Sigma(\omega)V^H(e^{j\omega}). \tag{11.21}$$

In (11.21), U and V are matrices containing the left and right singular vectors of G, satisfying $U \in \mathbb{C}^{p \times p}$, $U^H U = UU^H = I_p$, $V \in \mathbb{C}^{m \times m}$, $V^H V = VV^H = I_m$, and

$$\Sigma(\omega) = \text{diag}\{\sigma_1(\omega), \dots, \sigma_m(\omega)\}, \tag{11.22}$$

is the matrix of singular values of G, sometimes referred to as the *principal gains* of the system. Note that $\sigma_1(\omega) = \|G(e^{j\omega})\|_2$.

We next state a simple, though important, result, which is key to the discussion that we will present in the following section.

Lemma 11.4.1 *Let $\bar{G}(z)$ be the two-sided Z-transform of the sequence $\{\bar{h}_k : k = -\infty, \dots, \infty\}$ embedded in $\bar{\Gamma}$ as shown in (11.10). Then $\bar{G}(z)$ is given by*

$$\bar{G}(z) = zG(z), \tag{11.23}$$

where $G(z)$ is the system transfer function defined in (11.18). Moreover, the region of convergence of $\bar{G}(z)$ is given by

$$\max\{|\lambda_i(A_s)|\} < |z| < \min\{|\lambda_i(A_u)|\},$$

where $\{\lambda_i(\cdot)\}$ is the set of eigenvalues of the corresponding matrix.

Proof. The two-sided Z-transform $\bar{G}(z)$ is, by definition,

$$\bar{G}(z) = \sum_{k=-\infty}^{\infty} \bar{h}_k z^{-k} = \sum_{k=-\infty}^{-1} \bar{h}_k z^{-k} + \sum_{k=0}^{\infty} \bar{h}_k z^{-k}.$$

By replacing the corresponding values of $\{\bar{h}_k\}$ as per (11.11) and (11.12), we obtain

$$\bar{G}(z) = \underbrace{- \sum_{k=-\infty}^{-1} C_u A_u^k B_u z^{-k}}_{a} + \underbrace{\sum_{k=0}^{\infty} C_s A_s^k B_s z^{-k}}_{b}. \qquad (11.24)$$

Expressing $G_s(z)$ in (11.19) as

$$G_s(z) = \sum_{k=0}^{\infty} C_s A_s^k B_s z^{-(k+1)},$$

we see that the "b" term in (11.24) is

$$b = z G_s(z) \qquad (11.25)$$

and the region of convergence is $|z| > \max\{|\lambda_i(A_s)|\}$. For the "$a$" term in (11.24) we can let $l = -k$ to obtain

$$a = - \sum_{l=1}^{\infty} C_u A_u^{-l} B_u z^l. \qquad (11.26)$$

Define

$$\tilde{G}(z) \triangleq C_u(zI - A_u^{-1})^{-1} A_u^{-1} B_u = \sum_{k=1}^{\infty} C_u A_u^{-k} B_u z^{-k}. \qquad (11.27)$$

Then we have, comparing (11.26) and (11.27), that

$$a = -\tilde{G}(z^{-1}), \qquad (11.28)$$

with region of convergence

$$|z| < \frac{1}{\max\{|\lambda_i(A_u^{-1})|\}} = \min\{|\lambda_i(A_u)|\}.$$

Next, note from (11.27) and (11.20), that

$$-\tilde{G}(z^{-1}) = -z\, C_u(I - A_u^{-1}z)^{-1} A_u^{-1} B_u$$
$$= -z\, C_u(A_u - z)^{-1} B_u$$
$$= z\, G_u(z).$$

Substituting the above into (11.28), and combining with (11.25), we obtain

$$\bar{G}(z) = a + b = z[G_s(z) + G_u(z)],$$

which, using (11.18), yields the final result (11.23). $\qquad \square$

We observe that the region of convergence of the Z-transform $\bar{G}(z)$ includes the unit circle, and this allows us to refer to the frequency response of $\bar{G}(z)$ by taking $z = e^{j\omega}$. The frequency domain approach to the infinite horizon optimisation problem is the topic of the next section.

11.5 Connecting the Singular Values of the Regularised Sub-Hessian to the System Frequency Response

Here we explore another feature arising from the use of the regularised sub-Hessian. In particular, we will see that there is a direct connection between the singular values of $\bar{H}_N$ in (11.15), for large N, and the system frequency response.

Consider the sequence $\{\bar{h}_k : k = -\infty, \dots, \infty\}$ embedded in $\bar{\Gamma}$ as shown in (11.10). The result of Lemma 11.4.1 ensures that given any $\varepsilon > 0$ there exists $k_0 > 0$ such that

$$\left| \|\bar{G}(e^{j\omega})\|_2^2 - \left\| \sum_{k=-k_0}^{k_0} \bar{h}_k e^{-j\omega k} \right\|_2^2 \right| < \varepsilon \qquad \text{for all } w \in [-\pi, \pi], \qquad (11.29)$$

since the sequence $\{\bar{h}_k\}$ contains only exponentially decaying terms. The above is equivalent to saying that, for $k > k_0$ and $k < -k_0$, the terms of $\{\bar{h}_k\}$ are negligible, that is, we can effectively assume a finite sequence of length $2k_0 + 1$. (Actually, the result is also readily provable in the infinite sequence case if we bound the terms outside a finite interval. However, we adopt the finite impulse response approximation for clarity of exposition.) As a result, the autocorrelation matrix of the sequence $\{\bar{h}_k : k = -\infty, \dots, \infty\}$ can be approximated by

$$\Psi_\ell \approx \sum_{k=-k_0}^{k_0-\ell} \bar{h}_k^{\mathrm{T}} \bar{h}_{k+\ell} \qquad \text{for } 0 \leq \ell \leq 2k_0,$$

$$\Psi_\ell \approx 0 \qquad \text{for } \ell > 2k_0, \qquad (11.30)$$

$$\Psi_{-\ell} = \Psi_\ell^{\mathrm{T}}.$$

Recalling the structure of $\bar{\Gamma}$ in (11.10) and using the definition of Ψ_ℓ given in (11.30), we see that the regularised sub-Hessian (11.15) can be written as

$$\bar{H}_N = \begin{bmatrix} X_1 & | & 0 & \cdots & 0 \\ \hline \Psi_{-2k_0} \cdots \Psi_0 & \cdots & \Psi_{2k_0} & 0 & \cdots & 0 \\ 0 & \ddots & & \ddots & & \ddots & \vdots \\ \vdots & \ddots & \ddots & & \ddots & & 0 \\ 0 & \cdots & 0 & \Psi_{-2k_0} & \cdots & \Psi_0 & \cdots & \Psi_{2k_0} \\ \hline 0 & \cdots & 0 & | & & X_2 \end{bmatrix} + \rho I, \qquad (11.31)$$

provided $N \geq (4k_0 + 1)m$. Also, X_1 and X_2 are appropriate submatrices.

Finally, consider the $Nm \times m$ matrix

$$E_{N,\omega} \triangleq \begin{bmatrix} e_{N,\omega}^1 & \cdots & e_{N,\omega}^m \end{bmatrix} \triangleq \frac{1}{\sqrt{N}} \bar{E}_{N,\omega} V(e^{j\omega}), \tag{11.32}$$

where

$$\bar{E}_{N,\omega} \triangleq \begin{bmatrix} I_m \\ e^{-j\omega} I_m \\ \vdots \\ e^{-j(N-1)\omega} I_m \end{bmatrix}, \qquad \omega = \frac{2\pi}{N}\ell \quad \text{for } \ell \in \{0, \ldots, N-1\}.$$

We then have the following result.

Theorem 11.5.1 *Consider the discrete time linear system* (11.2), *and the SVD of its frequency response given by* (11.21)–(11.22). *Let $k_0 > 0$ be such that* (11.29) *holds. Consider $\bar{H}_N$ in* (11.31) *for $N \geq 4k_0 + 1$ and $\rho = 0$, which is the regularised sub-Hessian of the quadratic objective function V_N in* (11.13) *with $\rho = 0$. Then, for every given frequency $\omega_0 = \frac{2\pi}{N_0}\ell_0$, $\ell_0 \in \{0, \ldots, N_0 - 1\}$, we have that*

$$\lim_{\frac{N}{N_0} \to \infty} \left\| \bar{H}_N \, e_{N,\omega_0}^i \right\|_2 = \sigma_i^2(\omega_0) \quad \text{for } i = 1, \ldots, m,$$

where e_{N,ω_0}^i is the ith column of E_{N,ω_0} defined in (11.32), *and $N/N_0 \in \{1, 2, \ldots\}$.*

Proof. Let

$$\Psi(\omega) = \sum_{k=-\infty}^{\infty} \Psi_k e^{-j\omega k} \approx \sum_{k=-2k_0}^{2k_0} \Psi_k e^{-j\omega k} \quad \text{for } \omega \in [-\pi, \pi], \tag{11.33}$$

be the discrete time Fourier transform of the autocorrelation sequence in (11.30). Note that

$$\Psi(e^{j\omega}) = \bar{G}(e^{j\omega})^H \bar{G}(e^{j\omega}),$$

and, by means of Lemma 11.4.1,

$$\Psi(e^{j\omega}) = G(e^{j\omega})^H e^{-j\omega} e^{j\omega} G(e^{j\omega}) = G(e^{j\omega})^H G(e^{j\omega}).$$

Using (11.21) we have that

$$\Psi(e^{j\omega}) V(e^{j\omega}) = V(e^{j\omega}) \Sigma^2(\omega). \tag{11.34}$$

Next, consider $N_0 > 0$, $\ell_0 \in \{0, \ldots, N_0 - 1\}$, and $\omega_0 = \frac{2\pi}{N_0}\ell_0$. By direct calculation, we can write:

$$H_{N_0}E_{N_0,\omega_0} = \frac{1}{\sqrt{N_0}} \begin{bmatrix} W_1 \\ \cdots\cdots\cdots\cdots\cdots\cdots\cdots\cdots\cdots\cdots\cdots \\ \bar{E}_{N_0,\omega_0}[2k_0m+1:N_0m-2k_0m,:]\,\Psi(e^{j\omega_0})V(e^{j\omega_0}) \\ \cdots\cdots\cdots\cdots\cdots\cdots\cdots\cdots\cdots\cdots\cdots \\ W_2 \end{bmatrix},$$

$$(11.35)$$

where $\bar{E}_{N_0,\omega_0}[2k_0m+1:N_0m-2k_0m,:]$ denotes the section of the matrix $\bar{E}_{N_0,\omega_0}$ that starts at the $(2k_0m+1)$th row and finishes at the (N_0m-2k_0m)th row. In addition, W_1 and W_2 are some submatrices of dimension $2k_0m \times m$.

Replacing (11.34) in (11.35) and using the definition (11.32), we have

$$H_{N_0}E_{N_0,\omega_0} = \begin{bmatrix} \frac{1}{\sqrt{N_0}}W_1 \\ \cdots\cdots\cdots\cdots\cdots\cdots\cdots\cdots\cdots\cdots \\ E_{N_0,\omega_0}[2k_0m+1:N_0m-2k_0m,:]\,\Sigma^2(\omega_0) \\ \cdots\cdots\cdots\cdots\cdots\cdots\cdots\cdots\cdots\cdots \\ \frac{1}{\sqrt{N_0}}W_2 \end{bmatrix},$$

which can be considered columnwise as follows:

$$H_{N_0}\mathbf{e}^i_{N_0,\omega_0} = \begin{bmatrix} \frac{1}{\sqrt{N_0}}c_1 \\ \cdots\cdots\cdots\cdots\cdots\cdots\cdots\cdots\cdots\cdots \\ \sigma^2_i(\omega_0)\mathbf{e}^i_{N_0,\omega_0}[2k_0m+1:N_0m-2k_0m] \\ \cdots\cdots\cdots\cdots\cdots\cdots\cdots\cdots\cdots\cdots \\ \frac{1}{\sqrt{N_0}}c_2 \end{bmatrix} \quad \text{for } i = 1,\ldots,m,$$

where $\mathbf{e}^i_{N_0,\omega_0}[2k_0m+1:N_0m-2k_0m]$ denotes the section of the vector $\mathbf{e}^i_{N_0,\omega_0}$ that starts at the $(2k_0m+1)$th element and finishes at the (N_0m-2k_0m)th element. Subtracting $\sigma^2_i(\omega_0)\mathbf{e}^i_{N_0,\omega_0}$ from both sides of the above equation yields

$$H_{N_0}\mathbf{e}^i_{N_0,\omega_0} - \sigma^2_i(\omega_0)\mathbf{e}^i_{N_0,\omega_0} = d_{N_0,\omega_0}, \qquad (11.36)$$

where

$$d_{N_0,\omega_0} \triangleq \frac{1}{\sqrt{N_0}}\begin{bmatrix} d_1 \\ 0_{N_0-4k_0} \\ d_2 \end{bmatrix}. \qquad (11.37)$$

Both d_1 and d_2 are column vectors of the same dimension of vectors c_1 and c_2, respectively, and $0_{N_0-4k_0}$ is a column vector with zero entries and length $N_0 - 4k_0$. It can be easily shown that the norms of both vectors d_1 and d_2 are bounded. They are determined by the entries of submatrices X_1 and X_2 in (11.31) (which remain unchanged whenever the prediction horizon N is increased), the fixed value $\sigma^2_i(\omega_0)$ and the entries of vector $\mathbf{e}^i_{N_0,\omega_0}$ which are bounded. As a result, we can find $T_{\omega_0} > 0$ such that

$$\|d_{N_0,\omega_0}\|_2 = \frac{1}{\sqrt{N_0}}\sqrt{\|d_1\|^2_2 + \|d_2\|^2_2} \leq \frac{1}{\sqrt{N_0}}T_{\omega_0}. \qquad (11.38)$$

We now choose $N = LN_0$ and $\ell = L\ell_0$, for some $L \in \{1,2,\ldots\}$, such that

$$\omega = \frac{2\pi\ell}{N} = \frac{2\pi\ell_0}{N_0} = \omega_0. \tag{11.39}$$

Then, evaluating (11.36) at N instead of N_0 and $\omega = \omega_0$, that is,

$$\bar{H}_N e^i_{N,\omega_0} - \sigma_i^2(\omega_0) e^i_{N,\omega_0} = d_{N,\omega_0}, \tag{11.40}$$

and bounding from below, we have

$$\left| \left\| \bar{H}_N e^i_{N,\omega_0} \right\|_2 - \sigma_i^2(\omega_0) \right| \leq \left\| \bar{H}_N e^i_{N,\omega_0} - \sigma_i^2(\omega_0) e^i_{N,\omega_0} \right\|_2 = \left\| d_{N,\omega_0} \right\|_2,$$

where we have used the fact that $\left\| e^i_{N,\omega_0} \right\|_2 = 1$. Combining the above inequality with the bound derived in (11.38) (with N instead of N_0), and noting that $N = LN_0$, we obtain

$$\left| \left\| \bar{H}_N e^i_{N,\omega_0} \right\|_2 - \sigma_i^2(\omega_0) \right| \leq \left\| d_{N,\omega_0} \right\|_2 \leq \frac{1}{\sqrt{LN_0}} T_{\omega_0}.$$

Hence, for every $\varepsilon_0 > 0$, there exists L^* (for example, take the nearest integer towards infinity to $T_{\omega_0}^2 / (N_0 \varepsilon_0^2)$) such that

$$\left| \left\| \bar{H}_N e^i_{N,\omega_0} \right\|_2 - \sigma_i^2(\omega_0) \right| < \varepsilon_0 \text{ for all } N = LN_0 > L^* N_0 \text{ and } L \in \{1, 2, \dots\}.$$

The result then follows.   □

Corollary 11.5.2 *Consider the same conditions of Theorem 11.5.1 and choose $\rho > 0$ in (11.31). Then*

$$\lim_{\frac{N}{N_0} \to \infty} \left\| \bar{H}_N e^i_{N,\omega_0} \right\|_2 = \sigma_i^2(\omega_0) + \rho \quad for \ i = 1, \dots, m, \tag{11.41}$$

where $N/N_0 \in \{1, 2, \dots\}$.

Proof. Directly from Theorem 11.5.1.   □

The importance of the above result is that it establishes a direct link between the eigenvalues of the regularised sub-Hessian (for large N) and the principal gains of the system frequency response.

In the single input-single output case, (11.41) takes the simple form

$$\lim_{\frac{N}{N_0} \to \infty} \left\| \bar{H}_N E_{N,\omega_0} \right\|_2 = \left| G(e^{j\omega_0}) \right|^2 + \rho,$$

where E_{N,ω_0} is defined as in (11.32) for $m = 1$ and $V = 1$, and G is the transfer function (11.18).

The following example illustrates the result.

Example 11.5.1. Let (11.2) be a 2 input-2 output system with stable and unstable modes defined via the following matrices.

$$A_s = \begin{bmatrix} 1.442 & -0.64 \\ 1 & 0 \end{bmatrix}, \quad B_s = \begin{bmatrix} 1 & 0 \\ 0 & 0 \end{bmatrix}, \quad C_s = \begin{bmatrix} 0.721 & -0.64 \\ -0.36 & 0.32 \end{bmatrix},$$

and

$$A_u = 2, \quad B_u = \begin{bmatrix} 0 & 1 \end{bmatrix}, \quad C_u = \begin{bmatrix} -0.1 \\ -0.1 \end{bmatrix}.$$

Consider the finite horizon optimal control problem (11.6)–(11.7) and select

$$Q = \begin{bmatrix} C_s & C_u \end{bmatrix}^{\mathrm{T}} \begin{bmatrix} C_s & C_u \end{bmatrix}, \quad R = 0, \quad P = Q.$$

We construct the regularised sub-Hessian $\bar{H}_N$ as described in Section 11.3 solving the stable modes in forward time and the unstable modes in reverse time. We then compute the singular values of $\bar{H}_N$ and compare them with the square of the principal gains of the system. The results are presented in Figure 11.3 for $N = 61$ and in Figure 11.4 for $N = 401$.

We observe that, as the prediction horizon N is increased, the singular values of the regularised sub-Hessian converge to the continuous line representing the square of the principal gains of the system as predicted by Theorem 11.5.1. Also, observe that some singular values are very small. This property will be exploited in Section 11.6.   ○

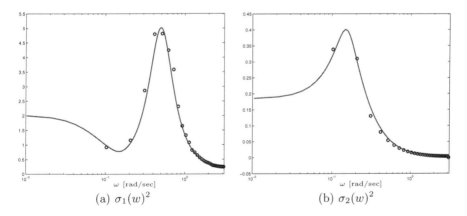

(a) $\sigma_1(w)^2$ (b) $\sigma_2(w)^2$

Figure 11.3. Singular values of the regularised sub-Hessian $\bar{H}_N$ (circles) with $N = 61$. The continuous lines represent the square of the two principal gains of the system.

11.6 Suboptimal Algorithms

We have seen above that the singular values of the regularised sub-Hessian are well-defined for large prediction horizons. Here we explore other possible ap-

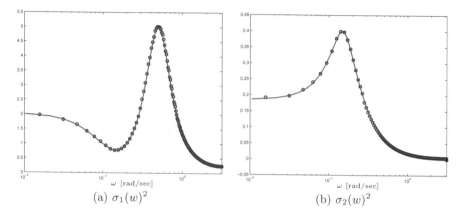

Figure 11.4. Singular values of the regularised sub-Hessian $\bar{H}_N$ (circles) with $N = 401$. The continuous lines represent the square of the two principal gains of the system.

plications of the formulation of the optimisation problem using the regularised sub-Hessian.

The idea of using an SVD decomposition of the (regularised sub-) Hessian of the QP leads to interesting possibilities in constructing suboptimal algorithms. We explore two such possibilities below. In Section 11.6.1 we show how the SVD decomposition can be used in conjunction with any QP solver (including active set and interior point methods) to obtain a suboptimal algorithm. In Section 11.6.2, we describe a related, but even simpler, strategy based on a line search.

11.6.1 A Restricted SVD-Based Strategy for RHC

For simplicity, we first consider the case of a stable system. Note that, in this case, the standard Hessian is equal to the regularised sub-Hessian.

Consider the system

$$x_{k+1} = Ax_k + Bu_k, \tag{11.42}$$

where $x_k \in \mathbb{R}^n$ is the state vector and $u_k \in \mathbb{R}^m$ is the input vector. Assume that the pair (A, B) is stabilisable. Consider the objective function as in (5.19) of Chapter 5 (here we take $\mathbf{d} = 0$ and $N = M$). We can rewrite this objective function as

$$V_N(x, \mathbf{w}) \triangleq V_{N,M} = \frac{1}{2}\mathbf{w}^\mathsf{T} H \mathbf{w} + \mathbf{w}^\mathsf{T} F(x - x_s) + \text{ constant}, \tag{11.43}$$

where $x = x_0$ is the current state (or an estimate of it) and

$$\mathbf{w} \triangleq \mathbf{u} - \mathbf{u}_s \triangleq \left[u_0^{\mathrm{T}} \ u_1^{\mathrm{T}} \ \dots \ u_{N-1}^{\mathrm{T}} \right]^{\mathrm{T}} - \left[u_s^{\mathrm{T}} \ u_s^{\mathrm{T}} \ \dots \ u_s^{\mathrm{T}} \right]^{\mathrm{T}}. \tag{11.44}$$

In (11.43), the Hessian H and the matrix F are computed from the problem data ($H = \Gamma^{\mathrm{T}} Q \Gamma + \mathbf{R}$, $F = \Gamma^{\mathrm{T}} Q \Omega$, with Γ, Ω, $\mathbf{Q}$, $\mathbf{R}$, defined as in (5.14) and (5.16) of Chapter 5, using $M = N$ and $C = I$). The matrices $Q = Q^{\mathrm{T}}$ and $R = R^{\mathrm{T}}$ in (5.16) are chosen to be positive definite, and P in (5.16) is chosen as the solution of the algebraic Riccati equation:

$$P = Q + A^{\mathrm{T}} P A - K^{\mathrm{T}} (R + B^{\mathrm{T}} P B) K, \tag{11.45}$$
$$K = (R + B^{\mathrm{T}} P B)^{-1} B^{\mathrm{T}} P A. \tag{11.46}$$

Suppose that we require the input and states of (11.42) to satisfy the constraints

$$\begin{aligned} u_k &\in \mathbb{U} \quad \text{for } k = 0, \dots, N - 1, \\ x_k &\in \mathbb{X} \quad \text{for } k = 1, \dots, N, \\ x_N &\in \mathbb{X}_f \subset \mathbb{X}, \end{aligned} \tag{11.47}$$

where $\mathbb{U}$, $\mathbb{X}$ and $\mathbb{X}_f$ are some sets that contain the origin. The constraints in (11.47) can be expressed, using the vector (11.44), as

$$\begin{aligned} \mathbf{w} + \mathbf{u}_s &\in \mathbb{U}^N, \\ \Omega x + \Gamma(\mathbf{w} + \mathbf{u}_s) &\in \mathbb{X}^N, \\ A^N x + \left[A^{N-1} B \ \dots \ AB \ B \right] (\mathbf{w} + \mathbf{u}_s) &\in \mathbb{X}_f, \end{aligned} \tag{11.48}$$

where Ω and Γ are defined in (5.14) of Chapter 5 (with $M = N$), and where $\mathbb{U}^N = \mathbb{U} \times \dots \times \mathbb{U}$, $\mathbb{X}^N = \mathbb{X} \times \dots \times \mathbb{X}$ are N-Cartesian products.

Let

$$H = V S V^{\mathrm{T}} \tag{11.49}$$

be the SVD of the Hessian $H = H^{\mathrm{T}} > 0$ of the objective function $V_N(x, \mathbf{u})$ in (11.43). In (11.49), $V \in \mathbb{R}^{Nm \times Nm}$ contains the singular vectors of H and $S \in \mathbb{R}^{Nm \times Nm}$ is the diagonal matrix

$$S = \text{diag} \left\{ \sigma_1, \sigma_2, \dots, \sigma_{Nm} \right\}, \tag{11.50}$$

with the singular values of H arranged in decreasing order. Observe that the left and right singular vectors of H are identical since H is symmetric, and that the singular values are strictly positive since H is positive definite.

Using the coordinate transformation $\mathbf{w} = V \tilde{\mathbf{w}}$, the vector of control moves $\mathbf{w}$ in (11.44) can be expressed as a linear combination of the singular vectors of H:

$$\mathbf{w} = V \tilde{\mathbf{w}} = \sum_{j=1}^{Nm} \mathbf{v}_j \tilde{\mathbf{w}}(j),$$

where $\mathbf{v}_j$, $j \in \{1, \ldots, Nm\}$, are the columns of V and $\tilde{\mathbf{w}}(j)$ are the components of the vector $\tilde{\mathbf{w}}$. We can then express the objective function (11.43) in the new $\tilde{\mathbf{w}}$ variables as

$$V_N(x, \tilde{\mathbf{w}}) = \frac{1}{2}\tilde{\mathbf{w}}^{\mathsf{T}}S\tilde{\mathbf{w}} + \tilde{\mathbf{w}}^{\mathsf{T}}V^{\mathsf{T}}F(x - x_s) + \text{ constant}, \qquad (11.51)$$

whose unconstrained minimum is

$$\tilde{\mathbf{w}}_{\mathrm{UC}}^{\mathrm{OPT}} = -S^{-1}V^{\mathsf{T}}F(x - x_s). \qquad (11.52)$$

Using (11.52) and (11.50) in (11.51), the objective function can be expressed as

$$V_N(x, \tilde{\mathbf{w}}) = \frac{1}{2}\sum_{j=1}^{Nm} \sigma_j[\tilde{\mathbf{w}}(j) - \tilde{\mathbf{w}}_{\mathrm{UC}}^{\mathrm{OPT}}(j)]^2 + \text{constant}, \qquad (11.53)$$

where $\tilde{\mathbf{w}}_{\mathrm{UC}}^{\mathrm{OPT}}(j)$ is the jth component of $\tilde{\mathbf{w}}_{\mathrm{UC}}^{\mathrm{OPT}}$.

We then have the following result.

Lemma 11.6.1 *For some $j \in \{1, \ldots, Nm\}$, fix $\tilde{\mathbf{w}}(k)$ for $k \neq j$. Then*

(i) *the change in objective function achieved by changing $\tilde{\mathbf{w}}(j)$ from 0 to $\tilde{\mathbf{w}}_{\mathrm{UC}}^{\mathrm{OPT}}(j)$ is*

$$\Delta V_j = \frac{-1}{2\sigma_j}[\tilde{x}(j)]^2;$$

(ii) *the change in input "energy", that is, $\tilde{\mathbf{w}}(j)^2$, resulting from changing $\tilde{\mathbf{w}}(j)$ from 0 to $\tilde{\mathbf{w}}_{\mathrm{UC}}^{\mathrm{OPT}}(j)$ is*

$$\Delta E_j = \frac{1}{\sigma_j^2}[\tilde{x}(j)]^2,$$

where $\tilde{x}(j)$ is the jth component of $V^{\mathsf{T}}F(x - x_s)$.

Proof. We note from (11.53) that

$$\Delta V_j = \frac{1}{2}\sigma_j[\tilde{\mathbf{w}}_{\mathrm{UC}}^{\mathrm{OPT}}(j)]^2.$$

The result is then immediate on noting, from (11.52), that

$$\tilde{\mathbf{w}}_{\mathrm{UC}}^{\mathrm{OPT}}(j) = -\frac{1}{\sigma_j}\tilde{x}(j),$$

where $\tilde{x}(j)$ is the jth component of $V^{\mathsf{T}}F(x - x_s)$. $\qquad \square$

We can think of

$$\frac{|\Delta V_j|}{\Delta E_j} = \frac{\sigma_j}{2} \qquad (11.54)$$

as a ratio between "benefit" (in terms of objective function reduction) and "cost" (in terms of input energy). This suggests that a "near optimal" strategy may be to restrict the optimisation so as not to utilise the singular vectors associated with small singular values since these singular directions are associated with poor benefit-to-cost ratio. (A more compelling reason for doing this in the context of RHC will be given in Section 11.7.)

Limiting the optimisation using only a subset (I, say) of singular vectors, leads to the following restricted optimisation problem:

$$\text{minimise} \quad \left\{ V_N'(x, \tilde{\mathbf{w}}) = \frac{1}{2} \sum_{j \in I} \sigma_j [\tilde{\mathbf{w}}(j) - \tilde{\mathbf{w}}_{\text{UC}}^{\text{OPT}}(j)]^2 \right\}, \qquad (11.55)$$

subject to (11.48), expressed in the $\tilde{\mathbf{w}}$-coordinates,

$$V\tilde{\mathbf{w}} + \mathbf{u}_s \in \mathbb{U}^N,$$
$$\Omega x + \Gamma(V\tilde{\mathbf{w}} + \mathbf{u}_s) \in \mathbb{X}^N, \qquad (11.56)$$
$$A^N x + \left[A^{N-1} B \ \dots \ AB \ B \right] (V\tilde{\mathbf{w}} + \mathbf{u}_s) \in \mathbb{X}_f,$$

where

$$\tilde{\mathbf{w}} = \left[\tilde{\mathbf{w}}(1) \ \dots \ \tilde{\mathbf{w}}(Nm) \right]^{\text{T}}$$

and

$$\tilde{\mathbf{w}}(j) = 0 \quad \text{for } j \notin I. \qquad (11.57)$$

Of course, the set I must be chosen so that there exists a control sequence satisfying (11.57) that is feasible. (More will be said about feasibility in Section 11.6.2.) Subject to this caveat, we can apply any of the standard QP solvers to this restricted problem. It has the following features:

(i) The number of free variables is reduced to $\{\tilde{\mathbf{w}}(j) \text{ for } j \in I\}$. (The cardinality of this set could be much less than the dimension of the original problem.)
(ii) The number of constraints remains the same.

An important practical reason for restricting the set of singular values used in the optimisation will be discussed in Section 11.7.2.

Remark 11.6.1. The above restricted optimisation problem (11.55)–(11.57) can be used in a receding horizon fashion to compute the current control action for the current state x by simply applying the first m components of the resulting vector $\mathbf{u} = V\tilde{\mathbf{w}} + \mathbf{u}_s$ at each time step. If this controller is combined with an observer as described in Section 5.5 of Chapter 5, it is easy to see that integral action is retained. The key idea presented in Section 5.5 to achieve integral action was that the observer for the *disturbance* only stops integrating when the observer output, that is, $C\hat{x}_k + \hat{d}_k$ (see (5.33) in Chapter 5), converges to the true system output y_k^{REAL}. It then follows that, provided the control law ensures the observer output be brought (in steady state) to $y^* - \bar{\hat{d}}$, then the true system output will be brought to y^* (see Lemma 5.5.1 in Chapter 5). Hence, zero steady state error depends on the control law being consistent with the observer (irrespective of the nature of the true system). We thus utilise $u_s = u_{s,k}$ and $x_s = x_{s,k}$ based on the *observer*, exactly as was done in (5.34) and (5.35) of Chapter 5. Hence, the steady state values are $\bar{u}_s$ and $\bar{x}_s$, given by (5.38) and (5.39), respectively. However, we also need to ensure that the *control law* has $\bar{u}_s$ and $\bar{x}_s$ as consistent steady state values. In the case of full QP, equation (11.52) gives $u = u_s$ when $x = x_s$. This is also the case for the restricted optimisation problem (11.55)–(11.57), since, when $x = x_s$, we have that $\tilde{\mathbf{w}}_{\mathrm{UC}}^{\mathrm{OPT}} = 0$ in (11.52), and hence $\tilde{\mathbf{w}} = 0$ minimises (11.55)–(11.57) in this case. It follows that $\mathbf{w} = V\tilde{\mathbf{w}} = 0$ and, from (11.44), $\mathbf{u} = \mathbf{u}_s$. Thus, u_s and x_s are consistent solutions of the control problem (solved in the restricted singular value space) as required to ensure integral action.   ∘

Remark 11.6.2. In the case where there are unstable modes and the objective function used in the optimisation is formulated as (11.13) (where $\mathbf{u}_s = 0$ for simplicity), one way to proceed is to limit the singular value restriction to the variables $\mathbf{u}$ whilst always including μ in the optimisation problem. We then use the transformation

$$\begin{bmatrix} \mathbf{u} \\ \mu \end{bmatrix} = \begin{bmatrix} \bar{V} & 0 \\ 0 & I \end{bmatrix} \begin{bmatrix} \tilde{\mathbf{u}} \\ \mu \end{bmatrix},$$

where $\bar{V}$ is the matrix of singular vectors corresponding to the SVD of the regularised sub-Hessian, that is,

$$\bar{H}_N = \bar{V}\bar{S}\bar{V}^{\mathrm{T}}.$$

We can then limit the optimisation using only a subset of the singular vectors contained in $\bar{V}$.   ∘

11.6.2 A Line Search Algorithm Based on SVDs

Here we consider, for simplicity, the standard formulation, and thus the subsequent singular value analysis applies to the standard Hessian H.

The result presented in Lemma 11.6.1, leading to the "benefit-to-cost" ratio (11.54), motivates an algorithm that uses a line search in the singular vector space. Specifically, suppose, for simplicity, that $x_s = 0$ and $\mathbf{u}_s = 0$ in (11.43) and (11.44), and consider the unconstrained optimal solution (11.52), that is,

$$\tilde{\mathbf{w}}_{\mathrm{UC}}^{\mathrm{OPT}} = \tilde{\mathbf{u}}_{\mathrm{UC}}^{\mathrm{OPT}} = -S^{-1}V^{\mathrm{T}}Fx. \tag{11.58}$$

The key idea of the algorithm is to construct a suboptimal solution to the QP problem by approximating (11.58) using the basis vectors starting from the largest singular values and proceeding downwards until a constraint boundary is reached. Heuristically, this algorithm examines the components of the input having largest benefit-to-cost ratio first (see (11.54)), that is, it sequentially decides whether to use $\tilde{\mathbf{u}}_{\mathrm{UC}}^{\mathrm{OPT}}(j)$ or not, starting from $j = 1$. Thus, the SVD strategy constructs a control vector $\mathbf{u}^{\mathrm{SVD}} \in \mathbb{R}^{Nm}$ using the components of (11.58) from the first one onwards until a constraint boundary is reached.

Then the first m components of $\mathbf{u}^{\text{SVD}}$ are used at the current time, and the algorithm proceeds in an RHC fashion. More precisely:

Algorithm 11.6.1 (SVD–RHC Line Search Algorithm)

(i) At time k, let $x = x_k$.

(ii) SVD control vector construction:

   a) Calculate the unconstrained optimal solution $\tilde{\mathbf{u}}_{\text{UC}}^{\text{OPT}}$ defined in (11.58).

   b) Let $\tilde{\mathbf{u}}_i \in \mathbb{R}^{Nm}$ be a vector whose first i, $i \in \{1, \ldots, Nm\}$, components are equal to the first i components of $\tilde{\mathbf{u}}_{\text{UC}}^{\text{OPT}}$, and having the remaining components equal to zero, that is,

   $$\tilde{\mathbf{u}}_i \triangleq \begin{bmatrix} \tilde{\mathbf{u}}_{\text{UC}}^{\text{OPT}}(1) & \tilde{\mathbf{u}}_{\text{UC}}^{\text{OPT}}(2) \ldots \tilde{\mathbf{u}}_{\text{UC}}^{\text{OPT}}(i) & 0 \ldots 0 \end{bmatrix}^{\text{T}},$$

   where $\tilde{\mathbf{u}}_{\text{UC}}^{\text{OPT}}(j)$, $j = 1, \ldots, Nm$, are the components of $\tilde{\mathbf{u}}_{\text{UC}}^{\text{OPT}}$.

   c) Find the integer

   $$r \triangleq \max i \quad for \quad i \in \{1, \ldots, Nm\}, \tag{11.59}$$

   such that vector $\tilde{\mathbf{u}}_r$ is feasible. That is, the vector

   $$\mathbf{u}_r \triangleq V\tilde{\mathbf{u}}_r = \sum_{j=1}^{r} \mathbf{v}_j \tilde{\mathbf{u}}_{\text{UC}}^{\text{OPT}}(j)$$

   satisfies the control, state and terminal constraints (11.48), namely

   $$V\tilde{\mathbf{u}}_r \in \mathbb{U}^N,$$

   $$\Omega x + \Gamma V\tilde{\mathbf{u}}_r \in \mathbb{X}^N, \tag{11.60}$$

   $$A^N x + \begin{bmatrix} A^{N-1}B & \ldots & AB & B \end{bmatrix} V\tilde{\mathbf{u}}_r \in \mathbb{X}_f.$$

   d) Set

   $$\mathbf{u}^{\text{SVD}} \triangleq V\tilde{\mathbf{u}}_r = \sum_{j=1}^{r} \mathbf{v}_j \tilde{\mathbf{u}}_{\text{UC}}^{\text{OPT}}(j), \tag{11.61}$$

   and let

   $$\mathscr{U}^{\text{SVD}} = \{u_k^{\text{SVD}}, u_{k+1}^{\text{SVD}}, \ldots, u_{k+N-1}^{\text{SVD}}\}, \tag{11.62}$$

   be the associated control sequence. That is, element j, $j = 1, \ldots, N$, of (11.62) is equal to the vector formed by components $(j-1)m + 1$ to jm of (11.61).

(iii) Apply, as the current control move, the first element of the sequence (11.62), that is,

   $$u_k = \mathscr{K}^{\text{SVD}}(x) \triangleq u_k^{\text{SVD}}.$$

(iv) Set $k = k + 1$ and return to step (i).   ∘

The basic Algorithm 11.6.1 is not in a form where one can readily establish stability. We thus present below a simple modification that ensures closed loop stability, provided the initial state belongs to an admissible set, defined next.

Definition 11.6.1 (Admissible Set $\mathbb{S}_\mathbf{N}$) *Let $\mathbb{S}_N$ be the set of all initial states $x \in \mathbb{R}^n$ for which there exists an admissible control sequence $\mathscr{U}^{\text{SVD}}$ of the form (11.62), computed as in step (ii) of Algorithm 11.6.1.* ○

Note that if we start with an initial state $x \in \mathbb{S}_N$, then there exists an admissible control sequence $\mathscr{U}^{\text{SVD}}$ of the form (11.62) that steers the resulting system's state trajectory into the *terminal constraint set* $\mathbb{X}_f$ in N steps. Suppose that we choose $\mathbb{X}_f$ in (11.60) to be the maximal output admissible set $\mathcal{O}_\infty$ of system (11.42) with the unconstrained control law $u_k = -Kx_k$, where K is the gain defined in (11.46), and with respect to the input and state constraints given by the first two lines of (11.47) (as in (5.63) of Chapter 5), that is,

$$\mathcal{O}_\infty \triangleq \{x : K(A - BK)^k x \in \mathbb{U} \text{ and } (A - BK)^k x \in \mathbb{X} \text{ for } k = 0, 1, \ldots\}. \tag{11.63}$$

Note that for states inside $\mathcal{O}_\infty$, we can keep applying the control $u_k = -Kx_k$ and remain inside $\mathcal{O}_\infty$ since it is positively invariant. Hence, if the N-move sequence $\mathscr{U}^{\text{SVD}}$ is applied to the system (11.42) in open loop, and subsequently, the unconstrained control $u_k = -Kx_k$ is applied for $k = N, N + 1, \ldots$, then (i) the resulting sequence is feasible, and (ii) the resulting state trajectory will converge to the origin since $A - BK$ is a Hurwitz matrix. Moreover, the objective function

$$V_N = (\{x_k\}, \{u_k\}) = \frac{1}{2}x_N^\mathrm{T} P x_N + \frac{1}{2}\sum_{k=0}^{N-1}(x_k^\mathrm{T} Q x_k + u_k^\mathrm{T} R u_k), \tag{11.64}$$

with $Q > 0$, $R > 0$, can be used as a Lyapunov function $\mathscr{V}^*(x)$ that decreases along the resulting state trajectory.

An algorithm with provable stabilising properties can then be developed as follows. Starting with a state $x \in \mathbb{S}_N$ and choosing $\mathbb{X}_f = \mathcal{O}_\infty$, we apply Algorithm 11.6.1 once. At the next time step, we check if the successor state x^+ belongs to $\mathbb{S}_N$. If $x^+ \notin \mathbb{S}_N$, we apply the second move of the SVD sequence $\mathscr{U}^{\text{SVD}}$ obtained in the previous step. If $x^+ \in \mathbb{S}_N$, we compute a new SVD sequence $\mathscr{U}^{\text{SVD}+}$ using step (ii) of Algorithm 11.6.1 and check whether applying the first move of $\mathscr{U}^{\text{SVD}+}$ would decrease the value of the Lyapunov function $\mathscr{V}^*(x)$. If so, we apply the first control move of $\mathscr{U}^{\text{SVD}+}$. If not, we apply the second move of the sequence $\mathscr{U}^{\text{SVD}}$ obtained in the previous step. At each time step, the value of $\mathscr{V}^*(x)$ is updated and the procedure is repeated. The resulting algorithm ensures that the Lyapunov function $\mathscr{V}^*(x)$ decreases at each step.

We formalise the above procedure in the following:

Algorithm 11.6.2 (Stable SVD–RHC Line Search Algorithm)

(i) At time $k = 0$, and given an initial state $x = x_0 \in \mathbb{S}_N$:

a) *Compute* $\mathscr{U}^* = \mathscr{U}^{\text{SVD}} = \{u_0^*, u_1^*, \ldots, u_{N-1}^*\}$, *the sequence of control moves determined by the SVD strategy, based on step (ii) of Algorithm 11.6.1. Let $\mathscr{X}^*$ be the corresponding state sequence resulting from the application of $\mathscr{U}^*$ to system (11.42) with initial state $x_0 = x$.*

b) *Initialise the value function using the corresponding value of the objective function as per (11.64), that is,*

$$\mathscr{V}^*(x) = V_N(\mathscr{X}^*, \mathscr{U}^*).$$

c) *Apply as the initial control action, the first element of $\mathscr{U}^*$, that is,*

$$u_0 = \mathscr{K}^{\text{SVD}}(x) \triangleq u_0^*.$$

d) *Set $k = 1$, $x_k = Ax + Bu_0^*$, and go to step (ii).*

(ii) *At time k set $x = x_k$. Then:*

a) *If $x \notin \mathbb{S}_N$, go to step (iii).*

b) *If $x \in \mathbb{S}_N$, determine the sequence of control moves $\mathscr{U}^{\text{SVD}}$ generated by the SVD strategy following step (ii) of Algorithm 11.6.1. Let $\mathscr{X}^{\text{SVD}}$ be the corresponding state sequence.*

c) *If $V_N(\mathscr{X}^{\text{SVD}}, \mathscr{U}^{\text{SVD}}) \geq \mathscr{V}^*(x)$, go to step (iii).*

d) *If $V_N(\mathscr{X}^{\text{SVD}}, \mathscr{U}^{\text{SVD}}) \leq \mathscr{V}^*(x) - \frac{1}{2}x^{\text{T}}Qx$, set*

$$\mathscr{U}^* = \mathscr{U}^{\text{SVD}} = \{u_0^*, u_1^*, \ldots, u_{N-1}^*\}, \qquad (11.65)$$
$$\mathscr{X}^* = \mathscr{X}^{\text{SVD}},$$
$$\mathscr{V}^*(x) = V_N(\mathscr{X}^*, \mathscr{U}^*).$$

e) *Apply as the current control action, the first element of $\mathscr{U}^*$, that is,*

$$u_k = \mathscr{K}^{\text{SVD}}(x) \triangleq u_0^*. \qquad (11.66)$$

f) *Set $k = k + 1$, $x_k = Ax + Bu_0^*$, and return to step (ii).*

(iii) a) *Apply as the current control action the second element of $\mathscr{U}^*$ in (11.65), that is,*

$$u_k = \mathscr{K}^{\text{SVD}}(x) \triangleq u_1^*. \qquad (11.67)$$

b) *Update the sequence of control moves $\mathscr{U}^*$ by retaining the last $N-1$ elements of the previous $\mathscr{U}^*$ in (11.65) and adding, as the last element, the unconstrained optimal control move, that is,*

$$\mathscr{U}^* = \{u_1^*, u_2^*, \ldots, u_{N-1}^*, -Kx_N^*\}, \qquad (11.68)$$

where K is the optimal feedback gain defined in (11.46).

c) *Update the value of the objective function $\mathscr{V}^*(x)$ using the updated control sequence $\mathscr{U}^*$ in (11.68), and the corresponding state sequence $\mathscr{X}^*$, that is, $\mathscr{V}^*(x) = V_N(\mathscr{X}^*, \mathscr{U}^*)$.*

d) *Set $k = k + 1$, $x_k = Ax + Bu_1^*$, and return to step (ii).* ○

The above algorithm ensures that the closed loop trajectories converge to the origin, as proved in the following theorem.

Theorem 11.6.2 (Stability of the SVD–RHC Strategy) *Let the terminal constraint set* $\mathbb{X}_f \subset \mathbb{X}$ *in (11.60) be the maximal output admissible set* $\mathcal{O}_\infty$ *defined in (11.63). Then the SVD–RHC strategy described in Algorithm 11.6.2 ensures attractivity of the origin for all initial states in* $\mathbb{S}_N$. *Moreover, the origin is an exponentially stable equilibrium point for the closed loop system* $x^+ = Ax + B\mathcal{K}^{\text{SVD}}(x)$ *in* $\mathcal{O}_\infty$.

Proof. We will show that the function $\mathcal{V}^*(x)$ is a Lyapunov function for the SVD–RHC strategy.

At any time k, with state $x = x_k$, Algorithm 11.6.2 generates a control sequence $\mathcal{U}^*$ that is given either by (11.65) or by (11.68). Note that in both cases the sequence $\mathcal{U}^*$ is feasible by construction. The corresponding value of the objective function is $\mathcal{V}^*(x) = V_N(\mathcal{X}^*, \mathcal{U}^*)$, where $\mathcal{X}^*$ is the corresponding state sequence, and the controller applied to the system is $u_k = \mathcal{K}^{\text{SVD}}(x)$, given either by (11.66) or (11.67).

Let $x^+ = Ax + Bu_k$ be the successor state. If $x^+ \in \mathbb{S}_N$ and the control sequence $\mathcal{U}^{*+}$ is computed as in (11.65), we have

$$\mathcal{V}^*(x^+) \le \mathcal{V}^*(x) - \frac{1}{2}x^\mathsf{T}Qx$$

by construction. On the other hand, if the control sequence $\mathcal{U}^{*+}$ is computed as in (11.68), we obtain, by direct calculation:

$$
\begin{aligned}
\mathcal{V}^*(x^+) &= V_N(\mathcal{X}^{*+}, \mathcal{U}^{*+}) \\
&= V_N(\mathcal{X}^*, \mathcal{U}^*) - \frac{1}{2}x^\mathsf{T}Qx - \frac{1}{2}u_0^{*\mathsf{T}}Ru_0^* - \frac{1}{2}x_N^{*\mathsf{T}}Px_N^* \\
&\quad + \frac{1}{2}x_N^{*\mathsf{T}}[(A-BK)^\mathsf{T}P(A-BK) + K^\mathsf{T}RK + Q]x_N^* \\
&= \mathcal{V}^*(x) - \frac{1}{2}x^\mathsf{T}Qx - \frac{1}{2}u_0^{*\mathsf{T}}Ru_0^*,
\end{aligned}
$$

where x_N^* is the last element of the state sequence $\mathcal{X}^*$, and where we have used $P = (A-BK)^\mathsf{T}P(A-BK) + K^\mathsf{T}RK + Q$, which follows from (11.45)–(11.46). Since $R > 0$, we then have that

$$\mathcal{V}^*(x^+) \le \mathcal{V}^*(x) - \frac{1}{2}x^\mathsf{T}Qx.$$

Since $Q > 0$, the objective function $\mathcal{V}^*(x)$ decreases along the trajectories of the system under the SVD–RHC strategy according to property (i) in Theorem 4.3.1 of Chapter 4. Also, $\mathcal{V}^*(x) \ge x^\mathsf{T}Qx$, and hence $\mathcal{V}^*(x)$ satisfies property (ii) in Theorem 4.3.1. The attractivity result then follows from similar arguments to those used in the proof of Theorem 4.3.1.

To show exponential stability, note that inside the terminal constraint set $\mathbb{X}_f = \mathcal{O}_\infty$, the SVD–RHC control coincides with the unconstrained control law $u_k = -Kx_k$, that is, $\mathscr{K}^{\mathrm{SVD}}(x) = -Kx$ for all $x \in \mathcal{O}_\infty$. Since $\mathcal{O}_\infty$ is positively invariant for $x^+ = Ax + B\mathscr{K}^{\mathrm{SVD}}(x) = (A - BK)x$, it follows that the origin is exponentially stable in $\mathcal{O}_\infty$ for the closed loop system since $A - BK$ is a Hurwitz matrix. $\qquad\square$

We observe that it is possible, depending on the initial state x_0, that the SVD strategy of Algorithm 11.6.1 might be used only once, to initialise Algorithm 11.6.2. This would imply that the control sequence $\mathscr{U}^*$ computed in step (i) be applied entirely in open loop. However, the control moves in $\mathscr{U}^*$, followed by the unconstrained control law $u_k = -Kx_k$ for $k = N, N+1, \ldots$, are guaranteed to stabilise system (11.42), as discussed before, and proved in Theorem 11.6.2.

When the system (11.42) is stable, and the state constraint set $\mathbb{X}$ in (11.47) is the whole space $\mathbb{R}^n$, the region of attraction of the SVD–RHC strategy can be trivially extended to be $\mathbb{R}^n$. Outside the region $\mathbb{S}_N$ we can simply apply $u_k = 0$, in which case the system's states would evolve asymptotically towards the origin. Once the state trajectory enters the admissible set $\mathbb{S}_N$, stability is ensured by means of Theorem 11.6.2.

Characterisation of the admissible set $\mathbb{S}_N$

For Algorithm 11.6.2 to be useful, we need a practical characterisation of the admissible set $\mathbb{S}_N$. This is done in the following result.

Lemma 11.6.3 *The set of admissible initial conditions $\mathbb{S}_N$ is given by*

$$\mathbb{S}_N = \bigcup_{r=1}^{Nm} C_r, \tag{11.69}$$

where C_r is the set

$$C_r \triangleq \left\{ x \in \mathbb{R}^n : V\tilde{\mathbf{u}}_r = \sum_{j=1}^{r} \mathbf{v}_j \tilde{\mathbf{u}}_{\mathrm{UC}}^{\mathrm{OPT}}(j) \text{ satisfies constraints (11.60)} \right\}, \tag{11.70}$$

and where $\tilde{\mathbf{u}}_{\mathrm{UC}}^{\mathrm{OPT}}(j)$ is the linear function of x given by the jth component of vector $\tilde{\mathbf{u}}_{\mathrm{UC}}^{\mathrm{OPT}}$ defined in (11.58).

Proof. If $x \in \mathbb{S}_N$, then by Definition 11.6.1 of the admissible set $\mathbb{S}_N$, there exists a sequence $\mathscr{U}^{\mathrm{SVD}}$ which satisfies the constraints in (11.47). However, this is equivalent to the fact that the corresponding vector of control moves (11.61) satisfies the constraints in (11.60). Thus $x \in C_r$ for the value of r determined using Algorithm 11.6.1. From this we conclude that $x \in \bigcup_{j=1}^{Nm} C_j$.

Conversely, suppose that $x \in C_i$ for some $i \in \mathcal{I}_{Nm}$. Then i is feasible for the maximisation in (11.59), and thus there exists a sequence $\mathscr{U}^{\mathrm{SVD}}$ which satisfies the constraints in (11.60). Thus, $x \in \mathbb{S}_N$, and the result follows. $\qquad\square$

It is useful to notice that, if $\mathbb{X}_f$ in (11.60) is chosen to be the maximal output admissible set $\mathcal{O}_\infty$ defined in (11.63), then the admissible region $\mathbb{S}_N$ is, at least, as large as $\mathcal{O}_\infty$. This is established in the following lemma:

Lemma 11.6.4 *Let $\mathbb{X}_f$ in (11.60) be the maximal output admissible set $\mathcal{O}_\infty$ defined in (11.63). Then $\mathcal{O}_\infty \subset \mathbb{S}_N$.*

Proof. Assume $x \in \mathcal{O}_\infty$. Compute the unconstrained optimal control sequence that minimises $V_N(\{x_k\}, \{u_k\})$. It is easy to verify that this sequence has the form $\{-Kx, -K(A - BK)x, \ldots, -K(A - BK)^{N-1}x\}$. By the definition of $\mathcal{O}_\infty$, this sequence is feasible with respect to the constraints (11.47), and so it is a feasible control sequence $\mathcal{U}^{\text{SVD}}$ corresponding to $r = Nm$ in (11.59). Thus, $x \in \mathbb{S}_N$, and the proof is completed. $\qquad\qquad\square$

Note that whenever the set of constraints (11.60) can be expressed via a finite set of linear inequalities, the same is true for the sets C_r in (11.70). Hence, the test $x \in \mathbb{S}_N$ required in Algorithm 11.6.1 is computationally simple. This is illustrated in Example 11.6.1 below.

Example 11.6.1. Consider the discrete time resonant system defined by the following state space matrices:

$$A = \begin{bmatrix} 1.5293 & -0.7408 \\ 1 & 0 \end{bmatrix}, \quad B = \begin{bmatrix} 0.5 \\ 0 \end{bmatrix}, \quad C = \begin{bmatrix} 0.2222 & 0.2009 \end{bmatrix}.$$

In the objective function (11.64) we use

$$Q = C^{\mathrm{T}}C, \quad R = 1, \quad N = 5,$$

and we assume that the system is subject to input constraints and a terminal state constraint as follows:

$$u_k \in \mathbb{U} \triangleq [-1, 1] \quad \text{for } k = 0, 1, \ldots, N - 1, \tag{11.71}$$
$$x_N \in \mathcal{O}_\infty.$$

The maximal output admissible set $\mathcal{O}_\infty$ is defined in terms of a finite set of inequalities

$$\mathcal{O}_\infty = \left\{ x \in \mathbb{R}^n : -LK(A - BK)^j x \leq W \quad \text{for } j = 0, 1, \ldots, t^* \right\},$$

where

$$L = \begin{bmatrix} 1 \\ -1 \end{bmatrix}, \quad W = \begin{bmatrix} 1 \\ 1 \end{bmatrix},$$

are defined based on (11.71) and t^* is the *output admissibility index* (Gilbert and Tan 1991). For this particular example, it can be shown that $t^* = 3$, that is, $\mathcal{O}_\infty$ is characterised via a set of eight inequalities.

The sets C_r, $r = 1, 2, \ldots, 5$, defined in Lemma 11.6.3 are given by:

$$C_r = \left\{ x \in \mathbb{R}^n : \bar{L}x \leq \bar{W} \right\} \quad \text{for } r = 1, \dots, 5, \tag{11.72}$$

where

$$\bar{L} = \begin{bmatrix} VP_jS^{-1}V^{\mathrm{T}}F \\ -VP_jS^{-1}V^{\mathrm{T}}F \\ -LK\bar{I}(\Omega - \Gamma VP_jS^{-1}V^{\mathrm{T}}F) \\ -LK(A-BK)\bar{I}(\Omega - \Gamma VP_jS^{-1}V^{\mathrm{T}}F) \\ -LK(A-BK)^2\bar{I}(\Omega - \Gamma VP_jS^{-1}V^{\mathrm{T}}F) \\ -LK(A-BK)^3\bar{I}(\Omega - \Gamma VP_jS^{-1}V^{\mathrm{T}}F) \end{bmatrix}, \quad \bar{W} = \begin{bmatrix} 1_{N\times1} \\ 1_{N\times1} \\ W \\ W \\ W \\ W \end{bmatrix},$$

and

$$\bar{I} = \begin{bmatrix} 0_{n\times(N-1)n} & I_n \end{bmatrix}, \quad P_j = \mathrm{diag}\{I_j,\ 0_{(N-j)\times(N-j)}\}.$$

The union of all five sets C_r in (11.72) determines the set of admissible initial conditions $\mathbb{S}_N$. The resulting $\mathbb{S}_N$ is depicted in Figure 11.5 where we also include the maximal output admissible set $\mathcal{O}_\infty$ for comparison. As we proved

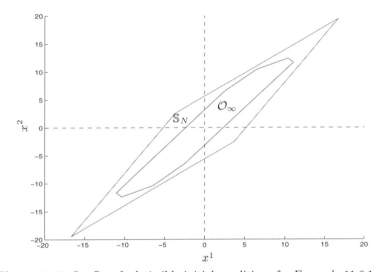

Figure 11.5. Set $\mathbb{S}_N$ of admissible initial conditions for Example 11.6.1.

in Lemma 11.6.4, $\mathcal{O}_\infty \subset \mathbb{S}_N$. Moreover, for the particular example presented here, we see that $\mathbb{S}_N$ is clearly larger than $\mathcal{O}_\infty$.

Now, let us consider the initial condition $x_0 = [16.66 \quad 19.43]^{\mathrm{T}} \in \mathbb{S}_N$. Figure 11.6 shows the corresponding state space trajectory for the closed loop system using the SVD–RHC strategy of Algorithm 11.6.2. We observe that the trajectory converges to the origin. Also observe that the trajectory is outside the set $\mathbb{S}_N$ during two steps in which no control sequence $\mathscr{U}^{\mathrm{SVD}}$ of

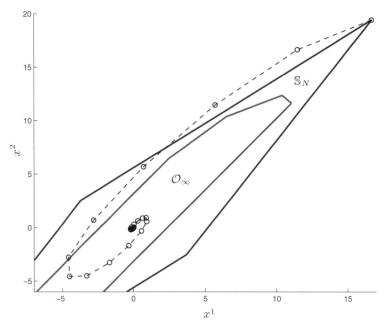

Figure 11.6. Closed loop state space trajectory using the SVD–RHC strategy of Algorithm 11.6.2 and $x_0 = [16.66 \quad 19.43]^{\mathrm{T}}$.

the form (11.62) is feasible. However, the SVD–RHC strategy steers the state trajectory towards the origin by successively using the elements of the control sequence $\mathscr{U}^{\mathrm{SVD}}$ (step (iii) in Algorithm 11.6.2) computed when initialising the algorithm.

Figure 11.7(a) shows the time evolution of the number of components r used in the basis function expansion of $\mathbf{u}^{\mathrm{SVD}}$ in (11.61). Note that at time $k = 0$ we have $r = 1$ which corresponds to $x_0 \in C_1$ (see (11.70)). At times $k = 1$ and $k = 2$ there is no feasible solution to (11.59), therefore no values for r are plotted in Figure 11.7(a). This is in agreement with the observation that $x_k \notin \mathbb{S}_N$ for $k = 1$ and $k = 2$ as shown in Figure 11.6.

Figure 11.7(b) shows the time evolution of $\mathscr{V}^*(x_k)$. We observe that the objective function is strictly decreasing along the state trajectory of the closed loop system.

We finally compare the closed loop performance obtained using the SVD–RHC strategy with the closed loop performance obtained using the standard, QP-based, RHC solution. Figure 11.8 shows that, although the inputs applied to the system are dissimilar, the time evolution of the output of the system is very similar in both cases.

○

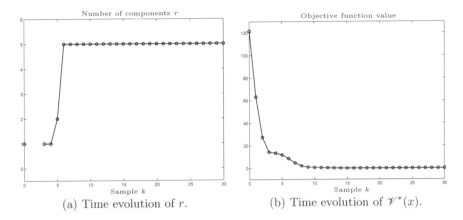

(a) Time evolution of r.
(b) Time evolution of $\mathscr{V}^*(x)$.

Figure 11.7. Time evolution of the number of components r used in the basis function expansion of $\mathbf{u}^{\mathrm{SVD}}$ and time evolution of $\mathscr{V}^*(x)$ using the SVD–RHC strategy of Algorithm 11.6.2 for $x_0 = [16.66 \quad 19.43]^{\mathrm{T}}$.

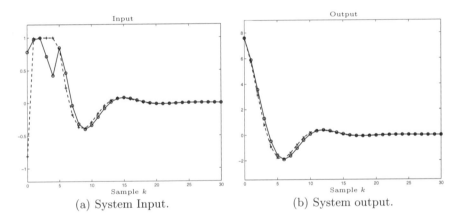

(a) System Input.
(b) System output.

Figure 11.8. Comparison between standard RHC (plus-dashed line) and the SVD–RHC strategy of Algorithm 11.6.2 (circle-solid line) for $x_0 = [16.66 \quad 19.43]^{\mathrm{T}}$.

11.7 Further Insights from the SVD Structure of the Regularised Sub-Hessian

Here we investigate further insights arising from the SVD decomposition of the regularised sub-Hessian. Our development here is heuristic[2] in nature since we will blur the distinction between the exact SVD representation obtained for a given value of the prediction horizon N and the associated asymptotic

---

[2] The problem with "heuristics" is that they are like Argentinian mate (a drink)—a matter of individual taste.

frequency domain results. However, in view of the results in Section 11.5, this blurring is well-justified for large prediction horizons.

Indeed, the motivation for the ideas presented here arises from the limiting results in Section 11.5. There it was shown that the singular values of the regularised sub-Hessian approach the principal gains of the system frequency response. This connection is not surprising in view of the fact that the objective function is a time-domain ℓ_2 design criterion. Therefore, subject to suitable regularity condition, this criterion can be evaluated in the frequency domain for large N.

We restrict our arguments here to the single input-single output case but the results can be extended to the multivariable case using the idea of principal gains. We assume that the system has input $u_k \in \mathbb{R}$ and output $y_k \in \mathbb{R}$. We take the system transfer function as $G(z)$ and we assume an objective function of the form

$$V_N = \sum_{k=0}^{N} \left[(y_k - y_k^s)^2 + R(u_k - u_k^s)^2 \right], \tag{11.73}$$

where y_k^s, u_k^s are given reference trajectories.

Assuming that the appropriate regularity conditions hold, we can use Parseval's theorem to express the asymptotic (that is, $N \to \infty$) objective function in the frequency domain as

$$V_\infty = \int_0^\pi \left\{ |Y(e^{j\omega}) - Y^s(e^{j\omega})|^2 + R|U(e^{j\omega}) - U^s(e^{j\omega})|^2 \right\} d\omega, \tag{11.74}$$

where $Y(e^{j\omega})$, $U(e^{j\omega})$, $Y^s(e^{j\omega})$, $U^s(e^{j\omega})$ denote the Fourier transforms of y_k, y_k^s, u_k, u_k^s, respectively. We also note that the Z-transform of y_k can be written as

$$Y(z) = G(z)U(z) + Y_0(z), \tag{11.75}$$

where $Y_0(z)$ is the Z-transform of the initial condition response. Since we have assumed that the Fourier transform of y_k exists (although not necessarily that of the initial condition response), we can set $z = e^{j\omega}$ in (11.75). Substituting into (11.74) gives

$$V_\infty = \int_0^\pi \left\{ |G(e^{j\omega})U(e^{j\omega}) + Y_0(e^{j\omega}) - Y^s(e^{j\omega})|^2 \right.$$

$$\left. + R|U(e^{j\omega}) - U^s(e^{j\omega})|^2 \right\} d\omega. \tag{11.76}$$

If we approximate the integral in (11.76) by a summation, then the objective function becomes

$$V_\infty \approx \sum_i (G_i U_i + Y_{0i} - Y_i^s)^* (G_i U_i + Y_{0i} - Y_i^s) + R(U_i - U_i^s)^*(U_i - U_i^s),$$

where * denotes conjugate. We then see that the "square term" in the *input* has weighting $|G_i|^2 + R$, which is consistent with the earlier rigorous analysis of the singular values of the regularised sub-Hessian.

Further insights arising from (11.76) are discussed below.

11.7.1 Achieved Closed Loop Bandwidth

There appear to be two ways one could achieve a desired closed loop bandwidth. These options are:

(i) **Adjusting R.**

Equation (11.76) suggests that the value of R should determine the resultant closed loop bandwidth as follows: (a) For those frequencies such that $|G(e^{j\omega})|^2 \gg R$, then the first term in (11.76) will dominate the optimisation problem and near perfect reference tracking should occur. (b) Similarly, for those frequencies such that $|G(e^{j\omega})|^2 \ll R$, then the second term in (11.76) will dominate the optimisation problem and the input should then be set to the corresponding steady state value. This argument suggests that the closed loop bandwidth will approximately occur at the frequency where $|G(e^{j\omega})|^2 = R$. Of course, between these two conditions, there will be a transition. The nature of the transition will depend on the system, especially the relationship between the critical frequency where $|G(e^{j\omega})|^2 = R$ and the system nonminimum phase zeros and unstable poles.

(ii) **Limiting the set of singular values used in the optimisation.**

An alternative direct way of limiting the closed loop bandwidth would be to set $R = 0$ in the objective function but then restrict the range of singular values used in the optimisation to include only those frequencies up to the desired bandwidth.

11.7.2 Robustness

It is usually the case that the model of a system gives only an approximate description. Moreover, the model error between the true system and the model typically grows with frequency. On the other hand, it is well-known (Goodwin et al. 2001) that closed loop stability cannot be guaranteed if the nominal closed loop bandwidth exceeds the frequency where the magnitude of the relative model error exceeds unity. With this as background, then the SVD frequency domain insights outlined in this chapter suggest that there are two ways that robustness can be achieved:

(i) **Via choice of R.**

We can choose R sufficiently large so that the resulting closed loop bandwidth is less than the frequency $\tilde{\omega}$ where the magnitude of the relative model error approaches unity. (Based on the arguments presented in Section 11.7.1, this implies that R needs to be greater than the magnitude squared of the model frequency response at the frequency $\tilde{\omega}$.)

(ii) Via limits on the set of singular values used in the optimisation.

A more direct way of restricting the bandwidth is simply to limit the set of singular values employed in the optimisation (whether constrained or unconstrained) to include only those "frequencies" where the magnitude of the relative model error is significantly below unity. (This is readily achieved using the algorithms described in Section 11.6.)

We illustrate the above heuristic ideas by a simple example.

Example 11.7.1. Consider the system

$$\dot{x}(t) = -x(t) + u(t),$$

sampled with period $T_s = 0.025$ sec. We develop the example in several steps.

(i) Closed loop bandwidth issues:

- Unconstrained response:

  We use the objective function (11.73) with $y_k = x_k$, $R = 0.01$, and horizon $N = 251$ (that is, 6.25 sec).

  Figure 11.9 shows the magnitude (in dB) of the open loop system frequency response. We see that this cuts the line $20 \log_{10} \sqrt{R}$ with $R = 0.01$ at $\omega = 10$ rad/sec. The line of argument presented in Section 11.7.1 suggests that using $R = 0.01$ should yield a closed loop bandwidth of approximately 10 rad/sec.

  This is confirmed in Figure 11.10, which shows the resultant closed

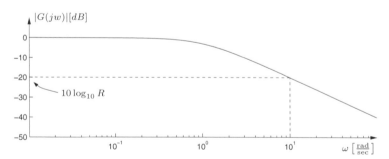

Figure 11.9. Frequency response of the open loop system.

loop responses for the input u and the output $y = x$ with a reference $y^s(t) = 1$. Indeed we see that a closed loop time constant of the order of 0.1 sec has resulted.

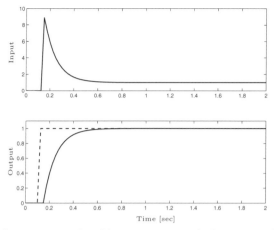

Figure 11.10. Input-output closed loop response with $R = 0.01$ and no constraints.

- Adding input constraints:

  We see that the input in Figure 11.10 reaches a maximum value of 9. We next add a mild input constraint of $|u_k| \leq 8$. The resulting input-output closed loop response is shown in Figure 11.11.

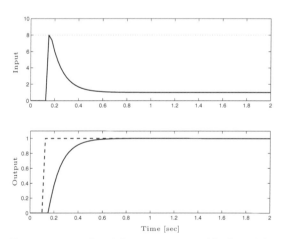

Figure 11.11. Input-output closed loop response with $R = 0.01$ and mild input constraints. RHC implementation using all singular values.

- RHC with restricted singular values:

  We next set $R = 0$ in (11.73), which would give a very large closed loop bandwidth in normal RHC if no other restrictions were added. However, utilising the link between singular values and frequency response, we anticipate that restricting the singular values to the most significant values should allow us to limit the closed loop bandwidth. In particular, Figure 11.12 shows the location of the 21st singular value in relation to the magnitude of the open loop system frequency response.

Comparing Figures 11.9 and 11.12 we see that the two critical frequen-

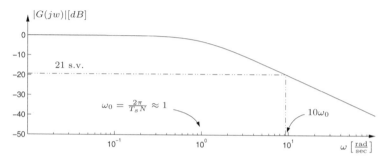

Figure 11.12. Frequency response of the open loop system with location of the 21st singular value.

cies are very close.

Restricting the optimisation to the first 21 singular values (and using $R = 0$) gives the response shown by the solid line in Figure 11.13. Also shown in this figure (in dotted lines) is the earlier response obtained using all of the singular values, but with $R = 0.01$. As predicted, the two responses are qualitatively the same, that is, essentially the same closed loop bandwidth has been achieved.

(ii) Robustness issues:

- Effect of R on robustness:

  We set R to the value 0.01. We then add a small unmodelled time delay of eight samples (that is, 0.2 sec) to the system. The relative model error is shown in Figure 11.14. We see that there is significant relative model error at $\omega = 10$ rad/sec. Thus, since $R = 0.01$ gives an effective closed loop bandwidth of about 10 rad/sec, we anticipate difficulties with the closed loop system when all singular values are utilised in the optimisation. This is confirmed in Figure 11.15, which shows that the closed loop system begins to oscillate. Note that the input signal essentially swings between the limits at ± 8.

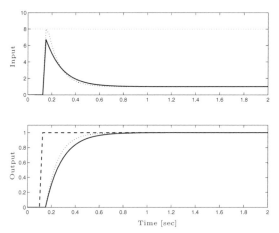

Figure 11.13. Input-output closed loop response with $R = 0$ and input constraints. RHC implementation with restricted singular values. The plots of Figure 11.11 are repeated in dotted lines.

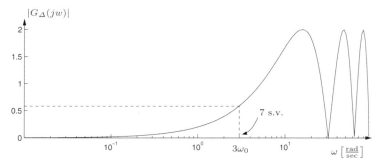

Figure 11.14. Frequency response of the relative model error.

- Effect of using a restricted set of singular values on robustness:

  Based again on the frequency domain link, we can see from Figure 11.14 that restricting to the first seven singular values should cause high feedback gain beyond about $\omega = 3$ rad/sec to be avoided, and intuitively this should yield robust stability for the given unmodelled dynamics. This hypothesis is confirmed in Figure 11.16, which shows that the response achieved with the restricted SVD algorithm using $R = 0.01$ and restricting the optimisation to the first seven singular values does indeed restore closed loop stability.

The above example confirms the link between closed loop bandwidth and robustness on the one hand and the choice of input weighting R and restrict-

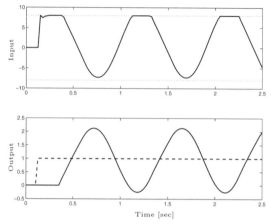

Figure 11.15. Input-output response with mild input constraint, $R = 0.01$ and plant-model mismatch. RHC implementation using all singular values.

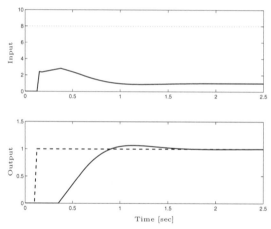

Figure 11.16. Input-output response with mild input constraint, $R = 0.01$ and plant-model mismatch. RHC implementation with restricted singular values.

ing the set of singular values in the optimisation on the other hand. These observations are consistent with the heuristic ideas outlined in this section.

11.8 Further Reading

For complete list of references cited, see References section at the end of book.

Section 11.6.2

More details regarding the SVD–RHC strategy described in Section 11.6.2 can be found in Rojas, Goodwin, Seron and Feuer (2004), and Rojas, Goodwin and Johnston (2002). The stability properties of the strategy are also addressed in Rojas, Seron and Goodwin (2003). The results regarding the singular values of the sub-Hessian for long prediction horizons were first derived for stable SISO plants and presented in Rojas et al. (2002).

Output Feedback Optimal Control
with Constraints

Contributed by Tristan Perez and Hernan Haimovich

12.1 Overview

In Chapters 1 through 9 of the book (with the exception of a brief discussion on observers and integral action in Section 5.5 of Chapter 5) we considered constrained optimal control problems for systems without uncertainty, that is, with no unmodelled dynamics or disturbances, and where the full state was available for measurement. More realistically, however, it is necessary to consider control problems for systems with uncertainty. This chapter addresses some of the issues that arise in this situation. As in Chapter 9, we adopt a stochastic description of uncertainty, which associates probability distributions to the uncertain elements, that is, disturbances and initial conditions. (See Section 12.6 for references to alternative approaches to model uncertainty.)

When incomplete state information exists, a popular observer-based control strategy in the presence of stochastic disturbances is to use the certainty equivalence [CE] principle, introduced in Section 5.5 of Chapter 5 for deterministic systems. In the stochastic framework, CE consists of *estimating* the state and then using these estimates as if they were the *true* state in the control law that results if the problem were formulated as a deterministic problem (that is, without uncertainty). This strategy is motivated by the unconstrained problem with a quadratic objective function, for which CE is indeed the optimal solution (Åström 1970, Bertsekas 1976).

One of the aims of this chapter is to explore the issues that arise from the use of CE in RHC in the presence of constraints. We then turn to the obvious question about the optimality of the CE principle. We show that CE is, indeed, not optimal in general.

We also analyse the possibility of obtaining truly optimal solutions for single input linear systems with input constraints and uncertainty related to output feedback and stochastic disturbances. We first find the optimal solution for the case of horizon $N = 1$, and then we indicate the complications that arise in the case of horizon $N = 2$. Our conclusion is that, for the case of linear constrained systems, the extra effort involved in the optimal feedback policy is probably not justified in practice. Indeed, we show by example that CE can give near optimal performance. We thus advocate this approach in real applications.

12.2 Problem Statement

We consider the following time-invariant, discrete time linear system with disturbances

$$
\begin{aligned}
x_{k+1} &= Ax_k + Bu_k + w_k, \\
y_k &= Cx_k + v_k,
\end{aligned}
\tag{12.1}
$$

where $x_k, w_k \in \mathbb{R}^n$ and $u_k, y_k, v_k \in \mathbb{R}$. The control u_k is constrained to take values in the set

$$
\mathbb{U} = \{u \in \mathbb{R} : -\Delta \leq u \leq \Delta\},
$$

for a given constant value $\Delta > 0$. The disturbances w_k and v_k are i.i.d. random vectors, with probability density functions (pdf) $p_w(\,\cdot\,)$ and $p_v(\,\cdot\,)$, respectively. The initial state, x_0, is characterised by a pdf $p_{x_0}(\,\cdot\,)$. We assume that the pair (A, B) is reachable and that the pair (A, C) is observable.

We further assume that, at time k, the value of the state x_k is not available to the controller. Instead, the following sets of past inputs and outputs, grouped as the *information vector* I^k, represent all the information available to the controller at the time instant k:

$$
I^k =
\begin{cases}
\{y_0\} & \text{if } k = 0, \\
\{y_0, y_1, u_0\} & \text{if } k = 1, \\
\{y_0, y_1, y_2, u_0, u_1\} & \text{if } k = 2, \\
\ \ \vdots & \quad \ \ \vdots \\
\{y_0, y_1, \ldots, y_{N-1}, u_0, u_1, \ldots u_{N-2}\} & \text{if } k = N - 1.
\end{cases}
$$

Then, $I^k \in \mathbb{R}^{2k+1}$, and also $I^{k+1} = \{I^k, y_{k+1}, u_k\}$, where $I^k \subset I^{k+1}$.

For system (12.1), under the assumptions made, we formulate the optimisation problem:

$$
\text{minimise } \mathbf{E}\left\{ F(x_N) + \sum_{k=0}^{N-1} L(x_k, u_k) \right\},
\tag{12.2}
$$

where

$$F(x_N) = x_N^{\mathrm{T}} P x_N,$$
$$L(x_k, u_k) = x_k^{\mathrm{T}} Q x_k + R u_k^2,$$

subject to the system equations (12.1) and the input constraint $u_k \in \mathbb{U}$, for $k = 0, \ldots, N-1$. Note that, under the stochastic assumptions, the expression $F(x_N) + \sum_{k=0}^{N-1} L(x_k, u_k)$ is a random variable. Hence, it is only meaningful to formulate the minimisation problem in terms of its statistics. A problem of practical interest is to minimise the expected value of this expression, which motivates the choice of the objective function in (12.2).

The result of the above minimisation problem will be a sequence of functions $\{\pi_0^{\mathrm{OPT}}(\,\cdot\,), \pi_1^{\mathrm{OPT}}(\,\cdot\,), \ldots, \pi_{N-1}^{\mathrm{OPT}}(\,\cdot\,)\}$ that enable the controller to calculate the desired optimal control action depending on the information available to the controller at each time instant k, that is, $u_k^{\mathrm{OPT}} = \pi_k^{\mathrm{OPT}}(I^k)$. These functions also must ensure that the constraints be always satisfied. We thus make the following definition.

Definition 12.2.1 (Admissible Policies for Incomplete State Information) *A policy Π_N is a finite sequence of functions $\pi_k(\,\cdot\,) : \mathbb{R}^{2k+1} \to \mathbb{R}$ for $k = 0, 1, \ldots, N-1$, that is,*

$$\Pi_N = \{\pi_0(\,\cdot\,), \pi_1(\,\cdot\,), \cdots, \pi_{N-1}(\cdot)\}.$$

A policy Π_N is called an admissible control policy *if and only if*

$$\pi_k(I^k) \in \mathbb{U} \quad \text{for all } I^k \in \mathbb{R}^{2k+1}, \quad \text{for } k = 0, \ldots, N-1.$$

Further, the class of all admissible control policies will be denoted by

$$\bar{\Pi}_N = \{\Pi_N : \Pi_N \text{ is admissible}\}.$$

○

Using the above definition, we can then state the optimal control problem of interest as follows.

Definition 12.2.2 (Stochastic Finite Horizon Optimal Control Problem) *Given the pdfs $p_{x_0}(\,\cdot\,)$, $p_w(\,\cdot\,)$ and $p_v(\,\cdot\,)$ of the initial state x_0 and the disturbances w_k and v_k, respectively, the problem considered is that of finding the control policy Π_N^{OPT}, called the* optimal control policy, *belonging to the class of all admissible control policies $\bar{\Pi}_N$, which minimises the objective function*

$$V_N(\Pi_N) = \operatorname*{\mathbf{E}}_{\substack{x_0, w_k, v_k \\ k=0,\ldots,N-1}} \left\{ F(x_N) + \sum_{k=0}^{N-1} L(x_k, \pi_k(I^k)) \right\}, \tag{12.3}$$

subject to the constraints

$$x_{k+1} = Ax_k + B \ \pi_k(I^k) + w_k,$$
$$y_k = Cx_k + v_k,$$
$$I^{k+1} = \{I^k, y_{k+1}, u_k\},$$

for $k = 0, \ldots, N - 1$. In (12.3) the terminal state weighting $F(\cdot)$ and the per-stage weighting $L(\cdot, \cdot)$ are given by

$$F(x_N) = x_N^{\mathrm{T}} P x_N,$$
$$L(x_k, \pi_k(I^k)) = x_k^{\mathrm{T}} Q x_k + R \pi_k^2(I^k),$$

(12.4)

with $P > 0$, $R > 0$ and $Q \geq 0$.

The optimal control policy is then

$$\Pi_N^{\mathrm{OPT}} = \arg \inf_{\Pi_N \in \bar{\Pi}_N} V_N(\Pi_N),$$

with the following resulting optimal objective function value

$$V_N^{\mathrm{OPT}} = \inf_{\Pi_N \in \bar{\Pi}_N} V_N(\Pi_N).$$

(12.5)

o

It is important to recognise that the optimisation problem of Definition 12.2.2 takes into account the fact that new information will be available to the controller at future time instants. This is called *closed loop optimisation*, as opposed to *open loop optimisation* where the control values $\{u_0, u_1, \ldots, u_{N-1}\}$ are selected all at once, at stage zero (Bertsekas 1976). For deterministic systems, in which there is no uncertainty, the distinction between open loop and closed loop optimisation is irrelevant, and the minimisation of the objective function over all sequences of controls or over all control policies yields the same result.

In what follows, and as in previous chapters, the matrix P in (12.4) will be taken to be the solution to the algebraic Riccati equation,

$$P = A^{\mathrm{T}} P A + Q - K^{\mathrm{T}} \bar{R} K,$$

(12.6)

where

$$K \triangleq \bar{R}^{-1} B^{\mathrm{T}} P A, \qquad \bar{R} \triangleq R + B^{\mathrm{T}} P B.$$

(12.7)

12.3 Optimal Solutions

The problem described in the previous section belongs to the class of the so-called sequential decision problems under uncertainty (Bertsekas 1976, Bertsekas 2000). A key feature of these problems is that an action taken at a particular stage affects all future stages. Thus, the control action has to be computed taking into account the future consequences of the current decision. The only general approach known to address sequential decision problems is dynamic programming.

The dynamic programming algorithm was introduced in Section 3.4 of Chapter 3 and was used in Chapters 6 and 7 to derive a closed form solution of a deterministic finite horizon optimal control problem. We next briefly show how this algorithm is used to solve the stochastic optimal control problem of Definition 12.2.2.

We define the functions

$$
\tilde{L}_{N-1}(I^{N-1}, \pi_{N-1}(I^{N-1})) = \mathbf{E}\left\{ F(x_N) + L(x_{N-1}, \pi_{N-1}(I^{N-1})) \right.
$$
$$
\left. | I^{N-1}, \pi_{N-1}(I^{N-1})) \right\},
$$
$$
\tilde{L}_k(I^k, \pi_k(I^k)) = \mathbf{E}\left\{ L(x_k, \pi_k(I^k)) | I^k, \pi_k(I^k) \right\} \quad \text{for } k = 0, \ldots, N-2.
$$

Then, the dynamic programming algorithm for the case of incomplete state information can be expressed via the following sequential optimisation (sub-) problems $[\mathcal{SOP}]$:

$$
\mathcal{SOP}_{N-1}: \quad J_{N-1}(I^{N-1}) = \inf_{u_{N-1} \in \mathbb{U}} \tilde{L}_{N-1}(I^{N-1}, u_{N-1}),
$$
$$
\text{subject to:} \tag{12.8}
$$
$$
x_N = Ax_{N-1} + Bu_{N-1} + w_{N-1},
$$

and, for $k = 0, \ldots, N-2$,

$$
\mathcal{SOP}_k: \quad J_k(I^k) = \inf_{u_k \in \mathbb{U}} \left[\tilde{L}_k(I^k, u_k) + \mathbf{E}\left\{ J_{k+1}(I^{k+1}) | I^k, u_k \right\} \right],
$$
$$
\text{subject to:}
$$
$$
x_{k+1} = Ax_k + Bu_k + w_k,
$$
$$
I^{k+1} = \{I^k, y_{k+1}, u_k\},
$$
$$
y_{k+1} = Cx_{k+1} + v_{k+1}.
$$

The dynamic programming algorithm starts at stage $N-1$ by solving $\mathcal{SOP}_{N-1}$ for all possible values of I^{N-1}. In this way, the law $\pi_{N-1}^{\text{OPT}}(\cdot)$ is obtained, in the sense that given the value of I^{N-1}, the corresponding optimal control is the value $u_{N-1} = \pi_{N-1}^{\text{OPT}}(I^{N-1})$, the minimiser of $\mathcal{SOP}_{N-1}$. The procedure then continues to solve the sub-problems $\mathcal{SOP}_{N-2}, \ldots, \mathcal{SOP}_0$ to obtain the laws $\pi_{N-2}^{\text{OPT}}(\cdot), \ldots, \pi_0^{\text{OPT}}(\cdot)$. After the last optimisation sub-problem is solved, the optimal control policy Π_N^{OPT} is obtained and the optimal objective function (see (12.5)) is

$$
V_N^{\text{OPT}} = V_N(\Pi_N^{\text{OPT}}) = \mathbf{E}\{J_0(I^0)\} = \mathbf{E}\{J_0(y_0)\}.
$$

12.3.1 Optimal Solution for N = 1

In the following proposition, we apply the dynamic programming algorithm to obtain the optimal solution of the problem in Definition 12.2.2 for the case $N = 1$.

Proposition 12.3.1 *For $N = 1$, the solution to the optimal control problem stated in Definition 12.2.2 is of the form $\Pi_1^{\mathrm{OPT}} = \{\pi_0^{\mathrm{OPT}}(\,\cdot\,)\}$, with*

$$u_0^{\mathrm{OPT}} = \pi_0^{\mathrm{OPT}}(I^0) = -\mathrm{sat}_\Delta(K\,\mathbf{E}\{x_0|I^0\}) \quad \textit{for all } I^0 \in \mathbb{R}, \tag{12.9}$$

where K is given in (12.7) and $\mathrm{sat}_\Delta : \mathbb{R} \to \mathbb{R}$ is the saturation function defined as

$$\mathrm{sat}_\Delta(z) = \begin{cases} \Delta & \textit{if } z > \Delta, \\ z & \textit{if } |z| \le \Delta, \\ -\Delta & \textit{if } z < -\Delta. \end{cases}$$

Moreover, the last step in the dynamic programming algorithm has the value

$$\begin{aligned} J_0(I^0) = \ &\mathbf{E}\left\{x_0^{\mathrm{T}}Px_0|I^0\right\} + \bar{R}\Phi_\Delta(K\,\mathbf{E}\{x_0|I^0\}) \\ &+ \mathrm{tr}(K^{\mathrm{T}}K\,\mathrm{cov}\{x_0|I^0\}) + \mathbf{E}\{w_0^{\mathrm{T}}Pw_0\}, \end{aligned} \tag{12.10}$$

where P and $\bar{R}$ are defined in (12.6) and (12.7), respectively, and where $\Phi_\Delta : \mathbb{R} \to \mathbb{R}$ is given by

$$\Phi_\Delta(z) = [z - \mathrm{sat}_\Delta(z)]^2. \tag{12.11}$$

Proof. For $N = 1$, the only optimisation sub-problem to solve is $\mathcal{SOP}_0$ (see (12.8)).

$$\begin{aligned} J_0(I^0) &= \inf_{u_0 \in \mathbb{U}} \mathbf{E}\left\{F(x_1) + L(x_0, u_0)|I^0, u_0\right\} \\ &= \inf_{u_0 \in \mathbb{U}} \mathbf{E}\left\{(Ax_0 + Bu_0 + w_0)^{\mathrm{T}}P(Ax_0 + Bu_0 + w_0) \right. \tag{12.12} \\ &\qquad\qquad \left. + x_0^{\mathrm{T}}Qx_0 + Ru_0^2|I^0, u_0\right\}. \end{aligned}$$

Using the fact that $\mathbf{E}\{w_0|I^0, u_0\} = E\{w_0\} = 0$ and that w_0 is neither correlated with the state x_0 nor correlated with the control u_0, (12.12) can be expressed, after distributing and grouping terms, as

$$\begin{aligned} J_0(I^0) = \ &\mathbf{E}\{w_0^{\mathrm{T}}Pw_0\} + \inf_{u_0 \in \mathbb{U}} \mathbf{E}\left\{x_0^{\mathrm{T}}(A^{\mathrm{T}}PA + Q)x_0 \right. \\ &\qquad\qquad \left. + 2u_0B^{\mathrm{T}}PAx_0(B^{\mathrm{T}}PB + R)u_0^2|I^0, u_0\right\}. \end{aligned}$$

Further, using (12.6) and (12.7), the above becomes

$$J_0(I^0) = \mathbf{E}\{w_0^{\mathrm{T}} P w_0\} + \inf_{u_0 \in \mathbb{U}} \mathbf{E} \left\{ x_0^{\mathrm{T}} P x_0 + \bar{R}(x_0^{\mathrm{T}} K^{\mathrm{T}} K x_0 \right.$$
$$\left. + 2u_0 K x_0 + u_0^2) | I^0, u_0 \right\}$$
$$= \mathbf{E}\{w_0^{\mathrm{T}} P w_0\} + \mathbf{E}\{x_0^{\mathrm{T}} P x_0 | I^0\} + \bar{R} \inf_{u_0 \in \mathbb{U}} \mathbf{E} \left\{ (u_0 + K x_0)^2 | I^0, u_0 \right\},$$

where we have used the fact that the conditional pdf of x_0 given I^0 and u_0 is equal to the pdf where only I^0 is given. Finally, using properties of the expected value of quadratic forms (see, for example, Åström 1970) the optimisation problem to solve becomes

$$J_0(I^0) = \mathbf{E}\{w_0^{\mathrm{T}} P w_0\} + \mathbf{E}\{x_0^{\mathrm{T}} P x_0 | I^0\} + \mathrm{tr}(K^{\mathrm{T}} K \, \mathrm{cov}\{x_0 | I^0\})$$
$$+ \bar{R} \inf_{u_0 \in \mathbb{U}} \left[(u_0 + K \, \mathbf{E}\{x_0 | I^0\})^2 \right]. \tag{12.13}$$

It is clear from (12.13) that the unconstrained minimum is attained at $u_0 = -K \, \mathbf{E}\{x_0 | I^0\}$. In the constrained case, equation (12.9) follows from the convexity of the quadratic function. The final value (12.10) is obtained by substituting (12.9) into (12.13). The result is then proved. □

Note that when $N = 1$ the optimal control law π_0^{OPT} depends on the information I^0 only through the *conditional expectation* $\mathbf{E}\{x_0 | I^0\}$. Therefore, this conditional expectation is a sufficient statistic in this case, that is, it provides all the necessary information to implement the control.

We observe that the control law given in (12.9) is also the optimal control law for the cases in which:

- the state is measured (complete state information) and the disturbance w_k is still acting on the system;
- the state is measured and w_k is set equal to a fixed value or to its mean value (see (6.17) in Chapter 6 for the case $w_k = 0$).

Therefore, CE is optimal for horizon $N = 1$, that is, the optimal control law is the same law that would result from an associated deterministic optimal control problem in which some or all uncertain quantities were set to a fixed value.

12.3.2 Optimal Solution for N = 2

We now consider the case where the optimisation horizon is $N = 2$.

Proposition 12.3.2 *For $N = 2$, the solution to the optimal control problem stated in Definition 12.2.2 is of the form $\Pi_2^{\mathrm{OPT}} = \{\pi_0^{\mathrm{OPT}}(\,\cdot\,), \pi_1^{\mathrm{OPT}}(\,\cdot\,)\}$, with*

$$u_1^{\mathrm{OPT}} = \pi_1^{\mathrm{OPT}}(I^1) = -\mathrm{sat}_\Delta (K \, \mathbf{E}\{x_1 | I^1\}) \quad \textit{for all } I^1 \in \mathbb{R}^3,$$

$$u_0^{\mathrm{OPT}} = \pi_0^{\mathrm{OPT}}(I^0) = \arg\inf_{u_0 \in \mathbb{U}} \left[\bar{R}(u_0 + K\,\mathbf{E}\{x_0|I^0\})^2 \right.$$

$$\left. + \bar{R}\,\mathbf{E}\left\{ \Phi_\Delta(K\,\mathbf{E}\{x_1|I^1\})|I^0, u_0 \right\} \right] \quad \text{for all } I^0 \in \mathbb{R},$$

$$(12.14)$$

where $\Phi_\Delta(\cdot)$ is given in (12.11).

Proof. The first step of the dynamic programming algorithm (see (12.8)) gives

$$J_1(I^1) = \inf_{u_1 \in \mathbb{U}} \mathbf{E}\left\{ F(Ax_1 + B\,u_1 + w_1) + L(x_1, u_1)|I^1, u_1 \right\}.$$

We see that $J_1(I^1)$ is similar to $J_0(I^0)$ in (12.12). By comparison, we readily obtain

$$\pi_1^{\mathrm{OPT}}(I^1) = -\mathrm{sat}_\Delta(K\,\mathbf{E}\{x_1|I^1\}),$$
$$J_1(I^1) = \mathbf{E}\{w_1^\mathsf{T} P w_1\} + \mathbf{E}\{x_1^\mathsf{T} P x_1|I^1\} + \mathrm{tr}(K^\mathsf{T} K \,\mathrm{cov}\{x_1|I^1\})$$
$$+ \bar{R}\Phi_\Delta(K\,\mathbf{E}\{x_1|I^1\}).$$

The second step of the dynamic programming algorithm proceeds as follows:

$$J_0(I^0) = \inf_{u_0 \in \mathbb{U}} \left[\mathbf{E}\{L(x_0, u_0)|I^0, u_0\} + \mathbf{E}\left\{J_1(I^1)|I^0, u_0\right\} \right],$$

subject to:

$$x_1 = Ax_0 + Bu_0 + w_0,$$
$$I^1 = \{I^0, y_1, u_0\},$$
$$y_1 = Cx_1 + v_1.$$

The objective function above can be written as

$$J_0(I^0) = \inf_{u_0 \in \mathbb{U}} \left[\mathbf{E}\{x_0^\mathsf{T} Q x_0 + R u_0^2 | I^0, u_0\} + \mathbf{E}\{w_1^\mathsf{T} P w_1\} \right.$$

$$+ \mathbf{E}\{\mathbf{E}\{x_1^\mathsf{T} P x_1 | I^1\}|I^0, u_0\} + \mathrm{tr}(K^\mathsf{T} K \,\mathrm{cov}\{x_1|I^1\}) \qquad (12.15)$$

$$\left. + \bar{R}\,\mathbf{E}\left\{\Phi_\Delta(K\,\mathbf{E}\{x_1|I^1\})|I^0, u_0\right\} \right].$$

Since $\{I^0, u_0\} \subset I^1$, using the properties of successive conditioning (Ash and Doléans-Dade 2000), we can express the third term inside the inf in (12.15) as

$$\mathbf{E}\{\mathbf{E}\{x_1^\mathsf{T} P x_1 | I^1\}|I^0, u_0\} = \mathbf{E}\{x_1^\mathsf{T} P x_1 | I^0, u_0\}$$
$$= \mathbf{E}\left\{ (Ax_0 + Bu_0 + w_0)^\mathsf{T} P(Ax_0 \right.$$
$$\left. + Bu_0 + w_0)|I^0, u_0 \right\}.$$

Using this, expression (12.15) becomes

$$J_0(I^0) = \inf_{u_0 \in \mathbb{U}} \Big[\mathbf{E}\{x_0^{\mathrm{T}} Q x_0 + R u_0^2 + (A x_0 + B u_0 + w_0)^{\mathrm{T}} P(A x_0$$
$$+ B u_0 + w_0) | I^0, u_0\} + \mathbf{E}\{w_1^{\mathrm{T}} P w_1\} + \mathrm{tr}(K^{\mathrm{T}} K \operatorname{cov}\{x_1 | I^1\})$$
$$+ \bar{R} \mathbf{E}\left\{\Phi_\Delta(K \mathbf{E}\{x_1 | I^1\}) | I^0, u_0\right\} \Big]. \tag{12.16}$$

Note that the first part of (12.16) is identical to (12.12). Therefore, using (12.13), expression (12.16) can be written as

$$J_0(I^0) = \mathbf{E}\{x_0^{\mathrm{T}} P x_0 | I^0\} + \sum_{j=0}^{1} \left[\mathrm{tr}(K^{\mathrm{T}} K \operatorname{cov}\{x_j | I^j\}) + \mathbf{E}\{w_j^{\mathrm{T}} P w_j\} \right]$$
$$+ \inf_{u_0 \in \mathbb{U}} \left[\bar{R}(u_0^2 + K \mathbf{E}\{x_0 | I^0\})^2 + \bar{R} \mathbf{E}\left\{\Phi_\Delta(K \mathbf{E}\{x_1 | I^1\}) | I^0, u_0\right\} \right], \tag{12.17}$$

where $\mathrm{tr}(K^{\mathrm{T}} K \operatorname{cov}\{x_1 | I^1\})$ has been left out of the minimisation because it is not affected by u_0 due to the linearity of the system equations (Bertsekas 1987, Bertsekas 2000). By considering only the terms that are affected by u_0, we find the result given in (12.14). $\qquad \square$

To obtain an explicit form for π_0^{OPT}, we would need to express $\mathbf{E}\{x_1 | I^1\} = \mathbf{E}\{x_1 | I^0, y_1, u_0\}$ explicitly as a function of I^0, u_0 and y_1. The optimal law $\pi_0^{\mathrm{OPT}}(\,\cdot\,)$ depends on I^0 not only through $\mathbf{E}\{x_0 | I^0\}$, as was the case for $N = 1$. Indeed, Haimovich, Perez and Goodwin (2003) have shown that, even for Gaussian disturbances, when input constraints are present, the optimal control law $\pi_0^{\mathrm{OPT}}(\,\cdot\,)$ depends also on $\operatorname{cov}\{x_0 | I^0\}$.

To calculate $\mathbf{E}\{x_1 | I^1\}$, we need to find the conditional pdf $p_{x_1 | I^1}(\,\cdot\,| I^1)$. At any time instant k, the conditional pdfs $p_{x_k | I^k}(\,\cdot\,| I^k)$ satisfy the Chapman–Kolmogorov equation and the observation update equation (see Section 9.8 in Chapter 9):

Time update

$$p_{x_k | I^{k-1}, u_{k-1}}(x_k | I^{k-1}, u_{k-1}) = \int_{\mathbb{R}^n} p_{x_k | x_{k-1}, u_{k-1}}(x_k | x_{k-1}, u_{k-1})$$
$$\times p_{x_{k-1} | I^{k-1}, u_{k-1}}(x_{k-1} | I^{k-1}, u_{k-1}) dx_{k-1}, \tag{12.18}$$

Observation update

$$p_{x_k | I^k}(x_k | I^k) = p_{x_k | I^{k-1}, y_k, u_{k-1}}(x_k | I^{k-1}, y_k, u_{k-1})$$
$$= \frac{p_{y_k | x_k}(y_k | x_k) p_{x_k | I^{k-1}, u_{k-1}}(x_k | I^{k-1}, u_{k-1})}{p_{y_k | I^{k-1}, u_{k-1}}(y_k | I^{k-1}, u_{k-1})}, \tag{12.19}$$

where

$$p_{y_k|I^{k-1},u_{k-1}}(y_k|I^{k-1},u_{k-1}) = \int_{\mathbb{R}^n} p_{y_k|x_k}(y_k|x_k)$$

$$\times p_{x_k|I^{k-1},u_{k-1}}(x_k|I^{k-1},u_{k-1})dx_k.$$

Remark 12.3.1. In general, depending on the pdfs of the initial state and the disturbances, it may be very difficult or even impossible to obtain an explicit form for the conditional pdfs that satisfy the recursion given by (12.18) and (12.19). If the pdfs of the initial state and the disturbances are Gaussian, however, all the conditional densities that satisfy (12.18) and (12.19) are also Gaussian. In this particular case, (12.18) and (12.19) lead to the well-known Kalman filter algorithm (see Section 9.6 in Chapter 9). The latter is a recursive algorithm in terms of the (conditional) expectation and covariance, which completely define any Gaussian pdf.   ○

Due to the way the information enters the conditional pdfs, it is, in general, very difficult to obtain an explicit form for the optimal control. On the other hand, even if the recursion given by (12.18) and (12.19) can be found explicitly, the implementation of such optimal control may also be complicated and computationally demanding. We illustrate this point by suggesting a way of implementing the optimal controller in such a case.

Let us first discretise the set $\mathbb{U}$ of admissible control values, and suppose that the discretised set $\mathbb{U}_d = \{u_{0i} \in \mathbb{U} : i = 1, 2, \ldots, r\}$ contains a finite number r of elements. We then approximate the optimal control as

$$u_0^{\text{OPT}} \approx \arg\inf_{u_0 \in \mathbb{U}_d} \left[\bar{R}(u_0 + K\,\mathbf{E}\{x_0|I^0\})^2 + \bar{R}\,\mathbf{E}\{\Phi_\Delta(K\,\mathbf{E}\{x_1|I^1\})|I^0, u_0\}\right].$$

(12.20)

Hence, to solve the above minimisation, only a finite number of values of u_0 have to be considered for a given $I^0 = y_0$. To do this, given the measurement $I^0 = y_0$ and for every value u_{0i}, $i = 1, \ldots, r$, of u_0, we can evaluate the expression between square brackets in (12.20) in the following way:

(i) Given y_0, we evaluate the function $W_2(\cdot)$ given by

$$W_2(y_0) = \mathbf{E}\{x_0|I^0\} = \mathbf{E}\{x_0|y_0\} = \int_{\mathbb{R}^n} x_0\, p_{x_0|y_0}(x_0|y_0)dx_0. \qquad (12.21)$$

(ii) Given y_0, $p_v(\cdot)$, $p_w(\cdot)$ and $p_{x_0}(\cdot)$, we obtain $p_{y_0|x_0}(\cdot|x_0)$ and $p_{x_1|y_0,y_1,u_0}(\cdot|y_0,y_1,u_0)$ in explicit form (as assumed) using the recursion (12.18) and (12.19) together with the system and measurement equations (12.1).

(iii) Using $p_{x_1|y_0,y_1,u_0}(\cdot|y_0,y_1,u_0)$, the expectation $\mathbf{E}\{x_1|y_0,y_1,u_0\}$ can be written as

$$h(y_0,y_1,u_0) = \mathbf{E}\{x_1|y_0,y_1,u_0\}$$
$$= \int_{\mathbb{R}^n} x_1 p_{x_1|y_0,y_1,u_0}(x_1|y_0,y_1,u_0)dx_1, \qquad (12.22)$$

which may not be expressible in explicit form even if (12.18) and (12.19) are so expressed. However, it may be evaluated numerically if y_0, y_1 and u_0 are given.

(iv) Using $p_{x_0|y_0}(\cdot|y_0)$, $p_v(\cdot)$, together with the measurement equation, we can obtain $p_{y_1|y_0,u_0}(\cdot|y_0,u_0)$. We can now express the expectation $\mathbf{E}\left\{\Phi_\Delta\left(Kh(y_0,y_1,u_0)\right)|I^0,u_0\right\}$ as

$$
\begin{aligned}
W_1(y_0,u_0) &= \mathbf{E}\left\{\Phi_\Delta\left(Kh(y_0,y_1,u_0)\right)|I^0,u_0\right\} \\
&= \int_{\mathbb{R}} \Phi_\Delta\left(Kh(y_0,y_1,u_0)\right)p_{y_1|y_0,u_0}(y_1|y_0,u_0)dy_1.
\end{aligned}
\tag{12.23}
$$

Note that, in order to calculate this integral numerically, the function $h(y_0,y_1,u_0)$ has to be evaluated for different values of y_1, even if u_0 and y_0 are given.

To find the value of the objective function achieved by *one* of the r values u_{0i}, expressions (12.21), (12.22) and (12.23) may need further discretisations (for x_0, x_1 and y_1, respectively).

From the previous comments, it is evident that the approximation of the optimal solution can be very computationally demanding depending on the discretisations performed. Note that in the above steps all the pdfs are assumed known in explicit form so that the integrals can be evaluated. As already mentioned in Remark 12.3.1, this may not always be possible.

As an alternative approach to brute force discretisations, we could use Markov chain Monte Carlo [MCMC] methods (Robert and Casella 1999). These methods approximate continuous pdfs by discrete ones by drawing samples from the pdfs in question or from other approximations. However, save for some very particular cases, the exponential growth in the number of computations as the optimisation horizon is increased seems to be unavoidable. We observe, in passing, that the application of MCMC methods to the recursion given by (12.18) and (12.19) gives rise to a special case of the, so-called, particle filters (Doucet et al. 2001).

The above discussion suggests that, not only does it seem impossible to analytically proceed with the optimisation for horizons greater than two but also the implementation of the optimal law (even for $N = 2$) appears to be quite intricate and computationally burdensome. This leads us to consider suboptimal solutions. In the next section, we analyse two alternative suboptimal strategies.

12.4 Suboptimal Strategies

12.4.1 Certainty Equivalent Control

As mentioned before, certainty equivalent control [CEC] uses the control law obtained as the solution of an associated *deterministic* control problem derived from the original problem by removing all uncertainty. Specifically, the associated problem is derived by setting the disturbance w_k to a fixed typical value (for example, $\bar{w} = \mathbf{E}\{w_k\}$) and by also assuming perfect state information. The resulting control law is a function of the true state. Then, the control is implemented using some estimate of the state $\hat{x}(I^k)$ in place of the true state.

For our problem, we first obtain the optimal policy for the deterministic problem

$$\Pi_N^{\text{DET}} = \{\pi_0^{\text{DET}}(\,\cdot\,), \ldots, \pi_{N-1}^{\text{DET}}(\,\cdot\,)\}, \tag{12.24}$$

where $\pi_k^{\text{DET}} : \mathbb{R}^n \to \mathbb{R}$ for $k = 0, 1, \ldots, N-1$. Then, the CEC evaluates the deterministic laws at the estimate of the state, that is,

$$u_k^{\text{CE}} = \pi_k^{\text{DET}}\left(\hat{x}(I^k)\right). \tag{12.25}$$

As we saw in Section 6.2 of Chapter 6, the associated deterministic problem for linear systems with a quadratic objective function is an example of a case where the control policy can be obtained explicitly for any finite optimisation horizon. The following example illustrates this for an optimisation horizon $N = 2$.

Example 12.4.1 (Closed Loop CEC). For $N = 2$, the deterministic policy $\Pi_2^{\text{DET}} = \{\pi_0^{\text{DET}}(\,\cdot\,), \pi_1^{\text{DET}}(\,\cdot\,)\}$ is given by (see Theorem 6.2.1 in Chapter 6):

$$\pi_1^{\text{DET}}(x) = -\text{sat}_\Delta(Kx) \quad \text{for all } x \in \mathbb{R}^n$$

$$\pi_0^{\text{DET}}(x) = \begin{cases} -\text{sat}_\Delta(Gx + h) & \text{if } x \in \mathbb{Z}^-, \\ -\text{sat}_\Delta(Kx) & \text{if } x \in \mathbb{Z}, \\ -\text{sat}_\Delta(Gx - h) & \text{if } x \in \mathbb{Z}^+. \end{cases}$$

K is given by (12.7) and

$$G = \frac{K + KBKA}{1 + (KB)^2}, \qquad h = \frac{KB}{1 + (KB)^2}\Delta.$$

The sets $\mathbb{Z}^-, \mathbb{Z}, \mathbb{Z}^+$ form a partition of $\mathbb{R}^n$, and are given by

$$\mathbb{Z}^- = \{x : K(A - BK)x < -\Delta\},$$
$$\mathbb{Z} = \{x : |K(A - BK)x| \leq \Delta\},$$
$$\mathbb{Z}^+ = \{x : K(A - BK)x > \Delta\}.$$

Therefore, a closed loop CEC applies the controls

$$u_0^{\text{CE}} = \pi_0^{\text{DET}}\left(\hat{x}(I^0)\right),$$
$$u_1^{\text{CE}} = \pi_1^{\text{DET}}\left(\hat{x}(I^1)\right),$$

where the estimate $\hat{x}(I^k)$ can be provided, for example, by the Kalman filter.

○

12.4.2 Partially Stochastic CEC

This variant of CEC uses the control law obtained as the solution to an associated problem that assumes perfect state information but takes stochastic disturbances into account. To actually implement the controller, the value of the state is replaced by its estimate $\hat{x}_k(I^k)$.

In our case, given a partially stochastic CEC [PS–CEC] admissible policy

$$\Lambda_N = \{\lambda_0(\,\cdot\,), \ldots, \lambda_{N-1}(\,\cdot\,)\}, \tag{12.26}$$

that is, a sequence of admissible control laws $\lambda_k(\,\cdot\,) : \mathbb{R}^n \to \mathbb{U}$ that map the (estimates of the) states into admissible control actions, the PS–CEC solves the following perfect state information problem.

Definition 12.4.1 (PS–CEC Optimal Control Problem) *Assuming that the state $\hat{x}_k$ will be available to the controller at time instant k to calculate the control, and given the pdf $p_w(\,\cdot\,)$ of the disturbances w_k, find the admissible control policy $\Lambda_N^{\mathrm{OPT}} = \{\lambda_0^{\mathrm{OPT}}(\,\cdot\,), \ldots, \lambda_{N-1}^{\mathrm{OPT}}(\,\cdot\,)\}$ that minimises the objective function*

$$\hat{V}_N(\Lambda_N) = \underset{\substack{w_k \\ k=0,\ldots,N-1}}{\mathbf{E}} \left\{ F(\hat{x}_N) + \sum_{k=0}^{N-1} L(\hat{x}_k, \lambda_k(\hat{x}_k)) \right\},$$

subject to $\hat{x}_{k+1} = A\hat{x}_k + B\lambda_k(\hat{x}_k) + w_k$ for $k = 0, \ldots, N-1$. ○

The optimal control policy for perfect state information thus found will be used, as in CEC, to calculate the control action based on the estimate $\hat{x}_k$ provided by the estimator; that is,

$$u_k = \lambda_k^{\mathrm{OPT}}(\hat{x}_k).$$

Next, we apply this suboptimal strategy to the problem of interest for horizons 1 and 2.

PS–CEC for N = 1.

Using the dynamic programming algorithm, we have

$$\hat{J}(\hat{x}_0) = \inf_{u_0 \in \mathbb{U}} \mathbf{E} \left\{ \hat{x}_1^{\mathrm{T}} P \hat{x}_1 + \hat{x}_0^{\mathrm{T}} Q \hat{x}_0 + R u_0^2 | \hat{x}_0, u_0 \right\}.$$

As with the true optimal solution for $N = 1$, the PS–CEC optimal control has the form

$$\hat{u}_0^{\mathrm{OPT}} = \lambda_0^{\mathrm{OPT}}(\hat{x}_0) = -\mathrm{sat}_\Delta(K\hat{x}_0),$$
$$\hat{J}_0(\hat{x}_0) = \hat{x}_0^{\mathrm{T}} P \hat{x}_0 + \bar{R}\Phi_\Delta(K\hat{x}_0) + \mathbf{E}\{w_0^{\mathrm{T}} P w_0\}.$$

We can see that if $\hat{x}_0 = \mathbf{E}\{x_0|I^0\}$ then the PS–CEC for $N = 1$ coincides with the optimal solution.

PS–CEC for N = 2.

The first step of the dynamic programming algorithm yields

$$\hat{u}_1^{\text{OPT}} = \lambda_1^{\text{OPT}}(\hat{x}_1) = -\text{sat}_\Delta(K\hat{x}_1),$$
$$\hat{J}_1(\hat{x}_1) = \hat{x}_1^{\text{T}} P \hat{x}_1 + \bar{R}\Phi_\Delta(K\hat{x}_1) + \mathbf{E}\{w_1^{\text{T}} P w_1\}.$$

For the second step, we have, after some algebra, that

$$\hat{J}_0(\hat{x}_0) = \inf_{u_0 \in \mathbb{U}} \left[\mathbf{E}\{L(\hat{x}_0, u_0) + \hat{J}_1(\hat{x}_1)|\hat{x}_0, u_0\} \right],$$

subject to:

$$\hat{x}_1 = A\hat{x}_0 + Bu_0 + w_0,$$

$$\hat{u}_0^{\text{OPT}} = \arg\inf_{u_0 \in \mathbb{U}} \left[\bar{R}(u_0 + K\hat{x}_0)^2 + \bar{R}\,\mathbf{E}\{\Phi_\Delta[K(A\hat{x}_0 + Bu_0 + w_0)]|\hat{x}_0, u_0\} \right]. \quad (12.27)$$

Comparing $\hat{u}_0^{\text{OPT}}$ with expression (12.14) for the optimal control, we can appreciate that, given $\hat{x}_0$, even if $\mathbf{E}\left\{\Phi_\Delta[K(A\hat{x}_0 + Bu_0 + w_0)]|\hat{x}_0, u_0\right\}$ cannot be found in explicit form as a function of u_0, the numerical implementation of this suboptimal control action is much less computationally demanding than its optimal counterpart.

12.5 Simulation Examples

In this section we compare the performance of the suboptimal strategies CEC and PS–CEC by means of simulation examples. The performance is assessed by computing the achieved value of the objective function. The objective function is defined as the expected value of a random variable, which is a quadratic function of the states and controls in our case. Hence, a comparison between the values of the objective function incurred by using different policies is only meaningful in terms of these expected values. To numerically compute values of the objective function for a given control policy, different realisations of the initial state plus process and measurement disturbances have to be obtained and a corresponding realisation of the objective function evaluated. Then, the expected value can be approximated by averaging over the different realisations.

The following examples are simulated for the system:

$$A = \begin{bmatrix} 0.9713 & 0.2189 \\ -0.2189 & 0.7524 \end{bmatrix}, \qquad B = \begin{bmatrix} 0.0287 \\ 0.2189 \end{bmatrix}, \qquad C = [0.3700 \quad 0.0600].$$

$$(12.28)$$

The disturbances w_k are assumed to have a uniform distribution with support on $[-0.5, 0.5] \times [-1, 1]$ and likewise for v_k with support on $[-0.1, 0.1]$. The initial state x_0 is assumed to have a Gaussian distribution with zero mean and covariance diag$\{300^{-1}, 300^{-1}\}$. A Kalman filter was implemented to provide the state estimates needed. Although this estimator is not the optimal one in this case because the disturbances are not Gaussian, it yields the best linear unbiased estimator for the state. The parameters for the Kalman filter were chosen as the true mean and covariance of the corresponding variables in the system. The saturation limit of the control was taken as $\Delta = 1$. The optimisation horizon is in both cases $N = 2$.

For PS–CEC, we discretise the set $\mathbb{U}$ so that only 500 values are considered, and the expected value in (12.27) is approximated by taking 300 samples of the pdf $p_w(\cdot)$ for every possible value of u_0 in the discretised set. For CEC, we implement the policy given in Example 12.4.1.

We simulated the closed loop system over two time instants and repeated the simulation a large number of times (between 2000 and 8000). For each simulation, a different realisation of the disturbances and the initial state was used. A realisation of the objective function was calculated for every simulation run for each one of the control policies applied (PS–CEC and CEC). The sample average of the objective function values achieved by each policy was computed, and the difference between them was always found to be less than 0.1%.

Although the examples are based on a simple simulated model, the comparison between the objective function values for the two control policies seems to indicate that the trade-off between better performance and computational complexity favours the CEC implementation over the PS–CEC.

It would be of interest, from a practical standpoint, to extend the optimisation horizon beyond $N = 2$. However, as we explained in a previous section and observed in the examples, due to computational issues this becomes very difficult. In order to achieve this extension, one is led to conclude that CEC may be, at this point, the only way forward.

Of course, the ultimate test for the suboptimal strategies would be to contrast them with the optimal one. It would be expected that, in this case, an appreciable difference in the objective function values may be obtained due to the fact that the optimal strategy takes into account the process and measurement disturbances in a unified manner, as opposed to the above mentioned suboptimal strategies, which use estimates provided by an estimator as if they were the true state.

12.6 Further Reading

For complete list of references cited, see References section at the end of book.

General

For more details on general sequential decision problems under uncertainty and the use of dynamic programming, the reader is referred to Bellman (1957), Bertsekas (1976) and Bertsekas (2000).

Section 12.3

The use of CE in RHC, due to its simplicity, has been advocated in the literature (Muske and Rawlings 1993) and reported in a number of applications (see, for example, Angeli, Mosca and Casavola 2000, Marquis and Broustail 1988, Perez, Goodwin and Tzeng 2000).

For RHC literature for uncertain systems using a stochastic uncertainty description, see Haimovich, Perez and Goodwin (2003) and Perez, Haimovich and Goodwin (2004) (on which this chapter is based). Also, in Filatov and Unbehauen (1995) output-feedback predictive control of nonlinear systems with uncertain parameters is addressed. The control is assumed unconstrained and only suboptimal solutions are considered. Batina, Stoorvogel and Weiland (2001) consider the RHC problem for the case of state feedback, input constraints and scalar disturbances. The optimal solution is approximated via a randomised algorithm (Monte Carlo sampling). Examples for an optimisation horizon of length 1 are presented. In Batina, Stoorvogel and Weiland (2002), the authors extend their previous result to the state constrained case.

An alternative approach to model uncertainty is via a set-membership description, which only gives information regarding the sets in which the uncertain elements take values. When addressing uncertain systems, the RHC literature has somewhat favoured the set-membership description; see, for example, Shamma and Tu (1998) and Lee and Kouvaritakis (2001). For example, Shamma and Tu (1998) propose an observer-based strategy that assumes unknown but bounded disturbances, and generates a set of possible states based on past input and output information. Then, to each estimated state the strategy associates a set of control values that meet the constraint requirements. The actual control applied to the system is selected to belong to the intersection of all the control value sets. As another example, Lee and Kouvaritakis (2001) present an extension of the dual-mode paradigm of Mayne and Michalska (1993), in which invariant sets of estimation errors are used for the case of unknown-but-bounded measurement noise and disturbances.

Finite Alphabet Controllers and Estimators

Contributed by Daniel Quevedo

13.1 Introduction

In this chapter we address the issue of control and estimation when the decision variables must satisfy a finite set constraint. We will distinguish between finite alphabet control and estimation problems. As in the case of convex constraints, the essential difference is whether or not the initial state is given or can be considered a decision variable.

Finite alphabet control occurs in many practical situations including: on-off control, relay control, control where quantisation effects are important (in principle this covers all digital control systems and control systems over digital communication networks), and switching control of the type found in power electronics.

Exactly the same design methodologies can be applied in other areas; for example, the following problems can be directly formulated as finite alphabet control problems:

- quantisation of audio signals for compact disc production;
- design of filters where the coefficients are restricted to belong to a finite set (it is common in digital signal processing to use coefficients that are powers of two to facilitate implementation issues);
- design of digital-to-analog [D/A] and analog-to-digital [A/D] converters.

Finite alphabet estimation problems are also frequently encountered in practice. Common examples are:

- estimation of transmitted signals in digital communication systems where the signals are known to belong to a finite alphabet (say ± 1);
- state estimation problems where a disturbance is known to take only a finite set of values (for example, either "on" or "off").

In this chapter we show how these problems can be formulated in the same general framework as described in earlier chapters. However, special care is needed to address the finite set nature of the constraints. In particular, this restriction gives rise to a hard combinatorial optimisation problem, which is exponential in the dimension of the problem. Thus, various approximation techniques are necessary to deal with optimisation problems in which the horizon is large. Commonly used strategies are variants of well-known branch and bound algorithms (Land and Doig 1960, Bertsekas 1998).

We will show how the receding horizon principle can be used in these problems. A key observation in this context is the fact that often the "first" decision variable is insensitive to increasing the optimisation horizon beyond some modest value (typically 3 to 10 in many real world problems).

Also, a *closed form* expression for the control law is derived by exploiting the geometry of the underlying optimisation problem. The solution can also be characterised by means of a partition of the state space, which is closely related to the partition induced by the *interval-constrained* solution,[1] as developed in Chapter 6. As a consequence, the controller can be implemented without relying upon on-line numerical optimisation. Furthermore, the insight obtained from this viewpoint into the nature of the control law can be used to study the dynamic behaviour of the closed loop system.

13.2 Finite Alphabet Control

Consider a linear system having a scalar input u_k and state vector $x_k \in \mathbb{R}^n$ described by

$$x_{k+1} = Ax_k + Bu_k. \tag{13.1}$$

(Here we treat only the scalar input case, but the extension to multiple inputs presents no additional conceptual difficulties.) A key consideration here is that the input is restricted to belong to the finite set

$$\mathbb{U} = \{s_1, s_2, \ldots, s_{n_\mathbb{U}}\}, \tag{13.2}$$

where $s_i \in \mathbb{R}$ and $s_i < s_{i+1}$ for $i = 1, 2, \ldots, n_\mathbb{U} - 1$.

We will formulate the input design problem as a receding horizon quadratic regulator problem with finite set constraints. Thus, given the state $x_k = x$, we seek the optimising sequence of present and future control inputs:

---

[1] The explicit solution in problems with "interval-type" constraints of the form $|u| \leq \Delta$.

$$\mathbf{u}^{\mathrm{OPT}}(x) \triangleq \arg\min_{\mathbf{u}_k \in \mathbb{U}^N} V_N(x, \mathbf{u}_k), \tag{13.3}$$

where

$$\mathbf{u}_k \triangleq \begin{bmatrix} u_k \\ u_{k+1} \\ \vdots \\ u_{k+N-1} \end{bmatrix}, \quad \mathbb{U}^N \triangleq \mathbb{U} \times \cdots \times \mathbb{U}. \tag{13.4}$$

As in previous chapters, in (13.3) V_N is the finite horizon quadratic objective function[2]

$$V_N(x, \mathbf{u}_k) \triangleq \|x_{k+N}\|_P^2 + \sum_{t=k}^{k+N-1} (\|x_t\|_Q^2 + \|u_t\|_R^2), \tag{13.5}$$

with $Q = Q^{\mathrm{T}} > 0$, $P = P^{\mathrm{T}} > 0$, $R = R^{\mathrm{T}} > 0$ and where $x_k = x$. Note that, as usual, the formulation of $V_N(\cdot, \cdot)$, uses predictions of future plant states.

Whilst we concentrate here upon plant state deviations from the origin, nonzero references can also be encompassed within this framework. In order to accomplish this, the objective function (13.5) needs to be modified by considering *shifted coordinates* as is common when dealing with nonzero constant references in standard receding horizon control schemes (see Chapter 5).

The minimisation of (13.5) subject to the finite set constraint on $\mathbf{u}_k$ and the plant dynamics expressed in (13.1) yields the optimal sequence $\mathbf{u}^{\mathrm{OPT}}(x)$. It is a function only of the current state value $x_k = x$.

Following the usual receding horizon principle (see Chapter 4), only the first control action, namely

$$u^{\mathrm{OPT}}(x) \triangleq \begin{bmatrix} 1 & 0 & \cdots & 0 \end{bmatrix} \mathbf{u}^{\mathrm{OPT}}(x), \tag{13.6}$$

is applied. At the next time instant, the optimisation is repeated with a new *initial* state and the finite horizon window shifted by one.

In the next section we present a closed form expression for $\mathbf{u}^{\mathrm{OPT}}(x)$. This is directly analogous to the geometric interpretation of the constrained solution developed in Chapter 6. This result will allow us to characterise the control law as a partition of the state space and provide a tool for studying the dynamic behaviour of the resulting closed loop system.

13.3 Nearest Neighbour Characterisation of the Solution

Since the constraint set $\mathbb{U}^N$ is finite, the optimisation problem (13.3) is non-convex. Indeed, it is a hard combinatorial optimisation problem whose solution requires a computation time that is exponential in the horizon length. Thus,

---

[2] $\|\nu\|_S^2$ denotes $\nu^{\mathrm{T}} S \nu$, where ν is any vector and S is a matrix.

one needs either to use a relatively small horizon or to resort to approximate solutions. We will adopt the former strategy based on the premise that, due to the receding horizon technique, the *first* decision variable is all that is of interest. Moreover, it is a practical observation that this first decision variable is often insensitive to increasing the horizon length beyond some relative modest value. To proceed, it is useful to *vectorise* the objective function (13.5) as follows:

Define

$$
\mathbf{x}_k \triangleq \begin{bmatrix} x_{k+1} \\ x_{k+2} \\ \vdots \\ x_{k+N} \end{bmatrix}, \quad
\Phi \triangleq \begin{bmatrix} B & 0 & \dots & 0 & 0 \\ AB & B & \dots & 0 & 0 \\ \vdots & \vdots & \ddots & \vdots & \vdots \\ A^{N-1}B & A^{N-2}B & \dots & AB & B \end{bmatrix}, \quad
\Lambda \triangleq \begin{bmatrix} A \\ A^2 \\ \vdots \\ A^N \end{bmatrix}, \quad (13.7)
$$

so that, given $x_k = x$ and by iterating (13.1), the predictor $\mathbf{x}_k$ satisfies

$$
\mathbf{x}_k = \Phi \mathbf{u}_k + \Lambda x. \tag{13.8}
$$

Hence, the objective function (13.5) can be re-written as

$$
V_N(x, \mathbf{u}_k) = \bar{V}_N(x) + \mathbf{u}_k^{\mathrm{T}} H \mathbf{u}_k + 2\mathbf{u}_k^{\mathrm{T}} F x, \tag{13.9}
$$

where

$$
H \triangleq \Phi^{\mathrm{T}} \mathbf{Q} \Phi + \mathbf{R} \in \mathbb{R}^{N \times N}, \quad F \triangleq \Phi^{\mathrm{T}} \mathbf{Q} \Lambda \in \mathbb{R}^{N \times n},
$$
$$
\mathbf{Q} \triangleq \mathrm{diag}\{Q, \dots, Q, P\} \in \mathbb{R}^{Nn \times Nn}, \quad \mathbf{R} \triangleq \mathrm{diag}\{R, \dots, R\} \in \mathbb{R}^{N \times N},
$$

and $\bar{V}_N(x)$ does not depend upon $\mathbf{u}_k$.

By direct calculation, it follows that the minimiser to (13.9), without taking into account any constraints on $\mathbf{u}_k$, is

$$
\mathbf{u}_{\mathrm{UC}}^{\mathrm{OPT}}(x) = -H^{-1} F x. \tag{13.10}
$$

Our subsequent development will utilise a nearest neighbour vector quantiser in order to characterise the constrained optimiser. This is defined as follows:

Definition 13.3.1 (Nearest Neighbour Vector Quantiser)   *Given a countable (not necessarily finite) set of nonequal vectors* $\mathbb{B} = \{b_1, b_2, \dots\} \subset \mathbb{R}^{n_\mathbb{B}}$*, the nearest neighbour quantiser is defined as a mapping* $q_\mathbb{B} \colon \mathbb{R}^{n_\mathbb{B}} \to \mathbb{B}$ *that assigns to each vector* $c \in \mathbb{R}^{n_\mathbb{B}}$ *the closest element of* $\mathbb{B}$ *(as measured by the Euclidean norm), that is,* $q_\mathbb{B}(c) = b_i \in \mathbb{B}$ *if and only if* c *belongs to the region*

$$
\{c \in \mathbb{R}^{n_\mathbb{B}} : \|c - b_i\|^2 \le \|c - b_j\|^2 \text{ for all } b_j \ne b_i,\ b_j \in \mathbb{B}\}
$$
$$
\setminus \{c \in \mathbb{R}^{n_\mathbb{B}} : \text{there exists } j < i \text{ such that } \|c - b_i\|^2 = \|c - b_j\|^2\}. \tag{13.11}
$$

○

Note that in the special case, when $n_{\mathbb{B}} = 1$, this vector quantiser reduces to the standard scalar quantiser.

In the above definition, the zero measure set of points that satisfy (13.11) with equality have been arbitrarily assigned to the element having the smallest index. This is done in order to avoid ambiguity in the case of *frontier points*, that is, points equidistant to two or more elements of $\mathbb{B}$. If this aspect does not matter, then expression (13.11) can be simplified to

$$\left\{c \in \mathbb{R}^{n_B} : \|c - b_i\|^2 \le \|c - b_j\|^2 \text{ for all } b_j \ne b_i, \ b_j \in \mathbb{B}\right\}. \tag{13.12}$$

Given Definition 13.3.1, we can now restate the solution to (13.3). This leads to:

Theorem 13.3.1 (Closed Form Solution) *Let* $\mathbb{U}^N = \{v_1, v_2, \ldots, v_r\}$, *where* $r = (n_{\mathbb{U}})^N$. *Then the optimiser* $\mathbf{u}^{\mathrm{OPT}}(x)$ *in* (13.3) *is given by*

$$\mathbf{u}^{\mathrm{OPT}}(x) = H^{-1/2} q_{\tilde{\mathbb{U}}^N}(-H^{-\mathrm{T}/2} F x), \tag{13.13}$$

where the nearest neighbour quantiser $q_{\tilde{\mathbb{U}}^N}(\cdot)$ *maps* $\mathbb{R}^N$ *to* $\tilde{\mathbb{U}}^N$, *defined as*

$$\tilde{\mathbb{U}}^N \triangleq \{\tilde{v}_1, \tilde{v}_2, \ldots, \tilde{v}_r\}, \quad \tilde{v}_i = H^{1/2} v_i, \quad v_i \in \mathbb{U}^N. \tag{13.14}$$

Proof. For fixed x, the level sets of the objective function (13.9) are ellipsoids in the input sequence space $\mathbb{R}^N$. These are centred at the point $\mathbf{u}_{\mathrm{UC}}^{\mathrm{OPT}}(x)$ defined in (13.10). Thus, the optimisation problem (13.3) can be geometrically interpreted as the one where we find the point $\mathbf{u}_k \in \mathbb{U}^N$ that belongs to the smallest ellipsoid defined by (13.9) (that is, the point which provides the smallest objective function value whilst satisfying the constraints).

In order to simplify the problem, we introduce the same coordinate transformation utilised in Chapter 6, that is,

$$\tilde{\mathbf{u}}_k = H^{1/2} \mathbf{u}_k, \tag{13.15}$$

which transforms the constraint set $\mathbb{U}^N$ into $\tilde{\mathbb{U}}^N$ defined in (13.14). The optimiser $\mathbf{u}^{\mathrm{OPT}}(x)$ can be defined in terms of this auxiliary variable as

$$\mathbf{u}^{\mathrm{OPT}}(x) = H^{-1/2} \arg \min_{\tilde{\mathbf{u}}_k \in \tilde{\mathbb{U}}^N} J_N(x, \tilde{\mathbf{u}}_k), \tag{13.16}$$

where

$$J_N(x, \tilde{\mathbf{u}}_k) \triangleq \tilde{\mathbf{u}}_k^{\mathrm{T}} \tilde{\mathbf{u}}_k + 2 \tilde{\mathbf{u}}_k^{\mathrm{T}} H^{-\mathrm{T}/2} F x. \tag{13.17}$$

The level sets of J_N are spheres in $\mathbb{R}^N$, centred at

$$\tilde{\mathbf{u}}_{\mathrm{UC}}^{\mathrm{OPT}}(x) \triangleq -H^{-\mathrm{T}/2} F x. \tag{13.18}$$

Hence, the constrained optimiser (13.3) is given by the nearest neighbour to $\tilde{\mathbf{u}}_{\mathrm{UC}}^{\mathrm{OPT}}(x)$, namely

$$\arg \min_{\tilde{\mathbf{u}}_k \in \tilde{\mathbb{U}}^N} J_N(x, \tilde{\mathbf{u}}_k) = q_{\tilde{\mathbb{U}}^N}(-H^{-\mathrm{T}/2} F x). \tag{13.19}$$

The result (13.13) follows by substituting (13.19) into (13.16). $\quad\square$

We observe that, with $N > 1$, the optimiser $\mathbf{u}^{\mathrm{OPT}}(x)$ provided in Theorem 13.3.1 is, in general, different to the sequence obtained by direct quantisation of the unconstrained minimum (13.10), namely, $q_{\mathbb{U}^N}(\mathbf{u}_{\mathrm{UC}}^{\mathrm{OPT}}(x))$.

As a consequence of Theorem 13.3.1, the receding horizon controller (13.6) satisfies

$$u^{\mathrm{OPT}}(x) = \begin{bmatrix} 1 & 0 & \cdots & 0 \end{bmatrix} H^{-1/2} q_{\tilde{\mathbb{U}}^N}(-H^{-\mathrm{T}/2} Fx). \qquad (13.20)$$

This solution can be illustrated as the composition of the following transformations:

$$x \in \mathbb{R}^n \xrightarrow{\ -H^{-\frac{\mathrm{T}}{2}} F\ } \tilde{\mathbf{u}}_{\mathrm{UC}}^{\mathrm{OPT}} \in \mathbb{R}^N \xrightarrow{\ H^{-\frac{1}{2}} q_{\tilde{\mathbb{U}}^N}(\cdot)\ } \mathbf{u}^{\mathrm{OPT}} \in \mathbb{U}^N \xrightarrow{\ [1\,0\,\cdots\,0]\ } u^{\mathrm{OPT}} \in \mathbb{U}. \qquad (13.21)$$

It is worth noticing that $q_{\tilde{\mathbb{U}}^N}(\cdot)$ is a *memoryless nonlinearity*, so that (13.20) corresponds to a time-invariant nonlinear state feedback law. In a direct implementation, at each time step, the quantiser needs to perform $r - 1$ comparisons. However, in some cases, it is possible to exploit the nature of the problem to obtain more efficient search algorithms (Quevedo and Goodwin 2003a).

13.4 State Space Partition

Expression (13.11) partitions the domain of the quantiser into polyhedra, called *Voronoi partition*. Since the constrained optimiser $\mathbf{u}^{\mathrm{OPT}}(x)$ in (13.13) (see also (13.21)) is defined in terms of $q_{\tilde{\mathbb{U}}^N}(\cdot)$, an equivalent partition of the state space can be derived, as shown next.

Theorem 13.4.1 *The constrained optimising sequence $\mathbf{u}^{\mathrm{OPT}}(x)$ in (13.13) can be characterised as*

$$\mathbf{u}^{\mathrm{OPT}}(x) = v_i \iff x \in \mathcal{R}_i,$$

where

$$\mathcal{R}_i \triangleq \left\{ z \in \mathbb{R}^n : 2(v_i - v_j)^{\mathrm{T}} Fz \le \|v_j\|_H^2 - \|v_i\|_H^2 \text{ for all } v_j \ne v_i,\, v_j \in \mathbb{U}^N \right\}$$
$$\setminus \left\{ z \in \mathbb{R}^n : \text{there exists } j < i \text{ such that } 2(v_i - v_j)^{\mathrm{T}} Fz = \|v_j\|_H^2 - \|v_i\|_H^2 \right\}. \qquad (13.22)$$

Proof. From expressions (13.13) and (13.14) it follows that $\mathbf{u}^{\mathrm{OPT}}(x) = v_i$ if and only if $q_{\tilde{\mathbb{U}}^N}(-H^{-\mathrm{T}/2} Fx) = \tilde{v}_i$. On the other hand,

$$\| - H^{-\mathrm{T}/2} Fx - \tilde{v}_i\|^2 = \|H^{-\mathrm{T}/2} Fx\|^2 + \|\tilde{v}_i\|^2 + 2\tilde{v}_i^{\mathrm{T}} H^{-\mathrm{T}/2} Fx,$$

so that

$$\| - H^{-\mathrm{T}/2} Fx - \tilde{v}_i\|^2 \le \| - H^{-\mathrm{T}/2} Fx - \tilde{v}_j\|^2$$

holds if and only if

$$2(\tilde{v}_i - \tilde{v}_j)^\mathsf{T} H^{-\mathsf{T}/2} Fx \le \|\tilde{v}_j\|^2 - \|\tilde{v}_i\|^2.$$

This inequality together with expressions (13.14) and (13.11) shows that

$$q_{\tilde{\mathbb{U}}^N}(-H^{-\mathsf{T}/2}Fx) = \tilde{v}_i$$

if and only if x belongs to the region $\mathcal{R}_i$ defined in (13.22). This fact completes the proof. $\qquad\square$

The $n_{\tilde{\mathbb{U}}}^N$ regions $\mathcal{R}_i$ defined in (13.22) are polyhedra. Without taking into account constraint borders, we can write these in a compact form as

$$\mathcal{R}_i = \{x \in \mathbb{R}^n : D_i x \le h_i\},$$

where the rows of D_i are equal to all terms $2(v_i - v_j)^\mathsf{T} F$ as required, whilst the vector h_i contains the scalars $\|v_j\|_H^2 - \|v_i\|_H^2$.

Some of the inequalities in (13.22) may be redundant. In these cases, the corresponding regions do not share a common edge, that is, they are not adjacent. This phenomenon is illustrated in Figure 13.4 of Example 13.8.1, where the regions $\mathcal{R}_1$ and $\mathcal{R}_4$ are not adjacent. The inequality separating them is redundant.

Also, depending upon the matrix $H^{-\mathsf{T}/2}F$, some of the regions $\mathcal{R}_i$ may be empty. This might happen, in particular, if $N > n$. In this case, the rank of F is smaller than N and the transformation $H^{-\mathsf{T}/2}F$ does not span the entire space $\mathbb{R}^N$. Figure 13.1 illustrates this for the case $n = 1$, $n_{\mathbb{U}} = 2$ and $N = 2$. As can be seen from this figure, depending on the *unconstrained optimum locus* given by the (dashed) line $-H^{-\mathsf{T}/2}Fx$, $x \in \mathbb{R}$, there exist situations in which some sequences $\tilde{v}_j$ will never be optimal, thus yielding empty regions in the state space.

On the other hand, if the pair (A, B) is completely controllable and A is invertible, then the rank of F is equal to $\min(N, n)$. In this case, if $n \ge N$, then $H^{-\mathsf{T}/2}F$ is onto, so that for every $\tilde{v}_j \in \tilde{\mathbb{U}}^N$ there exists at least one x such that $q_{\tilde{\mathbb{U}}^N}(-H^{-\mathsf{T}/2}Fx) = \tilde{v}_j$ and none of the regions $\mathcal{R}_i$ are empty.

13.5 The Receding Horizon Case

In the receding horizon law (13.20), only $n_{\mathbb{U}}$ instead of (at most) $(n_{\mathbb{U}})^N$ regions are needed to characterise the control law. Each of these $n_{\mathbb{U}}$ regions is given by the union of all regions $\mathcal{R}_i$ that correspond to vertices v_i having the same first element. The appropriate extension of Theorem 13.4.1 is presented below. This result follows directly from Theorem 13.4.1.

Corollary 13.5.1 (State Space Partition) *Let the constraint set $\mathbb{U}$ be given in (13.2) and consider the partition into equivalence classes*

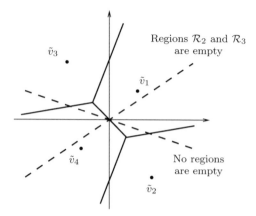

Figure 13.1. Partition of the transformed input sequence space with $N = 2$ (solid lines) and two examples of $-H^{-\mathrm{T}/2} F x$, $x \in \mathbb{R}$ (dashed lines).

$$\mathbb{U}^N = \bigcup_{i=1,\ldots,n_{\mathbb{U}}} \mathbb{U}_i^N,$$

where

$$\mathbb{U}_i^N \triangleq \{v \in \mathbb{U}^N : [1 \ 0 \cdots 0] v = s_i\}.$$

Then, the receding horizon control law (13.20) is equivalent to

$$u^{\mathrm{OPT}}(x) = s_i, \quad \text{if } x \in \mathcal{X}_i, \quad i = 1, 2, \ldots n_{\mathbb{U}}. \tag{13.23}$$

Here, the polyhedra $\mathcal{X}_i$ are given by

$$\mathcal{X}_i \triangleq \bigcup_{j : \, v_j \in \mathbb{U}_i^N} \mathcal{X}_{ij},$$

where

$$\mathcal{X}_{ij} \triangleq \left\{z \in \mathbb{R}^n : 2(v_j - v_k)^{\mathrm{T}} F z \leq \|v_k\|_H^2 - \|v_j\|_H^2 \text{ for all } v_k \in \mathbb{U}^N \backslash \mathbb{U}_i^N\right\}$$
$$\backslash \left\{z \in \mathbb{R}^n : \text{there exists } v_k \in \mathbb{U}^N \backslash \mathbb{U}_i^N, \ k < j, \text{ such that}\right.$$
$$\left. 2(v_j - v_k)^{\mathrm{T}} F z = \|v_k\|_H^2 - \|v_j\|_H^2\right\}.$$

It should be emphasised that this description requires evaluation of less inequalities than the direct calculation of the union of all $\mathcal{R}_j$ (as defined in (13.22)) with $v_j \in \mathbb{U}_i^N$, since inequalities corresponding to *internal borders* are not evaluated. Moreover, the definition of $\mathcal{X}_i$ (and of $\mathcal{R}_i$) can be simplified if the *ambiguity problem* is not addressed.

The state space partition obtained can be calculated off-line so that on-line computational burden can be reduced. The partition induced is related to the partition that characterises the interval-constrained case, as detailed in the following section.

13.6 Relation to Interval Constraints

If in the setup described above, the input is not constrained to belong to a finite set $\mathbb{U}$, but instead, needs to satisfy the interval-type constraint

$$-\Delta \le u_k \le \Delta \quad \text{for all } k, \tag{13.24}$$

where $\Delta \in \mathbb{R}$ is fixed, then a convex optimisation problem is obtained.

For this case, as shown in Chapter 6, the control law can be finitely parameterised and calculated off-line. The state space is partitioned into polytopes in which the receding horizon controller is piecewise affine in the state.

The partition of the $\tilde{\mathbf{u}}$-space using the transformation (13.15) and a geometric argument similar to the one used in the proof of Theorem 13.3.1 is sketched in Figure 13.2 for the case $N = 2$ and the restriction (13.24). In Figure 13.2, the polytope Θ_0 is obtained by applying the transformation (13.15) to the region in which the constraints are not active. It is the *allowed set*. The regions denoted as Θ_{si} are adjacent to a face of Θ_0.

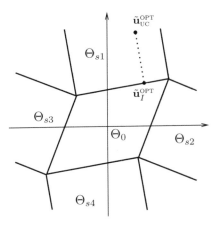

Figure 13.2. Partitions of $\tilde{\mathbf{u}}$-space with the interval constraint set (13.24).

As shown in Section 13.3, in the finite set-constrained case, the constrained solution $\mathbf{u}^{\text{OPT}}$ is related to $\tilde{\mathbf{u}}_{\text{UC}}^{\text{OPT}}$ by means of a nearest neighbour quantiser as stated in (13.13). (For ease of notation, the dependence on x of this and other vectors to follow has not been explicitly included.) A similar result holds in the interval-constrained case. Given (13.24), the constrained optimiser, denoted here as $\tilde{\mathbf{u}}_I^{\text{OPT}}$, is related to $\tilde{\mathbf{u}}_{\text{UC}}^{\text{OPT}}$ via a minimum Euclidean distance projection to the allowed set. This result was shown earlier in Chapter 6 and can be summarised as follows:

Remark 13.6.1. (Projection in the Interval-Constrained Case) If $\tilde{\mathbf{u}}_{\text{UC}}^{\text{OPT}}$ lies inside of Θ_0, then it holds that $\tilde{\mathbf{u}}_I^{\text{OPT}} = \tilde{\mathbf{u}}_{\text{UC}}^{\text{OPT}}$. On the other hand, if $\tilde{\mathbf{u}}_{\text{UC}}^{\text{OPT}} \notin$

Θ_0, then the constrained solution is obtained by the minimum Euclidean distance projection (see, for example, (10.15) in Chapter 10) onto the border of Θ_0. In particular, if the unconstrained solution lies in any of the regions adjacent to a face of Θ_0, then $\tilde{\mathbf{u}}_I^{\mathrm{OPT}}$ is obtained by an orthogonal projection onto the nearest face (as illustrated in Figure 13.2 by means of a dotted line).

○

As a consequence of the foregoing discussion, we obtain the following theorem, which establishes a connection between the partition of the $\tilde{\mathbf{u}}$-space in the interval-constrained case and the Voronoi partition of the quantiser defining the solution with a special finite set constraint.

Theorem 13.6.1 (Relationship Between the Binary and Interval-Constrained Cases) *Consider the binary constraint set $\mathbb{U} = \{-\Delta, \Delta\}$ and the region outside of Θ_0. Then, the borders of the Voronoi partition of the quantiser in (13.13) are parallel and equidistant to the borders of those regions of the interval-constrained case, which are adjacent to an $(N-1)$-dimensional face of Θ_0. (These regions are denoted in Figure 13.2 as Θ_{si}.)*

Proof. From (13.18) it follows that the solution (13.13) can be stated alternatively as $\mathbf{u}^{\mathrm{OPT}}(x) = H^{-1/2} q_{\tilde{\mathbb{U}}^N}(\tilde{\mathbf{u}}_{\mathrm{UC}}^{\mathrm{OPT}}(x))$. The result is a consequence of the fact that, as can be seen in Figure 13.3, the borders of the regions Θ_{si} are formed by orthogonal projections to $\tilde{v}_i$, and that the Voronoi partition is formed by equidistant hyperplanes, which are also orthogonal to the corresponding $(N-1)$-dimensional face of Θ_0. $\qquad\square$

This result is illustrated in Figure 13.3, where the Voronoi partition is depicted via dashed lines. Due to linearity of the mapping $H^{-\mathrm{T}/2} F$ in (13.21), the induced partition of the state space given the constraint (13.24) and the partition defined in (13.22) are similarly related.

13.7 Stability

In Chapters 4 and 5, we found that, for the case of interval-type constraints, one could utilise the value function of the optimal control problems as a candidate Lyapunov function to establish stability. The situation in the finite alphabet case is more difficult. Indeed, asymptotic stability is, in general, too strong a requirement for the finite alphabet problem. In this section we explore various stability issues associated with this case.

The closed loop that results when controlling the plant (13.1) with the receding horizon law (13.23) is described via the following piecewise-affine map, which follows from Corollary 13.5.1:

$$
\begin{aligned}
x_{k+1} &= g(x_k), \\
g(x_k) &\triangleq A x_k + B s_i, \quad \text{if } x_k \in \mathcal{X}_i, \quad i = 1, 2, \ldots n_{\mathbb{U}}.
\end{aligned}
\tag{13.25}
$$

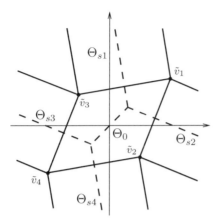

Figure 13.3. Relationship between partitions induced by binary constraints (dashed line) and interval-type constraints (solid line).

Piecewise-affine maps are *mixed mappings* and also form a special class of hybrid systems with underlying discrete time dynamics (see, for example, Bemporad, Ferrari-Trecate and Morari (2000), Heemels, De Schutter and Bemporad (2001) and the references therein). They also appear in connection with some signal processing problems, namely arithmetic overflow of digital filters (Chua and Lin 1988) and $\Sigma\Delta$-modulators (Feely 1997, Norsworthy, Schreier and Temes 1997) and have also been studied in a more theoretical mathematical context (see, for example, Adler, Kitchens and Tresser 2001, Wu and Chua 1994).

Since there exist fundamental differences in the dynamic behaviour of (13.25), depending on whether the plant (13.1) is open loop stable or unstable, that is, on whether the matrix A is Hurwitz or not, it is convenient to divide the discussion that follows accordingly.

13.7.1 Stable Plants

If the plant (13.1) is stable, then its states are always bounded when controlled by means of any finite set constraint law. This follows directly from the fact that $\mathbb{U}$ is always bounded.

Moreover, it can also be shown that all state trajectories[3] of (13.25) either converge towards a fixed point or towards a limit cycle (see, for example, Wu and Chua 1994, Ramadge 1990).

The properties stated so far apply to general systems described by (13.25), where $\mathcal{X}_i$ defines any partition of the state space. In contrast, the following theorem is more specific. It utilises the fact that the control law $u^{\mathrm{OPT}}(x)$ is optimising in a receding horizon sense in order to establish a stronger result.

---

[3] Exceptions are limited to trajectories that emanate from initial conditions belonging to a zero-measure set.

Theorem 13.7.1 (Asymptotic Stability) *If A is Hurwitz, $0 \in \mathbb{U}$ and $P = P^{\mathrm{T}} > 0$ satisfies the Lyapunov equation $A^{\mathrm{T}}PA + Q = P$, then the closed loop (13.25) has a globally attractive, locally asymptotically stable, equilibrium point at the origin.*

Proof. The proof follows standard techniques used in the receding horizon control framework as summarised in Chapter 4 (see also Section 5.6.1 in Chapter 5). In particular, we will use Theorem 4.4.2 of Chapter 4. We choose $\mathbb{X}_f = \mathbb{R}^n$ and $\mathcal{K}_f(x) = 0$ for all $x \in \mathbb{X}_f$. Clearly conditions **B1**, **B3**, **B4** and **B5** hold and $\mathbb{S}_N = \mathbb{R}^n$.

Direct calculation yields that $F(x) = x^{\mathrm{T}}Px$ satisfies

$$
\begin{aligned}
F(f(x, \mathcal{K}_f(x))) - F(x) + L(x, \mathcal{K}_f(x)) &= (Ax + B\mathcal{K}_f(x))^{\mathrm{T}}P(Ax + B\mathcal{K}_f(x)) \\
&\quad - x^{\mathrm{T}}Px + x^{\mathrm{T}}Qx + (\mathcal{K}_f(x))^{\mathrm{T}}R\mathcal{K}_f(x) \\
&= x^{\mathrm{T}}(A^{\mathrm{T}}PA + Q - P)x \\
&= 0 \quad \text{for all } x \in \mathbb{X}_f,
\end{aligned}
$$

so that condition **B2** is also satisfied. Global attractivity of the origin then follows from Theorem 4.4.2.

Next, note that there exists a region containing an open neighbourhood of the origin where $u^{\mathrm{OPT}}(x) = 0$, hence local asymptotic stability of the origin follows since A is Hurwitz. □

As can be seen, if the conditions of this theorem are satisfied, then the receding horizon law (13.6) ensures that the origin is not only a fixed point, but also that it has region of attraction $\mathbb{R}^n$.

It should be emphasised here that, in a similar manner, it can be shown that a finite alphabet control law can steer the plant state asymptotically to any point $x^{\star}$, such that there exist $s_i \in \mathbb{U}$ that allow one to write $x^{\star} = (I - A)^{-1}Bs_i$.

13.7.2 Unstable Plants

In case of strictly unstable plants (13.1), the situation becomes more involved. Although fixed points and periodic sequences may be admissible, they are basically nonattractive.

Moreover, with control signals that are limited in magnitude, as is the case with finite set constraints (and also with interval constraints), there always exists an unbounded region, such that initial states contained in it lead to unbounded state trajectories. This does not mean that every state trajectory of (13.25) is unbounded. Despite the fact that the unstable open loop dynamics (as expressed in A) makes neighbouring trajectories diverge locally, under certain circumstances the control law may keep the state trajectory bounded.

As a consequence of the highly nonlinear (non-Lipschitz) dynamics resulting from the quantiser defining the control law (13.13), in the bounded case

the resulting closed loop trajectories may be quite complex. In order to analyse them without exploring their fine geometrical structure, it is useful that
we relax the usual notion of asymptotic stability of the origin. A more useful
characterisation here is that of *ultimate boundedness* of state trajectories. This
notion refers to convergence towards a bounded region of $\mathbb{R}^n$, instead of to a
point or a specific periodic orbit (Blanchini 1999). (Ultimate boundedness has
also been considered in Li and Soh (1999), and by several other authors in the
context of *practical stability*.) We refer the reader to the literature, especially
Quevedo, De Doná and Goodwin (2002), where these more detailed issues are
discussed and analysed for the case of finite alphabet receding horizon control
of unstable open loop plants.

13.8 Examples

13.8.1 Open Loop Stable Plant

Consider an open loop stable plant described by

$$x_{k+1} = \begin{bmatrix} 0.1 & 2 \\ 0 & 0.8 \end{bmatrix} x_k + \begin{bmatrix} 0.1 \\ 0.1 \end{bmatrix} u_k, \tag{13.26}$$

and the binary constraint set $\mathbb{U} = \{-1, 1\}$. The receding horizon control law
with $R = 0$ and

$$P = Q = \begin{bmatrix} 1 & 0 \\ 0 & 1 \end{bmatrix}, \tag{13.27}$$

partitions the state space into the regions depicted in Figure 13.4, for constraint horizons $N = 2$ and $N = 3$. In this figure x_k^1 and x_k^2 denote the two
components of the state vector x_k.

The receding horizon control law is

$$u^{\text{OPT}}(x) = \begin{cases} -1 & \text{if } x \in \mathcal{X}_1, \\ 1 & \text{if } x \in \mathcal{X}_2, \end{cases}$$

where

$$\mathcal{X}_1 = \bigcup_{i=2^{N-1}+1, 2^{N-1}+2, \dots, 2^N} \mathcal{R}_i, \qquad \mathcal{X}_2 = \bigcup_{i=1, 2, \dots, 2^{N-1}} \mathcal{R}_i.$$

13.8.2 Open Loop Unstable Plant

We next analyse a situation when the plant is open loop unstable. For that
purpose, consider

$$x_{k+1} = \begin{bmatrix} 1.02 & 2 \\ 0 & 1.05 \end{bmatrix} x_k + \begin{bmatrix} 0.1 \\ 0.1 \end{bmatrix} u_k, \tag{13.28}$$

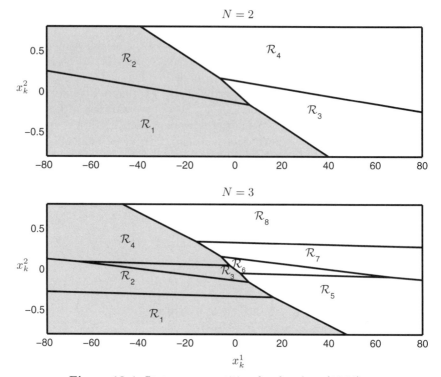

Figure 13.4. State space partition for the plant (13.26).

controlled with a receding horizon controller with parameters $\mathbb{U}$, P, Q and R as in Example 13.8.1 above. The constraint horizon is chosen to be $N = 2$.

Figure 13.5 illustrates the induced state space partition and a closed loop trajectory, which starts at $x = [-10 \quad 0]^{\mathrm{T}}$. As can be seen, due to the limited control action available, the trajectory becomes unbounded.

The situation is entirely different when the initial condition is chosen as $x = [0.7 \quad 0.2]^{\mathrm{T}}$. As depicted in Figure 13.6, the closed loop trajectory now converges to a bounded region, which contains the origin in its interior. Within that region, the behaviour is not periodic, but appears to be random, despite the fact that the system is deterministic. Neighbouring trajectories diverge due to the action of the unstable poles of the plant. However, the control law manifests itself by maintaining the plant state *ultimately bounded*. As already mentioned in Section 13.7.2, the *complex* dynamic behaviour obtained is a consequence of the insertion of a nonsmooth nonlinearity in the feedback loop.

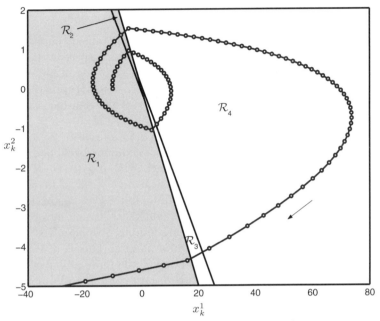

Figure 13.5. State trajectories of the controlled plant (13.28) with initial condition $x = [-10 \quad 0]^{\mathrm{T}}$.

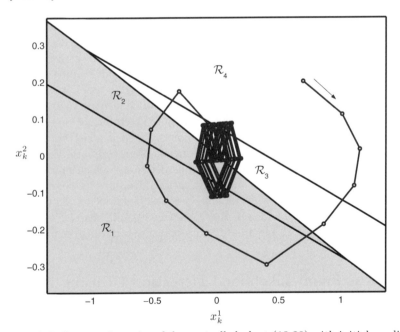

Figure 13.6. State trajectories of the controlled plant (13.28) with initial condition $x = [0.7 \quad 0.2]^{\mathrm{T}}$.

13.9 Finite Alphabet Estimation

As we have seen in Chapter 9, the problems of constrained control and estimation are very similar (differing essentially only with respect to the nature of
the boundary conditions). Here we give a brief description of finite alphabet
estimation. To fix ideas, we refer to the specific problem of estimating a signal
drawn from a given finite alphabet that has been transmitted over a noisy
dispersive communication channel.

   This problem, which is commonly referred to as one of *channel equalisation*, can be formulated as a fixed-delay maximum likelihood detection
problem. The resultant detector estimates each symbol based upon the entire
sequence received to a point in time and hence constitutes, in principle, a
growing memory structure. In the case of finite impulse response [FIR] channels, the Viterbi algorithm can be used for solving the resultant optimisation
problem. However, for more general infinite impulse response [IIR] channels,
the complexity of the Viterbi Algorithm is infinite. This is a direct consequence of the requirement to take into account the finite alphabet nature of
the transmitted signal, which makes this a hard combinatorial optimisation
problem.

   In order to address this problem, various simplified detectors of fixed memory and complexity have been proposed. The simplest such scheme is the decision feedback equaliser [DFE] (Qureshi 1985), which is a symbol-by-symbol detector. It basically corresponds to the scheme depicted in Figure 1.10 in Chapter 1. It is a feedback loop comprising linear filters and a scalar quantiser. The
DFE is extended and outperformed by more complex multistep detector structures, which estimate channel inputs based upon blocks of sampled outputs
of fixed size (see, for example, Williamson, Kennedy and Pulford 1992, Duel-
Hallen and Heegard 1989).

   In these schemes, decision feedback (also called *genie-aided feedback*) is
used to overcome the growing memory problem. The information contained in
the sampled outputs received before the block where the constraints are taken
into account explicitly is summarised by means of an estimate of the channel
state. This estimate is based upon previous decisions, which are assumed to
be correct. Not taking into account that various decisions may contain errors
can lead to error propagation problems (see also Cantoni and Butler 1976).

   Here we show how the idea of the "benevolent genie" can be extended by
means of an *a priori* state estimate and a measure of its degree of belief.

13.10 Maximum Likelihood Detection Utilising An A Priori State Estimate

Consider a linear channel (which may include a whitening matched filter and
any other pre-filter) with scalar input u_k drawn from a finite alphabet $\mathbb{U}$. The
channel output y_k is scalar and is assumed to be perturbed by zero-mean

additive white Gaussian noise n_k of variance r, denoted by $n_k \sim N(0, r)$, yielding the state space model

$$
\begin{aligned}
x_{k+1} &= Ax_k + Bu_k, \\
y_k &= Cx_k + Du_k + n_k,
\end{aligned}
\tag{13.29}
$$

where $x_k \in \mathbb{R}^n$. The above model may equivalently be expressed in transfer function form as

$$
y_k = H(\rho)u_k + n_k, \quad H(\rho) = D + C(\rho I - A)^{-1}B = h_0 + \sum_{i=1}^{\infty} h_i \rho^{-i},
$$

where[4]

$$
h_0 = D, \quad h_i = CA^{i-1}B, \quad i = 1, 2, \dots. \tag{13.30}
$$

We incorporate an a priori state estimate into the problem formulation. This is achieved as follows:

As described in Section 9.9 of Chapter 9, we fix integers $L_1 \geq 0$, $L_2 \geq 1$ and suppose, for the moment, that

$$
x_{k-L_1} \sim N(z_{k-L_1}, P), \tag{13.31}
$$

that is, z_{k-L_1} is a given a priori estimate for x_{k-L_1} which has a Gaussian distribution. The matrix P^{-1} reflects the degree of belief in this a priori state estimate. Absence of prior knowledge of x_{k-L_1} can be accommodated by using $P^{-1} = 0$, and decision feedback is achieved by taking $P = 0$, which effectively locks x_{k-L_1} at z_{k-L_1}.

Additionally, we define the vectors

$$
\begin{aligned}
\mathbf{u}_k &\triangleq \begin{bmatrix} u_{k-L_1} & u_{k-L_1+1} & \cdots & u_{k+L_2-1} \end{bmatrix}^{\mathrm{T}}, \\
\mathbf{y}_k &\triangleq \begin{bmatrix} y_{k-L_1} & y_{k-L_1+1} & \cdots & y_{k+L_2-1} \end{bmatrix}^{\mathrm{T}}.
\end{aligned}
$$

The vector $\mathbf{y}_k$ gathers time samples of the channel output and $\mathbf{u}_k$ contains channel inputs, which are the decision variables of the estimation problem considered here.

The *maximum a posteriori* [MAP] sequence detector, which at time $t = k$ provides an estimate of $\mathbf{u}_k$ and x_{k-L_1} based upon the received data contained in $\mathbf{y}_k$, maximises the probability density function (see Chapter 9 for further discussion)[5]

$$
p\left(\begin{bmatrix} \mathbf{u}_k \\ x_{k-L_1} \end{bmatrix} \Big| \mathbf{y}_k \right) = \frac{p\left(\mathbf{y}_k \Big| \begin{bmatrix} \mathbf{u}_k \\ x_{k-L_1} \end{bmatrix}\right) p\left(\begin{bmatrix} \mathbf{u}_k \\ x_{k-L_1} \end{bmatrix}\right)}{p(\mathbf{y}_k)}, \tag{13.32}
$$

---

[4] ρ denotes the forward shift operator, $\rho v_k = v_{k+1}$, where $\{v_k\}$ is any sequence.

[5] For ease of notation, in what follows we will denote all (conditional) probability density functions by p. The specific function referred to will be clear from the context.

where we have utilised Bayes' rule.

Note that only the numerator of this expression influences the maximisation. Assuming that $\mathbf{u}_k$ and x_{k-L_1} are independent (which is a consequence of (13.29) if u_k is white), it follows that

$$p\left(\begin{bmatrix} \mathbf{u}_k \\ x_{k-L_1} \end{bmatrix}\right) = p\left(x_{k-L_1}\right) p\left(\mathbf{u}_k\right).$$

Hence, if all finite alphabet-constrained symbol sequences $\mathbf{u}_k$ are equally likely (an assumption that we make in what follows), then the MAP detector that maximises (13.32) is equivalent to the following *maximum likelihood* sequence detector

$$\begin{bmatrix} \hat{\mathbf{u}}_k \\ \hat{x}_{k-L_1} \end{bmatrix} \triangleq \underset{\mathbf{u}_k, x_{k-L_1}}{\arg\max} \left\{ p\left(\mathbf{y}_k \middle| \begin{bmatrix} \mathbf{u}_k \\ x_{k-L_1} \end{bmatrix}\right) p\left(x_{k-L_1}\right) \right\}. \tag{13.33}$$

Here,

$$\hat{\mathbf{u}}_k \triangleq \begin{bmatrix} \hat{u}_{k-L_1} & \hat{u}_{k-L_1+1} & \cdots & \hat{u}_k & \cdots & \hat{u}_{k+L_2-1} \end{bmatrix}^\mathsf{T}, \tag{13.34}$$

and $\mathbf{u}_k$ needs to satisfy the constraint

$$\mathbf{u}_k \in \mathbb{U}^N, \quad \mathbb{U}^N \triangleq \mathbb{U} \times \cdots \times \mathbb{U}, \quad N \triangleq L_1 + L_2, \tag{13.35}$$

in accordance with the restriction $u_k \in \mathbb{U}$. Our working assumption (see (13.31)) is that the initial channel state x_{k-L_1} has a Gaussian probability density function

$$p\left(x_{k-L_1}\right) = \frac{1}{(2\pi)^{n/2}(\det P)^{1/2}} \exp\left\{ \frac{-\|x_{k-L_1} - z_{k-L_1}\|_{P^{-1}}^2}{2} \right\}. \tag{13.36}$$

In order to derive analytic expressions for the other probability density functions in (13.33), we rewrite the channel model (13.29) at time instants $t = k - L_1, k - L_1 + 1, \ldots, k + L_2 - 1$ in block form as

$$\mathbf{y}_k = \Psi \mathbf{u}_k + \Gamma x_{k-L_1} + \mathbf{n}_k.$$

Here,

$$\mathbf{n}_k \triangleq \begin{bmatrix} n_{k-L_1} \\ n_{k-L_1+1} \\ \vdots \\ n_{k+L_2-1} \end{bmatrix}, \quad \Gamma \triangleq \begin{bmatrix} C \\ CA \\ \vdots \\ CA^{N-1} \end{bmatrix}, \quad \Psi \triangleq \begin{bmatrix} h_0 & 0 & \ldots & 0 \\ h_1 & h_0 & \ddots & \vdots \\ \vdots & \ddots & \ddots & 0 \\ h_{N-1} & \ldots & h_1 & h_0 \end{bmatrix}.$$

The entries of Ψ obey (13.30), that is, its columns contain truncated impulse responses of the model (13.29).

Since the noise n_k is assumed Gaussian with variance r, it follows that

$$p\left(\mathbf{y}_k \middle| \begin{bmatrix} \mathbf{u}_k \\ x_{k-L_1} \end{bmatrix}\right) = \frac{1}{(2\pi)^{N/2}(\det R)^{1/2}} \exp\left\{\frac{-\|\mathbf{y}_k - \Psi\mathbf{u}_k - \Gamma x_{k-L_1}\|_{R^{-1}}^2}{2}\right\},$$

$$(13.37)$$

where the matrix $R \triangleq \text{diag}\{r, \ldots, r\} \in \mathbb{R}^{N \times N}$.

After substituting expressions (13.36) and (13.37) into (13.33) and applying the natural logarithm, one obtains the sequence detector

$$\begin{bmatrix} \hat{\mathbf{u}}_k \\ \hat{x}_{k-L_1} \end{bmatrix} = \arg\min_{\mathbf{u}_k, x_{k-L_1}} V(\mathbf{u}_k, x_{k-L_1}), \quad (13.38)$$

subject to the constraint (13.35). In (13.38), the objective function V is defined as

$$V(\mathbf{u}_k, x_{k-L_1}) \triangleq \|x_{k-L_1} - z_{k-L_1}\|_{P^{-1}}^2 + \|\mathbf{y}_k - \Psi\mathbf{u}_k - \Gamma x_{k-L_1}\|_{R^{-1}}^2$$

$$= \|x_{k-L_1} - z_{k-L_1}\|_{P^{-1}}^2 + r^{-1}\sum_{j=k-L_1}^{k+L_2-1}(y_j - C\check{x}_j - Du_j)^2,$$

$$(13.39)$$

and the vectors $\check{x}_j$ denote predictions of the channel states x_j. They satisfy (13.29), that is,

$$\check{x}_{j+1} = A\check{x}_j + Bu_j \quad \text{for } j = k - L_1, \ldots, k + L_2 - 1,$$
$$\check{x}_{k-L_1} = x_{k-L_1}.$$

$$(13.40)$$

Remark 13.10.1. (Notation) Since $\hat{\mathbf{u}}_k$ and $\hat{x}_{k-L_1}$ in (13.33) are calculated using data up to time $t = k + L_2 - 1$, they could perhaps be more insightfully denoted as $\hat{\mathbf{u}}_{k|k+L_2-1}$ and $\hat{x}_{k-L_1|k+L_2-1}$, respectively (see Chapter 9). However, in order to keep the notation simple, we will here avoid double indexing, in anticipation that the context will always allow for correct interpretation.

○

As a consequence of considering the joint probability density function (13.32), the objective function (13.39) includes a term which allows one to obtain an a posteriori state estimate $\hat{x}_{k-L_1}$ which differs from the a priori estimate z_{k-L_1} as permitted by the confidence matrix P^{-1}.

13.11 Information Propagation

Having set up the fixed horizon estimator as the finite alphabet optimiser (13.38)–(13.40), in this section we show how this information can be utilised as part of a moving horizon scheme. Here we essentially follow the methodology outlined in Section 9.9 of Chapter 9.

13.11.1 Moving Horizon Implementation

Minimisation of the objective function V in (13.39) yields the entire optimising sequence $\hat{\mathbf{u}}_k$ defined in (13.38). However, following our usual procedure, we will utilise a moving horizon approach in which only the *present* value[6]

$$\hat{u}_k^{\mathrm{OPT}} \triangleq \begin{bmatrix} 0_{L_1} & 1 & 0_{L_2-1} \end{bmatrix} \hat{\mathbf{u}}_k, \tag{13.41}$$

will be delivered at the output of the detector.

At the next time instant the optimisation is repeated, providing $\hat{u}_{k+1}^{\mathrm{OPT}}$ and so on. Thus, the data window "slides" (or moves) forward in time. The scheme previews $L_2 - 1$ samples, hence has a decision-delay of $L_2 - 1$ time units.

The *window length* $N = L_1 + L_2$ fixes the complexity of the computations needed in order to minimise (13.39). It is intuitively clear that good performance of the detector can be ensured if N is sufficiently large. However, in practice, there is a strong incentive to use small values for L_1 and L_2, since large values give rise to high complexity in the associated computations to be performed at each time step.

13.11.2 Decision-Directed Feedback

The provision of an a priori estimate, z_{k-L_1}, together with an associated degree of belief via the term $\|x_{k-L_1} - z_{k-L_1}\|_{P^{-1}}^2$ in (13.39) provides a means of propagating the information contained in the data received before $t = k - L_1$. Consequently, an *information horizon* of growing length is effectively obtained in which the computational effort is fixed by means of the window length N.

One possible approach to choose the a priori state estimate is as follows: Each optimisation step provides estimates for the channel state and input sequence (see (13.38)). These decisions can be re-utilised in order to formulate a priori estimates for the channel state x_k. We propose that the estimates be propagated in blocks according to[7]

$$z_k = A^N \hat{x}_{k-N} + M \hat{\mathbf{u}}_{k-L_2},$$

where $M \triangleq \begin{bmatrix} A^{N-1}B & A^{N-2}B & \dots & AB & B \end{bmatrix}$. In this way, the estimate obtained in the previous block is rolled forward. Indeed, in order to operate in a moving horizon manner, it is necessary to store N a priori estimates. This is depicted graphically in Figure 13.7.

13.11.3 The Matrix P as a Design Parameter

Since channel states depend on the finite alphabet input, one may well question the assumption made in Section 13.10 that x_{k-L_1} is Gaussian. (This

---

[6] The row vector $0_m \in \mathbb{R}^{1 \times m}$ contains only zeros.

[7] Since z_k is based upon channel outputs up to time $k - 1$, it could alternatively be denoted as $\hat{x}_{k|k-1}$; see also Remark 13.10.1.

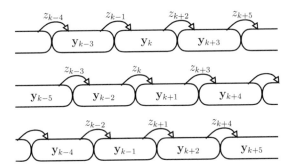

Figure 13.7. Information propagation with parameters $L_1 = 1$ and $L_2 = 2$.

situation is similar to that of other detectors that utilise Gaussian approximations; (see, for example, Lawrence and Kaufman 1971, Thielecke 1997, Baccarelli, Fasano and Zucchi 2000). However, we could always use this structure by interpreting the matrix P in (13.39) as a design parameter.

As a guide for tuning P, we recall that in the unconstrained case, where the channel input and initial state are Gaussian, that is, $u_k \sim N(0, Q)$ and $x_0 \sim N(\mu_0, P_0)$, the Kalman filter provides the minimum variance estimate for x_{k-L_1} (see, for example, Anderson and Moore 1979). Its covariance matrix P_{k-L_1} obeys the Riccati difference equation (see Chapter 9),

$$P_{k+1} = AP_k A^{\mathrm{T}} - K_k(CP_k C^{\mathrm{T}} + r + DQD^{\mathrm{T}})K_k^{\mathrm{T}} + BQB^{\mathrm{T}}, \qquad k \geq 0, \quad (13.42)$$

where $K_k \triangleq (AP_k C^{\mathrm{T}} + BQD^{\mathrm{T}})(CP_k C^{\mathrm{T}} + r + DQD^{\mathrm{T}})^{-1}$.

A further simplification occurs if we replace the recursion (13.42) by its steady state equivalent. In particular, it is well-known (Goodwin and Sin 1984) that, under reasonable assumptions, P_k converges to a steady state value P as $k \to \infty$. The matrix P satisfies the following algebraic Riccati equation:

$$P = APA^{\mathrm{T}} - K(CPC^{\mathrm{T}} + r + DQD^{\mathrm{T}})K^{\mathrm{T}} + BQB^{\mathrm{T}}, \qquad (13.43)$$

where $K = (APC^{\mathrm{T}} + BQD^{\mathrm{T}})(CPC^{\mathrm{T}} + r + DQD^{\mathrm{T}})^{-1}$. Of course, the Gaussian assumption on u_k is not valid in the constrained case. However, the choice (13.43) may still provide good performance. Alternatively, one may simply use P as a design parameter and test different choices via simulation studies.

13.12 Closed Loop Implementation of the Finite Alphabet Estimator

Here we follow similar arguments to those used with respect to finite alphabet control in Section 13.3 to obtain a closed form expression for the solution to the finite alphabet estimation problem. This closed form expression utilises a vector quantiser as defined earlier in Definition 13.3.1.

For general recursive channels, it is useful to assume that, whilst the input is always constrained to a finite alphabet, the channel state x_k in (13.29) is left unconstrained. In this case, the optimisers (13.38) are characterised as follows:

Lemma 13.12.1 (Closed Form Solution) *The optimisers corresponding to (13.38) given the constraint $\mathbf{u}_k \in \mathbb{U}^N$ are given by*

$$\hat{\mathbf{u}}_k = \Omega^{-1/2} q_{\tilde{\mathbb{U}}^N}\left(\Omega^{-1/2}(\Lambda_1 \mathbf{y}_k - \Lambda_2 z_{k-L_1})\right), \tag{13.44}$$

$$\hat{x}_{k-L_1} = \Upsilon\left(P^{-1} z_{k-L_1} + \Gamma^{\mathsf{T}} R^{-1} \mathbf{y}_k - \Gamma^{\mathsf{T}} R^{-1} \Psi \hat{\mathbf{u}}_k\right), \tag{13.45}$$

where

$$\begin{aligned}
\Omega &= \Psi^{\mathsf{T}}\left(R^{-1} - R^{-1}\Gamma\Upsilon\Gamma^{\mathsf{T}} R^{-1}\right)\Psi, \qquad \Omega^{\mathsf{T}/2}\Omega^{1/2} = \Omega, \\
\Upsilon &= (P^{-1} + \Gamma^{\mathsf{T}} R^{-1}\Gamma)^{-1}, \\
\Lambda_1 &= \Psi^{\mathsf{T}}\left(R^{-1} - R^{-1}\Gamma\Upsilon\Gamma^{\mathsf{T}} R^{-1}\right), \\
\Lambda_2 &= \Psi^{\mathsf{T}} R^{-1}\Gamma\Upsilon P^{-1}.
\end{aligned} \tag{13.46}$$

The nonlinear function $q_{\tilde{\mathbb{U}}^N}(\cdot)$ is the nearest neighbour vector quantiser described in Definition 13.3.1. The image of this mapping is the set

$$\tilde{\mathbb{U}}^N = \Omega^{1/2}\mathbb{U}^N \triangleq \{\tilde{v}_1, \tilde{v}_2, \dots, \tilde{v}_r\} \subset \mathbb{R}^N, \text{ with } \tilde{v}_i = \Omega^{1/2} v_i, \ v_i \in \mathbb{U}^N. \tag{13.47}$$

Proof. The objective function (13.39) can be expanded as

$$\begin{aligned}
V(\mathbf{u}_k, x_{k-L_1}) = {} & \|x_{k-L_1}\|_{\Upsilon^{-1}}^2 + \|z_{k-L_1}\|_{P^{-1}}^2 + \|\mathbf{y}_k\|_{R^{-1}}^2 \\
& + \|\mathbf{u}_k\|_{\Psi^{\mathsf{T}} R^{-1}\Psi}^2 + \mathbf{u}_k^{\mathsf{T}}\Psi^{\mathsf{T}} R^{-1}\Gamma x_{k-L_1} + x_{k-L_1}^{\mathsf{T}}\Gamma^{\mathsf{T}} R^{-1}\Psi \mathbf{u}_k \\
& - 2\left[\mathbf{u}_k^{\mathsf{T}}\Psi^{\mathsf{T}} R^{-1}\mathbf{y}_k + x_{k-L_1}^{\mathsf{T}}\left(P^{-1} z_{k-L_1} + \Gamma^{\mathsf{T}} R^{-1}\mathbf{y}_k\right)\right], \tag{13.48}
\end{aligned}$$

with Υ defined in (13.46). This expression can be written as

$$\begin{aligned}
V(\mathbf{u}_k, x_{k-L_1}) = {} & \alpha(\mathbf{u}_k, \mathbf{y}_k, z_{k-L_1}) + \|x_{k-L_1}\|_{\Upsilon^{-1}}^2 \\
& - 2 x_{k-L_1}^{\mathsf{T}}\left[P^{-1} z_{k-L_1} + \Gamma^{\mathsf{T}} R^{-1}\mathbf{y}_k - \Gamma^{\mathsf{T}} R^{-1}\Psi \mathbf{u}_k\right],
\end{aligned}$$

where $\alpha(\mathbf{u}_k, \mathbf{y}_k, z_{k-L_1})$ does not depend upon x_{k-L_1}.

Since x_{k-L_1} is assumed unconstrained, it follows that, for every fixed value of $\mathbf{u}_k$, the objective function is minimised by means of $x_{\mathrm{UC}}^{\mathrm{OPT}} = \Upsilon\left(P^{-1} z_{k-L_1} + \Gamma^{\mathsf{T}} R^{-1}\mathbf{y}_k - \Gamma^{\mathsf{T}} R^{-1}\Psi \mathbf{u}_k\right)$ from where (13.45) follows.

In order to obtain the constrained optimiser $\hat{\mathbf{u}}_k \in \mathbb{U}^N$, observe that

$$\hat{\mathbf{u}}_k = \arg\min_{\mathbf{u}_k \in \mathbb{U}^N} J(\mathbf{u}_k), \tag{13.49}$$

where $J(\mathbf{u}_k) \triangleq V(\mathbf{u}_k, x_{\mathrm{UC}}^{\mathrm{OPT}})$. Substitution of (13.45) into (13.48) yields

$$J(\mathbf{u}_k) = \beta(\mathbf{y}_k, z_{k-L_1}) + \mathbf{u}_k^\mathrm{T}\Omega\mathbf{u}_k - 2\mathbf{u}_k^\mathrm{T}(\Lambda_1\mathbf{y}_k - \Lambda_2 z_{k-L_1}), \qquad (13.50)$$

where Ω, Λ_1 and Λ_2 are defined in (13.46) and $\beta(\mathbf{y}_k, z_{k-L_1})$ does not depend upon $\mathbf{u}_k$.

As in the proof of Theorem 13.3.1, it is useful to introduce the coordinate transformation $\tilde{\mathbf{u}}_k \triangleq \Omega^{1/2}\mathbf{u}_k$. This transforms $\mathbb{U}^N$ into $\tilde{\mathbb{U}}^N$ defined in (13.47). Equation (13.50) then allows one to rewrite (13.49) as

$$\hat{\mathbf{u}}_k = \Omega^{-1/2} \arg\min_{\tilde{\mathbf{u}}_k \in \tilde{\mathbb{U}}^N} \tilde{J}(\tilde{\mathbf{u}}_k), \qquad (13.51)$$

with $\tilde{J}(\tilde{\mathbf{u}}_k) \triangleq \tilde{\mathbf{u}}_k^\mathrm{T}\tilde{\mathbf{u}}_k - 2\tilde{\mathbf{u}}_k^\mathrm{T}\Omega^{-1/2}(\Lambda_1\mathbf{y}_k - \Lambda_2 z_{k-L_1})$. The level sets of $\tilde{J}$ are spheres in $\mathbb{R}^N$, centred at $\Omega^{-1/2}(\Lambda_1\mathbf{y}_k - \Lambda_2 z_{k-L_1})$. Hence,

$$\arg\min_{\tilde{\mathbf{u}}_k \in \tilde{\mathbb{U}}^N} \tilde{J}(\tilde{\mathbf{u}}_k) = q_{\tilde{\mathbb{U}}^N}\left(\Omega^{-1/2}(\Lambda_1\mathbf{y}_k - \Lambda_2 z_{k-L_1})\right),$$

which, after substituting into (13.51) yields (13.44).   □

13.13 Example

Consider an FIR channel described by

$$H(z) = 1 + 2z^{-1} + 2z^{-2}. \qquad (13.52)$$

In order to illustrate the performance of the multistep optimal equaliser presented, we carry out simulations of this channel with an input consisting of 10000 independent and equiprobable binary digits drawn from the alphabet $\mathbb{U} = \{-1, 1\}$. The system is affected by Gaussian noise with different variances. The following detection architectures are used: direct quantisation of the channel output, decision feedback equalisation and moving horizon estimation, with parameters $(L_1, L_2) = (1, 2)$ and also with $(L_1, L_2) = (2, 3)$.

Figure 13.8 documents the results. It contains the empirical probabilities of symbol errors obtained at several noise levels. It can be appreciated how moving horizon estimation clearly outperforms direct quantisation of the channel output and also the DFE for this example.

13.14 Conclusions

In this chapter we have presented an approach that addresses control and estimation problems where the decision variables are constrained to belong to a finite alphabet.

It turns out that concepts introduced in previous chapters, namely receding horizon optimisation, exploration of the geometry of the underlying

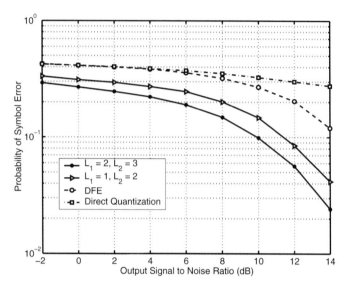

Figure 13.8. Bit error rates of the communication systems simulated.

optimisation problem and information propagation can be readily utilised. It is also apparent that some aspects, such as dynamics and stability of the closed loop, demand for other, more specialised, tools.

Bearing in mind the wide range of applications that can be cast as finite alphabet-constrained control and estimation problems, we invite the reader to apply the acquired expertise in these *nontraditional* areas. The cross-fertilisation of ideas gives new insight and may lead to improved design methodologies in various realms of application.

13.15 Further Reading

General

A more detailed presentation of the ideas outlined in this chapter, including several application studies can be found in Quevedo and Goodwin (2004c). More information on computational complexity of combinatorial optimisation problems can be found in Garey and Johnson (1979). Vector quantisers and Voronoi partitions are described thoroughly in Gersho and Gray (1992), Gray and Neuhoff (1998).

Finite Set Control

In relation to finite set-constrained control problems, Quevedo, Goodwin and De Doná (2004) forms the basis of our presentation. Alternative views are given, for example, in Brockett and Liberzon (2000), Ishii and Francis (2003), Richter and Misawa (2003), Sznaier and Damborg (1989), and Bicchi, Marigo and Piccoli (2002).

Channel Equalisation

Channel equalisation is an important problem in digital communications. It is described in standard textbooks, such as Proakis (1995) and also in the survey papers Qureshi (1985), Tugnait et al. (2000). The presentation given in this chapter follows basically Quevedo, Goodwin and De Doná (2003) and is also related to the multistep estimation schemes described in Williamson et al. (1992), Duel-Hallen and Heegard (1989).

Other Application Areas

Design of networked control systems based upon the ideas presented in this chapter can be found, for example, in Quevedo, Goodwin and Welsh (2003), Goodwin, Haimovich, Quevedo and Welsh (2004), Kiihtelys (2003) and also in Chapter 16. Other interesting references include Bushnell (ed.) (2001), Wong and Brocket (1999), Ishii and Francis (2002), Zhivoglyadov and Middleton (2003), Hristu and Morgansen (1999), Elia and Mitter (2001). The related problem of state estimation with quantised measurements has also been treated in Curry (1970) and Delchamps (1989), Haimovich, Goodwin and Quevedo (2003). (See also Chapter 16.)

Applications to the design of FIR filters with finite set constrained coefficients can be found in Quevedo and Goodwin (2003b), Goodwin, Quevedo and De Doná (2003). These problems have been studied extensively in the signal-processing literature, see, for example, Evangelista (2002), Lim and Parker (1983b), Kodek (1980), Lim and Parker (1983a).

Audio quantisation and A/D conversion can be dealt with in a way similar to finite alphabet-constrained control problems, as detailed in Goodwin, Quevedo and McGrath (2003), Quevedo and Goodwin (2003a), Quevedo and Goodwin (2004b). Other references include Norsworthy et al. (1997), Lipschitz, Vanderkooy and Wannamaker (1991) and the collection Candy and Temes (1992).

Applications of finite set-constrained control in power electronics abound. One particular case resides in the design of the switching signal for switch-mode power supplies, as described in Quevedo and Goodwin (2004a). The book Rashid (1993) is a good introductory level textbook on power electronics.

Part III

Case Studies

14

Rudder Roll Stabilisation of Ships

Contributed by Tristan Perez

14.1 Overview

In this chapter, we present a case study of control system design for rudder-based stabilisers of ships using RHC. The rudder's main function is to correct the heading of a ship; however, depending on the type of ship, the rudder may also be used to produce, or correct, roll motion. Rudder roll stabilisation consists of using rudder-induced roll motion to reduce the roll motion induced by waves. When this technique is employed, an automatic control system is necessary to provide the rudder command based on measurements of ship motion. The RHC formulation provides a unified framework to address many of the difficulties associated with this control system design problem.

14.2 Ship Roll Stabilisation

The success or failure of a ship's mission (fishing, landing a helicopter on deck, serving meals during transit, and so on) is judged by comparing the ship's performance indices with levels that are deemed satisfactory for the particular mission, type of ship and sea environment considered. To accomplish missions successfully, and to improve the performance, marine vehicles are often equipped with sophisticated devices and control systems. Amongst the many different control systems encountered on board a marine vehicle, there is often the so-called *roll stabilising system*, or simply *stabiliser*, whose function is to reduce undesirable roll motion.

Reduced roll motion is important for it can affect the performance of the ship, as indicated in the following considerations.

- Transverse accelerations that occur due to roll interrupt tasks performed by the crew. This increases the amount of time required to complete a mission.
- Roll accelerations may produce cargo damage, for example, on soft loads such as fruit.
- Roll motion increases hull resistance.
- Large roll angles limit crew capability to handle equipment on board, and/or to launch and recover systems.

Several type of stabilisers and stabilisation techniques have been developed and are commonly used: bilge keels, water tanks, fins and rudder (see Sellars and Martin (1992) for a description and benefits of each of these stabilisers).

Amongst the different types of stabilisers, rudder-based stabilisation is a very attractive technique. The reasons for this are that almost every ship has a rudder (thus no extra equipment may be necessary), and also this technique can be used in conjunction with other stabilisers (such as water tanks and fins) to improve performance under various conditions. In this chapter, we will focus on the control system design of rudder-based stabilisers. As we shall see, this design problem is far from trivial.

14.3 A Challenging Control Problem

Using the rudder for simultaneous course keeping and roll reduction is not a simple task. The ability to accomplish this depends on the dynamic characteristics of the ship, and also on the control strategy used to command the rudder. The design of such a control strategy must then be performed so as to best deal with the following issues:

- **Underactuated System**. One control action (rudder force) achieves two control objectives: *roll reduction* and *low heading (yaw) interference*. A key fact that must be understood for the successful application of this technique is that the dynamics associated with the rudder-induced roll motion are faster than the dynamics associated with the rudder-induced yaw motion. This phenomenon depends on the shape of the hull and the location of the rudder and the centre of gravity of the ship. The difference in dynamic response between roll and yaw is characterised by the location of a nonminimum phase zero [NMP] associated with the rudder to roll response. The closer the NMP zero is to the imaginary axis, the faster the roll response to the rudder will be with respect to the response in yaw; and, thus, the better the potential for successful application of rudder roll stabilisation (Roberts 1993). Nevertheless, this effect of the NMP zero will not, per se, guarantee good performance in all conditions; see Perez (2003) for details.

- **Uncertainty**. There are three sources of uncertainty associated with the control problem. First, there is incomplete state information available to implement the control law. Although, complete measurement of the state is possible, the necessary sensors can be very expensive. Second, there are disturbances from the environment (wave-induced motion) that, in principle, cannot be known a priori. Third, in the case of model-based control (such as RHC), there is uncertainty associated with the accuracy of the model.

- **Disturbance Rejection with a Nonminimum Phase System**. As already mentioned, the response of roll due to rudder action presents nonminimum phase dynamics. This imposes fundamental limitations and trade-offs regarding disturbance rejection and achievable roll reduction in different *sailing conditions* (ship speed and heading relative to the waves). The energy of the disturbance shifts in frequency according to the sea state and sailing conditions. Because these changes can be significant, roll amplification can be induced if the controller is not adapted to the changes in the disturbance characteristics (see, for example, Blanke, Adrian, Larsen and Bentsen 2000, Perez 2003). This is a consequence of reducing the sensitivity close to the frequency of the NMP zero.

- **Input constraints**. The rudder action demanded by the controller should satisfy *rate* and *magnitude* constraints. Rate constraints are associated with safety and reliability. By imposing rate constraints on the rudder command, we ensure an adequate lifespan of the hydraulic actuators and avoid their saturation. However, this produces time delays that could lead to stability problems. Magnitude constraints are associated with performance and economy. Large rudder angles induce flow separation (loss of actuation and poor performance), and a significant increase in drag (resistance). Also, it is desirable to reduce the maximum rudder action at higher speeds to reduce the mechanical loads on the rudder and the steering machinery.

- **Output constraints**. Since the rudder affects the ship heading, it may be necessary to include constraints on the maximum heading deviations allowed when the rudder is used to reduce roll.

- **Unstable plant**. The response of yaw to rudder action is marginally unstable: there is an integrator. Indeed, if the rudder is offset from its central position with a step-like command, there will be a ramp-like increase in the heading angle. Some vessels are even directionally unstable, requiring permanent rudder offset to keep a heading.

Based on the above considerations, it is evident that the problem of rudder roll stabilisation of ships is a challenging one and, as such, the chosen control strategy plays an important role in achieving high performance. In what follows, we will further describe the different effects that give rise to the issues mentioned above, define the performance criteria and carry out the design.

14.4 Ship Motion Description and Modelling for Control

The motion of a marine vehicle can be considered in six degrees of freedom. The motion components are resolved into translation components in three directions: *surge*, *sway* and *heave*, and rotation components about three axis: *roll*, *pitch* and *yaw*. Table 14.1 shows the notation used to describe the different motion components.

translation	surge	sway	heave		rotation	roll	pitch	yaw
position	x	y	z		angle	ϕ	θ	ψ
linear rate	u	v	w		angular rate	p	q	r

Table 14.1. Adopted nomenclature for the description of ship motion.

To describe the motion of a vehicle, two reference frames are considered: an *inertial* frame and a *body-fixed* frame. Figure 14.1 shows the two reference frames together with the variables often expressed relative to these frames. This figure also indicates the adopted positive convention.

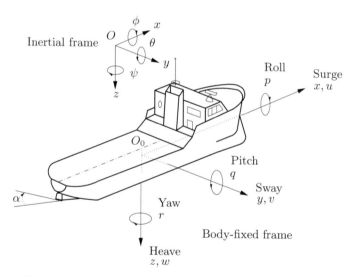

Figure 14.1. Notation, reference frames and sign conventions for ship motion description.

For marine vehicles position and orientation are described relative to the inertial reference frame, whilst linear and angular velocities are expressed in the body-fixed frame. This choice is convenient since some of these magnitudes are measured on board, and thus, relative to the body-fixed frame. In addition, by choosing the appropriate location of the body-fixed frame, the resulting equations of motion that describe the dynamic behaviour of the vessel are simplified—see Fossen (1994) for details.

A convenient abstraction to obtain a mathematical model that captures the different effects that give rise to ship motion, is to separate the motion due to control action (rudder motion) from the motion due to the waves. This abstraction results in two models that can be combined using superposition (Faltinsen 1990).

The first part of the model can be obtained using Newton's laws. This approach yields a nonlinear model that describes the motion components in terms of the forces and moments acting on the hull (Fossen 1994). By linearising this model and incorporating a linear approximation of the forces and moments generated by the rudder, we can obtain a linear state space model that describes the ship response due to the rudder (control) action. A discrete time version of this model can be expressed as

$$x_{k+1}^c = A_c x_k^c + B_c u_k, \tag{14.1}$$

where the state x_k^c and control u_k, are given by (see Figure 14.1 and Table 14.1)

$$x_k^c = \begin{bmatrix} v_k^c & p_k^c & r_k^c & \phi_k^c & \psi_k^c \end{bmatrix}^{\mathrm{T}} \quad \text{and} \quad u_k = \alpha_k, \tag{14.2}$$

with α_k being the current rudder angle.

For the particular motion control problem being considered, it is a common practice to decouple the vertical motion components of pitch and heave and to consider a constant forward speed. Hence, the surge equation can also be decoupled leaving a model that captures the couplings between roll, sway and yaw, that is, the state indicated in (14.2).

The parameters of the model (14.1), the values of the entries of the matrices A_c and B_c, will vary with the forward speed of the vessel. However, this variation is such that constant values can be considered for different speed ranges; usually close to the nominal speed of the vessel. Because of this, system identification techniques and data collected from tests in calm water can be used to estimate the parameters for different speed ranges; see, for example, Zhou, Cherchas and Calisal (1994). The different sets of parameters can then be used in a gain scheduling-like approach to update the model. This helps to minimise model uncertainty.

The second part of the model incorporates the motion induced by the waves. The sea surface elevation can be described in stochastic terms by its power spectral density; or, simply, *sea spectrum*. The ship motion induced by the waves can be interpreted as a filtering process made by the ship's hull, which has a selected response to certain frequencies and attenuates others. The frequency response of the hull due to wave excitation is called the *ship response operator*. The total effect can be incorporated into our model as a coloured noise output disturbance. The roll motion induced by the waves will thus be modelled with a shaping filter:

$$x_{k+1}^w = A_w x_k^w + w_k,$$
$$y_k^w = x_k^w + v_k,$$

(14.3)

where $x^w = [p^w \quad \phi^w]^\mathsf{T}$, and w_k and v_k are sequences of i.i.d. Gaussian vectors with appropriate dimensions.

For a given hull shape, the filtering characteristics of the hull depend on the forward speed of the ship U, and the heading angle relative to the waves: the *encounter angle* χ, which is defined as indicated in Figure 14.2. The variations in the characteristics of the motion response due to speed and encounter angle are the consequence of a Doppler-like effect. Indeed, if the ship is moving with a forward speed, the wave frequency observed from the ship is, in general, different from that observed from a fixed zero-speed reference frame. The frequency observed from the ship is called the *encounter frequency* ω_e. Expression (14.4) shows the relationship between the wave frequency ω_w (observed from a fixed-reference frame), and the encounter frequency:

$$\omega_e = \omega_w - \frac{\omega_w^2 U}{g} \cos \chi.$$

(14.4)

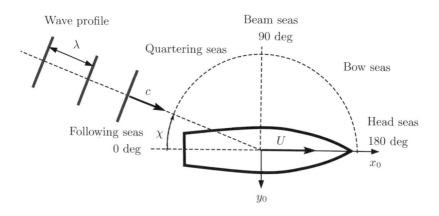

Figure 14.2. Encounter angle definition and usual denomination for sailing conditions.

The encounter effect produces significant variations in the motion response of the ship even for the same sea state, hereby defined by the sea spectrum. Consequently, the values of the parameters of the model (14.3) should be

updated for changes in different sea states and sailing conditions. We will
address this point in a latter section.

The complete state space model can then be represented via state aug-
mentation as

$$x_{k+1} = Ax_k + Bu_k + Jw_k,$$
$$y_k = Cx_k + n_k,$$

where $x^{\mathrm{T}} = [x_k^c \quad x_k^w]$. The following measurements are assumed available to
implement the control:

$$y_k = \left[p_k \; r_k \; \phi_k \; \psi_k \right]^{\mathrm{T}} + n_k$$
$$= \left[(p_k^c + p_k^w) \; r_k^f \; (\phi_k^c + \phi_k^w) \; \psi_k^f \right]^{\mathrm{T}} + n_k, \quad (14.5)$$

where, n_k is noise introduced by the sensors, and r_k^f and ψ_k^f are the filtered
yaw rate and yaw angle respectively.

The ramifications of an output disturbance model shall be evident when
we estimate the parameters of the disturbance part of the model. Note also
that we have only considered the wave disturbance affecting the roll angle
and the roll rate but not the yaw. The reason for this is that, in conventional
autopilot design, the yaw is filtered and only low frequency yaw motion is
corrected. This is done to avoid the autopilot making corrections to account
for the first-order (sinusoidal) wave-induced yaw motion. Therefore, in the
problem we are assuming that the yaw is measured after the yaw wave filter;
see Fossen (1994) and Blanke et al. (2000) for details.

14.5 Control Problem Definition

The basic control objectives for the particular motion control problem being
addressed here are as follows:

(i) minimise the roll motion, which includes roll angle and accelerations;
(ii) produce low interference with yaw;
(iii) satisfy input constraints.

In a discrete time framework, all the above objectives are captured in the
following optimisation problem.

**Definition 14.5.1 (Output Feedback Control Problem with Input
Constraints)** *Find the feedback control command* $u_k = \mathcal{K}(y_k)$ *that minimises
the objective function*

$$V = \lim_{N \to \infty} \frac{1}{N} \mathbf{E} \left\{ \sum_{k=0}^{N} y_k^{\mathrm{T}} Q y_k + (y_{k+1} - y_k)^{\mathrm{T}} S(y_{k+1} - y_k) + u_k^{\mathrm{T}} R u_k \right\} \quad (14.6)$$

subject to the system equations

$$x_{k+1} = Ax_k + Bu_k + Jw_k,$$
$$y_k = Cx_k + n_k,$$

and the input constraints

$$|u_k| \leq u_{max} \qquad \text{and} \qquad |u_{k+1} - u_k| \leq \delta u_{max},$$

with y_k given in (14.5).   ○

Choosing the matrices Q and S as:

$$Q = \text{diag}\{0, Q_p, 0, Q_\phi, Q_\psi\},$$
$$S = \text{diag}\{0, S_p, 0, 0, 0\},$$

the objective function (14.6) becomes (assuming no sensor noise)

$$V = \lim_{N \to \infty} \frac{1}{N} \mathbf{E}\left\{\sum_{k=0}^{N} [Q_p p_k^2 + Q_\phi \phi_k^2 + S_p(p_{k+1} - p_k)^2] + Q_\psi \psi_k^2 + R\alpha_k^2\right\},$$
(14.7)

which can be interpreted as

$$V \propto Q_p \, \mathbf{var}[p] + Q_\phi \, \mathbf{var}[\phi] + S_p \, \mathbf{var}[\dot{p}] + Q_\psi \, \mathbf{var}[\psi] + R \, \mathbf{var}[\alpha].$$

The objective function (14.7) is a discrete time version of the objective function proposed by van Amerongen, van der Klught and Pieffers (1987). The function (14.7), however, incorporates an extra term that weights the roll accelerations via the difference $p_{k+1} - p_k$. The reason for incorporating this extra term is that both roll angle and roll acceleration affect the performance of the ship; these are directly related to the criteria often used to evaluate ship performance in the marine environment (see Lloyd 1989, Graham 1990).

We next show how the above problem can be cast in the RHC framework.

14.6 A Receding Horizon Control Solution

As discussed in Chapter 12, the problem defined above is not easy to solve due to the presence of constraints.

We will approximate its solution using the certainty equivalent solution of an associated finite horizon problem, together with a receding horizon implementation.

In this context, we define the following associated finite horizon problem.

Definition 14.6.1 (Finite Horizon Optimal Control Problem) *Given the initial condition $\check{x}_0$, we seek the sequence of control moves*

$$\{\check{u}_0^{\text{OPT}}(\check{x}_0), \dots, \check{u}_{N-1}^{\text{OPT}}(\check{x}_0)\}$$
(14.8)

that minimises the objective function

$$V_N \triangleq \frac{1}{2}\check{x}_N^{\mathsf{T}}\check{P}\check{x}_N + \sum_{j=0}^{N-1}\frac{1}{2}(\check{x}_j^{\mathsf{T}}\check{Q}\check{x}_j + \check{u}_j^{\mathsf{T}}\check{R}\check{u}_j + \check{u}_j^{\mathsf{T}}\check{T}\check{x}_j + \check{x}_j^{\mathsf{T}}\check{T}^{\mathsf{T}}\check{u}_j), \qquad (14.9)$$

subject to

$$\begin{aligned} \check{x}_{j+1} &= A\check{x}_j + B\check{u}_j, \\ \check{y}_j &= C\check{x}_j, \end{aligned} \qquad (14.10)$$

and the constraints

$$|\check{u}_j| \le u_{max} \qquad \text{and} \qquad |\check{u}_{j+1} - \check{u}_j| \le \delta |u_{max}|.$$

The augmented state $\check{x}$ in (14.10) is given by

$$\check{x} = \begin{bmatrix} \check{v}^c & \check{p}^c & \check{r}^c & \check{\phi}^c & \check{\psi}^c & \check{\phi}^w & \check{p}^w \end{bmatrix}^{\mathsf{T}}.$$

NB. The notation $\check{x}$ is used here to distinguish the predicted state (predicted using the model (14.10)) from the true state x.

The matrices in the objective function are

$$\begin{aligned} \check{Q} &= (A - I)^{\mathsf{T}}(C^{\mathsf{T}}SC)(A - I) + C^{\mathsf{T}}QC, \\ \check{R} &= B^{\mathsf{T}}(C^{\mathsf{T}}SC)B + R, \\ \check{T} &= B^{\mathsf{T}}(C^{\mathsf{T}}SC)(A - I). \end{aligned}$$

The matrices Q, S and R are the parameters defining the objective function (14.6), and the matrices A, B describing the augmented system are

$$A = \begin{bmatrix} A_c & 0 \\ 0 & A_w \end{bmatrix}, \qquad B = \begin{bmatrix} B_c \\ 0 \end{bmatrix}, \qquad (14.11)$$

where the zeros denote zero matrices of appropriate dimensions. The matrix $\check{P}$ in (14.9) is taken as the solution of the following discrete time algebraic Riccati equation:

$$\check{P} = A^{\mathsf{T}}\check{P}A + \check{Q} - K^{\mathsf{T}}\bar{R}K.$$

with $K = \bar{R}^{-1}B^{\mathsf{T}}\check{P}A$ and $\bar{R} = \check{R} + B^{\mathsf{T}}\check{P}B.$ ○

The cross terms in the objective function (14.9), which were not considered in the earlier RHC formulation given in Chapter 5 (see (5.9)), appear due to the terms in the objective function penalising the difference $p_{k+1} - p_k$. These cross terms only affect the matrices that define the associated quadratic program [QP]. The QP solution of the above problem is given by (see Section 5.3 in Chapter 5)

$$\mathbf{u}^{\mathrm{OPT}}(\check{x}_0) = \arg\min_{L\mathbf{u}\le W}\frac{1}{2}\mathbf{u}^{\mathsf{T}}(H_1 + H_2)\mathbf{u} + \mathbf{u}^{\mathsf{T}}(F_1 + F_2)\check{x}_0, \qquad (14.12)$$

where

$$H_1 = \Gamma^{\mathrm{T}} \mathbf{Q} \Gamma + \mathbf{R}, \qquad\qquad H_2 = \mathbf{T}\bar{\Gamma} + \bar{\Gamma}^{\mathrm{T}} \mathbf{T}^{\mathrm{T}},$$
$$F_1 = \Gamma^{\mathrm{T}} \mathbf{Q} \Omega, \qquad\qquad F_2 = \mathbf{T}\bar{\Omega},$$

$$\mathbf{Q} = \mathrm{diag}\{\check{Q}, \ldots, \check{Q}, \check{P}\},$$
$$\mathbf{R} = \mathrm{diag}\{\check{R}, \ldots, \check{R}\},$$
$$\mathbf{T} = \mathrm{diag}\{\check{T}, \ldots, \check{T}\},$$

and

$$\Omega = \begin{bmatrix} A \\ A^2 \\ \vdots \\ A^N \end{bmatrix}, \qquad \bar{\Omega} = \begin{bmatrix} I \\ A \\ \vdots \\ A^{N-1} \end{bmatrix},$$

$$\Gamma = \begin{bmatrix} B & 0 & \cdots & 0 \\ AB & B & \cdots & 0 \\ \vdots & \vdots & \ddots & \vdots \\ A^{N-1}B & A^{N-2}B & \cdots & B \end{bmatrix} \qquad \bar{\Gamma} = \begin{bmatrix} 0 & 0 & \cdots & 0 \\ B & 0 & \cdots & 0 \\ \vdots & \vdots & \ddots & \vdots \\ A^{N-2}B & A^{N-3}B & \cdots & 0 \end{bmatrix}.$$

The matrices L and W that define the constraint set in (14.12) are given by (see Section 5.3.2 in Chapter 5)

$$L = \begin{bmatrix} I \\ E \\ -I \\ -E \end{bmatrix}; \quad W = \begin{bmatrix} \bar{M}_{\mathrm{mag}} \\ \bar{M}_{\mathrm{rate}} \\ \bar{M}_{\mathrm{mag}} \\ \bar{M}_{\mathrm{rate}} \end{bmatrix}$$

where I is the $N \times N$ identity matrix and E is the $N \times N$ matrix

$$E = \begin{bmatrix} 1 & \cdots & & 0 \\ -1 & 1 & & 0 \\ \vdots & \ddots & \ddots & \vdots \\ 0 & \cdots & -1 & 1 \end{bmatrix},$$

and

$$\bar{M}_{\mathrm{mag}} = \begin{bmatrix} u_{\max} \\ \vdots \\ u_{\max} \end{bmatrix}; \qquad \bar{M}_{\mathrm{rate}} = \begin{bmatrix} u_{-1} + \delta u_{\max} \\ \delta u_{\max} \\ \vdots \\ \delta u_{\max} \end{bmatrix}.$$

The above problem is be solved on line, and the implicit receding horizon feedback control law is implemented, that is,

$$u_k = \mathcal{K}_N(\check{x}_0(y_k)) = \check{u}_0^{\mathrm{OPT}}(\check{x}_0(y_k)),$$

the first element of the optimal sequence (14.8).

Using the certainty equivalence principle as described in Chapter 12 the initial condition for solving the above problem is provided by a Kalman filter (see Theorem 9.6.2 in Chapter 9). That is, at each step k we take $\check{x}_0 = \hat{x}_{k|k}$, where:

Prediction:

$$\begin{aligned}
\hat{x}_{k|k-1} &= A\hat{x}_{k-1|k-1} + Bu_{k-1}, \\
\Sigma_{k|k-1} &= A\Sigma_{k-1|k-1}A^{\mathrm{T}} + R_w.
\end{aligned} \tag{14.13}$$

Measurement update:

$$\begin{aligned}
L_k &= \Sigma_{k|k-1}C^{\mathrm{T}}(C\Sigma_{k|k-1}C^{\mathrm{T}} + R_v)^{-1}, \\
\hat{x}_{k|k} &= \hat{x}_{k|k-1} + L_k(y_k - C\hat{x}_{k|k-1}), \\
\Sigma_{k|k} &= (I_n - L_kC)\Sigma_{k|k-1},
\end{aligned} \tag{14.14}$$

The predictions $\check{x}_j$ used in the finite horizon problem (see (14.10)) are then j-step predictions given the measurement y_k at the time instant k, that is,

$$\check{x}_j = \hat{x}_{k+j|k}.$$

To summarise the proposed control strategy, the following steps are envisaged at each sampling instant:

(i) Take measurements, that is, obtain y_k (see (14.5)) and the previous control action u_{k-1}.
(ii) Update the state prediction (14.13) and estimate the state $\hat{x}_{k|k}$ using (14.14) and the measured output.
(iii) Using u_{k-1} and the initial condition $\check{x}_0 = \hat{x}_{k|k}$ solve the QP (14.12) to obtain the sequence of controls (14.8).
(iv) Update the control command $u_k^c = \check{u}_0^{\mathrm{OPT}}(\check{x}_0)$.

Thus far, we have defined the control and the estimation problem. The only missing element of the proposed strategy is a method to update the parameters of the model that describes the wave-induced motion, that is, the matrix A^w in (14.11).

14.7 Disturbance Model Parameter Estimation

The solution proposed in the previous section assumes that a model is available to predict the output disturbance; namely, (14.3). Here, we present a simple approach to estimate the parameters based on the control scheme shown in Figure 14.3. If the stabiliser control loop is open (see Figure 14.3)—that is,

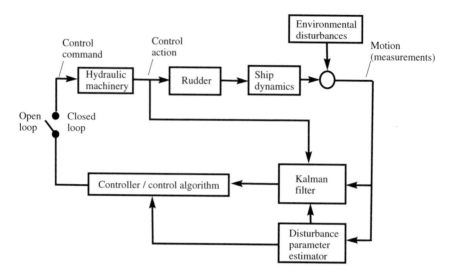

Figure 14.3. Block diagram of the control system architecture used for design.

the rudder is kept to zero angle, and only minor corrections are applied to the rudder to keep the course—we can then use the roll angle and roll rate measurements to estimate the parameters of a second order shaping filter. Under these conditions, the measurements coincide with the state of the following shaping filter:

$$\begin{bmatrix} \phi_{k+1}^w \\ p_{k+1}^w \end{bmatrix} = \begin{bmatrix} \theta_{11} & \theta_{12} \\ \theta_{21} & \theta_{22} \end{bmatrix} \begin{bmatrix} \phi_k^w \\ p_k^w \end{bmatrix} + \begin{bmatrix} w_k^\phi \\ w_k^p \end{bmatrix}.$$

By defining the vector

$$\theta_k = \begin{bmatrix} \theta_{11}(k) & \theta_{12}(k) & \theta_{21}(k) & \theta_{22}(k) \end{bmatrix}^\mathsf{T},$$

we can express the available measurements as

$$\theta_{k+1} = \theta_k + \theta_{wk},$$
$$\begin{bmatrix} \phi_k^w \\ p_k^w \end{bmatrix} = \begin{bmatrix} \phi_{k-1}^w & p_{k-1}^w & 0 & 0 \\ 0 & 0 & \phi_{k-1}^w & p_{k-1}^w \end{bmatrix} \theta_k + v_k. \tag{14.15}$$

The system (14.15) is in a form that we can apply a Kalman filter (see (14.14) and (14.13)) to estimate $\hat{\theta}_{k|k}$ from the measurements ϕ_k^w and p_k^w. The variable θ_{wk} represents a small random variable that accounts for unmodelled dynamics, so the Kalman filter does not assume a perfect model and eventually stops incorporating the information provided by the measurements. This method is a recursive implementation of the least-squares estimation method; see, for example, Goodwin and Sin (1984).

   To assess the proposed method to estimate the parameters of the filter and the quality of the prediction using the above model, a series of simulations were performed generating the roll angle and roll rate as a sum of regular components using parameters corresponding to different sea states and sailing conditions.

   The parameters to describe the sea state that were used are the average wave period T and the significant wave height H_s (average of the highest one third of the wave heights). These parameters are used in the International Towing Tank Conference (ITTC, 1952, 1957) recommended model for the wave power spectral density, commonly termed *ITTC spectrum* in the marine literature (Lloyd 1989):

$$\mathbf{S}_{\zeta\zeta}(\omega_w) = \frac{172.75 H_s^2}{T^4 \omega_w^5} \exp\left\{ \frac{-691}{T^4 \omega_w^4} \right\} \quad (\text{m}^2\text{sec/rad}). \qquad (14.16)$$

The above sea state description is combined with the ship roll response operator to obtain the roll power spectral density and then simulate the time series used to estimate the parameters. The process for obtaining the ship roll power spectral density is indicated via an example in Figure 14.4. The first two plots

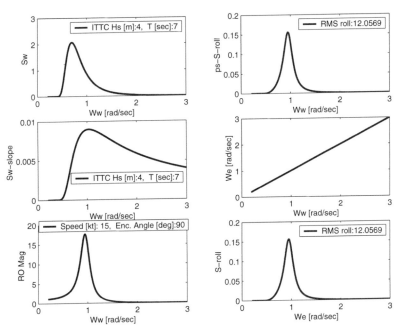

Figure 14.4. Roll motion spectral density used in the simulations.

on the left hand side show the sea elevation spectrum (ITTC spectrum) and the sea slope spectrum for the adopted sea state. The third plot on the left hand side shows the ship roll response operator from a particular vessel for the adopted sailing conditions (see Perez (2003) for this particular vessel model). The wave slope spectrum is filtered by the hull, and this effect is depicted in the first plot on the right hand side. This plot is called a *pseudo-spectrum* because it is in the wave frequency domain and it is the product of the wave slope spectrum and the roll response operator. The roll power spectral density is finally obtained by transforming the pseudo-spectrum to the encounter frequency domain according to (14.4). This transformation is depicted in the second plot on the right hand side. The roll power spectral density is depicted in the last plot on the right hand side of Figure 14.4.

Once the roll power spectral density has been obtained, the roll motion realisations (time series) can be computed as

$$\phi(t) = \sum_i \bar{\phi}_i \sin(\omega_{ei} t + \theta_i), \qquad (14.17)$$

where the phases are chosen randomly with a uniform distribution in $[-\pi, \pi]$ and the amplitudes calculated from

$$\bar{\phi}_i^2 = 2\mathbf{S}_{\phi\phi}(\omega_{ei}) \Delta_{\omega i}, \qquad (14.18)$$

where $\mathbf{S}_{\phi\phi}(\omega_e)$ represents the roll power spectral density (S-roll in Figure 14.4). The number of regular (sinusoidal) components used to simulate the time series is normally between 500 and 1000 to avoid pattern repetition depending on the total simulation time. This procedure for simulating time series for a ship response is a standard practice in naval architecture and marine engineering; see, for example, Faltinsen (1990).

The measurements taken from one of the realisations were used to estimate the parameters. Figure 14.5 shows the evolution of the estimates of the parameters for the model at a particular sea state and sailing condition. A sampling period of 0.25 sec was adopted based on the value of the roll natural period of the vessel (approximately 7 sec). Finally, Figure 14.6, shows the roll angle and roll rate predictions for two other different realisations using 5 and 10 step-ahead predictions. This consists of using the measured state as initial condition for the model (14.3) with the noises set to their mean values and then running this model forward in time.

From the above example we can see that the filter converges relatively quickly, and the quality of predictions for the behaviour of the ship can be deemed satisfactory. We note that, for the chosen sailing conditions, the use of a second order disturbance model seems to give good results. For other sea conditions, it may be necessary to resort to higher order models, but this shall not be considered here.

This process should be performed before closing the control loop. Then, the proposed control strategy can be considered a quasiadaptive control strategy. That is, if the sailing condition (heading and speed) or the sea state changes,

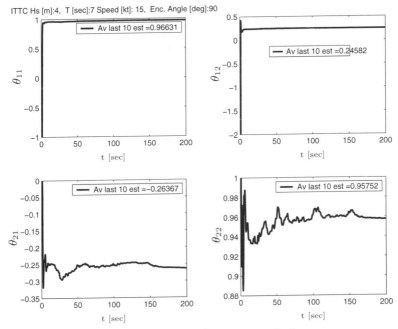

Figure 14.5. Estimated parameters for beam seas.

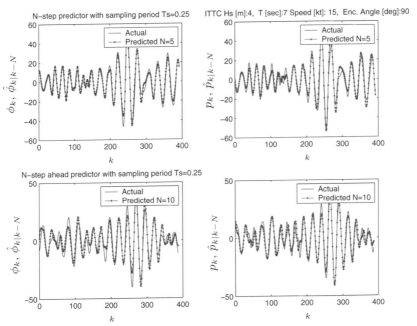

Figure 14.6. Roll angle and roll rate predictions in beam seas.

it may be necessary to open the loop and re-estimate the parameters of the disturbance model to avoid significant degradation in the closed loop performance.

14.8 Constrained Predictive Control of Rudder-Based Stabilisers

In this section, we will present simulation results aimed at assessing the performance of a rudder-based stabiliser designed according to the proposed strategy. In our simulations we will use as a ship a high fidelity nonlinear (calibration) model of a naval vessel adapted from Blanke and Christensen (1993). This model is a very comprehensive model that includes features that are not captured by the simple model used for control system design. However, these features have a direct bearing on the ship dynamic response description. In this fashion, we preserve the degree of uncertainty present in the real application. For the complete model see Perez (2003).

We have selected a speed of 15 kts for the simulations. This is the nominal speed of the vessel and also the speed at which this vessel performs its missions most of the time. The performance will be assessed for the following scenarios:

- **Case A**: Beam seas ($\chi = 90$ deg), $H_s = 2.5$ m, $T = 7.5$ sec;
- **Case B**: Quartering seas ($\chi = 45$ deg), $H_s = 2.5$ m $T = 7.5$ sec;
- **Case C**: Bow seas ($\chi = 135$ deg), $H_s = 4$ m, $T = 9.5$ sec.

The wave heights (H_s) have been chosen to represent moderate and rough conditions under which a vessel the size of this naval vessel can be expected to perform. The particular wave average periods T are the most probable periods for the chosen wave heights in ocean areas around Australia (Perez 2003). The control action will be updated with a sampling rate of 0.25 sec. Finally, the maximum rudder angle will be limited to 25 deg, and maximum rudder rate will be limited to 20 deg/sec.

We will assess the performance via

(i) percentage of reduction in roll angle variance and RMS value;
(ii) yaw angle RMS value induced by the rudder;
(iii) percentage of reduction of motion induced interruptions [MII].

MII is an index that depends on the roll angle and roll acceleration and yields the number of interruptions per minute that a worker can expect due to tipping and loss of balance. The value of this index depends on the particular location on the ship at which it is evaluated. It is a commonly used index to evaluate ship performance in the marine environment (Graham 1990). We will consider a location 7 m above the vertical centre of gravity [VCG] such that we can neglect the effect of vertical motion on the MII. Thus we can simplify the calculations and consider only roll motion. This location coincides with the rear part of the bridge of the vessel.

14.8.1 Tuning

The tuning was performed in beam seas, and then the parameters of the controller were fixed for the rest of the simulation scenarios. The only part of the controller that changed with each sailing condition was the model for the disturbance, which was estimated prior to closing the loop, as indicated in Section 14.7.

Different prediction horizons were tested. As expected, for short prediction horizons ($N = 1$ and $N = 2$), the performance was poorer than for longer horizons ($N = 5$ and $N = 10$). For a horizon of 10 samples periods, an improvement of 10% in roll reduction was achieved with respect to that obtained with $N = 1$. For horizons longer than 10 sample periods there was no significant improvement. Therefore, this is the horizon that was adopted so as to limit the size of the QP problem.

We next present some simulation results. For each case we present a table with the data calculated from the simulated time series and, also, a plot of the corresponding time series.

Case A: Beam Seas

The data for the case of beam seas are shown in Table 14.2.

```
CASE A. Rudder Roll Stabilisation Simulation Report.
        Perez (2003), 14-Jul-2003.
RHC parameters:
Prediction horizon           --------------------> 10      samples
Sampling time                --------------------> 0.25    [sec]
Rudder magnitude constraint --------------------> 25      [deg]
Rudder rate constraint       --------------------> 20      [deg/sec]
Controller tuning parameters:
Sp=1000; Qp=0; Qr=1; Qphi=100; Qpsi=1; R=0.25


Sailing Conditions:
Wave spectrum                --------------------> ITTC
Significant wave height       --------------------> 2.5     [m]
Average Wave period          --------------------> 7.5     [sec]
Encounter angle              --------------------> 90      [deg]

Yaw due to stabiliser (RMS) --------------------> 1.7283  [deg]
Roll open loop (RMS)         --------------------> 7.2533  [deg]
Roll closed loop (RMS)       --------------------> 2.7845  [deg]
Reduction   (RMS)            --------------------> 61.6111 %
Reduction   (VAR)            --------------------> 85.2629 %

Motion Induced Interruptions @ bridge (7m above VCG):
MII open loop                --------------------> 4.9124   [per min]
MII closed loop              --------------------> 0.28575  [per min]
Reduction                    --------------------> 94.1831 %
```

Table 14.2. Data from the simulated time series for case A: beam seas.

The time series corresponding to the data in Table 14.2 are shown in Figure 14.7. This case is close to the worst condition that the ship can experience

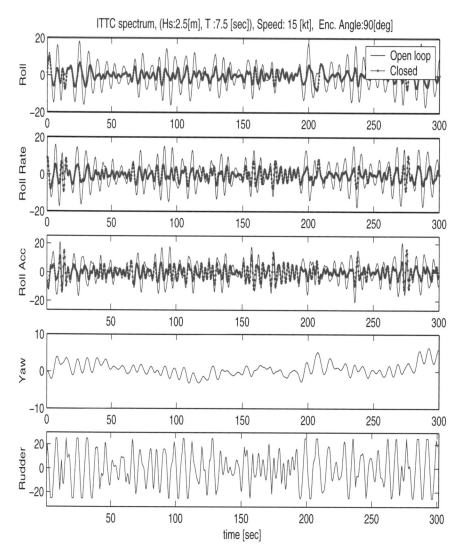

Figure 14.7. Case A. Simulation in beam seas.

in regard to roll motion for the assumed sea state: beam seas. The wave period in this case is close to the natural roll period, which is approximately 7 sec. Therefore, the roll excitation due to the waves is close to resonance. Notwithstanding this, we can still observe good performance.

Results obtained from over 20 different realisations indicate roll reductions on the order of 55–60% for roll RMS values. A significant improvement is achieved, however, in regard to MII: 80–90% reduction.

The inclusion of the term in the objective function that weights roll acceleration $\dot{p}$ yields a smoother control action and a better MII reduction with respect to the case that does not consider this term (between 5–15% higher reduction for the MII), and it yields only a small improvement in the roll angle reduction (less than 5%). This seems to indicate that weighting the roll accelerations in the objective function can be beneficial for operations (missions) that require low MII.

From the rudder action depicted in Figure 14.7, we can see that the controller generates a command that satisfies the magnitude constraints.

Case B: Quartering Seas

The data for the case of quartering seas are shown in Table 14.3.

```
CASE B. Rudder Roll Stabilisation Simulation Report.
       Perez (2003), 14-Jul-2003.
RHC parameters:
Prediction horizon          --------------------> 10    samples
Sampling time               --------------------> 0.25  [sec]
Rudder magnitude constraint --------------------> 25    [deg]
Rudder rate constraint      --------------------> 20    [deg/sec]
Controller tuning parameters:
Sp= 1000; Qp=0; Qr=1; Qphi=100; Qpsi=1; R=0.25


Sailing Conditions:
Wave spectrum               --------------------> ITTC
Significant wave height      --------------------> 2.5   [m]
Average Wave period         --------------------> 7.5   [sec]
Encounter angle             --------------------> 45    [deg]


Yaw due to stabiliser (RMS) --------------------> 6.8057 [deg]
Roll open loop (RMS)        --------------------> 4.0204 [deg]
Roll closed loop (RMS)      --------------------> 2.6248 [deg]
Reduction   (RMS)           --------------------> 34.7126 %
Reduction   (VAR)           --------------------> 57.3755 %


Motion Induced Interruptions @ bridge (7m above VCG):
MII open loop               ------------------> 0.38515   [per min]
MII closed loop             ------------------> 0.0016905 [per min]
Reduction                   ------------------> 99.5611 %
```

Table 14.3. Data from the simulated time series for case B: quartering seas.

The time series corresponding the data in Table 14.3 are shown in Figure 14.8. The performance in quartering seas decreases significantly. This is expected due to the low encounter frequency of the disturbance. In these conditions, the fundamental limitations associated with the NMP zero and the

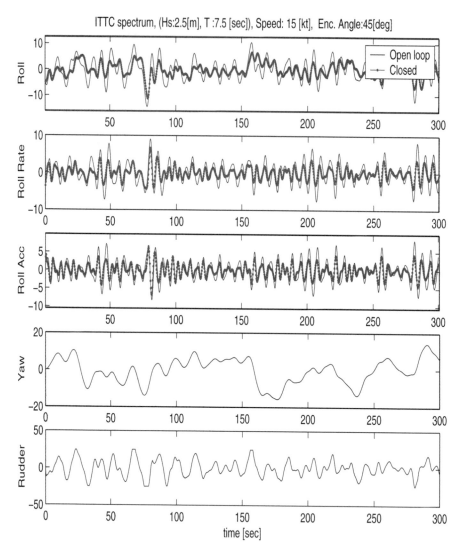

Figure 14.8. Case B. Simulation in quartering seas.

underactuated nature of the system swamp the limitations imposed by the input constraints. Note that the rudder angle depicted in Figure 14.8 rarely hits the constraints. The analysis of the performance in these conditions is beyond the scope of this chapter. The interested reader is encouraged to examine the broader discussion given in Perez, Goodwin and Skelton (2003) and in Perez (2003).

As depicted in Figure 14.8, due to the high interference with yaw for sailing conditions having low encounter frequency, it may be necessary to incorporate an output constraint in order to limit the maximum heading deviation. The low encounter frequency, here appearing in quartering seas, may also be present in other sailing conditions if the sea state is given by very low period waves produced in severe storms.

Case C: Bow Seas

The data for the case of bow seas are shown in Table 14.4.

```
CASE C. Rudder Roll Stabilisation Simulation Report.
        Perez (2003), 14-Jul-2003.
RHC parameters:
Prediction horizon          ------------------> 10    samples
Sampling time               ------------------> 0.25 [sec]
Rudder magnitude constraint ------------------> 25   [deg]
Rudder rate constraint      ------------------> 20   [deg/sec]
Controller tuning parameters:
Sp=1000; Qp=0; Qr=1; Qphi=100; Qpsi=1; R=0.25


Sailing Conditions:
Wave spectrum               ------------------> ITTC
Significant wave height      ------------------> 4    [m]
Average Wave period          ------------------> 9.5  [sec]
Encounter angle              ------------------> 135  [deg]


Yaw due to stabiliser (RMS) ------------------> 1.1673 [deg]
Roll open loop (RMS)         ------------------> 4.2398 [deg]
Roll closed loop (RMS)       ------------------> 1.4856 [deg]
Reduction   (RMS)            ------------------> 64.9593 %
Reduction   (VAR)            ------------------> 87.7215 %


Motion Induced Interruptions @ bridge (7m above VCG):
MII open loop                ------------------> 1.4513 [per min]
MII closed loop              ------------------> 9.6e-5 [per min]
Reduction                    ------------------> 99.9934 %
```

Table 14.4. Data from the simulated time series for case C: bow seas.

The time series corresponding the data in Table 14.4 are shown in Figure 14.9. This case presents the best performance despite the more severe sea state: 4 m waves. If we compare the RMS of roll in open loop with that of case B, we can see that these are similar. However, due to the higher encounter frequency of the disturbances in case C, the roll reduction is significantly better. The reason for this is the relative location of the NMP zero (which appears in the response of roll due to rudder) with respect to the bulk of energy of the disturbance (Perez 2003).

14.8.2 The Role of Adaptation

Table 14.5 shows how the adaptation improves the performance of the proposed control strategy for a particular example in which changes in course

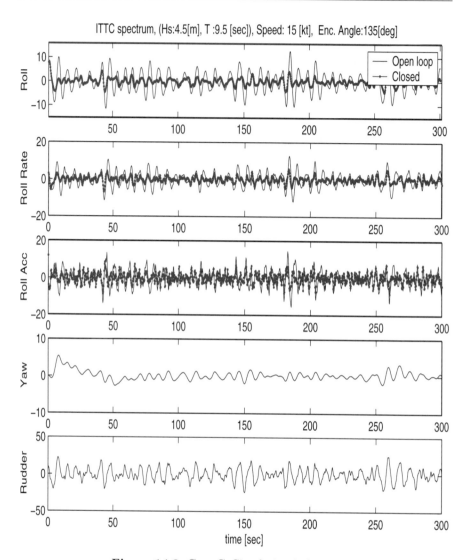

Figure 14.9. Case C. Simulation in bow seas.

from quartering seas to beam seas (χ [deg]: 45→90) and from beam seas to bow seas (χ [deg]: 90→135) were simulated. The second column shows the performance obtained if the disturbance model is not adapted (NA) after the change in course. The third column shows the performance after opening the loop to re-estimate the disturbance model, that is, the model is adapted (A).

We can see a significant improvement in performance due to the adaptation for the course change from quartering to beam seas. However, there is a small improvement for the course change from beam to bow seas. The reason for this is that the first course change produces more variation in the roll response of the vessel than the second. Therefore, after re-estimating the parameters of the model the performance improves significantly. For the second course change the nonadapted model is still good for the predictions; hence, after adaptation, only a small improvement is obtained.

Although we have only showed, as an illustrative example, variations due to a course change, one should bear in mind that changes in speed and, more importantly, in the sea, determine the characteristics of roll motion. Adaptation plays an important role for the proposed control strategy.

ITTC, Hs=2.5m,T=7.5s	χ [deg]: 45→90 (NA)	χ [deg]: 45→90 (A)
Roll red %	41.5	61.2
MII red %	81.2	99.1
Yaw rms	0.70	0.86
Rudder rms	6.2	10.1
ITTC, Hs=2.5m,T=7.5s	χ [deg]: 90→135 (NA)	χ [deg]: 90→135 (A)
Roll red %	62.0	66. 4
MII red %	100	100
Yaw rms	0.33	0.4
Rudder rms	4.2	5.2

Table 14.5. Performance after a change in course from quartering to beam seas (χ [deg]: 45→90) and from beam to bow seas (χ [deg]: 90→135) with no adaptation (NA) and after adapting the disturbance predictor (A).

14.9 Summary and Discussion

In this chapter, we have presented a case study and control system design for a problem of significant practical importance. The simplifying assumptions under which we performed the design have been kept to a minimum. Hence, almost all aspects of the design process have been addressed to some degree, including choosing the appropriate model, adopting the type of disturbance description and selecting the performance criteria.

The RHC formulation offers a unified framework to address many of the difficulties associated with the control system design for this particular problem: multivariable nature, constraints, uncertainty, stochastic disturbance rejection. The simulations presented illustrate the performance of RHC and suggest that the methodology should be successful in practical applications.

14.10 Further Reading

For complete list of references cited, see References section at the end of book.

General

Further information on rudder roll stabilisation of ships can be found in, for example, Perez (2003) and van Amerongen, van der Klugt and van Nauta Lemke (1990).

For more information on ship dynamics and control, see, for example, Fossen (1994) and Fossen (2002).

15

Cross-Directional Control

Contributed by Osvaldo J. Rojas

15.1 Overview

In this chapter we describe a practical application of receding horizon control to a common industrial problem, namely *web-forming processes*. Web-forming processes represent a wide class of industrial processes with relevance in many different areas such as paper making, plastic film extrusion, steel rolling, coating and laminating.

In a general set up, web processes (also known as *film and sheet forming processes*) are characterised by raw material entering one end of the process machine and a thin web or film being produced in (possibly) several stages at the other end of the machine. The raw material is fed to the machine in a continuous or semi-continuous fashion and its flow through the web-forming machine is generally referred to as the machine direction [MD].

Sheet and film processes are effectively two-dimensional spatially distributed processes with several of the properties of the sheet of material varying in both the machine direction and in the direction across the sheet known as the cross direction [CD].

The main objective of the control applied to sheet and film processes is to maintain both the MD and CD profiles of the sheet as flat as possible, in spite of disturbances such as variations in the composition of the raw material fed to the machine, uneven distribution of the material in the cross direction, and deviations in the cross-directional profile. The weight, moisture and calliper of the sheet are the most commonly controlled properties of the web. Usually their average values are controlled in the machine direction whilst their deviations from the mean are controlled in the cross direction (Featherstone, VanAntwerp and Braatz 2000).

Figure 15.1 shows a typical web-forming process. In order to control the cross-directional profile of the web, several actuators are evenly distributed along the cross direction of the sheet. The number of actuators can vary from only 30 up to as high as 300. The film properties, on the other hand, are either measured via an array of sensors placed in a downstream position or via a scanning sensor that moves back and forth in the cross direction. The number of measurements taken by a single scan of the sensor can be up to 1000.

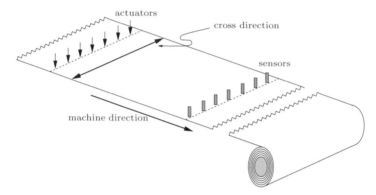

Figure 15.1. Generic web-forming process.

It is generally recognised that the cross-directional control problem in web-forming processes is much more challenging than the machine direction control problem. This is due to several difficulties; some of these difficulties are:

- the high dimensionality of the cross-directional system;
- the high cross-direction spatial interaction between actuators;
- the uncertainty in the model;
- the limited control authority of the actuators.

We present this application problem as an example to illustrate the concepts and approaches studied in previous chapters.

15.2 Problem Formulation

It is generally the case that web-forming processes can be effectively modelled by assuming a decoupled spatial and dynamical response (see, for example, Laughlin, Morari and Braatz 1993, Bergh and MacGregor 1987). This is equivalent to saying that the effect of one single actuator movement is almost instantaneous in the cross direction whilst its effect in the machine direction shows a certain dynamic behaviour. This behaviour, as measured by the array of sensors in the downstream position, is commonly assumed to be identical across the web.

These observations allow one to consider a general model for a cross-directional system of the form

$$y_k = q^{-d}h(q)\bar{B}u_k + d_k, \tag{15.1}$$

where q^{-1} is the unitary shift operator.

The vector $y_k \in \mathbb{R}^m$ represents the sensor measurements whilst $u_k \in \mathbb{R}^m$ is the vector of control signals. The vector $d_k \in \mathbb{R}^m$ represents an output disturbance. Note that the model (15.1) is assumed to have the same number of inputs and outputs. In practice, the sensor measurements often outnumber the actuators. In that case, the sensor data can be filtered in the spatial domain to reduce its order whilst minimising the loss of controllable information (Stewart, Gorinevsky and Dumont 1998).

It is assumed that the system dynamics are the same across the machine and thus $h(q)$ can be taken to be a scalar transfer function. In addition, $h(q)$ is typically taken to be a low order, stable and minimum-phase transfer function. A typical model is a simple first-order system with unit gain (Featherstone et al. 2000):

$$h(q) = \frac{(1 - \alpha)}{q - \alpha}. \tag{15.2}$$

The transport delay q^{-d} in (15.1) accounts for the physical separation that exists between the actuators and the sensors in a typical cross-directional process application (see Figure 15.1).

The matrix $\bar{B}$ is the normalised steady state *interaction matrix* and represents the spatial influence of each actuator on the system outputs. In most applications it is reasonably assumed that the steady state cross-directional profile generated by each actuator is identical. As a result, the interaction matrix $\bar{B}$ usually has the structure of a Toeplitz symmetric matrix (Featherstone et al. 2000, Featherstone and Braatz 1997). In other applications the structure of $\bar{B}$ is that of a circulant symmetric matrix. This is the case, for example, in paper machines where edge effects are neglected, in dyes for plastic films and in multizone crystal growth furnaces (Hovd, Braatz and Skogestad 1997).

As the system model (15.1)–(15.2) suggests, the main difficulties in dealing with cross-directional control problems are related to the spatial interaction between actuators and not so much to the complexity of dynamics, which could reasonably be regarded as benign.

A key feature is that a single actuator movement not only affects a single sensor measurement in the downstream position but also influences sensors placed in nearby locations. Indeed, the interaction matrix $\bar{B}$ is typically poorly conditioned in most cases of practical importance.

The poor conditioning of $\bar{B}$ can be quantified via a singular value decomposition

$$\bar{B} = USV^T \tag{15.3}$$

where $S, U, V \in \mathbb{R}^{m \times m}$. $S = \mathrm{diag}\{\sigma_1, \sigma_2, \ldots, \sigma_m\}$ is a diagonal matrix with positive singular values arranged in decreasing order, and U and V are orthogonal matrices such that $UU^{\mathrm{T}} = U^{\mathrm{T}}U = I_m$ and $VV^{\mathrm{T}} = V^{\mathrm{T}}V = I_m$, where I_m is the $m \times m$ identity matrix. If $\bar{B}$ is symmetric then $U = V$.

If $\bar{B}$ is poorly conditioned then the last singular values on the diagonal of S are very small compared to the singular values at the top of the chain $\{\sigma_i\}_{i=1}^m$. This characteristic implies that the control directions associated with the smallest singular values are more difficult to control than those associated with the biggest singular values, in the sense that a larger control effort is required to compensate for disturbances acting in directions associated with small σ_i. (See also the discussion in Section 11.6.1 of Chapter 11.)

This constitutes not only a problem in terms of the limited control authority usually available in the array of actuators, but it is also an indication of the sensitivity of the closed loop to uncertainties in the spatial components of the model in (15.1).

The control objective in cross-directional control systems is usually stated as the requirement to minimise the variations of the output profile subject to input constraints. This can be stated in terms of minimising the following objective function:

$$V_\infty = \sum_{k=0}^{\infty} \|y_k\|_2^2$$

subject to input constraints

$$\|u_k\|_\infty \leq u_{\max}. \tag{15.4}$$

Another type of constraint typical of CD control systems is a second-order bending constraint defined as[1]

$$\|\Delta u_k^{i+1} - \Delta u_k^i\|_\infty \leq b_{\max} \quad \text{for } i = 1, \ldots, m, \tag{15.5}$$

where $\Delta u_k^i = u_k^i - u_k^{i-1}$ is the deviation between adjacent actuators in the input profile at a given time instant k. Constraints of this type are necessary to prevent damage to the array of actuators, in particular, in paper making applications where excessive variation between adjacent actuators can compromise the physical integrity of the slice lip (Kristinsson and Dumont 1996).

15.3 Simulation Results

To illustrate the ideas involved in cross-directional control, we consider a 21-by-21 interaction matrix $\bar{B}$ with a Toeplitz symmetric structure and exponential profile:

---

[1] The superscript indicates the actuator number.

$$b_{ij} = e^{-0.2|i-j|} \quad \text{for } i, j = 1, \ldots, 21, \tag{15.6}$$

where b_{ij} are the entries of the matrix $\bar{B}$. For illustrative purposes we assume that the transport delay in the model (15.1) is negligible and we consider the transfer function

$$h(q) = \frac{1 - e^{-0.2}}{q - e^{-0.2}}, \tag{15.7}$$

which is a discretised version of the first-order system $\dot{y}(t) = -y(t) + u(t)$ with sampling period $T_s = 0.2$ sec. (Note that a nonzero delay can be treated as in Section 5.5.1 of Chapter 5.)

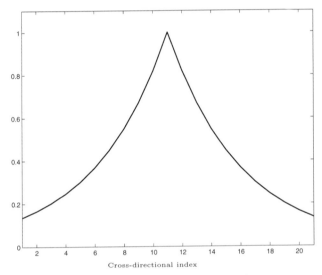

Cross-directional index

Figure 15.2. Cross-directional profile for a unit step in actuator number 11.

Figure 15.2 shows the steady state CD profile created by a unit step in actuator number 11. Although the chosen profile is not necessarily representative of real sheet and film processes, it captures the main characteristics of such systems, namely the wide spatial influence of one actuator over the entire output profile.

Figure 15.3 shows the singular values of the interaction matrix $\bar{B}$. We observe that there exists a significant difference between the largest singular value σ_1 and the smallest singular value σ_{21}, indicating that the matrix is poorly conditioned. Dealing with the poor conditioning of $\bar{B}$ is one of the main challenges in CD control problems as we will show later.

In order to estimate the states of the system and the output disturbance d_k, a Kalman filter is implemented as described in Section 5.5 of Chapter 5 for an extended system (see (5.32)) that includes the dynamics of a constant output disturbance:

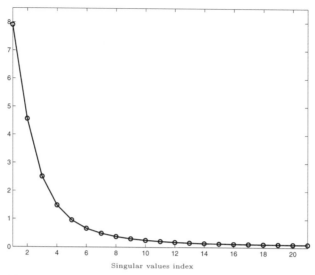

Figure 15.3. Singular values of the interaction matrix $\bar{B}$.

$$x_{k+1} = Ax_k + Bu_k,$$
$$d_{k+1} = d_k,$$
$$y_k = Cx_k + d_k,$$

where in our case

$$A = \text{diag}\{e^{-0.2}, \ldots, e^{-0.2}\},$$
$$B = (1 - e^{-0.2})\bar{B},$$
$$C = I_m.$$

The state noise covariance

$$Q_n = \begin{bmatrix} I_m & 0 \\ 0 & 100I_m \end{bmatrix},$$

and output noise covariance

$$R_n = I_m,$$

were considered in the design of the Kalman filter (see, for example, Goodwin et al. 2001).

In what follows we will illustrate how some of the constrained control strategies described in previous chapters perform on a large scale control problem such as the one described here. We will consider the finite horizon quadratic objective function of the form (5.9) in Chapter 5 with both prediction and control horizons set equal to one, that is

$$V_{1,1} = \frac{1}{2}(y_0^{\mathsf{T}}Qy_0 + u_0^{\mathsf{T}}Ru_0 + x_1^{\mathsf{T}}Px_1). \tag{15.8}$$

We have chosen a "one-step ahead" prediction horizon owing to the simplicity of the first-order dynamics of the system. However, it is perhaps worth noting that the minimisation of the objective function $V_{1,1}$ subject to the constraints described in (15.4) and (15.5) could still be a computationally intensive problem considering the high dimensionality of a general CD control system.

In the objective function (15.8) we use the weighting matrices

$$Q = I_m, \quad R = 0.1I_m. \tag{15.9}$$

The first control strategy that we try on the problem is a linear quadratic Gaussian [LQG] controller designed with the same weighting matrices as in (15.9). This design clearly does not take into consideration the constraints imposed on the input profile. As might be expected, the application of such a blind (or serendipitous) approach to the problem would, in general, not achieve satisfactory performance (see comments made in Chapter 1).

To illustrate, let us assume the system is subject to physical constraints on the inputs of the form:

$$|u_k^i| \le 1 \quad \text{for all } k, i = 1, \ldots, 21.$$

We then apply the controller

$$u_k = -K\hat{x}_k,$$

where $\hat{x}_k$ is the state estimated by the Kalman filter and K is the optimal feedback gain.

The system is allowed to settle after a first transient in order to adequately reduce the state estimation error. After that, a specified constant disturbance profile is applied to the output of the system. The disturbance is selected to contain large components in the directions of the small spatial singular values.

Figure 15.4 shows the time response of the input-output pair number 10 compared to the response achieved when no constraints are imposed on the system. We observe that the closed loop response with the unconstrained controller is faster than the constrained response and, perhaps more important, that the controller subject to hard input constraints is not able to compensate the disturbance in steady state due to its limited control authority.

In Figure 15.5 we can observe the achieved steady state input and output profiles across the strip. The dotted lines in Figure 15.5 (a) represent the input constraint levels whilst the dotted line in Figure 15.5 (b) is the applied constant disturbance profile d. Figure 15.5 (a) illustrates a phenomenon that is well known in the area of cross-directional control, namely alternate inputs across the strip converge to large alternate values, that is, "input picketing" occurs. We will see below, when we test alternative design methods, that this picketing effect can be avoided by careful design leading to significantly improved disturbance compensation.

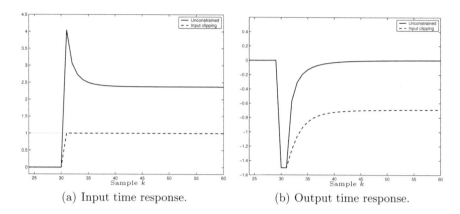

(a) Input time response.   (b) Output time response.

Figure 15.4. Time response of the input clipping strategy (dashed line) compared to the unconstrained LQG response (solid line).

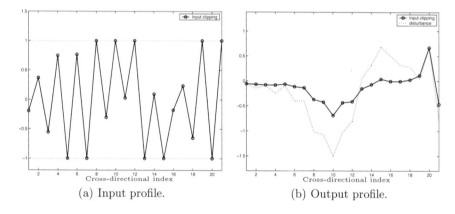

(a) Input profile.   (b) Output profile.

Figure 15.5. Input-output steady state profiles with input clipping.

We next try RHC considering initially only input constraints (see Chapter 5). The achieved steady state input and output profiles are presented in Figure 15.6 where we have also included, for comparison, the profiles obtained with the input clipping approach.

We observe, perhaps surprisingly, that the steady state response achieved with RHC does not seem to have improved significantly compared with the result obtained by just clipping the control in the LQG controller. In addition, the input profile obtained with RHC seems to be dominated by the same high spatial frequency modes as those that resulted from the input clipping approach. This is an indication of the difficulties inherent in dealing with CD

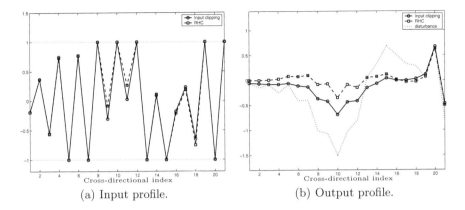

(a) Input profile.   (b) Output profile.

Figure 15.6. Input-output steady state profiles using RHC (square-dashed line) and input clipping (circle-solid line).

control systems which, in turn, arise from the strong spatial "interference" between actuators.

However, this is a reasonably well understood difficulty in CD control systems: The "picket fence" profile in the input arises from the controller trying to compensate for the components of the disturbance in the high spatial modes which, in turn, require bigger control effort, driving the inputs quickly into saturation. With limited control authority no controller can completely compensate for a disturbance that contains high spatial frequency components. In addition, Figure 15.6 shows that even when the compensation of high frequency disturbance modes is performed optimally (since the RHC strategy solves a QP at each step) large and usually unacceptable deviations between adjacent actuators occur.

The commonly accepted solution to this inherent difficulty is to let the controller seek disturbance compensation only in the low spatial frequencies (Heath 1996, Kristinsson and Dumont 1996). Note that this is the spatial version of the algorithm described in Section 11.6.2 of Chapter 11.

This is indeed how the SVD–RHC line search algorithm described in Section 11.6.2 generates a control signal that meets the constraints. If the prediction horizon chosen is $N = 1$ then in the vector formulation of the quadratic optimisation problem described in Chapter 5 we can write

$$\Gamma = B$$

and the Hessian of the objective function (see (5.20) in Chapter 5) is simply

$$H = B^{\mathrm{T}} \mathbf{Q} B + \mathbf{R}$$
$$= \bar{B}^{\mathrm{T}} \bar{B} + \mathbf{R}.$$

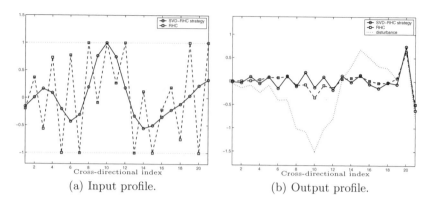

(a) Input profile.   (b) Output profile.

Figure 15.7. Input and output steady state profiles using the SVD–RHC strategy (circle-solid line) and RHC (square-dashed line).

This implies that the singular values of the Hessian are simply the singular values squared of the interaction matrix $\bar{B}$ shifted by the weighting in the input $\mathbf{R}$ (recall the analysis presented in Chapter 11).

Figure 15.7 shows the steady state input and output profiles obtained by using the SVD–RHC strategy when input constraints of the type in (15.4) are considered. We observe that in this case the picket fence profile has disappeared from the input whilst the output profile has not changed significantly. Clearly a slight degradation of the output variance is to be expected owing to the suboptimality of the strategy. The steady state profiles obtained with RHC are repeated in Figure 15.7 for comparison purposes. Note that the output profile obtained with the SVD–RHC strategy contains the higher spatial frequency components that the strategy avoids compensating. These components are less evident in the output profile obtained with RHC but they generate the undesirable input variations observed in Figure 15.7 (a).

Figure 15.8 shows the spatial components considered by the SVD–RHC strategy at each sampling time. We observe that, in steady state, the strategy only retains the first seven spatial modes completely whilst scaling the eighth mode and discarding the higher frequency modes.

Finally, we compare the performance obtained using RHC with the performance obtained using the SVD–RHC strategy when the second-order bending constraint in (15.5) is also considered. One might suspect that a constraint imposed on the maximum acceleration that neighbouring actuators are allowed to have will necessarily "smooth out" the input profile. This is in fact the case as it is shown in Figure 15.9.

Note that the input profiles obtained with both strategies are remarkably similar. Figure 15.9 (b) shows the second-order bending profile achieved in steady state. Both strategies meet the constraint imposed on the second-order

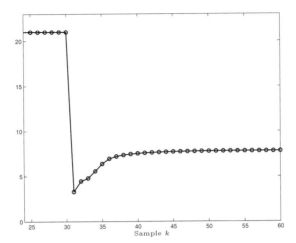

Figure 15.8. Time evolution of the number of singular values considered by the SVD–RHC strategy.

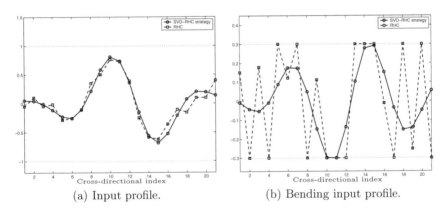

(a) Input profile.   (b) Bending input profile.

Figure 15.9. Steady state input and second-order bending input profiles using the SVD–RHC strategy (circle-solid line) and RHC (square-dashed line).

bending even though the SVD–RHC profile is once again smoother, for only low spatial modes are included in the solution.

Figure 15.10 shows the resultant steady state output profiles. As anticipated from the similarities already observed in the input profiles, the output profiles are very similar as well. When observing the spatial modes included in the output profiles (see Figure 15.10 (b)) the SVD–RHC strategy clearly achieves complete disturbance compensation in the first six spatial modes and significant attenuation in the seventh mode. However, no attenuation is ob-

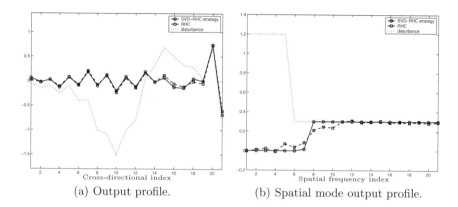

(a) Output profile.   (b) Spatial mode output profile.

Figure 15.10. Steady state output profile and spatial components profile using the SVD–RHC strategy (circle-solid line) and RHC (square-dashed line).

tained in any of the other higher spatial frequency modes. The RHC solution, on the other hand, gives the optimal balance that results when compensating the disturbance both in the low and high spatial frequency modes. In this particular case, it seems that no significant improvement is achieved when adopting a QP solution.

This observation may suggest that in certain specific applications, alternative sub-optimal strategies like the SVD–RHC can achieve a performance that is comparable to that of the full QP solution, but with only a fraction of the computational load.

15.4 Further Reading

For complete list of references cited, see References section at the end of book.

General

Further details regarding cross-directional control problems can be found in Featherstone et al. (2000); Wellstead, Zarrop and Duncan (2000); Stewart, Gorinevsky and Dumont (2003). In particular, Chapter 3 in Featherstone et al. (2000) presents a detailed literature review of cross-directional control. Web-forming processes are industrial applications that need high capital investments due to the complexity and scale of the machinery involved (Duncan 2002). As such, increased effort has been devoted to devise and implement more sophisticated controllers to improve the performance of cross-directional control systems and reduce operational costs. This interest has been reflected in a growing number of publications analysing the control and identification aspects of the problem. See, for example, two recent special issues on cross-directional control (Duncan 2002, Dochain, Dumont, Gorinevsky and Ogunnaike 2003).

Different representations of the steady state output profile have been used by several authors, including Gram polynomials (Kristinsson and Dumont 1996, Heath 1996), splines (Halouskova, Karny and Nagy 1993) and singular value decomposition (Duncan and Bryant 1997, Featherstone et al. 2000).

It is perhaps interesting to notice that although input constraints play an important role in the formulation of cross-directional control problems, it is only recently that constrained control formulations such as receding horizon control have been applied to such large scale systems (see, for example, Heath 1996, Rawlings and Chien 1996). The idea of using QP to obtain the optimal steady state cross-directional profile was however proposed by Boyle as early as 1977 (Boyle 1977). Constraint handling is also addressed in Dave, Willing, Kudva, Pekny and Doyle (1997), where a linear programming formulation is adopted.

In Bartlett, Biegler, Backstrom and Gopal (2002) different QP algorithms are compared in terms of their computational costs and a new fast QP algorithm is proposed particularly tailored to large scale CD control systems. A suboptimal solution to QP applied to cross-directional control problems is studied in VanAntwerp and Braatz (2000) based on an ellipsoidal approximation of the constraint polytope. A further interesting topic is the analysis of the achievable steady state performance in cross-directional control (Wills and Heath 2002).

Section 15.3

The main idea underpinning the SVD–RHC strategy used here and described in Section 11.6.2 of Chapter 11 was initially developed as an anti-windup strategy applied to cross-directional control problems, reported in Rojas et al. (2002).

16

Control over Communication Networks

Contributed by James Welsh, Hernan Haimovich and Daniel Quevedo

16.1 Overview

This chapter presents a case study of constrained control and estimation in the area of networked control. *Networked control* is a term used to describe control actions that take place over a networked communication system. A key issue that arises in this context is the need to quantise data so that it can be communicated over the network. Quantisation occurs in both time and space and is needed on both the "up-link" (between the plant and the control computer) and the "down-link" (between the control computer and the plant). In this chapter, we show how this problem can be formulated in the framework of constrained control and estimation.

16.2 Description of Networked Control Systems

Networked control systems are of substantial interest in a variety of applications. They are control systems in which controller and plant are connected via a communication channel. There are many practical applications of these systems. They have been made possible by technological developments, including the development of MEMS arrays, and may deploy wireless links (for example, Bluetooth or IEEE 802.11), Ethernet (for example, IEEE 802.3) or specialised protocols such as CAN. A special feature of networked control systems is the need for signal quantisation.

Due to the digital nature of the communication channel, every signal transmitted is expressed as a finite number of bits, hence the signal needs to be quantised to a finite set; see, for example, Ishii and Francis (2002).

The specific problem addressed in this chapter is the design of a methodology for minimising network traffic between a centralised controller and a multivariable plant or collection of plants. We split the design problem into two sub-problems, namely: *up-link* and *down-link*.

On the down-link side, we restrict the controller such that only one actuator can be addressed at any given time. Moreover, only one of a finite number of levels can be transmitted. The design of the resulting system is aimed at optimising performance subject to these constraints. In this context, we choose to send increments in the control signal, rather than the actual values. Between updates, all inputs are kept at their previous values. We utilise receding horizon optimisation to deal with the associated computational issues.

On the up-link side, we employ signal quantisation to minimise data rate requirements. In particular, at each sampling time, we transmit only data specifying the region in which the measurements lie. We propose two design strategies for dealing with these quantised measurements. First, we adopt a stochastic problem formulation and develop an approximate minimum variance state estimator. Second, by adopting a deterministic formulation, we propose a set-valued observer that is able to approximate the smallest region of the state space in which the state to be estimated has to lie. For both estimators, we resort to moving horizon techniques in order to limit computational complexity.

As outlined above, the design methodology described here divides the problem into the up-link and down-link problems as depicted in Figure 16.1. Implicit in this division is the assumption that we will utilise certainty equivalence (see Chapter 12), that is, we base the down-link design on the state estimates provided by the up-link state estimator. The scheme described here

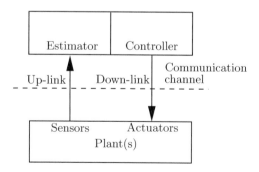

Figure 16.1. System overview.

is particularly suited to protocols where the message size can be manipulated.

16.3 Down-Link

Here we assume knowledge of the system states (the state estimation problem will be addressed in Section 16.4) and use x_k to denote the current state. We consider optimisation over the future time interval $k, k + 1, \ldots, k + N$, and use $\mathbf{y}_{k:k+N}$ to denote future output predictions.

We consider a linear time-invariant noiseless plant with m inputs and s outputs. We assume a given and fixed reference vector $r^\star$. We also assume that the link between the controller and actuators is characterised by a known and fixed time-delay and that data are sent at a bounded rate. This last assumption is achieved by imposing the following two communication constraints on the design:

Constraint 1 *The data sent from the controller to each actuator is restricted to belong to a (small and fixed) finite set of scalars, $\mathbb{U}$.*   ○

Constraint 2 *Only data corresponding to one input of the plant can be transmitted at a time. Between updates (which may be separated by several sampling periods), all plant inputs are held at their previous values.*   ○

The delay between controller and plant can be incorporated into the noiseless[1] predicted plant response

$$
\begin{aligned}
x'_{j+1} &= Ax'_j + Bu_j \quad \text{for } j \geq k, \\
y'_j &= Cx'_j,
\end{aligned}
\tag{16.1}
$$

with initial condition

$$
x'_k = x_k.
\tag{16.2}
$$

In (16.1), $x'_j \in \mathbb{R}^n$, $A \in \mathbb{R}^{n \times n}$, $B \in \mathbb{R}^{n \times m}$, $C \in \mathbb{R}^{s \times n}$, $u_j \in \mathbb{R}^m$ and $y'_j \in \mathbb{R}^s$. For future reference, we note that u_j can be expanded as:

$$
u_j = \left[(u_1)_j \; (u_2)_j \; \ldots \; (u_m)_j \right]^{\mathrm{T}}.
\tag{16.3}
$$

The design problem can thus be stated as that of developing a control strategy, which regulates the model (16.1) to the constant reference $r^\star$, whilst not violating Constraints 1 or 2. Thus, the control strategy for the networked system is characterised by choosing, at each time step, *which* of the m inputs in (16.3) to access and *what* to send, that is, the controller needs to divide its attention between all plant inputs. We will see that this can be formulated as a receding horizon finite alphabet control problem.

Note that the above formulation encompasses the problem of controlling a collection of geographically separated plants. Simply note that a set of p plants, each described by

$$
\begin{aligned}
(x_i)_{k+1} &= A_i(x_i)_k + B_i(u_i)_k, \\
(y_i)_k &= C_i(x_i)_k \quad \text{for } i = 1, \ldots, p,
\end{aligned}
\tag{16.4}
$$

---

[1] Compare with (16.21).

can be put into the form (16.1) by defining x as the overall state

$$x_k \triangleq \left[((x_1)_k)^{\mathrm{T}} \; \ldots \; ((x_p)_k)^{\mathrm{T}} \right]^{\mathrm{T}}.$$

With this, the matrices in the realisation (16.1) are given by

$$A = \mathrm{diag}\{A_1, \ldots, A_p\},$$
$$B = \mathrm{diag}\{B_1, \ldots, B_p\},$$
$$C = \mathrm{diag}\{C_1, \ldots, C_p\}.$$

Rather than sending the control signals directly, we propose to send their increments

$$\Delta(u_i)_k \triangleq (u_i)_k - (u_i)_{k-1} \tag{16.5}$$

when such increments are nonzero. This generally requires fewer bits to specify the control signal. The pair $(\Delta(u_i)_k, i)$ is received at the actuator node specified by the index i. The actual signal $(u_i)_k$ is readily reconstructed by discrete time integration as shown in Figure 16.2.[2]

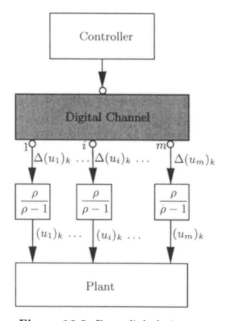

Figure 16.2. Down-link design.

The communication Constraints 1 and 2 can be summarised by means of a simple *finite set constraint* on the increments (16.5). More precisely, at every time instant k, the vector

---

[2] ρ denotes the forward shift operator, $\rho v_k = v_{k+1}$, where $\{v_k\}$ is any sequence.

$$\Delta u_k \triangleq \left[\Delta(u_1)_k \ \Delta(u_2)_k \ \ldots \ \Delta(u_m)_k\right]^{\mathrm{T}} \in \mathbb{R}^m$$

is restricted to belong to the set $\mathbb{V}$, which is defined as

$$\mathbb{V} \triangleq \left\{v \in \mathbb{R}^m \text{ such that there exists } \xi \in \mathbb{U}\colon v = \begin{bmatrix} 0 \ldots 0 \ \xi \ 0 \ldots 0 \end{bmatrix}^{\mathrm{T}}\right\}.$$

As depicted in Figure 16.3 for $\mathbb{V}$ in $\mathbb{R}^2$, this set contains all column vectors formed by one element of $\mathbb{U}$, whilst all its other components are zero.

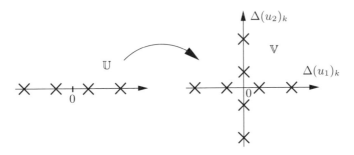

Figure 16.3. Construction of the set $\mathbb{V}$.

The performance of the model (16.1) over a finite horizon N starting at time k is quantified by means of the quadratic objective function

$$V_N \triangleq \sum_{j=k+1}^{k+N} \|y_j' - r^\star\|^2 + \sum_{j=k}^{k+N-1} \|u_j - u^\star\|_R^2, \tag{16.6}$$

where $R = R^{\mathrm{T}} > 0$ and $u^\star$ is defined via[3]

$$r^\star = C(I_n - A)^{-1} B u^\star. \tag{16.7}$$

In (16.6), y_j' and u_j are predicted trajectories satisfying (16.1).

The control increments Δu_j are to be sent down the channel. Thus, the decision variables in the minimisation of (16.6) are the predicted control increments

$$\Delta u_j \triangleq u_j - u_{j-1} \quad \text{for } j = k, \ldots, k + N - 1. \tag{16.8}$$

These can be gathered into the vector

$$\Delta \mathbf{u}_{k:k+N-1} \triangleq \left[(\Delta u_k)^{\mathrm{T}} \ (\Delta u_{k+1})^{\mathrm{T}} \ \ldots \ (\Delta u_{k+N-1})^{\mathrm{T}}\right]^{\mathrm{T}} \in \mathbb{R}^{Nm}, \tag{16.9}$$

which needs to satisfy the finite set constraint $\Delta \mathbf{u}_{k:k+N-1} \in \mathbb{V}^N$, where

$$\mathbb{V}^N \triangleq \mathbb{V} \times \cdots \times \mathbb{V}.$$

In the remainder of this section, we will write $V_N(\Delta \mathbf{u}_{k:k+N-1})$ in order to make explicit the dependence of the objective function (16.6) on the control increments.

---

[3] We assume that A has no eigenvalue at one.

Remark 16.3.1. The framework can be extended to plants whose inputs u_k are affected by saturation. In this case, the predicted increments $\Delta\mathbf{u}_{k:k+N-1}$ can be restricted to belong to a subset of $\mathbb{V}^N$, which keeps predicted inputs u_{k+i}, $i = 0, \ldots, N-1$, within the unsaturated region. Naturally, the resulting subsets are time-varying, since they depend upon the value of the last control input, u_{k-1}.

$\circ$

Based upon the objective function (16.6), we propose, as in Chapter 13, to utilise a receding horizon scheme. Therefore, at each time step, we solve the finite set constrained optimisation problem

$$\Delta\mathbf{u}_{k:k+N-1}^{\text{OPT}} \triangleq \arg\min_{\Delta\mathbf{u}_{k:k+N-1}\in\mathbb{V}^N} V_N(\Delta\mathbf{u}_{k:k+N-1}). \tag{16.10}$$

Instead of implementing the entire control sequence contained in the vector $\Delta\mathbf{u}_{k:k+N-1}^{\text{OPT}}$, we only utilise its first m components (where m is the number of actuators):

$$\Delta u_k^{\text{OPT}} \triangleq L^{\text{T}}\Delta\mathbf{u}_{k:k+N-1}^{\text{OPT}}, \tag{16.11}$$

where[4]

$$L^{\text{T}} \triangleq \begin{bmatrix} I_m & 0_m & \ldots & 0_m \end{bmatrix} \in \mathbb{R}^{m\times Nm}. \tag{16.12}$$

The vector (16.11) contains the data corresponding to all m inputs. Since $\Delta u_k^{\text{OPT}} \in \mathbb{V}$, not more than one of its components is nonzero. Only this value is sent to the plant input determined by its index (see Figure 16.2). The other $m-1$ inputs are left unchanged as prescribed in Constraint 2. It can be seen that the strategy consists of only sending the most relevant control increment, as quantified by the objective function (16.6). Note that not only the control increment to be applied at each time is provided by solving (16.10), but also the question of which input to access is addressed.

As shown in Chapter 13, the control scheme can be implemented by utilising linear filters and a nearest neighbour vector quantiser.

For that purpose, it is useful to define the vectors

$$\mathbf{u}_{k:k+N-1} \triangleq \begin{bmatrix} u_k \\ u_{k+1} \\ \vdots \\ u_{k+N-1} \end{bmatrix}, \quad \mathbf{y}'_{k+1:k+N} \triangleq \begin{bmatrix} y'_{k+1} \\ y'_{k+2} \\ \vdots \\ y'_{k+N} \end{bmatrix}, \quad \mathbf{r}^\star \triangleq \begin{bmatrix} r^\star \\ r^\star \\ \vdots \\ r^\star \end{bmatrix}, \quad \mathbf{u}^\star \triangleq \begin{bmatrix} u^\star \\ u^\star \\ \vdots \\ u^\star \end{bmatrix},$$

and to note that (16.6) can be re-written in vector form as

$$V_N(\Delta\mathbf{u}_{k:k+N-1}) = \|\mathbf{y}'_{k+1:k+N} - \mathbf{r}^\star\|^2 + \|\mathbf{u}_{k:k+N-1} - \mathbf{u}^\star\|_{\bar{R}}^2, \tag{16.13}$$

where

$$\bar{R} \triangleq \text{diag}\{R, \ldots, R\}.$$

---

[4] I_m denotes the $m \times m$ identity matrix and $0_m \triangleq 0 \cdot I_m$.

Furthermore, iteration of (16.1) yields

$$\mathbf{y}'_{k+1:k+N} = \Phi \mathbf{u}_{k:k+N-1} + \Lambda x_k, \tag{16.14}$$

where

$$\Phi \triangleq \begin{bmatrix} CB & 0 & \cdots & 0 \\ CAB & CB & \ddots & \vdots \\ \vdots & & \ddots & \ddots & 0 \\ CA^{N-1}B & \cdots & CAB & CB \end{bmatrix}, \quad \Lambda \triangleq \begin{bmatrix} CA \\ CA^2 \\ \vdots \\ CA^N \end{bmatrix}.$$

In order to include the predicted increments (16.9) in these expressions, notice that

$$\Delta \mathbf{u}_{k:k+N-1} = E \mathbf{u}_{k:k+N-1} - L u_{k-1}, \tag{16.15}$$

where

$$E \triangleq \begin{bmatrix} I_m & \cdots & & \cdots & 0_m \\ -I_m & I_m & & & \vdots \\ 0_m & \ddots & \ddots & & \\ \vdots & \ddots & \ddots & \ddots & \vdots \\ 0_m & \cdots & 0_m & -I_m & I_m \end{bmatrix} \in \mathbb{R}^{Nm \times Nm},$$

and L is defined in (16.12).

Using the nearest neighbour vector quantiser (see Definition 13.3.1 in Chapter 13) we can describe the control law as follows:

Theorem 16.3.1 (Closed Form Solution) *The optimiser in (16.10) satisfies*

$$\Delta \mathbf{u}^{\text{OPT}}_{k:k+N-1} = H^{-1/2} q_{\widetilde{\mathbb{V}}^N}(-H^{-\text{T}/2} f), \tag{16.16}$$

where

$$\begin{aligned} H &\triangleq E^{-\text{T}} \left(\Phi^\text{T} \Phi + \bar{R} \right) E^{-1}, \\ f &\triangleq H L u_{k-1} - E^{-\text{T}} \left[\bar{R} \mathbf{u}^\star + \Phi^\text{T} \left(\mathbf{r}^\star - \Lambda x_k \right) \right], \end{aligned} \tag{16.17}$$

and the square matrix $H^{1/2}$ is defined via $H^{\text{T}/2} H^{1/2} = H$.

The nonlinear mapping $q_{\widetilde{\mathbb{V}}^N}(\cdot)$ is the nearest neighbour quantiser described in Definition 13.3.1 of Chapter 13. The image of this mapping is the set:

$$\widetilde{\mathbb{V}}^N \triangleq \left\{ \widetilde{\mathbf{v}} \in \mathbb{R}^{Nm} \text{ such that there exists } \mathbf{v} \in \mathbb{V}^N : \widetilde{\mathbf{v}} = H^{1/2} \mathbf{v} \right\}. \tag{16.18}$$

Proof. The proof follows closely that of Quevedo et al. (2004).

Given expressions (16.13) to (16.15) it follows that

$$V_N(\Delta\mathbf{u}_{k:k+N-1}) = \|\Phi\mathbf{u}_{k:k+N-1} + \Lambda x_k - \mathbf{r}^\star\|^2 + \|\mathbf{u}_{k:k+N-1} - \mathbf{u}^\star\|_{\bar{R}}^2$$
$$= \|\Phi E^{-1}(\Delta\mathbf{u}_{k:k+N-1} + Lu_{k-1}) + \Lambda x_k - \mathbf{r}^\star\|^2$$
$$+ \|E^{-1}(\Delta\mathbf{u}_{k:k+N-1} + Lu_{k-1}) - \mathbf{u}^\star\|_{\bar{R}}^2$$
$$= \|\Delta\mathbf{u}_{k:k+N-1}\|_H^2 + 2(\Delta\mathbf{u}_{k:k+N-1})^\mathrm{T}f + \bar{V}_N(x_k, \mathbf{r}^\star, u_{k-1}),$$
$$(16.19)$$

where $\bar{V}_N(x_k, \mathbf{r}^\star, u_{k-1})$ does not depend upon $\Delta\mathbf{u}_{k:k+N-1}$, and f and H are defined in (16.17).

We introduce the change of variables,

$$\Delta\tilde{\mathbf{u}}_{k:k+N-1} = H^{1/2}\Delta\mathbf{u}_{k:k+N-1}.$$

This transforms $\mathbb{V}^N$ into $\widetilde{\mathbb{V}}^N$ defined in (16.18). Expression (16.19) then allows us to rewrite the optimiser (16.10) as

$$\Delta\mathbf{u}_{k:k+N-1}^{\mathrm{OPT}} = H^{-1/2} \quad \underset{\Delta\tilde{\mathbf{u}}_{k:k+N-1}\in\widetilde{\mathbb{V}}^N}{\arg\min} \quad \varphi(\Delta\tilde{\mathbf{u}}_{k:k+N-1}), \qquad (16.20)$$

$$\varphi(\Delta\tilde{\mathbf{u}}_{k:k+N-1}) \triangleq (\Delta\tilde{\mathbf{u}}_{k:k+N-1})^\mathrm{T}\Delta\tilde{\mathbf{u}}_{k:k+N-1} + 2(\Delta\tilde{\mathbf{u}}_{k:k+N-1})^\mathrm{T}H^{-\mathrm{T}/2}f.$$

The level sets of $\varphi(\cdot)$ are spheres in $\mathbb{R}^{Nm}$, centred at $-H^{-\mathrm{T}/2}f$. Hence, the constrained optimiser is given by the nearest neighbour:

$$\underset{\Delta\tilde{\mathbf{u}}_{k:k+N-1}\in\widetilde{\mathbb{V}}^N}{\arg\min} \quad \varphi(\Delta\tilde{\mathbf{u}}_{k:k+N-1}) = q_{\widetilde{\mathbb{V}}N}(-H^{-\mathrm{T}/2}f).$$

Using the above in (16.20) then establishes the result given in (16.16).   □

As a consequence of Theorem 16.3.1, the scheme can be characterised by means of the closed loop depicted in Figure 16.4.

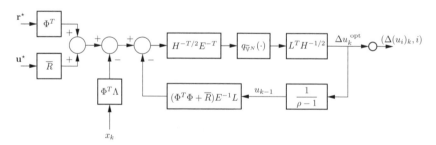

Figure 16.4. Implementation of the receding horizon networked controller.

The term f in the solution (16.16) contains the previous control value u_{k-1}. We propose to calculate it directly in the controller by integrating all previous increments as illustrated in Figure 16.4.

16.4 Up-Link

The up-link problem contains some novel challenges not met elsewhere in the book. The key issue here is that of *quantised measurements*. This is a form of constraint on how we "look at" the system rather than the system itself. Thus, the methodologies needed differ somewhat from those described in Chapter 9, where we treated constrained estimation with constraints that were integral to the system operation. In this section we will describe two approaches to deal with quantised measurements. In the first approach we use a stochastic problem formulation and design an approximate minimum variance state estimator. In the second approach we adopt a deterministic formulation with unknown-but-bounded description of uncertainty and design a recursive set-valued observer. Here we give a brief outline of the algorithms and refer the reader to Haimovich, Goodwin and Quevedo (2003) and Haimovich, Goodwin and Welsh (2004) for details.

We take k as the current sample time and use $\mathbf{y}_{a:b}$ to denote past output data from sample time a to b inclusive.

16.4.1 Stochastic Formulation

We consider a noisy discrete time, linear time-invariant plant described by

$$\begin{aligned} x_{j+1} &= Ax_j + Bu_j + w_j \\ y_j &= Cx_j + v_j, \end{aligned} \tag{16.21}$$

where $x_j, w_j \in \mathbb{R}^n$, $u_j \in \mathbb{R}^m$, $y_j, v_j \in \mathbb{R}^s$, and w_j and v_j are i.i.d. samples of white Gaussian noise processes having covariance matrices Q and R, respectively. We further assume that the measurements are *quantised* prior to transmission:

$$y_j^q = \sigma(y_j), \tag{16.22}$$

where y_j^q is the measured variable. Here, the finite range mapping $\sigma(\cdot)$ corresponds to a general quantiser, defined as

$$\sigma(y_j) = \begin{cases} \sigma_1 & \text{if } y_j \in \mathcal{R}_{\sigma_1}, \\ \sigma_2 & \text{if } y_j \in \mathcal{R}_{\sigma_2}, \\ \vdots & \vdots \quad \vdots \\ \sigma_{\text{L}} & \text{if } y_j \in \mathcal{R}_{\sigma_{\text{L}}}, \end{cases} \tag{16.23}$$

where L is the number of quantisation regions and $\{\mathcal{R}_{\sigma_i} : i = 1, \ldots, \text{L}\}$ is a given partition of $\mathbb{R}^s$.

Note that we have assumed that (quantised) measurements are available at each discrete time-step. However, it would be a straightforward extension to also use time-stamped data with event-based sampling to further limit transmission rates so that we only transmit when a measurement changes zones.

The statistical behaviour of the system is described by

$$x_0 \sim N(x_{00}, P_{00}), \quad w_i \sim N(0, Q), \quad v_i \sim N(0, R), \tag{16.24}$$

$$\mathbf{E}(w_i w_j^\mathsf{T}) = 0 \text{ and } \mathbf{E}(v_i v_j^\mathsf{T}) = 0 \quad \text{if } i \neq j, \tag{16.25}$$

$$\mathbf{E}(w_i v_j^\mathsf{T}) = 0, \quad \mathbf{E}(w_i x_0^\mathsf{T}) = 0, \quad \mathbf{E}(v_i x_0^\mathsf{T}) = 0. \tag{16.26}$$

The input u_j is assumed to be known at the controller.

We are interested in determining the *conditional mean* of the current state, x_k, given the collection of measurements from time instant 1 to the present time instant k. Let $\mathbf{y}_{1:k}^q$ denote the vector containing all the measurements up to the current time instant. Then, the required state estimate at time instant k is given by the following expected value:

$$\mathbf{E}(x_k | \mathbf{y}_{1:k}^q) = \mathbf{E}(x_k | y_1^q, \dots, y_k^q) = \mathbf{E}(x_k | y_1 \in \mathcal{R}_{y_1^q}, \dots, y_k \in \mathcal{R}_{y_k^q}). \tag{16.27}$$

The estimate given by (16.27) can be obtained in a conceptually simple way by taking advantage of the linear structure of the system between input, state and the unavailable (unmeasured) variable y_k. Using a property of the expectation operator, we can write

$$\mathbf{E}(x_k | \mathbf{y}_{1:k}^q) = \mathbf{E}(x_k | y_1^q, \dots, y_k^q) = \mathbf{E}(\mathbf{E}(x_k | y_1, \dots, y_k) | y_1^q, \dots, y_k^q)$$
$$= \mathbf{E}(\mathbf{E}(x_k | \mathbf{y}_{1:k}) | \mathbf{y}_{1:k}^q), \tag{16.28}$$

where $\mathbf{y}_{1:k}$ denotes the collection of variables $y_1, \dots, y_k$. The inner expected value on the right hand side of (16.28) is provided by the equations of the Kalman filter (see Theorem 9.6.2 in Chapter 9), which are summarised below in equations (16.29) to (16.35):[5]

$$\mathbf{E}(x_0) \triangleq x_{00}, \tag{16.29}$$

$$P_{0|0} \triangleq P_{00}, \tag{16.30}$$

$$\mathbf{E}(x_k | \mathbf{y}_{1:k-1}) = A \, \mathbf{E}(x_{k-1} | \mathbf{y}_{1:k-1}) + B u_{k-1}, \tag{16.31}$$

$$P_{k|k-1} = A P_{k-1|k-1} A^\mathsf{T} + Q, \tag{16.32}$$

$$K_k = P_{k|k-1} C^\mathsf{T} (C P_{k|k-1} C^\mathsf{T} + R)^{-1}, \tag{16.33}$$

$$\mathbf{E}(x_k | \mathbf{y}_{1:k}) = \mathbf{E}(x_k | \mathbf{y}_{1:k-1}) + K_k \left[y_k - C \, \mathbf{E}(x_k | \mathbf{y}_{1:k-1}) \right], \tag{16.34}$$

$$P_{k|k} = (I_n - K_k C) P_{k|k-1}. \tag{16.35}$$

The analysis that follows depends on interpreting $\mathbf{E}(x_k | \mathbf{y}_{1:k})$ as a *function of the random variable* $\mathbf{y}_{1:k}$ rather than as a function of a given realisation of the output, $\{y_j\}$. Once this function is determined, the outer expected value in (16.28) has to be evaluated as follows:

---

[5] I_n denotes the $n \times n$ identity matrix.

$$\mathbf{E}(x_k|y_1^q,\ldots,y_k^q) = \frac{\displaystyle\int_{\rho_{1:k}} p_{y_1,\ldots,y_k}(\gamma_1,\ldots,\gamma_k)\,\mathbf{E}(x_k|y_1{=}\gamma_1,\ldots,y_k{=}\gamma_k)d\gamma_1\ldots d\gamma_k}{\displaystyle\int_{\rho_{1:k}} p_{y_1,\ldots,y_k}(\gamma_1,\ldots,\gamma_k)d\gamma_1\ldots d\gamma_k},$$

(16.36)

where $\rho_{1:k}$ is the region $\mathcal{R}_{y_1^q} \times \cdots \times \mathcal{R}_{y_k^q}$ (in $\mathbb{R}^{s\times k}$), and $p_{y_1,\ldots,y_k}$ is the joint probability density function (pdf) of the random variables $y_1,\ldots,y_k$. This joint pdf can be expressed as in the following lemma.

Lemma 16.4.1 *The joint pdf of the random variables $y_1,\ldots,y_k$ is Gaussian and given by*

$$p_{y_1,\ldots,y_k}(\gamma_1,\ldots,\gamma_k) = \frac{\exp\{-\frac{1}{2}[\Gamma - \mathbf{E}(\mathbf{y}_{1:k})]^{\mathrm{T}}[\mathrm{cov}(\mathbf{y}_{1:k})]^{-1}[\Gamma - \mathbf{E}(\mathbf{y}_{1:k})]\}}{(2\pi)^{sk/2}|\mathrm{cov}(\mathbf{y}_{1:k})|^{1/2}},$$

(16.37)

with $\Gamma = (\gamma_1^{\mathrm{T}},\ldots,\gamma_k^{\mathrm{T}})^{\mathrm{T}}$, and where $\mathbf{E}(\mathbf{y}_{1:k})$ and $\mathrm{cov}(\mathbf{y}_{1:k})$ are defined as

$$\mathbf{E}(\mathbf{y}_{1:k}) = \mathbf{\Lambda}_k\,\mathbf{E}(x_0) + \mathbf{\Phi}_k\mathbf{u}_{0:k-1} = \mathbf{\Lambda}_k x_{00} + \mathbf{\Phi}_k\mathbf{u}_{0:k-1},$$

(16.38)

$$\mathrm{cov}(\mathbf{y}_{1:k}) = \begin{bmatrix} \mathrm{cov}(y_1,y_1) & \mathrm{cov}(y_1,y_2) & \cdots & \mathrm{cov}(y_1,y_k) \\ \mathrm{cov}(y_2,y_1) & \mathrm{cov}(y_2,y_2) & \cdots & \mathrm{cov}(y_2,y_k) \\ \vdots & \vdots & \ddots & \vdots \\ \mathrm{cov}(y_k,y_1) & \mathrm{cov}(y_k,y_2) & \cdots & \mathrm{cov}(y_k,y_k) \end{bmatrix},$$

$$\mathrm{cov}(y_i,y_j) = CA^i P_{00}(CA^j)^{\mathrm{T}} + \sum_{l=1}^{\min\{i,j\}} CA^{i-l}Q(CA^{j-l})^{\mathrm{T}} + \delta_{ij}R,$$

$$i,j = 1,\ldots,k, \quad (16.39)$$

with δ_{ij} the Kronecker delta, $\mathbf{u}_{0:k-1} = \begin{bmatrix} u_0^{\mathrm{T}} & u_1^{\mathrm{T}} & \ldots & u_{k-1}^{\mathrm{T}} \end{bmatrix}^{\mathrm{T}}$, and for suitable constant matrices $\mathbf{\Lambda}_k$ and $\mathbf{\Phi}_k$.

Proof. By iterating (16.21), we can write

$$\begin{bmatrix} y_1 \\ y_2 \\ \vdots \\ y_k \end{bmatrix} = \begin{bmatrix} CA \\ CA^2 \\ \vdots \\ CA^k \end{bmatrix} x_0 + \begin{bmatrix} CB & 0 & \cdots & 0 \\ CAB & CB & \cdots & 0 \\ \vdots & \vdots & \ddots & \vdots \\ CA^{k-1}B & CA^{k-2}B & \cdots & CB \end{bmatrix} \begin{bmatrix} u_0 \\ u_1 \\ \vdots \\ u_{k-1} \end{bmatrix}$$

$$+ \begin{bmatrix} C & 0 & \cdots & 0 \\ CA & C & \cdots & 0 \\ \vdots & \vdots & \ddots & \vdots \\ CA^{k-1} & CA^{k-2} & \cdots & C \end{bmatrix} \begin{bmatrix} w_0 \\ w_1 \\ \vdots \\ w_{k-1} \end{bmatrix} + \begin{bmatrix} v_1 \\ v_2 \\ \vdots \\ v_k \end{bmatrix},$$

(16.40)

which can be succinctly written as

$$\mathbf{y}_{1:k} = \mathbf{\Lambda}_k x_0 + \mathbf{\Phi}_k \mathbf{u}_{0:k-1} + \mathbf{\Omega}_k \mathbf{w}_{0:k-1} + \mathbf{v}_{1:k}, \qquad (16.41)$$

with the obvious definitions for $\mathbf{\Lambda}_k$, $\mathbf{\Phi}_k$, $\mathbf{u}_{0:k-1}$, $\mathbf{\Omega}_k$, $\mathbf{w}_{0:k-1}$ and $\mathbf{v}_{1:k}$. It can be appreciated that $\mathbf{y}_{1:k}$ results from a linear combination of (jointly) Gaussian random variables and hence it is also Gaussian. Its mean and covariance (equations (16.38) to (16.39)) can be straightforwardly obtained from (16.40) and (16.41), noting that $\mathrm{cov}(x_0, x_0) = P_{00}$. □

Thus, the desired estimate could, in principle, be obtained by performing the multivariate integration shown in (16.36). However, direct application of (16.36) becomes impractical for the following reasons:

(i) The multiple integration in (16.36) has to be solved numerically since, typically, no explicit closed form solution exists.
(ii) As new measurements become available, the complexity of the integration in (16.36) increases since it is performed over the region $\rho_{1:k}$ in $\mathbb{R}^{s \times k}$ with growing k (recall that k denotes the current time).

A practical approach to this problem is achieved by bringing together ideas from Monte Carlo sampling and moving horizon estimation.

To provide a numerical solution to (16.36), we propose the use of Monte Carlo sampling. Denote by S_k a set of N_s samples drawn from the Gaussian density $p_{y_1,\dots,y_k}$ at time instant k. Define $S_{\rho_{1:k}} \subset S_k$ as the set of samples that fall inside the region $\rho_{1:k}$, that is,

$$S_{\rho_{1:k}} \triangleq S_k \cap \rho_{1:k}. \qquad (16.42)$$

Equation (16.36) can then be approximated by

$$\mathbf{E}(x_k | \mathbf{y}_{1:k}^q) = \mathbf{E}(x_k | y_1^q, \dots, y_k^q) \approx \frac{1}{N_{\rho_{1:k}}} \sum_{\mathbf{z} \in S_{\rho_{1:k}}} \mathbf{E}(x_k | \mathbf{y}_{1:k} = \mathbf{z}), \qquad (16.43)$$

where $N_{\rho_{1:k}}$ is the number of samples that fall inside the region $\rho_{1:k}$, that is, the cardinality of the set $S_{\rho_{1:k}}$. The closeness of the Monte Carlo approximation to the true value of the estimate will directly depend on $N_{\rho_{1:k}}$: Larger values of this number yield better approximations.

As far as the growing complexity is concerned, we adopt a *moving horizon* approach as in Section 9.9 of Chapter 9. At time instant k, we will explicitly take into account only the last N_e measurements available, $\mathbf{y}_{k-N_e+1:k}^q$, whereas the older measurements, $\mathbf{y}_{1:k-N_e}^q$, will be dealt with by summarising the history of the system before time instant $k - N_e$ by adopting a suitable conditional distribution for the state x_{k-N_e}. Consequently, at time instant k, the state x_{k-N_e} will be regarded as the initial state. The parameter N_e will be called the *estimation horizon* and will allow us to trade performance for complexity.

At the outset, the distribution ascribed to the initial state x_0 given in (16.24) could be interpreted as the summary of the system history prior to time instant 1. This Gaussianity assumption for the initial state, together with the corresponding ones for the process and measurement noises and the linearity of the system defined in (16.21) rendered the equations of the Kalman filter suitable for determining the expected value $\mathbf{E}(x_k|\mathbf{y}_{1:k})$ needed for the two-step evaluation procedure described in (16.28).

It follows from the Chapman–Kolmogorov equation (see Section 9.8 in Chapter 9) that knowledge of the conditional probability density function of the state x_{k-N_e} of the form $p(x_{k-N_e}|\mathbf{y}^q_{1:k-N_e})$ is all that is needed in order to summarise the history of the system if we are interested only in subsequent estimates of the form

$$\mathbf{E}(x_{k-N_e+i}|\mathbf{y}^q_{1:k-N_e+i}) \quad \text{for } i = 1, 2, \ldots. \tag{16.44}$$

Due to the quantised nature of the measurements, the conditional density of the state x_{k-N_e} given the measurements $\mathbf{y}^q_{1:k-N_e}$ is not Gaussian. However, approximating this density as Gaussian enables us to utilise the two-step procedure shown in (16.28) because the inner expected value can again be provided by the equations of the Kalman filter, regarding x_{k-N_e} as the initial state.

The proposed approach works as follows. The initial estimates $\mathbf{E}(x_k|\mathbf{y}^q_{1:k})$ for $k = 1, \ldots, N_e$ are found exactly as described so far in this section and the multivariate integrals in (16.36) are approximated with the aid of Monte Carlo sampling. From time instant $N_e + 1$ on, the conditional densities $p(x_{k-N_e}|\mathbf{y}^q_{1:k-N_e})$ are approximated by $N(x'_{k-N_e}, P'_{k-N_e})$, where x'_{k-N_e} and P'_{k-N_e} are design parameters. Equations (16.29) and (16.30) are replaced by

$$\mathbf{E}(x_{k-N_e}|\mathbf{y}_{1:k-N_e}) \triangleq x'_{k-N_e}, \tag{16.45}$$

$$P_{k-N_e|k-N_e} \triangleq P'_{k-N_e}, \tag{16.46}$$

whilst equations (16.31) to (16.35) remain unchanged. The approximate estimate of $\mathbf{E}(x_k|\mathbf{y}^q_{1:k})$ is then computed as

$$\mathbf{E}(x_k|\mathbf{y}^q_{1:k}) \approx \frac{\mathcal{I}_1}{\mathcal{I}_2}, \tag{16.47}$$

where

$$\mathcal{I}_1 \triangleq \int_{\rho_{k-N_e+1:k}} p_{y_{k-N_e+1},\ldots,y_k}(\gamma_1, \ldots, \gamma_{N_e})$$

$$\times \mathbf{E}(x_k|y_{k-N_e+1} = \gamma_1, \ldots, y_k = \gamma_{N_e})d\gamma_1 \ldots d\gamma_{N_e},$$

$$\mathcal{I}_2 \triangleq \int_{\rho_{k-N_e+1:k}} p_{y_{k-N_e+1},\ldots,y_k}(\gamma_1, \ldots, \gamma_{N_e})d\gamma_1 \ldots d\gamma_{N_e},$$

and where $\rho_{k-N_e+1:k}$ is the region $\mathcal{R}_{y^q_{k-N_e+1}} \times \cdots \times \mathcal{R}_{y^q_k}$ in $\mathbb{R}^{s \times N_e}$. The multivariate integrals $\mathcal{I}_1$ and $\mathcal{I}_2$ are performed with the aid of Monte Carlo sampling. The density $p_{y_{k-N_e+1},\ldots,y_k}$ is now the joint pdf of the variables $y_{k-N_e+1}, \ldots, y_k$ under the Gaussianity assumption for x_{k-N_e}. This density is quantified in the following lemma.

Lemma 16.4.2 *The joint pdf of the random variables* $y_{k-N_e+1}, \ldots, y_k$, *under the assumption that the distribution of the state* x_{k-N_e} *is Gaussian, is given by*

$$p_{y_{k-N_e+1},\ldots,y_k}(\gamma_1, \ldots, \gamma_{N_e}) =$$

$$\frac{\exp\left\{ -\tfrac{1}{2}[\Gamma_{1:N_e} - \mathbf{E}(\mathbf{y}_{k-N_e+1:k})]^{\mathrm{T}}[\mathrm{cov}(\mathbf{y}_{k-N_e+1:k})]^{-1}[\Gamma_{1:N_e} - \mathbf{E}(\mathbf{y}_{k-N_e+1:k})] \right\}}{(2\pi)^{sN_e/2}|\mathrm{cov}(\mathbf{y}_{k-N_e+1:k})|^{1/2}},$$

with $\Gamma_{1:N_e} = (\gamma_1^{\mathrm{T}}, \ldots, \gamma_{N_e}^{\mathrm{T}})^{\mathrm{T}}$, *and where* $\mathbf{E}(\mathbf{y}_{k-N_e+1:k})$ *and* $\mathrm{cov}(\mathbf{y}_{k-N_e+1:k})$ *are given by*

$$\mathbf{E}(\mathbf{y}_{k-N_e+1:k}) = \mathbf{\Lambda}_{N_e} \mathbf{E}(x_{k-N_e}) + \mathbf{\Phi}_{N_e}\mathbf{u}_{k-N_e:k-1}$$
$$= \mathbf{\Lambda}_{N_e}x'_{k-N_e} + \mathbf{\Phi}_{N_e}\mathbf{u}_{k-N_e:k-1}, \qquad (16.48)$$

$$\mathrm{cov}(\mathbf{y}_{k-N_e+1:k}) =$$
$$\begin{bmatrix} \mathrm{cov}(y_{k-N_e+1}, y_{k-N_e+1}) & \mathrm{cov}(y_{k-N_e+1}, y_{k-N_e+2}) & \cdots & \mathrm{cov}(y_{k-N_e+1}, y_k) \\ \mathrm{cov}(y_{k-N_e+2}, y_{k-N_e+1}) & \mathrm{cov}(y_{k-N_e+2}, y_{k-N_e+2}) & \cdots & \mathrm{cov}(y_{k-N_e+2}, y_k) \\ \vdots & \vdots & \ddots & \vdots \\ \mathrm{cov}(y_k, y_{k-N_e+1}) & \mathrm{cov}(y_k, y_{k-N_e+2}) & \cdots & \mathrm{cov}(y_k, y_k) \end{bmatrix},$$
$$(16.49)$$

$$\mathrm{cov}(y_{k-N_e+i}, y_{k-N_e+j}) = CA^i P'_{k-N_e}(CA^j)^{\mathrm{T}} + \sum_{l=1}^{\min\{i,j\}} CA^{i-l}Q(CA^{j-l})^{\mathrm{T}}$$
$$+ \delta_{ij}R, \qquad i,j = 1,\ldots,N_e,$$

where δ_{ij} *is the Kronecker delta and* $\mathbf{u}_{k-N_e:k-1}$, $\mathbf{\Lambda}_{N_e}$ *and* $\mathbf{\Phi}_{N_e}$ *are as in Lemma 16.4.1.*

Proof. The result follows as in Lemma 16.4.1, by iterating (16.21), starting from time instant $k - N_e$, and under the Gaussianity assumption for the state x_{k-N_e}. $\qquad \square$

A graphical representation of the suggested strategy is depicted in Figure 16.5.

The values adopted for the parameters x'_{k-N_e} and P'_{k-N_e} will affect the approximate estimate sought. The quantity x'_{k-N_e} can be adopted as the (approximate) estimate of x_{k-N_e} found N_e time instants before. As regards P'_{k-N_e}, different options can be considered:

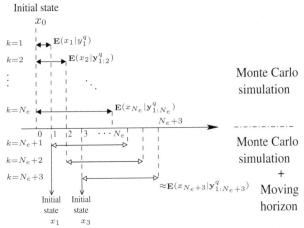

Figure 16.5. Moving horizon Monte Carlo state estimation strategy.

(i) $P'_{k-N_e} = 0$. In this case, perfect knowledge of x_{k-N_e} is assumed.

(ii) $P'_{k-N_e} \to \infty$. This corresponds to assuming very little precision of the estimate of the initial state.

(iii) $0 < P'_{k-N_e} < \infty$, arbitrary. Here, P'_{k-N_e} is regarded as a tuning parameter.

(iv) $P'_{k-N_e} \approx \text{cov}(x_{k-N_e}|\mathbf{y}^q_{1:k-N_e})$. Here, P'_{k-N_e} is set equal to an approximation of the true conditional covariance of the initial state.

Option (i) is analogous to the idea of decision feedback equalisers in digital communications; see for example, Proakis (1995).

The procedure described above can be summarised in the following algorithm, which holds for $k > N_e$.

Algorithm 16.4.1 (Moving Horizon Monte Carlo State Estimation)

(i) *Adopt the parameters for the initial state distribution* $(x'_{k-N_e}$ *and* $P'_{k-N_e})$.

(ii) *Calculate the mean and covariance of* $\mathbf{y}_{k-N_e+1:k}$ *as in* (16.48) *and* (16.49).

(iii) *Draw* N_s *samples from* $N(\mathbf{E}(\mathbf{y}_{k-N_e+1:k}), \text{cov}(\mathbf{y}_{k-N_e+1:k}))$ *and determine which ones fall inside the region* $\rho_{k-N_e+1:k}$.

(iv) *Approximate* $\mathbf{E}(x_k|\mathbf{y}^q_{1:k})$ *as in* (16.47) *and* (16.43).

(v) *Set* $k \leftarrow k + 1$ *and go to step* (i).   ○

16.4.2 Deterministic Formulation

Here we consider an alternative discrete time, deterministic linear time-invariant system with quantised state measurements

$$x_{k+1} = Ax_k + Bu_k + w_k, \qquad (16.50)$$

$$y^q_k = \sigma(x_k + v_k), \qquad (16.51)$$

where $x_k \in \mathbb{R}^n$, $u_k \in \mathbb{R}^m$, and y_k^q is the measured variable. $\{v_k\}$ and $\{w_k\}$ are unknown disturbance sequences that comply with

$$||w_k||_p < \delta \quad \text{for } k = 0, 1, \ldots, \tag{16.52}$$
$$||v_k||_p < \epsilon \quad \text{for } k = 0, 1, \ldots, \tag{16.53}$$

where ϵ and δ are known constants, and p is either 1 or ∞. (Although assumed deterministic, we will refer to v_k and w_k as measurement noise and process noise, respectively.) In (16.51), $\sigma(\cdot)$ is a finite range mapping that defines the state quantiser

$$\sigma(x_k) = \begin{cases} \sigma_1 & \text{if } x_k \in \mathcal{Y}_1, \\ \sigma_2 & \text{if } x_k \in \mathcal{Y}_2, \\ \vdots & \vdots \quad \vdots \\ \sigma_L & \text{if } x_k \in \mathcal{Y}_L, \end{cases} \tag{16.54}$$

where L is the number of quantisation regions.

We make the following assumptions:

A1. $\mathcal{Y}_i$, $i = 1, \ldots, L$, are dense in $\mathbb{R}^n$, nonempty, convex and polytopic.[6]
A2. The regions $\{\mathcal{Y}_i : i = 1, \ldots, L\}$ form a partition of a given convex polytope $\mathcal{H}$, defined in $\mathbb{R}^n$.
A3. $x_k \in \mathcal{H}$, $\quad k = 0, 1, \ldots$.

Assumptions A2 and A3 imply that at any time instant, the state will belong to the domain of definition of the quantiser function.

Note that the system model defined in (16.50), (16.51) and (16.54) is completely deterministic. There exists, however, uncertainty from the observer point of view due to the presence of the disturbance sequences $\{w_k\}$ and $\{v_k\}$, and since all that is known about the initial state x_0 is that is must belong to the region $\mathcal{H}$.

At the current time instant k we would like to combine the information provided by the available measurements in order to obtain the smallest region of the state space where the state x_{k+1} has to lie. We will henceforth refer to this region as the *set-valued state estimate*.

If no measurement noise were present ($v_k \equiv 0$), the event of receiving the quantised measurement $y_k^q = \sigma_i$ would imply that $x_k \in \mathcal{Y}_i$ (see (16.54)). However, measurement noise could cause a measurement indicating a different region to be received whenever the state x_k is closer than ϵ (see (16.53)) to the border of region $\mathcal{Y}_i$. We overcome this problem in the following way. Whenever a measurement $y_k^q = \sigma_i$ is received, the observer assumes that $x_k \in \mathrm{Expand}(\mathcal{Y}_i, \epsilon) \cap \mathcal{H}$, where $\mathrm{Expand}(\cdot, \cdot)$ is defined by

$$\mathrm{Expand}(\mathcal{Y}, \epsilon) \triangleq \{x \in \mathbb{R}^n : x = z + v, \text{ for some } z \in \mathcal{Y}$$
$$\text{and some } v \text{ such that } ||v||_p \leq \epsilon\}.$$

---

[6] A polytopic set, or polytope, is a bounded polyhedral set.

If $p = 1$ or ∞ and $\mathcal{Y}_i$ is polytopic, we can readily see that $\mathrm{Expand}(\mathcal{Y}_i, \epsilon)$ is also polytopic. We thus define

$$\mathcal{Y}_i^{\mathrm{OBS}} \triangleq \mathrm{Expand}(\mathcal{Y}_i, \epsilon) \cap \mathcal{H} \quad \text{for } i = 1, \ldots, \mathrm{L}. \tag{16.55}$$

Hence, $x_k \in \mathcal{Y}_i^{\mathrm{OBS}}$ whenever the measurement $y_k^q = \sigma_i$ is received.

We will next describe how the set-valued estimate is obtained. We start by explaining the initialisation procedure and how the first iteration is performed. Then we develop a recursive algorithm to be utilised by the observer at each successive time instant. In what follows, let $\mathcal{X}_{j+1}$ denote the set-valued state estimate of x_{j+1} based on measurements y_0^q to y_j^q.

Initialisation. By assumption A3, the initial state, x_0, belongs to the region $\mathcal{H}$. Since no additional information is available at time $k = 0$, the set-valued estimate at time $k = 0$ is just $\mathcal{X}_0 \triangleq \mathcal{H}$.

First Iteration. The measurement $y_0^q = \sigma_i$ becomes available. From (16.55), it is now known that $x_0 \in \mathcal{Y}_i^{\mathrm{OBS}}$. It is also known that $x_0 \in \mathcal{X}_0$. Hence,

$$x_0 \in \mathcal{Y}_i^{\mathrm{OBS}} \cap \mathcal{X}_0.$$

We now define the region

$$\mathcal{X}_0^{\mathrm{M}} \triangleq \mathcal{Y}_i^{\mathrm{OBS}} \cap \mathcal{X}_0,$$

where the superscript ^{M} is used to denote that the region has been updated using the information provided by the last available measurement.

To obtain the set-valued estimate $\mathcal{X}_1$, for the state at time instant $k = 1$, we must transform the region $\mathcal{X}_0^{\mathrm{M}}$ by means of the system equation (16.50). This can be performed in two steps:

$$\tilde{\mathcal{X}}_1 = \{x_1 \in \mathbb{R}^n : x_1 = Ax_0 + Bu_0, \text{for some } x_0 \in \mathcal{X}_0^{\mathrm{M}}\}, \tag{16.56}$$

$$\mathcal{X}_1 = \mathrm{Expand}(\tilde{\mathcal{X}}_1, \delta) \cap \mathcal{H}. \tag{16.57}$$

The region $\mathcal{X}_1$ is the set-valued estimate of the state x_{k+1}. All available information has already been used in order to calculate it. We will denote the calculation in equation (16.56) by

$$\tilde{\mathcal{X}}_1 = A\mathcal{X}_0^{\mathrm{M}} + Bu_0.$$

The procedure described above obtains the set-valued estimate of the state at time $k + 1 = 1$ based on all the information available at time $k = 0$. We describe the procedure at an arbitrary time instant $k > 0$ in the following recursive algorithm.

Algorithm 16.4.2 (Set-valued State Estimate) *The set-valued state estimate at time $k + 1$ based on the information available at time k requires the following steps:*

(i) Measurement update: $\mathcal{X}_k^{\mathrm{M}} = \mathrm{Expand}(\sigma^{-1}(y_k^q), \epsilon) \cap \mathcal{X}_k$.
(ii) Time update: $\tilde{\mathcal{X}}_{k+1} = A\mathcal{X}_k^{\mathrm{M}} + Bu_k$.
(iii) Noise correction: $\mathcal{X}_{k+1} = \mathrm{Expand}(\tilde{\mathcal{X}}_{k+1}, \delta) \cap \mathcal{H}$.   ∘

The following properties of Algorithm 16.4.2 readily follow.

Proposition 16.4.3 (Convexity of Set-valued Estimates) *The regions* $\mathcal{X}_k$, $k = 0, 1, \ldots$, *calculated by Algorithm 16.4.2 are either empty or convex polytopes.*

Proof. By induction on k. By assumption A2, $\mathcal{X}_0 = \mathcal{H}$ is a convex polytope. Assume that the region $\mathcal{X}_k$ is a convex polytope. By assumption A1, $\sigma^{-1}(y_k^q)$ is also a convex polytope. Since the expansion operation preserves the convex-polytopic nature of the set, step (i) of Algorithm 16.4.2 then shows that $\mathcal{X}_k^{\mathrm{M}}$ is either a convex polytope or empty (since it is the result of intersecting two convex polytopes, see for example, Grünbaum (1967)). From step (ii), $\tilde{\mathcal{X}}_{k+1}$ is obtained via an affine transformation of a convex-polytopic or empty region and hence it also must be either convex-polytopic or empty (Grünbaum 1967). step (iii) is similar to step (i). The result then follows.   □

Proposition 16.4.4 (Empty Set-valued Estimates) *If, whilst iterating Algorithm 16.4.2, an empty region is obtained as the result, then all the following regions obtained will also be empty.*

Proof. By induction. Assume $\mathcal{X}_k = \emptyset$. Then, by step (i) of Algorithm 16.4.2, $\mathcal{X}_k^{\mathrm{M}} = \emptyset$. By this and steps (ii) and (iii), we have that $\mathcal{X}_k = \emptyset \Rightarrow \mathcal{X}_{k+1} = \emptyset$. The result then follows.   □

The occurrence of an empty region as the result of Algorithm 16.4.2 implies that the measurements obtained up to that moment could not have been generated by the system defined in (16.50) and (16.51) under the assumptions made. This implies that the set-valued observer is unable to supply any useful information regarding the location of the state. We will henceforth assume that this situation can only be caused by a disturbance that may instantly change the position of the state to any point in the region $\mathcal{H}$. Accordingly, to overcome the empty-region difficulty, we add the following modification to the observer strategy: *Whenever an empty region is obtained, the observer will reset, discarding all previous information, except for the very last measurement, and begin again from that time instant.*

By including this modification, we obtain the final version of the observer algorithm.

Algorithm 16.4.3 (Set-valued Observer Algorithm)

(i) Measurement update: $\mathcal{X}_k^{\mathrm{M}} = \mathrm{Expand}(\sigma^{-1}(y_k^q), \epsilon) \cap \mathcal{X}_k$.
(ii) Disturbance detection: If $\mathcal{X}_k^{\mathrm{M}} = \emptyset$, set $\mathcal{X}_k^{\mathrm{M}} = \mathrm{Expand}(\sigma^{-1}(y_k^q), \epsilon)$.
(iii) Time update: $\tilde{\mathcal{X}}_{k+1} = A\mathcal{X}_k^{\mathrm{M}} + Bu_k$.
(iv) Noise correction: $\mathcal{X}_{k+1} = \mathrm{Expand}(\tilde{\mathcal{X}}_{k+1}, \delta) \cap \mathcal{H}$.   ∘

To use the observer in combination with a state feedback controller (such as the receding horizon controller described in Section 16.3), we require a point state estimate for the current state x_k. Whenever the set-valued estimate $\mathcal{X}_k$ is nonempty, any point contained in it is a possible value for the state x_k. We propose to use a Chebyshev centre of $\mathcal{X}_k$ as the required state estimate. A Chebyshev centre of $\mathcal{X}_k$ is defined as a point inside $\mathcal{X}_k$ that is farthest from the exterior of $\mathcal{X}_k$. Note that, by definition, a Chebyshev centre of a convex region is a point that belongs to the region and may not be unique.

Computational Issues. A crucial aspect in the computational implementation of Algorithm 16.4.2 is bounding the complexity of the polytopic regions obtained. Complexity may increase in every iteration of the algorithm due to the need for intersecting polytopes, that is, the polytope $\mathcal{X}_{k+1}$ may contain a higher number of vertices than $\mathcal{X}_k$.

Complexity may be dealt with by deploying a *block moving horizon* strategy. This strategy consists of running N_0 parallel observers at every time instant and combining them in a moving horizon manner. We propose a set-valued observer strategy where no less than the last $N_e \cdot (N_0 - 1)/N_0$, but no more than the last N_e, measurements received are explicitly considered in order to calculate the region. As in the stochastic formulation, previous measurements are suitably summarised. The summarising step is carried out by bounding the polytopic region obtained with an axis-aligned hypercube. This bounding step is responsible for reducing the worst-case computational complexity of all the polytopic regions obtained.

Figure 16.6 shows the suggested scheme when only two observers are utilised. Each observer bounds the polytopic region obtained after running

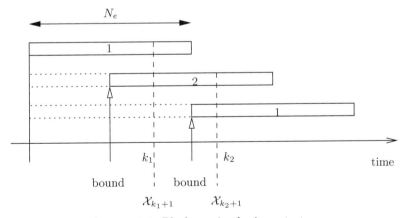

Figure 16.6. Block moving horizon strategy.

for a period of N_e time instants. At any time instant, both observers calculate a set-valued estimate. The required set-valued estimate is provided by the observer that performed the region-bounding step the longest time ago. In the figure, $\mathcal{X}_{k_1+1}$ is provided at time k_1 by observer number 1 whereas at time k_2, observer number 2 provides $\mathcal{X}_{k_2+1}$.

In a similar way, a strategy comprising more than two parallel observers could be devised. Note that the parameters N_e and N_0 allow one to trade off observer performance for computational complexity. If $N_0 = N_e$, the suggested strategy becomes the exact deterministic counterpart to the stochastic observer designed in Section 16.4.1.

16.5 Experimental Results

We consider a practical experiment undertaken on a pilot-scale plant where level control for four similar tanks was configured. We study both the up-link and down-link problems of controlling the plants over a limited bandwidth channel.

In each of the four tanks we seek to maintain a constant level by controlling the inflow by adjusting the voltage applied to a pump, where outflow is through an orifice of constant size. Control is required to reject disturbances from unmeasured inflows and increased outflows. In our setup, a single tank having the fluid level as output and, as input, the pump voltage, is described around the chosen operating point by the discrete time model (16.4) with a sampling period of 2 sec and matrices

$$A_i = 0.9826, \quad B_i = 0.0452, \quad C_i = 1.$$

On the up-link, we transmit only one bit of information per tank corresponding to whether the tank level is above or below the set-point, that is, $Y = \{-1, 1\}$ (Note that this is deliberately specified as a very poorly quantised system to highlight performance issues). Since we have 4 tanks and a 2-second sampling period, we need only send 4 bits over a 2-second period, that is, the required up-link data rate is 2 bits/sec.

We use the deterministic set-valued observer described in Section 16.4.2 to provide the state estimates. (We have also tested the alternative Monte Carlo scheme described in Section 16.4.1 and found it to be more computationally intensive than the deterministic algorithm, although it did actually give more accurate state estimates.)

Noise (var = 0.04) is deliberately added to the measured signal prior quantisation to provide a small dithering signal. To account for this noise, a region expansion of $\epsilon = 0.6$ is utilised in the estimator. The estimation horizon is set to 100 and we run two parallel observers to reduce the computational complexity.

On the down-link side, we emphasise the quantised nature of the problem and choose the (small) set $\mathbb{U} = \{-1, 0, 1\}$. This requires only 1 bit in each sampling interval corresponding to an increase or decrease of the input signal since $0 \in \mathbb{U}$ can be specified by sending nothing. We also need to send address information for the tank whose input is to be changed. This requires 2 bits for our application. Hence, the down-link data rate will be 1.5 bits/sec.

We control the four tanks by means of the receding horizon networked controller with horizon $N = 2$ based on the algorithm described in Section 16.3. We remove from the possible control actions the possibility of sending no control action to any plant. This provides the system with some degree of excitation when the plant output is close to the setpoint.

Once the tanks have reached the desired level (0 V) after the initial start-up, unmeasured inflow disturbances were introduced as follows:

(i) a disturbance inflow was applied to Tank 4 at approximately 250 sec and removed at approximately 800 sec;

(ii) a disturbance inflow was applied simultaneously to Tanks 1 and 2 at approximately 1080 sec and removed at 1950 sec.

We first focus on the disturbance into Tank 4 applied at 250 sec and removed at 800 sec. The effect of this can be clearly seen in the actual tank level shown in Figure 16.7 (Note that the data shown in Figure 16.7 were *not* available to the control law, but they are presented here to highlight the resulting performance). Figure 16.8 shows the estimated tank levels. It can be seen that the observer detects the disturbance in Tank 4 at approximately 370 sec. Figure 16.9 shows the estimated disturbance. Once the disturbance is detected the controller begins to act. This can be observed in Figure 16.10, which shows the control signal increments sent to each plant. Note that Tank 4 begins to receive attention once the disturbance is detected. The control signal applied to the actuators is shown in Figure 16.11.

We next consider the disturbances applied to Tanks 1 and 2 at approximately 1080 sec and removed at 1950 sec. The effect of these disturbances can be seen in the actual levels shown in Figure 16.7. Again the estimator can be seen (Figure 16.9) to lock onto the disturbances. The controller then pays attention to the necessary tanks, in this case Tanks 1 and 2 (see the control increments shown in Figure 16.10).

The attention-sharing facility by the controller is also apparent in the stem plots given in Figure 16.12. We see that most control attention ($\Delta u_i = \pm 1$) was applied to those tanks to which the disturbances were applied.

16.6 Further Reading

For complete list of references cited, see References section at the end of book.

General

Further information on networked control systems can be found in the book Ishii and Francis (2002). The current chapter is based on references Goodwin,

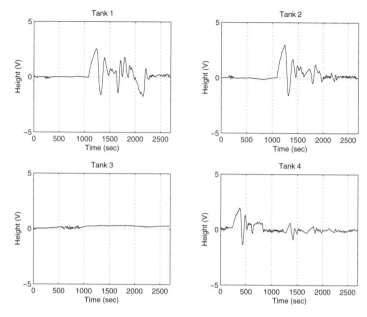

Figure 16.7. Actual levels.

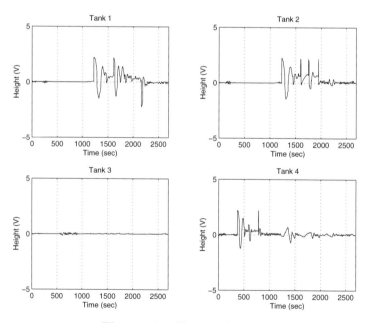

Figure 16.8. Estimated levels.

Haimovich, Quevedo and Welsh (2004), Haimovich, Goodwin and Quevedo (2003), Quevedo, Goodwin and Welsh (2003).

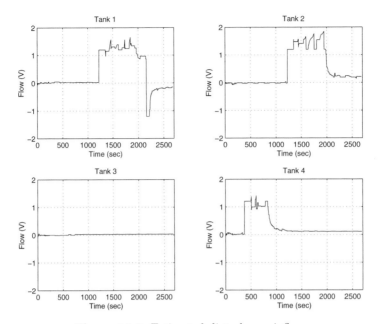

Figure 16.9. Estimated disturbance inflow.

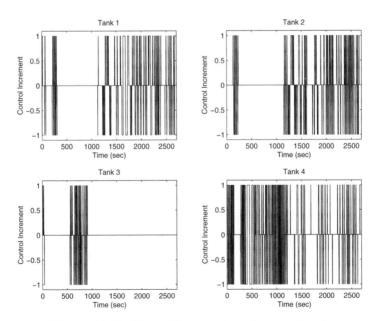

Figure 16.10. Control increments sent over network.

In relation to the estimation problem, in Curry (1970) different estimators for quantised discrete time linear systems affected by noise are pre-

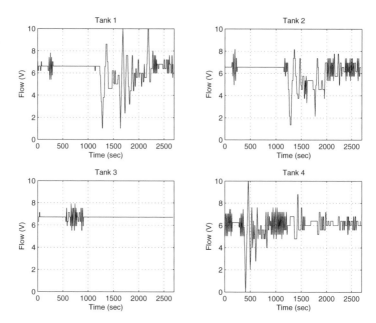

Figure 16.11. Control signal at actuators.

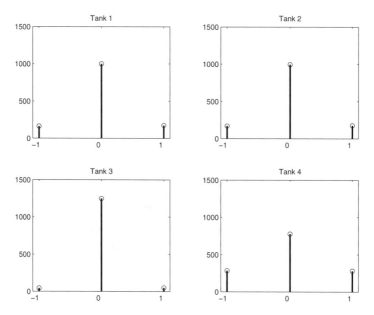

Figure 16.12. Stem Plots of Control Increments.

sented. In some situations the quantisation effect may be neglected or approximated by white noise (Gray and Neuhoff 1998, Curry 1970, Uchino, Ohta and

Hatakeyama 1991). However, this is not appropriate when only a few quantisation levels are available (see, for example, Delchamps 1989, Delchamps 1990, Koplon and Sontag 1993, Lunze 1994, Lunze 1998). It is also known (Feely 1997, Baillieul 2002) that the introduction of quantisers has a strong impact on closed loop dynamics. As well, the relationship between sampling rate and the size of the quantised sets has been explored by several authors, such as Tatikonda, Sahai and Mitter (1998), Wong and Brocket (1999), Nair and Evans (2003), Elia and Mitter (2001), Kao and Venkatesh (2002), Ishii and Francis (2002).

Another consequence of utilising channels with limited bandwidth is the introduction of time delays. Since the medium is data rate-limited, signals may have to *queue* before being transmitted (Lian, Moyne and Tilbury 2001), leading to delays in the *up-link* (from sensors to controller) and in the *down-link* (from controller to actuator). Without taking into account quantisation effects, several problems have been studied in the literature (see, for example, Liou and Ray (1991), Nilsson, Bernhardsson and Wittenmark (1998), Lincoln and Bernhardsson (2000), which address control system design with random time delays; and Branicky, Phillips and Zhang (2000)). In many cases it is possible to *time-stamp* the up-link data (Chan and Özgüner 1995), so that the delays are known at the receiving end. Thus (without taking into account quantisation issues), plant state estimates can be obtained at the controller side by simply solving a standard Kalman filtering problem (Goodwin and Sin 1984, Anderson and Moore 1979), see also Beldiman and Walsh (2000), Bauer, Sichitiu, Lorand and Premaratne (2001) and Zhivoglyadov and Middleton (2003). One can deploy event-based (nonuniform) sampling of the plant outputs, as described in Åström (1999). Sensor data should be sent only when needed (see also Bauer et al. 2001).

The down-link design problem considered in this chapter includes the issue of allocating communication time and thus has connections to *limited communication control* as treated in Brockett (1995), Hristu and Morgansen (1999), Hristu (1999), Hristu (2001), Rehbinder and Sanfridson (2000), see also Lincoln and Bernhardsson (2002), Palopoli, Bicchi and Sangiovanni-Vincentelli (2002). Allocation of different forms of network resources has also been explored in Xiao, Johansson, Hindi, Boyd and Goldsmith (2003), Walsh, Ye and Bushnell (2002), Walsh, Beldiman and Bushnell (2002), Branicky, Phillips and Zhang (2002), Park, Kim, Kim and Kwon (2002), and Hong (1995).

References

Abadie, J., ed. (1967), *Nonlinear Programming*, North Holland, Amsterdam, The Netherlands.

Adler, R., Kitchens, B. and Tresser, C. (2001), Dynamics of non-ergodic piecewise affine maps of the torus, *Ergodic theory and dynamical systems* **21**, 959–999.

Alamir, M. and Bornard, G. (1994), On the stability of receding horizon control of nonlinear discrete-time systems, *Systems and Control Letters* **23**, 291–296.

Allgöwer, F. and Zhen, A., eds (2000), *Nonlinear Model Predictive Control*, Vol. 26 of *Progress in Systems and Control Theory*, Birkhauser, Boston.

Anderson, T. W. (1958), *An Introduction to Multivariate Statistical Analysis*, Wiley, New York.

Anderson, B. D. O. and Moore, J. B. (1979), *Optimal Filtering*, Prentice Hall.

Anderson, B. D. O. and Moore, J. B. (1989), *Optimal Control. Linear Quadratic Methods*, Prentice Hall.

Angeli, D., Mosca, E. and Casavola, A. (2000), Output predicitive control of linear plants with saturated and rate limited actuators, in *Proceedings of the American Control Conference*, Chicago, IL, USA, pp. 272–276.

Ash, R. B. and Doléans-Dade, C. (2000), *Probability and Measure Theory*, Harcourt Academic Press.

Åström, K. J. (1970), *Introduction to Stochastic Control Theory*, Academic Press, New York and London.

Åström, K. J. (1999), Riemann and Lebesgue sampling, in *Advances in System Theory: A Symposium in Honor of Sanjoy K. Mitter*, Cambridge, MA, USA.

Åström, K. J. and Wittenmark, B. (1990), *Computer-Controlled Systems: Theory and Design*, 2nd edn, Prentice Hall, Englewood Cliffs, NJ.

Athans, M. and Falb, P. L. (1966), *Optimal Control: An Introduction to the Theory and its Applications*, McGraw-Hill, New York.

Baccarelli, E., Fasano, A. and Zucchi, A. (2000), A reduced-state soft-statistics-based MAP/DF equalizer for data transmission over long ISI channels, *IEEE Transactions on Communications* **48**(9), 1441–1446.

Baillieul, J. (2002), Feedback coding for information-based control: Operating near the data-rate limit, in *Proceedings of the 41st IEEE Conference on Decision and Control*, Las Vegas, NV, USA.

Bartlett, R. A., Biegler, L. T., Backstrom, J. and Gopal, V. (2002), Quadratic programming algorithms for large-scale model predictive control, *Journal of Process Control* **12**(7), 775–795.

Batina, I., Stoorvogel, A. A. and Weiland, S. (2001), Stochastic disturbance rejection in model predictive control by randomized algorithms, in *Proceedings of the American Control Conference*, Arlington, VA, USA.

Batina, I., Stoorvogel, A. A. and Weiland, S. (2002), Optimal control of linear, stochastic systems with state and input constraints, in *Proceedings of the 41st IEEE Conference on Decision and Control*, Las Vegas, NV, USA.

Bauer, P. H., Sichitiu, M., Lorand, C. and Premaratne, K. (2001), Total delay compensation in LAN control systems and implications for scheduling, in *Proceedings of the American Control Conference*, Arlington, VA, USA, pp. 4300–4305.

Bazaraa, M. S., Sherali, H. D. and Shetty, C. M. (1993), *Nonlinear Programming: Theory and Algorithms*, 2nd. edn, Wiley, New York.

Beale, E. M. L. (1955), On minimizing a convex function subject to linear inequalities, *Journal of the Royal Statistical Society, Ser. B* **17**, 173–184.

Beldiman, O. and Walsh, G. C. (2000), Predictors for networked control systems, in *Proceedings of the American Control Conference*, Chicago, IL, USA, pp. 2347–2351.

Bellman, R. (1957), *Dynamic Programming*, Princeton University Press.

Bemporad, A., Chisci, L. and Mosca, E. (1995), On the stabilizing property of SIORHC, *Automatica* **30**, 2013–2015.

Bemporad, A., Ferrari-Trecate, G. and Morari, M. (2000), Observability and controllability of piecewise affine and hybrid systems, *IEEE Transactions on Automatic Control* **45**(10), 1864–1876.

Bemporad, A. and Filippi, C. (2001), Suboptimal explicit MPC via approximate multiparametric quadratic programming, in *Proceedings of the 40th IEEE Conference on Decision and Control*, Orlando, FL, USA.

Bemporad, A. and Morari, M. (1999), Robust model predictive control: A survey, in A. Garulli, A. Tesi and A. Vicino, eds, Robustness in Identification and Control, number 245 in Lecture Notes in Control and Information Sciences, Springer, London, pp. 207–226.

Bemporad, A., Morari, M., Dua, V. and Pistikopoulos, E. N. (2002), The explicit linear quadratic regulator for constrained systems, *Automatica* **38**(1), 3–20.

Bergh, L. G. and MacGregor, J. F. (1987), Spatial control of film and sheet forming processes, *The Canadian Journal of Chemical Engineering* **65**, 148–155.

Bertsekas, D. P. (1976), *Dynamic Programming and Stochastic Control*, Vol. 125 of *Mathematics in science and engineering*, Academic Press, New York.

Bertsekas, D. P. (1987), *Dynamic Programming: Deterministic and stochastic models*, Prentice Hall.

Bertsekas, D. P. (1998), *Network Optimization: Continuous and Discrete Models*, Athena Scientific, Belmont, MA, USA.

Bertsekas, D. P. (2000), *Dynamic Programming and Optimal Control*, Vol. 1,2 of *Optimization and Computation Series*, Athena Scientific, Belmont, MA, USA.

Bicchi, A., Marigo, A. and Piccoli, B. (2002), On the reachability of quantized control systems, *IEEE Transactions on Automatic Control* **47**(4), 546–562.

Bitmead, R., Gevers, M. and Wertz, V. (1990), *Adaptive Optimal Control, The Thinking Man's GPC*, Prentice Hall.

Blanchini, F. (1999), Set invariance in control, *Automatica* **35**, 1747–1767.

Blanke, M., Adrian, J., Larsen, K. and Bentsen, J. (2000), Rudder roll damping in coastal region sea conditions, in *Proceedings of 5th IFAC Conference on Manoeuvering and Control of Marine Craft, MCMC'00*, Aalborg, Denmark.

Blanke, M. and Christensen, A. (1993), Rudder roll damping autopilot robustness to sway-yaw-roll couplings, Proceedings of 10th Ship Control System Symposium SCSS'93 pp. 93–119.

Borrelli, F. (2003), *Constrained Optimal Control of Linear and Hybrid Systems*, Springer, Heidelberg.

Boyd, S. and Vandenberghe, L. (2003), *Convex Optimization*, Cambridge University Press. Available on the web: http://www.stanford.edu/~boyd/cvxbook.html.

Boyle, T. J. (1977), Control of cross-direction variations in web forming machines, *The Canadian Journal of Chemical Engineering* **55**, 457–461.

Branicky, M. S., Phillips, S. M. and Zhang, W. (2000), Stability of networked control systems: Explicit analysis of delay, in *Proceedings of the American Control Conference*, Chicago, IL, USA, pp. 2352–2357.

Branicky, M. S., Phillips, S. M. and Zhang, W. (2002), Scheduling and feedback co-design for networked control systems, in *Proceedings of the 41st IEEE Conference on Decision and Control*, Las Vegas, NV, USA, pp. 1211–1217.

Brockett, R. W. (1995), Stabilization of motor networks, in *Proceedings of the IEEE Conference on Decision and Control*, New Orleans, LA, USA, pp. 1484–1488.

Brockett, R. W. and Liberzon, D. (2000), Quantized feedback stabilization of linear systems, *IEEE Transactions on Automatic Control* **45**(7), 1279–1289.

Bryson, A. E. and Ho, Y. C. (1969), *Applied Optimal Control*, Blaisdell, Waltham, Mass.

Bryson, A. E. J. and Frazier, M. (1963), Smoothing for linear and nonlinear dynamic systems, Technical Report ASD–TDR 63–119, Aero. Syst. Div., Wright-Patterson Air Force Base, Ohio.

Bushnell (ed.), L. (2001), Special section on networks and control, *IEEE Control Systems Magazine* **21**(1), 57–99.

Camacho, E. F. and Bordons, C. (1999), *Model Predictive Control*, Advanced Textbooks in Control and Signal Processing, Springer, London.

Candy, J. C. and Temes, G. C., eds (1992), *Oversampling Delta-Sigma Data Converters Theory, Design and Simulation*, IEEE Press., New York.

Canon, M. D., Cullum, C. D. and Polak, E. (1970), *Theory of Optimal Control and Mathematical Programming*, McGraw-Hill.

Cantoni, A. and Butler, P. (1976), Stability of decision feedback inverses, *IEEE Transactions on Communications* **24**(9), 970–977.

Chan, H. and Özgüner, Ü. (1995), Closed-loop control of systems over a communications network with queues, *International Journal of Control* **62**(3), 493–510.

Chen, C. and Shaw, L. (1982), On receding horizon feedback control, *Automatica* **18**, 349–352.

Chmielewski, D. and Manousiouthakis, V. (1996), On constrained infinite-time linear quadratic optimal control, *Systems and Control Letters* **29**, 121–129.

Chua, L. O. and Lin, T. (1988), Chaos in digital filters, *Theoretical Computer Science* **35**(6), 648–658.

Clarke, D. and Scattolini, R. (1981), Constrained receding horizon predictive control, *IEE Proceedings Control Theory and Applications* **138**, 347–354.

Cox, H. (1964), Estimation of state variables via dynamic programming, in *Proceedings of the Joint Automatic Control Conference*, Stanford, CA, USA, pp. 376–381.

Curry, R. E. (1970), *Estimation and Control with Quantized Measurements*, M.I.T. Press, Cambridge, MA.

Dave, P., Willing, D. A., Kudva, G. K., Pekny, J. F. and Doyle, F. J. (1997), LP methods in MPC of large scale systems: Application to paper-machine CD control, *AIChE Journal* **43**, 1016–1031.

De Doná, J. A. (2000), Input Constrained Linear Control, PhD thesis, School of Electrical Engineering and Computer Science, The University of Newcastle, Australia.

De Doná, J. A. and Goodwin, G. C. (2000), Elucidation of the state-space regions wherein model predictive control and anti-windup strategies achieve identical control policies, in *Proceedings of the American Control Conference*, Chicago, IL, USA.

De Doná, J. A., Goodwin, G. C. and Seron, M. M. (2000), Anti-windup and model predictive control: Reflections and connections, *European Journal of Control* **6**(5), 467–477.

De Doná, J. A., Seron, M. M., Goodwin, G. C. and Mayne, D. Q. (2002), Enlarged terminal sets guaranteeing stability of receding horizon control, *Systems and Control Letters* **47**(1), 57–63.

De Nicolao, G., Magni, L. and Scattolini, R. (1996), Stabilizing nonlinear receding horizon control via a nonquadratic penalty, in *Proceedings of the IMACS Multiconference for Computer Engineering in Systems Applications CESA*, Vol. 1, Lille, France, pp. 185–187.

Delchamps, D. F. (1989), Extracting state information from a quantized output record, *Systems and Control Letters* **13**, 365–372.

Delchamps, D. F. (1990), Stabilizing a linear system with quantized state feedback, *IEEE Transactions on Automatic Control* **35**, 916–924.

den Hertog, D. (1994), *Interior Point Approach to Linear, Quadratic, and Convex Programming*, Kluwer Academic Publishers, Dordrecht; Boston.

Dochain, D., Dumont, G., Gorinevsky, D. M. and Ogunnaike, T. (2003), Editorial. Special issue on control of industrial spatially distributed processes, *IEEE Transactions on Control Systems Technology* **11**(5), 609–611.

Doucet, A., de Freitas, N. and Gordon, N., eds (2001), *Sequential Monte Carlo Methods in Practice*, Springer, New York.

Dua, V. and Pistikopoulos, E. N. (2000), An algorithm for the solution of multiparametric mixed integer linear programming problems, *Annals of Operations Research* **99**, 123–139.

Duel-Hallen, A. and Heegard, C. (1989), Delayed decision-feedback sequence estimation, *IEEE Transactions on Communications* **37**(5), 428–436.

Duncan, S. R. (2002), Editorial. Special section: Cross directional control, *IEE Proceedings Control Theory and Applications* **149**(5), 412–413.

Duncan, S. R. and Bryant, G. F. (1997), The spatial bandwidth of cross-directional control systems for web processes, *Automatica* **33**(2), 139–153.

Elia, N. and Mitter, S. K. (2001), Stabilization of linear systems with limited information, *IEEE Transactions on Automatic Control* **46**(9), 1384–1400.

Evangelista, G. (2002), Design of optimum high-order finite-wordlength digital FIR filters with linear phase, *Signal Processing* **82**(2), 187–194.

Faltinsen, O. M. (1990), *Sea Loads on Ships and Offshore Structures*, Cambridge University Press.

Featherstone, A. P. and Braatz, R. D. (1997), Control-oriented modeling of sheet and film processes, *AIChE Journal* **43**(8), 1989–2001.

Featherstone, A. P., VanAntwerp, J. G. and Braatz, R. D. (2000), *Identification and control of sheet and film processes*, Springer Verlag, London.

Feely, O. (1997), A tutorial introduction to non-linear dynamics and chaos and their application to Sigma-Delta modulators, *International Journal of Circuit Theory and Applications* **25**, 347–367.

Fiacco, A. V. (1983), *Introduction to Sensitivity and Stability Analysis in Nonlinear Programming*, Academic Press, London.

Fiacco, A. V. and McCormick, G. P. (1990), *Nonlinear Programming : Sequential Unconstrained Minimization Techniques*, Society for Industrial and Applied Mathematics, Philadelphia.

Filatov, N. M. and Unbehauen, H. (1995), Adaptive predictive control policy for non-linear stochastic systems, *IEEE Transactions on Automatic Control* **40**(11), 1943–1949.

Findeisen, R., Imsland, L., Allgöwer, F. and Foss, B. A. (2003), State and output feedback nonlinear model predictive control: An overview, *European Journal of Control* **9**, 190–206.

Fletcher, R. (1981), *Practical Method of Optimization. Volume 2 Constrained Optimization*, Wiley, New York.

Fletcher, R. (2000), *Practical methods of optimization*, 2nd edn, Wiley, New York.

Fletcher, R. and Jackson, M. P. (1974), Minimization of a quadratic function of many variables subject only to lower and upper bounds, *Journal of the Institute of Mathematics and its Applications* **14**, 159–174.

Floudas, C. A. (1995), *Nonlinear and Mixed-Integer Optimization: Fundamentals and Applications*, Oxford University Press, New York.

Fossen, T. I. (1994), *Guidance and Control of Ocean Marine Vehicles*, Wiley, New York.

Fossen, T. I. (2002), *Marine Control Systems, Guidance, Navigation and Control of Ships, Rigs and Underwater Vehicles*, Marine Cybernetics.

Garcia, C. E., Prett, D. M. and Morari, M. (1989), Model predictive control: Theory and practice—a survey, *Automatica* **25**(3), 335–348.

Garey, M. R. and Johnson, D. S. (1979), *Computers and Intractability: A guide to the Theory of NP-Completeness*, W. H. Freeman, San Francisco.

Gersho, A. and Gray, R. M. (1992), *Vector Quantization and Signal Compression*, Kluwer Academic.

Gilbert, E. G. and Tan, K. T. (1991), Linear systems with state and control constraints: The theory and application of maximal output admissible sets, *IEEE Transactions on Automatic Control* **36**, 1008–1020.

Gill, P. E., Murray, W. and Wright, M. H. (1981), *Practical Optimization*, Academic Press.

Goodwin, G. C., De Doná, J. A., Seron, M. M. and Zhuo, X. W. (2004), On the duality of constrained estimation and control, in *Proceedings of the American Control Conference*, Boston, MA, USA.

Goodwin, G. C., Graebe, S. F. and Salgado, M. E. (2001), *Principles of Control System Design*, Prentice Hall.

Goodwin, G. C., Quevedo, D. E. and De Doná, J. A. (2003), An application of receding horizon control to estimation with quantized measurements, in *Proceedings of the American Control Conference*, Denver, CO, USA.

Goodwin, G. C., Haimovich, H., Quevedo, D. E. and Welsh, J. S. (2004), A moving horizon optimization approach to networked control system design, *IEEE Transactions on Automatic Control* . To appear.

Goodwin, G. C., Quevedo, D. E. and McGrath, D. (2003), Moving-horizon optimal quantizer for audio signals, *Journal of the Audio Engineering Society* **51**(3), 138–149.

Goodwin, G. C. and Sin, K. S. (1984), *Adaptive Filtering, Prediction, and Control*, Prentice Hall.

Graham, R. (1990), Motion-induced interruptions as ship operability criteria, *Naval Engineers Journal* **103**(3).

Grancharova, A. and Johansen, T. A. (2002), Approximate explicit model predictive control incorporating heuristics, in *Proceedings of the IEEE Conference on Computer Aided Control Design*, Glasgow, Scotland, pp. 92–97.

Gray, R. M. and Neuhoff, D. L. (1998), Quantization, *IEEE Transactions on Information Theory* **44**(6), 2325–2383.

Grünbaum, B. (1967), *Convex Polytopes*, Wiley.

Haimovich, H., Goodwin, G. C. and Quevedo, D. E. (2003), Moving horizon Monte Carlo state estimation for linear systems with output quantization, in *Proceedings of the 42st IEEE Conference on Decision and Control*, Maui, HI, USA.

Haimovich, H., Goodwin, G. C. and Welsh, J. S. (2004), Set-valued observers for constrained state estimation of discrete-time systems with quantized measurements, in *Asian Control Conference*, Melbourne, Australia.

Haimovich, H., Perez, T. and Goodwin, G. C. (2003), On certainty equivalence in output feedback receding horizon control of constrained linear systems., in *Proceedings of the European Control Conference*, Cambridge, UK.

Halkin, H. (1966), A maximum principle of the Pontryagin type for systems described by nonlinear difference equations, *SIAM Journal on Control* **4**(1), 90–111.

Halouskova, A., Karny, M. and Nagy, I. (1993), Adaptive cross-direction control of paper basis weight, *Automatica* **29**(2), 425–429.

Heath, W. P. (1996), Orthogonal functions for cross-directional control of web forming processes, *Automatica* **32**(2), 183–198.

Heemels, W. P. M. H., De Schutter, B. and Bemporad, A. (2001), Equivalence of hybrid dynamical models, *Automatica* **37**(7), 1085–1091.

Holtzman, J. M. (1966a), Convexity and the maximum principle for discrete systems, *IEEE Transactions on Automatic Control* **11**(1), 30–35.

Holtzman, J. M. (1966b), On the maximum principle for nonlinear discrete-time systems, *IEEE Transactions on Automatic Control* **11**(14), 273–274.

Holtzman, J. M. and Halkin, H. (1966), Directional convexity and the maximum principle for discrete systems, *SIAM Journal on Control* **4**(2), 263–275.

Hong, S. H. (1995), Scheduling algorithm of data sampling times in the integrated communication and control systems, *IEEE Transactions On Control Systems Technology* **3**(2), 225–230.

Hovd, M., Braatz, R. D. and Skogestad, S. (1997), SVD controllers for H_2-, H_∞- and μ-optimal control, *Automatica* **33**(3), 433–439.

Hristu, D. (1999), Generalized inverses for finite-horizon tracking, in *Proceedings of the 38th IEEE Conference on Decision and Control*, Phoenix, AZ, USA, pp. 1397–1402.

Hristu, D. (2001), Feedback control systems as users of a shared network: Communication sequences that guarantee stability, in *Proceedings of the 40th IEEE Conference on Decision and Control*, Orlando, FL, USA, pp. 3631–3636.

Hristu, D. and Morgansen, K. (1999), Limited communication control, *Systems and Control Letters* **37**(4), 193–205.

Ishii, H. and Francis, B. A. (2002), *Limited Data Rate in Control Systems with Networks*, Springer, Heidelberg.

Ishii, H. and Francis, B. A. (2003), Quadratic stabilization of sampled-data systems with quantization, *Automatica* **39**, 1793–1800.

Jadbabaie, A., Persis, C. D. and Yoon, T.-W. (2002), A globally stabilizing receding horizon controller for neutrally stable linear systems with input constraints, in *Proceedings of the 41st IEEE Conference on Decision and Control*, Las Vegas, NV, USA.

Jadbabaie, A., Yu, J. and Hauser, J. (2001), Unconstrained receding-horizon control of nonlinear systems, *IEEE Transactions on Automatic Control* **46**(5), 776–783.

Jazwinski, A. H. (1970), *Stochastic processes and filtering theory*, New York, Academic Press.

Johansen, T. A. (2002), Structured and reduced dimension explicit linear quadratic regulators for systems with constraints, in *Proceedings of the 41st IEEE Conference on Decision and Control*, Las Vegas, NV, USA, pp. 3970–3975.

Johansen, T. A. and Grancharova, A. (2002), Approximate explicit model predictive control implemented via orthogonal search tree partitioning, in *Proceedings of the 15th IFAC World Congress*, Barcelona, Spain.

Johansen, T. A., Petersen, I. and Slupphaug, O. (2002), Explicit sub-optimal linear quadratic regulation with state and input constraints, *Automatica* **38**, 1099–1111.

Kailath, T., Sayed, A. H. and Hassibi, B. (2000), *Linear Estimation*, Prentice Hall, New York.

Kalman, R. E. (1960a), Contributions to the theory of optimal control, *Boletín Sociedad Matemática Mexicana* **5**, 102–119.

Kalman, R. E. (1960b), A new approach to linear filtering and prediction problems, *Transactions of the ASME, Journal of Basic Engineering* **82**, 34–45.

Kalman, R. E. and Bertram, J. E. (1960), Control system analysis and design via the second method of Lyapunov. II Discrete-time systems, *Transactions of the ASME, Journal of Basic Engineering*, pp. 394–400.

Kalman, R. E. and Bucy, R. S. (1961), New results in linear filtering and prediction theory, *Transactions of the ASME, Journal of Basic Engineering* **83-D**, 95–107.

Kao, C.-Y. and Venkatesh, S. R. (2002), Stabilization of linear systems with limited information – multiple input case, in *Proceedings of the American Control Conference*, Anchorage, AK, USA, pp. 2406–2411.

Keerthi, S. S. and Gilbert, E. G. (1988), Optimal infinite horizon feedback laws for a general class of constrained discrete time systems: Stability and moving-horizon approximations, *Journal of Optimization Theory and Applications* **57**, 265–293.

Khalil, H. (1996), *Nonlinear Systems*, 2nd. edn, Prentice Hall, New Jersey.

Kiihtelys, K. (2003), Bandwidth and quantization constraints in feedback control, Master's thesis, Department of Systems, Signals and Sensors (S3), Royal Institute of Technology (KTH), Stockholm, Sweden.

Kirk, D. E. (1970), *Optimal Control Theory. An Introduction*, Prentice Hall, Englewood Cliffs, NJ.

Kleinman, B. L. (1970), An easy way to stabilize a linear constant system, *IEEE Transactions on Automatic Control* **15**(12), 693.

Kodek, D. (1980), Design of optimal finite wordlength FIR digital filters using integer programming techniques, *IEEE Transactions on Acoustics Speech And Signal Processing* **28**(3), 304–308.

Koplon, R. and Sontag, E. D. (1993), Linear systems with sign-observations, *SIAM Journal on Control and Optimization* **31**(5), 1245–1266.

Kothare, M., Campo, P., Morari, M. and Nett, C. (1994), A unified framework for the study of anti-windup designs, *Automatica* **30**(12), 1869–1883.

Kouvaritakis, B., Cannon, M. and Rossiter, J. A. (2002), Who needs QP for linear MPC anyway?, *Automatica* **38**, 879–884.

Kristinsson, K. and Dumont, G. A. (1996), Cross-directional control on paper machines using Gram polynomials, *Automatica* **32**(4), 533–548.

Kwon, W., Bruckstein, A. and Kailath, T. (1983), Stabilizing state-feedback design via the moving horizon method, *International Journal of Control* **37**(3), 631–643.

Kwon, W. and Pearson, A. (1977), A modified quadratic cost problem and feedback stabilization of a linear system, *IEEE Transactions on Automatic Control* **22**(5), 838–842.

Land, A. H. and Doig, A. G. (1960), An automatic method for solving discrete programming problems, *Econometrica* **28**, 497–520.

Laughlin, D. L., Morari, M. and Braatz, R. D. (1993), Robust performance of cross-directional basis-weight control in paper machines, *Automatica* **29**(6), 1395–1410.

Lawrence, R. E. and Kaufman, H. (1971), The Kalman filter for the equalization of a digital communications channel, *IEEE Transactions on Communication Technology* **19**(6), 1137–1141.

Lee, E. B. and Markus, L. (1967), *Foundations of Optimal Control Theory*, Wiley, New York.

Lee, Y. I. and Kouvaritakis, B. (2001), Receding horizon output feedback control for linear systems with input saturation, *IEE Proceedings, Control Theory and Applications* **148**(2), 109–115.

Li, Z. and Soh, C. B. (1999), Lyapunov stability of discontinuous dynamic systems, *Ima Journal of Mathematical Control And Information* **16**, 261–274.

Lian, F. L., Moyne, J. R. and Tilbury, D. M. (2001), Performance evaluation of control networks, *IEEE Control Systems Magazine* **21**(1), 66–83.

Lim, Y. C. and Parker, S. R. (1983a), Discrete coefficient FIR digital filter design based upon an LMS criteria, *Theoretical Computer Science* **30**(10), 723–739.

Lim, Y. C. and Parker, S. R. (1983b), FIR filter design over a discrete powers-of-two coefficient space, *IEEE Transactions on Acoustics Speech And Signal Processing* **31**(3), 583–591.

Lincoln, B. and Bernhardsson, B. (2000), Optimal control over networks with long random delays, in *Proceedings of the International Symposium on Mathematical Theory of Networks and Systems*, Perpignan, France.

Lincoln, B. and Bernhardsson, B. (2002), LQR optimization of linear system switching, *IEEE Transactions on Automatic Control* **47**(10), 1701–1705.

Liou, L. W. and Ray, A. (1991), A stochastic regulator for integrated communication and control systems, *Transactions of the ASME* **113**, 604–619.

Lipschitz, S. P., Vanderkooy, J. and Wannamaker, R. A. (1991), Minimally audible noise shaping, *Journal of the Audio Engineering Society* **39**(11), 836–852.

Lloyd, A. R. J. M. (1989), *Seakeeping: Ship Behaviour in Rough Weather*, Ellis Horwood Series in Marine Technology, Ellis Horwood.

Luenberger, D. G. (1984), *Linear and Nonlinear Programming*, Addison-Wesley, Reading, Mass.

Luenberger, D. G. (1989), *Linear and Nonlinear Programming*, 2nd. edn, Addison–Wesley.

Lunze, J. (1994), Qualitative modelling of linear dynamical systems with quantized state measurements, *Automatica* **30**(3), 417–431.

Lunze, J. (1998), On the Markov property of quantised state measurement sequences, *Automatica* **34**(11), 1439–1444.

Maciejowski, J. M. (2002), *Predictive Control with Constraints*, Prentice Hall.

Magni, L., De Nicolao, G. and Scattolini, R. (2001), Output feedback and tracking of nonlinear systems with model predictive control, *Automatica* **37**(10), 1601–1607.

Magni, L., ed. (2003), Special Issue: Control of nonlinear systems with model predictive control, *International Journal of Robust and Nonlinear Control*, Wiley.

Marquis, P. and Broustail, J. (1988), SMOC, a bridge between state space and model predictive controllers: Application to the automation of hydrotreating unit, in *Proceedings of the IFAC workshop on model based process control*, Oxford: Pergamon Press, pp. 37–43.

Mayne, D. Q. and Michalska, H. (1990), Receding horizon control of nonlinear systems, *IEEE Transactions on Automatic Control* **35**(5), 814–824.

Mayne, D. Q. and Michalska, H. (1993), Robust receding horizon control of constrained nonlinear systems, *IEEE Transactions on Automatic Control* **38**(11), 1623–1633.

Mayne, D. Q., Rawlings, J. B., Rao, C. V. and Scokaert, P. O. M. (2000), Constrained model predictive control: Stability and optimality, *Automatica* **36**, 789–814.

Michalska, H. and Mayne, D. Q. (1995), Moving horizon observers and observer-based control, *IEEE Transactions on Automatic Control* **40**(6), 995–1006.

Mosca, E., Lemos, J. and Zhang, J. (1990), Stabilizing I/O receding horizon control, in *Proceedings of the 29th IEEE Conference on Decision and Control*, Honolulu, HI, USA, pp. 2518–2523.

Mosca, E. and Zhang, J. (1992), Stable redesign of predictive control, *Automatica* **28**(6), 1229–1233.

Muske, K. R., Meadows, E. S. and Rawlings, J. R. (1994), The stability of constrained receding horizon control with state estimation, in *Proceedings of the American Control Conference*, Baltimore, MD, USA.

Muske, K. R. and Rawlings, J. B. (1993), Model predictive control with linear models, *AIChE Journal* **39**(2), 262–287.

Nair, G. N. and Evans, R. J. (2003), Exponential stabilisability of finite-dimensional linear systems with limited data rates, *Automatica* **39**, 585–593.

Nash, S. G. and Sofer, A. (1996), *Linear and Nonlinear Programming*, McGraw-Hill.

Nesterov, Y. and Nemirovskii, A. (1994), *Interior-Point Polynomial Algorithms in Convex Programming*, Society for Industrial and Applied Mathematics, Philadelphia.

Nilsson, J., Bernhardsson, B. and Wittenmark, B. (1998), Stochastic analysis and control of real-time systems with random time delays, *Automatica* **34**(1), 57–64.

Nocedal, J. and Wright, S. J. (1999), *Numerical Optimization*, Springer, New York.

Norsworthy, S. R., Schreier, R. and Temes, G. C., eds (1997), *Delta–Sigma Data Converters: Theory, Design and Simulation*, IEEE Press, Piscataway, N.J.

Palopoli, L., Bicchi, A. and Sangiovanni-Vincentelli, A. (2002), Numerically efficient control of systems with communication constraints, in *Proceedings of the 41st IEEE Conference on Decision and Control*, Las Vegas, NV, USA.

Park, H. S., Kim, Y. H., Kim, D.-S. and Kwon, W. H. (2002), A scheduling method for network-based control systems, *IEEE Transactions on Automatic Control* **10**(3), 318–330.

Perez, T. (2003), Performance Analysis and Constrained Control of Fin and Rudder-Based Roll Stabilizers for Ships, PhD thesis, School of Electrical Engineering and Computer Science, The University of Newcastle, Australia.

Perez, T., Goodwin, G. C. and Skelton, R. E. (2003), Analysis of performance and applicability of rudder-based stabilizers, in *6th IFAC Conference on Manouvering and Control of Marine Craft MCMC'03*, Girona, Spain, pp. 185–190.

Perez, T., Goodwin, G. C. and Tzeng, C. Y. (2000), Model predictive rudder roll stabilization control for ships, in *Proceedings of the 5th IFAC Conference on Manoeuvering and Control of Marine Crafts MCMC'00*, Aalborg, Denmark.

Perez, T., Haimovich, H. and Goodwin, G. C. (2004), On optimal control of constrained linear systems with imperfect state information and stochastic disturbances, *International Journal of Robust and Nonlinear Control* **14**(4), 379–393.

Polak, E. (1971), *Computational Methods in Optimization; A Unified Approach*, Academic Press, New York.

Pontryagin, L. S. (1959), Optimal regulation processes, *Uspekhi Mat. Nauk (N.S.)* **14**, 3–20. English translation in Amer. Math. Soc. Transl. Ser. **2**(18), 1961, 321–339.

Pontryagin, L. S., Boltyanskii, V. G., Gamkrelidze, R. V. and Mischenko, E. F. (1962), *The Mathematical Theory of Optimal Processes*, Interscience Publishers, New York.

Proakis, J. G. (1995), *Digital Communications*, McGraw-Hill.

Qin, S. J. and Badgwell, T. A. (1997), An overview of industrial model predictive control technology, in *Chemical Process Control–V, CACHE, AIChE*, pp. 232–256.

Quevedo, D. E., De Doná, J. A. and Goodwin, G. C. (2002), On the dynamics of receding horizon linear quadratic finite alphabet control loops, in *Proceedings of the 41st IEEE Conference on Decision and Control*, Las Vegas, NV, USA.

Quevedo, D. E. and Goodwin, G. C. (2003a), Audio quantization from a receding horizon control perspective, in *Proceedings of the American Control Conference*, Denver, CO, USA, pp. 4131–4136.

Quevedo, D. E. and Goodwin, G. C. (2003b), Moving horizon design of discrete co-efficient FIR filters, Technical Report EE03027, School of Electrical Engineering and Computer Science, The University of Newcastle, NSW 2308, Australia.

Quevedo, D. E. and Goodwin, G. C. (2004a), Control of EMI from switch-mode power supplies via multi-step optimization, in *Proceedings of the American Control Conference*, Boston, MA, USA.

Quevedo, D. E. and Goodwin, G. C. (2004b), Multi-step optimal analog-to-digital conversion, *IEEE Transactions on Circuits and Systems I*. To appear.

Quevedo, D. E. and Goodwin, G. C. (2004c), Finite alphabet constraints in engineering. In preparation.

Quevedo, D. E., Goodwin, G. C. and De Doná, J. A. (2003), A multi-step detector for linear ISI-channels incorporating degrees of belief in past estimates, Technical Report EE03013, School of Electrical Engineering and Computer Science, The University of Newcastle, NSW 2308, Australia.

Quevedo, D. E., Goodwin, G. C. and De Doná, J. A. (2004), Finite constraint set receding horizon quadratic control, *International Journal of Robust and Nonlinear Control* **14**(4), 355–377.

Quevedo, D. E., Goodwin, G. C. and Welsh, J. S. (2003), Minimizing down-link traffic in networked control systems via optimal control techniques, in *Proceedings of the 42st IEEE Conference on Decision and Control*, Maui, HI, USA.

Qureshi, S. U. H. (1985), Adaptive equalization, *Proceedings of the IEEE* **73**(9), 1349–1387.

Ramadge, P. J. (1990), On the periodicity of symbolic observations of piecewise smooth discrete-time systems, *IEEE Transactions on Automatic Control* **35**(7), 807–813.

Rao, C. V. (2000), Moving Horizon Strategies for the Constrained Monitoring and Control of Nonlinear Discrete-Time Systems, PhD thesis, University of Wisconsin-Madison, USA.

Rao, C. V., Rawlings, J. B. and Lee, J. H. (2001), Constrained linear estimation—a moving horizon approach, *Automatica* **37**(1619–1628).

Rao, C. V., Rawlings, J. B. and Mayne, D. Q. (2003), Constrained state estimation for nonlinear discrete-time systems: Stability and moving horizon approximations, *IEEE Transactions on Automatic Control* **48**(2), 246 –258.

Rashid, M. H. (1993), *Power Electronics*, second edn, Prentice Hall, Englewood Cliffs, NJ.

Rawlings, J. B. and Chien, I.-L. (1996), Gage control of film and sheet-forming processes, *AIChE Journal* **42**(3), 753–766.

Rawlings, J. B. and Muske, K. R. (1993), Stability of constrained receding horizon control, *IEEE Transactions on Automatic Control* **38**(10), 1512–1516.

Rehbinder, H. and Sanfridson, M. (2000), Scheduling of a limited communication channel for optimal control, in *Proceedings of the 39th IEEE Conference on Decision and Control*, Sydney, Australia.

Richter, H. and Misawa, E. A. (2003), Stability of discrete-time systems with quantized input and state measurements, *IEEE Transactions on Automatic Control* **48**(8), 1453–1458.

Robert, C. R. and Casella, G. (1999), *Monte Carlo Statistical Methods*, Springer, New York.

Roberts, G. N. (1993), A note on the applicability of rudder roll stabilization for ships, in *Proceedings of the American Control Conference*, San Francisco, CA, USA, pp. 2403–2407.

Robertson, D. G. and Lee, J. H. (2002), On the use of constraints in least squares estimation and control, *Automatica* **38**(7), 1113–1123.

Rojas, O. J., Goodwin, G. C. and Johnston, G. V. (2002), Spatial frequency anti-windup strategy for cross directional control problems, *IEE Proceedings Control Theory and Applications* **149**(5), 414–422.

Rojas, O. J., Goodwin, G. C., Seron, M. M. and Feuer, A. (2004), An SVD based strategy for receding horizon control of input constrained linear systems, *International Journal of Robust and Nonlinear Control*. In press.

Rojas, O. J., Seron, M. M. and Goodwin, G. C. (2003), SVD based receding horizon control for constrained linear systems: Stability results, in *Proceedings of the 42st IEEE Conference on Decision and Control*, Maui, HI, USA.

Rosen, J. B. (1960), The gradient projection method for nonlinear programming, Part I, Linear constraints, *SIAM Journal of Applied Mathematics* **8**, 181–217.

Rossiter, J. A. (2003), *Model-Based Predictive Control*, CRC Press.

Rossiter, J. A., Kouvaritakis, B. and Rice, M. J. (1998), A numerically robust state-space approach to stable-predictive control strategies, *Automatica* **34**(1), 65–74.

Sastry, S. (1999), *Nonlinear Systems. Analysis, Stability and Control*, Springer, New York.

Scokaert, P. O. M. and Rawlings, J. B. (1998), Constrained linear quadratic regulation, *IEEE Transactions on Automatic Control* **43**(8), 1163–1169.

Scokaert, P. O. M., Rawlings, J. B. and Meadows, E. S. (1997), Discrete-time stability with perturbations: Aplication to model predictive control, *Automatica* **33**(3), 463–470.

Sellars, F. H. and Martin, J. P. (1992), Selection and evaluation of ship roll stabilization systems, *Marine Technology, SNAME* **29**(2), 84–101.

Seron, M. M., Goodwin, G. C. and De Doná, J. A. (2003), Characterisation of receding horizon control for constrained linear systems, *Asian Journal of Control* **5**(2), 271–286.

Shamma, J. S. and Tu, K. Y. (1998), Output feedback for systems with constraints and saturations: Scalar control case, *Systems and Control Letters* **35**, 265–293.

Stewart, G. E., Gorinevsky, D. M. and Dumont, G. A. (1998), Robust GMV cross directional control of paper machines, in *Proceedings of the American Control Conference*, Philadelphia, PA, USA, pp. 3002–3007.

Stewart, G. E., Gorinevsky, D. M. and Dumont, G. A. (2003), Feedback controller design for a spatially distributed system: The paper machine problem, *IEEE Transactions on Control Systems Technology* **11**(5), 612–628.

Sznaier, M. and Damborg, M. J. (1987), Suboptimal control of linear systems with state an control inequality constraints, in *Proceedings of the 26th Conference on Decision and Control*, Los Angeles, CA, pp. 761–762.

Sznaier, M. and Damborg, M. J. (1989), Control of constrained discrete time linear systems using quantized controls, *Automatica* **25**(4), 623–628.

Sznaier, M. and Damborg, M. J. (1990), Heuristically enhanced feedback control of discrete-time linear systems, *Automatica* **26**, 521–532.

Tatikonda, S., Sahai, A. and Mitter, S. (1998), Control of LQG systems under communication constraints, in *Proceedings of the 37th Conference on Decision and Control*, Tampa, FL, USA, pp. 1165–1170.

Teel, A. (1999), Anti-windup for exponentially unstable linear systems, *International Journal of Robust and Nonlinear Control* **9**(10), 701–716.

Thielecke, J. (1997), A soft-decision state-space equalizer for FIR channels, *IEEE Transactions on Communications* **45**(10), 1208–1217.

Thomas, Y. A. (1975), Linear quadratic optimal estimation and control with receding horizon, *Electronics Letters* **11**, 19–21.

Tøndel, P. and Johansen, T. A. (2002), Complexity reduction in explicit linear model predictive control, in *Proceedings of the 15th IFAC World Congress*, Barcelona, Spain.

Tøndel, P., Johansen, T. A. and Bemporad, A. (2002), Computation and approximation of piecewise affine control laws via binary search trees, in *Proceedings of the 41st IEEE Conference on Decision and Control*, Las Vegas, NV, USA, pp. 3144–3149.

Tugnait, J. K., Tong, L. and Ding, Z. (2000), Single-user channel estimation and equalization, *IEEE Signal Processing Magazine* **17**(3), 17–28.

Uchino, E., Ohta, M. and Hatakeyama, K. (1991), Various type digital filters for an arbitrary sound environmental system with quantized observation mechanism and its application, in *Proceedings of the International Conference on Industrial Electronics, Control and Instrumentation IECON'91*, pp. 2205–2210.

van Amerongen, J., van der Klught, P. G. M. and Pieffers, J. B. M. (1987), Rudder roll stabilization–controller design and experimental results, in *Proceedings of 8th International Ship Control System Symposium SCSS'87*, The Netherlands.

van Amerongen, J., van der Klugt, P. and van Nauta Lemke, H. (1990), Rudder roll stabilization for ships, *Automatica* **26**, 679–690.

VanAntwerp, J. G. and Braatz, R. D. (2000), Fast model predictive control of sheet and film processes, *IEEE Transactions on Control Systems Technology* **8**(3), 408–417.

Vidyasagar, M. (2002), *Nonlinear Systems Analysis*, second edn, SIAM Society for Industrial and Applied Mathematics.

Walsh, G. C., Beldiman, O. and Bushnell, L. G. (2002), Error encoding algorithms for networked control systems, *Automatica* **38**, 261–267.

Walsh, G. C., Ye, H. and Bushnell, L. G. (2002), Stability analysis of networked control systems, *IEEE Transactions on Automatic Control* **10**(3), 438–446.

Wellstead, P. E., Zarrop, M. B. and Duncan, S. R. (2000), Signal processing and control paradigms for industrial web and sheet manufacturing, *International Journal of Adaptive Control and Signal Processing* **14**(1), 51–76.

Williamson, D., Kennedy, R. A. and Pulford, G. W. (1992), Block decision feedback equalization, *IEEE Transactions on Communications* **40**(2), 255–264.

Wills, A. G. and Heath, W. P. (2002), Analysis of steady-state performance for cross-directional control, *IEE Proceedings Control Theory and Applications* **149**(5), 433–440.

Wolfe, P. (1963), Methods of nonlinear programming, in R. L. Graves and P. Wolfe, eds, Recent Advances in Mathematical Programming, McGraw-Hill, New York.

Wong, W. S. and Brocket, R. W. (1999), Systems with finite communication bandwidth constraints–Part II: Stabilization with limited information feedback, *IEEE Transactions on Automatic Control* **44**(5), 1049–1053.

Wright, S. (1997a), *Primal-Dual Interior-Point Methods*, SIAM Publications.

Wright, S. J. (1996), A path-following interior-point algorithm for linear and quadratic problems, *Annals of Operations Research* **62**, 103–130.

Wright, S. J. (1997b), Applying new optimization algorithms to model predictive control, in J. Kantor, C. Garcia and B. Carnahan, eds, *Proceedings of the Fifth International Conference on Chemical Process Control, CACHE, AIChE*, Vol. 93, pp. 147–155.

Wu, C. W. and Chua, L. O. (1994), Symbolic dynamics of piecewise-linear maps, *IEEE Transactions on Circuits and Systems II* **41**(6), 420–424.

Xiao, L., Johansson, M., Hindi, H., Boyd, S. and Goldsmith, A. (2003), Joint optimization of communication rates and linear systems, *IEEE Transactions on Automatic Control* **48**(1), 148–153.

Ye, Y. (1997), *Interior Point Algorithms: Theory and Analysis*, Wiley, New York.

Yoon, T.-W., Kim, J.-S., Jadbabaie, A. and Persis, C. D. (2003), A globally stabilizing model predictive controller for neutrally stable linear systems with input constraints, in *Proceedings of International Conference on Control, Automation and Systems ICCAS'03*, Gyeongju, Korea, pp. 1901–1904.

Zheng, A. and Morari, M. (1995), Stability of model predictive control with mixed constraints, *IEEE Transactions on Automatic Control* **40**(10), 1818–1823.

Zhivoglyadov, P. V. and Middleton, R. H. (2003), Networked control design for linear systems, *Automatica* **39**, 743–750.

Zhou, K., Doyle, J. C. and Glover, K. (1996), *Robust and Optimal Control*, Prentice Hall.

Zhou, W.-W., Cherchas, D. B. and Calisal, S. (1994), Identification of rudder-yaw
    and rudder-roll steering models using prediction error techniques, *Optimal Con-
    trol Applications and Methods* **15**, 101–114.

Index